VOLUME ONE HUNDRED AND SIXTEEN

ADVANCES IN PROTEIN CHEMISTRY AND STRUCTURAL BIOLOGY

Intracellular Signalling Proteins

VOLUME ONE HUNDRED AND SIXTEEN

ADVANCES IN PROTEIN CHEMISTRY AND STRUCTURAL BIOLOGY

Intracellular Signalling Proteins

Edited by

ROSSEN DONEV

MicroPharm Ltd,
United Kingdom

ELSEVIER

ACADEMIC PRESS

An imprint of Elsevier

Academic Press is an imprint of Elsevier
50 Hampshire Street, 5th Floor, Cambridge, MA 02139, United States
525 B Street, Suite 1650, San Diego, CA 92101, United States
The Boulevard, Langford Lane, Kidlington, Oxford OX5 1GB, United Kingdom
125 London Wall, London, EC2Y 5AS, United Kingdom

First edition 2019

ISBN: 978-0-12-815561-5
ISSN: 1876-1623

Publisher: Zoe Kruze
Acquisition Editor: Ashlie M. Jackman
Editorial Project Manager: Leticia M. Lima
Production Project Manager: James Selvam
Cover Designer: Christian J. Bilbow

Typeset by SPi Global, India

Contents

Contributors

Alessandro Allegra
Division of Hematology, Department of Human Pathology in Adulthood and Childhood "Gaetano Barresi", University of Messina, Messina, Italy

Andrea Gaetano Allegra
Division of Hematology, Department of Human Pathology in Adulthood and Childhood "Gaetano Barresi", University of Messina, Messina, Italy

Anne Biever
Max Planck Institute for Brain Research, Frankfurt am Main, Germany

Ilya Borisenko
St. Petersburg State University, St. Petersburg, Russia

Angela Bruzzaniti
Department of Anatomy and Cell Biology, Indiana University School of Medicine; Indiana Center for Musculoskeletal Health; Department of Biomedical and Applied Sciences, Indiana University School of Dentistry, Indianapolis, IN, United States

Miribane Dërmaku-Sopjani
Department of Chemistry, University of Prishtina, Prishtina, Kosova

Alexander V. Ereskovsky
St. Petersburg State University, St. Petersburg, Russia; Mediterranean Institute of Marine and Terrestrial Biodiversity and Ecology (IMBE), Aix Marseille University, CNRS, IRD, Avignon Université, Marseille, France

Nathalie Fuentes
Department of Pediatrics, The Pennsylvania State University College of Medicine, Hershey, PA, United States

Giuseppe Gangarossa
Unité de Biologie Fonctionnelle et Adaptative, CNRS UMR 8251, Université Paris Diderot, Sorbonne Paris Cité, Paris, France

Christoph Garbers
Department of Pathology, Otto-von-Guericke-University Magdeburg, Medical Faculty, Magdeburg, Germany

Kyung Ah Han
Department of Brain and Cognitive Sciences, Daegu Gyeongbuk Institute of Science and Technology (DGIST), Daegu, Korea

Vanessa Innao
Division of Hematology, Department of Human Pathology in Adulthood and Childhood "Gaetano Barresi", University of Messina, Messina, Italy

Jaewon Ko
Department of Brain and Cognitive Sciences, Daegu Gyeongbuk Institute of Science and Technology (DGIST), Daegu, Korea

Adelaine Kwun-Wai Leung
Department of Veterinary Biomedical Sciences, Western College of Veterinary Medicine, Saskatoon, SK, Canada

Jerrold S. Levine
Department of Medicine, Section of Nephrology, University of Illinois College of Medicine and Jesse Brown Veterans Affairs Medical Center, Chicago, IL, United States

Sabrina Lisi
Department of Basic Medical Sciences, Neurosciences and Sensory Organs (SMBNOS), Section of Human Anatomy and Histology, University of Bari "Aldo Moro", Bari, Italy

Juliane Lokau
Department of Pathology, Otto-von-Guericke-University Magdeburg, Medical Faculty, Magdeburg, Germany

Yinglong Miao
Center for Computational Biology and Department of Molecular Biosciences, University of Kansas, Lawrence, KS, United States

Caterina Musolino
Division of Hematology, Department of Human Pathology in Adulthood and Childhood "Gaetano Barresi", University of Messina, Messina, Italy

Lilian I. Plotkin
Department of Anatomy and Cell Biology, Indiana University School of Medicine; Indiana Center for Musculoskeletal Health; Roudebush Veterans Administration Medical Center, Indianapolis, IN, United States

Olga I. Podgornaya
St. Petersburg State University; Institute of Cytology RAS, St. Petersburg; Far Eastern Federal University, Vladivostok, Russia

Emma Puighermanal
Department of Cell Biology, Physiology and Immunology, Institute of Neuroscience, Autonomous University of Barcelona, Barcelona, Spain

Naresh Ramesh
Department of Veterinary Biomedical Sciences, Western College of Veterinary Medicine, Saskatoon, SK, Canada

Domenico Ribatti
Department of Basic Medical Sciences, Neurosciences and Sensory Organs (SMBNOS), Section of Human Anatomy and Histology, University of Bari "Aldo Moro", Bari, Italy

Patricia Silveyra
Department of Pediatrics, The Pennsylvania State University College of Medicine, Hershey, PA; The University of North Carolina at Chapel Hill, School of Nursing, Chapel Hill, NC, United States

Margherita Sisto
Department of Basic Medical Sciences, Neurosciences and Sensory Organs (SMBNOS), Section of Human Anatomy and Histology, University of Bari "Aldo Moro", Bari, Italy

Mentor Sopjani
Faculty of Medicine, University of Prishtina, Prishtina, Kosova

David S. Ucker
Department of Microbiology and Immunology, University of Illinois College of Medicine, Chicago, IL, United States

Ji Won Um
Department of Brain and Cognitive Sciences, Daegu Gyeongbuk Institute of Science and Technology (DGIST), Daegu, Korea

Suraj Unniappan
Department of Veterinary Biomedical Sciences, Western College of Veterinary Medicine, Saskatoon, SK, Canada

Emmanuel Valjent
IGF, CNRS, INSERM, University of Montpellier, Montpellier, France

Christine Vogel
Department of Biology, New York University, New York, NY, United States

Jinan Wang
Center for Computational Biology and Department of Molecular Biosciences, University of Kansas, Lawrence, KS, United States

CHAPTER ONE

Voices from the dead: The complex vocabulary and intricate grammar of dead cells

Jerrold S. Levine[a,*], David S. Ucker[b]

[a]Department of Medicine, Section of Nephrology, University of Illinois College of Medicine and Jesse Brown Veterans Affairs Medical Center, Chicago, IL, United States

[b]Department of Microbiology and Immunology, University of Illinois College of Medicine, Chicago, IL, United States

*Corresponding author: e-mail address: jslevine@uic.edu

Contents

Abstract

Of the roughly one million cells per second dying throughout the body, the vast majority dies by apoptosis, the predominant form of regulated cell death in higher organisms. Long regarded as mere waste, apoptotic cells are now recognized as playing a prominent and active role in homeostatic maintenance, especially resolution of inflammation, and in the sculpting of tissues during development. The activities associated with apoptotic cells are continually expanding, with more recent studies demonstrating their ability to modulate such vital functions as proliferation, survival, differentiation, metabolism,

Advances in Protein Chemistry and Structural Biology, Volume 116
ISSN 1876-1623
https://doi.org/10.1016/bs.apcsb.2019.02.004

migration, and angiogenesis. In each case, the role of apoptotic cells is active, exerting their effects via new activities acquired during the apoptotic program. Moreover, the capacity to recognize and respond to apoptotic cells is not limited to professional phagocytes. Most, if not all, cells receive and integrate an array of signals from cells dying in their vicinity. These signals comprise a form of biochemical communication. As reviewed in this chapter, this communication is remarkably sophisticated; each of its three critical steps—encoding, transmission, and decoding of the apoptotic cell's "message"—is endowed with exquisite robustness. Together, the abundance and intricacy of the variables at each step comprise the vocabulary and grammar of the language by which dead cells achieve their post-mortem voice. The combinatorial complexity of the resulting communication network permits dying cells, through the signals they emit and the responses those signals elicit, to partake of an expanded role in homeostasis, acting as both sentinels of environmental change and agents of adaptation.

1. Introduction

The importance of a biologic process or pathway can often be inferred from the extent of its redundancy and specialization, and/or from the intricacy of its regulation. As a general rule, the more vital a process or pathway is to the wellbeing of an organism, the greater the number of specialized alternatives or backups it will incorporate, and the more precise and nimble will be its level of control. Characteristic examples include such critical and complex processes as the coagulation system and cell cycle, each of which is endowed with an extensive network of positive and negative regulators, as well as various safeguards and substitutes. The same applies to the machinery overseeing cell death, not only the various sensors and integrative pathways aiding the cell in its choice between life and death (Evan & Vousden, 2001) but also the various forms and manifestations of death available to the cell (Galluzzi et al., 2018), as well as the phagocytic apparatus responsible for the rapid and efficient clearance of the roughly one million cellular corpses generated per second throughout the body (Arandjelovic & Ravichandran, 2015; Devitt & Marshall, 2011; Green, Oguin, & Martinez, 2016).

Remarkably, the consequences of cell death do not end with disposal of the corpse. The dead cell's influence ranges far beyond its mere demise, well into the afterlife, and affects most, if not all, live cells in the vicinity (Arandjelovic & Ravichandran, 2015; Birge & Ucker, 2008; Elliott, Koster, & Murphy, 2017; Hochreiter-Hufford & Ravichandran, 2013; Peter, Wesselborg, Herrmann, & Lauber, 2010). Although images of dead cells possessing many of the morphologic features associated with apoptosis, the predominant form of regulated cell death in higher organisms (Nagata, 2018), first appeared in the scientific literature in the late 1800s, recognition of the postmortem activities acquired by cells dying apoptotically lagged considerably behind (Lockshin & Zakeri, 2001; Wallach, Kang, & Kovalenko, 2014). As recently as the mid-1990s, apoptosis was regarded simply as a process for the elimination of unwanted and functionally inappropriate cells. The reason for this misapprehension relates in large part to the very rapidity and efficiency with which dead cells are removed (Arandjelovic & Ravichandran, 2015; Devitt & Marshall, 2011; Elliott et al., 2017). Indeed, in the absence of specific histologic labeling, apoptotic cell death often goes undetected, even in situations of heightened cell turnover, such as embryogenesis or within the immature thymus (Nagata, 2018; Surh & Sprent, 1994; Suzanne & Steller, 2013).

Despite being largely invisible under physiological conditions, apoptosis is by no means silent. A striking breadth and complexity characterize the flow of information from apoptotic cells to live cells in their vicinity (Elliott et al., 2017; Hochreiter-Hufford & Ravichandran, 2013; Peter et al., 2010). Indeed, recognition of the potent and pleiotropic activities acquired by cells as they die apoptotically has led to a recasting of the role of apoptosis as a vital participant in homeostatic maintenance (Arandjelovic & Ravichandran, 2015). Initial studies leading to this redefinition focused on the ability of apoptotic cells to suppress inflammation (Fadok et al., 1998; Voll et al., 1997). It should be emphasized that the absence of inflammation is not simply a passive avoidance, achieved by prompt clearance and prevention of cellular leakage, but instead an active and potent suppression exerted on the environment (Birge & Ucker, 2008; Cvetanovic et al., 2006; Cvetanovic & Ucker, 2004). Since this initial description, apoptotic cells have been shown also to affect a broad range of cellular functions, including such critical activities as survival (Patel et al., 2012; Pérez-Garijo, Fuchs, & Steller, 2013; Reddy et al., 2002), proliferation (Patel et al., 2012; Reddy et al., 2002), differentiation (Jiang et al., 2009; Miura et al., 2004; Zernecke et al., 2009), metabolism (Patel et al., 2015), angiogenesis (Li et al., 2010), and migration (Elliott et al., 2017; Hochreiter-Hufford & Ravichandran, 2013; Peter et al., 2010). Moreover, these effects are not limited to the professional phagocytes charged with the clearance of apoptotic cells, but extend also to virtually every living cell in the vicinity, regardless of its origin or lineage (Arandjelovic & Ravichandran, 2015; Birge & Ucker, 2008; Devitt & Marshall, 2011).

In this chapter, we will review the evidence that most, if not all, viable cells receive and integrate a vast array of signals from cells dying in their vicinity. This is true, regardless of the form of cell death, whether apoptotic or non-apoptotic, regulated or accidental, physiologic or pathologic. Continuous monitoring of these signals provides cells with a streaming summary of prevailing conditions and serves as an early warning system of local perturbations to the environment. The response that a cell makes after processing this information is cell-specific, genetically determined, and for the most part advantageous to the organ and organism in which the cell resides. Our focus here will be on the "message" itself, its nuance and robustness, and the wealth of extractable information it contains. As we will review, this vigor derives not only from the abundance of death-related variables with which the dying cell encodes its message but also the cell-specific decoding of that message, in which the same set of signals can elicit vastly different responses depending upon the lineage or differentiative stage of the responding cell. The combinatorial complexity of the resulting communication network permits dying cells, through the signals they emit and the responses those signals elicit, to act both as sentinels of environmental change and as agents of adaptation.

2. The language of cell death

Communication from dead cells to neighboring live cells occurs in a multitude of diverse ways. In general, these may be divided into two broad categories: direct transmission, via a physical interaction between dead and live cells, or indirect transmission, via intermediary agents, in the absence of physical contact between communicating cells. Direct effects occur most commonly via receptor-mediated recognition by live cells of adjacent dead cells or their fragments (Birge & Ucker, 2008; Hochreiter-Hufford & Ravichandran, 2013; Penberthy & Ravichandran, 2016), whereas indirect effects are most frequently the result of soluble mediators released from dying cells (Elliott et al., 2017; Peter et al., 2010). The distinction, however, is not always clear-cut. For example, information may be packaged within exosomes or microvesicles, with specific effects deriving not from receptor-mediated recognition of the exosomes or microvesicles themselves, but rather, in the manner of a Trojan horse, from uptake of their hidden cargo (Zernecke et al., 2009).

Regardless of the encoded message, dead cells, as we have increasingly come to realize, achieve their effects in sometimes marvelously sophisticated and ingenious ways (Ariel et al., 2006; Röszer, 2017; Uderhardt et al., 2012). Viewed as a language, dead cells possess an extensive vocabulary and an intricate grammar. The enormity and complexity of the information available may be appreciated from the following considerations. As mentioned previously, approximately one million cells per second die throughout the body as part of routine homeostatic maintenance (Arandjelovic & Ravichandran, 2015; Devitt & Marshall, 2011; Green et al., 2016). Locally this number can increase dramatically, for example, during infection or acute injury. If we consider these dead or dying cells, and the signals they emit, as elements or words in a vocabulary, then the scope of conveyable information is revealed through an analysis of the myriad available death-related variables. At a single cell level, these include, but are by no means limited to, the mode of cell death (Cvetanovic & Ucker, 2004; Krysko et al., 2006; Patel et al., 2006)—and, under certain conditions, even the particular inducer of that form of cell death (Obeid et al., 2007; Patel et al., 2012)—the elapsed time since induction of death, and the origin and lineage of the dying cell, as well as the large assortment of surface determinants and soluble mediators expressed by the cell as it undergoes its particular death routine (Galluzzi et al., 2018; Hochreiter-Hufford & Ravichandran, 2013; Medina & Ravichandran, 2016; Peter et al., 2010). A grammar arises from the rules governing the manifold ways in which the words represented by dead cells and their signals are combined, and encompasses such additional variables as the pattern, distribution, kinetics, rate, and extent of cell death (Bergmann & Steller, 2010; King & Newmark, 2012; Suzanne & Steller, 2013).

As if this were not enough, a further layer of complexity stems from the fact that the same death-related variables may be interpreted differently by different responding cells. Thus, the response of any given live cell hinges as much on the identity of the responding cell itself—its lineage (Patel et al., 2010; Reddy et al., 2002), organ of residence (Patel et al., 2012), and stage of differentiation (Bournazou et al., 2009; Esmann et al., 2010)—as on the specific properties of the dead or dying cells. Further, it is relatively easy to imagine that even live cells lacking direct interaction with dead cells or their released mediators may still be affected by the dynamic flow of information. A ripple effect may ensue, in which dead cells stimulate the synthesis and release of cytokines and mediators from live cells, and in turn these cytokines and mediators then modulate the activity of other live cells that reside at a distance, at the fringe of or beyond the area of augmented cell death, even within distal organs and tissues.

To date, decrypting of the language used by dead and dying cells to transmit information is greatest for cells that die by apoptosis. With the possible exception of necrosis (Peter et al., 2010; Zitvogel, Kepp, & Kroemer, 2010), a general term for the unregulated and catastrophic death of cells subjected to extreme physical or chemical stress (Galluzzi et al., 2018), knowledge of the effects of cells dying by other mechanisms is still at a preliminary stage. For this reason, and because apoptosis is the primary mechanism by which cells die under physiologic conditions (Nagata, 2018), we will focus primarily on this form of cell death. Whenever appropriate, we will contrast the effects of cells dying by apoptosis with those of cells dying by non-apoptotic means. While based thus far on limited evidence, it is very likely that the paradigm presented in this chapter will extend to most, if not all, other forms of cell death (Galluzzi et al., 2018), both unregulated and regulated, including, for example, necroptosis (Kaczmarek, Vandenabeele, & Krysko, 2013; Martinez et al., 2011; Zargarian et al., 2017), pyroptosis (Jorgensen & Miao, 2015), and ferroptosis (Stockwell et al., 2017).

3. An expanded view of the role of cell death in homeostasis

In the absence of significant environmental stress, the cells within a tissue or organ exist in a steady state, referred to as homeostasis, in which a large number of variables characterizing the population as a whole are dynamically balanced (Cannon & Cannon, 1967). From the perspective of this chapter, a useful definition of homeostasis is the tendency of cells to resist change in order to maintain a relatively stable external environment. Straightforward examples of variables under homeostatic control include temperature, pH, water balance, and delivery of oxygen. Homeostasis operates at multiple levels, from the whole body down to individual organs and even to cellular niches within tissues. Moreover, the set point for any given variable may vary from tissue to tissue or from cell type to cell type within a given tissue.

With respect to the role of cell death in homeostasis, most attention has been directed to a few areas in particular. These include tissue renewal (for example, the continuous shedding and replacement of the epithelial cell lining of organs such as the gastrointestinal tract) (Günther, Neumann, Neurath, & Becker, 2013), population control (for example, the restoration of a stable number of circulating lymphocytes within the blood stream following an adaptive immune response) (Rathmell & Thompson, 2002), and, especially, clearance of cellular corpses (for example, to cite an extreme instance, the removal within the testes of the millions of spermatogenic corpses generated daily) (Nakanishi & Shiratsuchi, 2004). In all but the last of these examples, the predominant direction of information flow, at least initially, while the cell's fate remains up in the air, is from the environment to cells. Only after the cell has committed to its death, and its clearance becomes an active issue, does the direction of information flow shift and begin to move outward, from the cell to its environment, thereby providing the cell a voice in the manner and consequences of its disposal. Regardless of the direction of information flow, however, the vital contribution of cell death to homeostasis in each of these examples is revealed by the organ pathology and disease resulting from disruption of the cell death machinery through single gene loss or mutation (Drappa, Vaishnaw, Sullivan, Chu, & Elkon, 1996; Elliott et al., 2010; Günther et al., 2013).

Maintenance of homeostasis across a multitude of cell types and variables necessarily involves an elaborate interplay of counterbalancing processes. To take a simple example, consider the distinct constraints in maintaining population size for two cell types, resident (non-migratory) and infiltrating (migratory). Steady state for resident cells is achieved primarily through a balancing of the rates of cell division and cell death, whereas that for infiltrating cells requires an additional balancing of the rates of cellular influx and efflux. These balances, in turn, are realized through a variety of mechanisms: competition for essential nutrients and factors, rate-limiting steps within key metabolic and signaling pathways, limitations of diffusion, spatial and other physical barriers, as well as multiple others (Green, Galluzzi, & Kroemer, 2014; Johnston, 2009; Marinari et al., 2012). While the most critical, and obvious, variables include the availability of growth and survival factors, the delivery of adequate oxygen and nutrients through the bloodstream, and the effective clearance of metabolic waste products and cellular debris, there are an assortment of other subtle and less evident environmental and intercellular cues that also play an important role, not the least of which is the type and quantity of cell death (King & Newmark, 2012; Suzanne & Steller, 2013; Patel et al., 2006; Johnston, 2009; Marinari et al., 2012; Green et al., 2014).

With respect to dead cell clearance, especially within higher organisms, studies on homeostatic maintenance and restoration have focused largely on the silencing of inflammation and the prevention of an immune response against dead cells and their contents. Both effects, while notable for the absence of something, are nonetheless active in nature, and dependent upon an ongoing constraint. An equally large body of work has explored the cell- and tissue-specific factors and pathways that modulate the susceptibility, extent, and nature of the death response to specific stressors. Remarkably fewer studies, however, have examined systematically the role of cell death after initial injury, in terms of both the attenuation of further damage (aside from that due to inflammation) and the promotion of healing. This is a notable neglect, since in addition to their anti-inflammatory and immunosuppressive properties, apoptotic cells also modulate such vital activities as proliferation, differentiation, and metabolism, all of which play an essential role in wound healing. We use the term "wound healing" here in its broadest sense of restoration of an injured tissue to homeostatic equilibrium, a view supported by a recent editorial on cellular networks in wound healing (Willenborg & Eming, 2018). Skin repair remains the best-studied model, probably because of its accessibility, but the basic events in healing are likely similar across most organs and tissues. These steps encompass a wide range of reparative activities, from dermal and endothelial mending of minor insults, to tissue remodeling after surgical intervention, to the extensive and complex multi-tissue healing following severe traumatic or ischemic injury. In this expanded view of the homeostatic role of cell death, dead and dying cells contribute to the restoration of homeostasis throughout all phases of wound healing (Byun & Gardner, 2013; Dvorak, 1986; Schäfer & Werner, 2008), not only the initial inflammatory phase, in which professional phagocytes such as macrophages are recruited to the tissue in response to the injury, but also the later proliferative and remodeling phases, in which the role of resident non-professional phagocytes like epithelial cells is at the forefront.

Much of our understanding of the ways in which dead or dying cells contribute to homeostasis derives from circumstances in which homeostasis is lost or absent. The two most important examples are developmental morphogenesis (Suzanne & Steller, 2013; Magali Suzanne et al., 2010; Yamaguchi et al., 2011; Cuervo & Covarrubias, 2004; Pérez-Garijo, 2018) and recovery following a severe environmental stress or injury, such as the amputation of an organ or structure (Bergmann & Steller, 2010; Li, Lindblad, Perez, Bergmann, & Fan, 2016; Li et al., 2010; King & Newmark, 2012; Tseng, Adams, Qiu, Koustubhan, & Levin, 2007; Chera et al., 2009; Goessling et al., 2009; Pérez-Garijo, 2018). Both are states of extreme flux, in which there is a clearly increased rate of cell death and/or loss, and the balances necessary for homeostasis are severely challenged. The prominence of cell death in these two situations has facilitated elucidation of the bidirectional flow of information, from the environment to individual cells deciding their fate, and from dead or dying cells back to live cells in the vicinity.

A third, perhaps surprising, source of insight into the expanded role of cell death in the control of homeostasis is cancer (Gregory, Ford, & Voss, 2016; Ucker & Levine, 2018). The markedly increased rate and persistence of cell death, which characterize most tumors, necessarily entail a magnification of the extrinsic, postmortem effects of dead cells (Gregory et al., 2016; Ucker & Levine, 2018). Tumors, insightfully described as "wounds that do not heal" (Dvorak, 1986), or even "wounds that do not stop repairing" (Gregory et al., 2016), are particularly susceptible to the mitogenic effects of apoptotic cells, including those induced extensively by chemotherapy and irradiation (Donato et al., 2014; Huang et al., 2011; Lauber et al., 2011; Mao, Smith, Xie, & Wang, 2013). This likely underlies the perplexing tumor-enhancing role of apoptosis in cancer, for which the extent of apoptosis has been found to correlate with the aggressiveness of the underlying malignancy across a broad spectrum of cancers (Gregory et al., 2016; Ucker & Levine, 2018).

The underlying tendency in morphogenesis and recovery from severe injury or environmental stress, as well as in cancer, is toward establishment of a new steady state. In the case of morphogenesis, the absence of homeostasis is part and parcel of an intricate and evolutionarily conserved program in which cell death is a pervasive feature, often occurring in surprisingly large numbers (M. Suzanne & Steller, 2013; Suzanne et al., 2010; Yamaguchi et al., 2011; Cuervo & Covarrubias, 2004). As its name in this context implies—*programmed* cell death (Galluzzi et al., 2018)—cell death is an essential and active participant during morphogenesis, helping to sculpt the ultimate shape and configuration of the nascent tissues in which it occurs (Suzanne & Steller, 2013).

The role of cell death during recovery from injury or stress is, in essence, similar to that during morphogenesis. Dead and dying cells again help to guide the organ or tissue toward a new state of homeostasis (Bergmann & Steller, 2010; Li et al., 2016; Li et al., 2010; King & Newmark, 2012; Tseng et al., 2007; Chera et al., 2009; Goessling et al., 2009). Unlike morphogenesis, however, the outcome is not predetermined. Indeed, salvage and restoration of homeostasis may not even be possible. If morphogenesis may be viewed as scripted, a dialog between live and dead cells that has been rehearsed and perfected over the eons of evolution, then recovery from an environmental stress can be seen as improvisation. Through an extemporized back and forth between live and dead cells, the stressed organ or tissue does its best to adapt and restore homeostasis.

To take a physiologically relevant example, we may consider the consequences of a reduction of blood flow to a segment of an organ. This could be the result of a partial arterial occlusion. Downstream cells are subjected to a reduced delivery of oxygen and nutrients. If the reduction is severe enough, there will be an increased rate of cell death. Recognition by live cells of this increased number of dead cells serves as a signal of environmental perturbation and danger. The response of individual viable cells within the affected zone will depend on multiple death-related factors, including the mode, extent, kinetics, and distribution of cell death. Adaptive responses might include measures that lead to a decrease in the organ's global consumption of oxygen and nutrients. This may be achieved through a further "voluntary" decrease of population size beyond that imposed by the original stress (e.g., via death receptors and the extrinsic pathway of apoptosis) (Feng et al., 2018; Galluzzi et al., 2018), but also through inhibition of proliferation and other energy-consuming anabolic processes in surviving cells (Patel et al., 2012, 2010, 2015). Some responding cells, particularly those recruited for cleanup and repair, may be permitted energy-consuming activities, or even delivered specific survival signals (Reddy et al., 2002). The tissue's overall response, comprising the integrated responses of its component cells, both resident and recruited, represents an attempt at adaptation toward a new steady state.

While insights from experimental models in which there is a severe departure from homeostasis underpin this chapter, its major aim is to highlight the more subtle role that dead or dying cells play under more everyday conditions, during homeostatic steady state and in the aftermath of less severe environmental fluctuations to which adaptation is possible. Because the responses of neighboring viable cells require either direct physical interaction with dead cells, or the binding of diffusible factors emanating from these cells, the effects of cellular corpses will tend to be short-range and local. Indeed, unless specifically sought, the alerting signals and ensuing adaptive responses may be all but undetectable, so streamlined and finely calibrated is the regulatory network under the purview of cell death. Examples of the remarkable breadth and critical role of the contribution of apoptotic cells to the maintenance and/or restoration of homeostasis are provided in Table 1.

Table 1 Extended homeostatic effects induced by apoptotic cells.[a]

Activity		Modulatory effect	Mechanism	References
Proliferation	Stimulation:	Epidermal keratinocytes, neural/ mesenchymal stem cells, hepatocytes	PGE_2 generated by apoptotic cells	Li et al. (2010)
		Endothelial cells (lung microvascular)	Induced synthesis and release by phagocyte of VEGF	Golpon et al. (2004)
		Endothelial cells (HUVEC, aortic)	CXCL12-dependent effect via delivery of microRNA-126 by apoptotic bodies	Zernecke et al. (2009)
	Inhibition:	Macrophages (bone marrow-derived)	Non-specific effect of phagocytosis (replicated by latex beads)	Reddy et al. (2002)
		Epithelial cells (kidney proximal tubule)	Recognition-dependent, unclear mechanism	Patel et al. (2010, 2012)
Apoptosis	Protection (pro-survival):	Endothelial cells (lung microvascular), epithelial cells (mammary)	Uncharacterized factor released by phagocyte	Golpon et al. (2004)
		Macrophage cell lines (RAW 264.7, THP-1)	S1P generated by apoptotic cells	Weigert et al. (2006)
		Endothelial cells (HUVEC, aortic)	CXCL12-dependent effect via delivery of microRNA-126 by apoptotic bodies	Zernecke et al. (2009)
	"Death-induced death" (anti-survival):	*Drosophila* wing disc (apoptosis in posterior compartment produces massive coordinated death in anterior compartment); coordinated apoptotic death of murine hair follicle cells	Release of TNF-α or Eiger (TNF-α homolog) by apoptotic cells	Pérez-Garijo et al. (2013)
		Epithelial cells (kidney proximal tubule, glomerular podocytes)	Recognition-dependent, caspase-8 dependent via Fas or other death receptor?	Patel et al., 2012, 2010, 2015, Linkermann et al. (2011), and Matsusaka et al. (2011)

Continued

Table 1 Extended homeostatic effects induced by apoptotic cells.[a]—cont'd

Activity		Modulatory effect	Mechanism	References
Metabolism	Inhibition:	Epithelial cells (kidney proximal tubule)	Recognition-dependent activation of AMPK and inhibition of mTORC1	Patel et al. (2015)
Differentiation	Induction:	Murine embryonic stem cells into $CD31^{+}$ endothelial cells	CXCL12-dependent mechanism, via uptake of microRNA-126-containing apoptotic bodies from apoptotic endothelial cells	Zernecke et al. (2009)
		Drosophila gut progenitor cells into mature enterocytes	Release of IL-6-like cytokines	Jiang et al. (2009)
		Osteogenic differentiation from bone marrow stromal stem cells	Caspase-3-dependent mechanism	Miura et al. (2004)
Wound healing/repair	Epithelium:	Accelerated closure/healing of excision skin wounds in mice	PGE_2 generated by apoptotic cells	Li et al. (2010)
		Accelerated closure of wounds in mouse mammary epithelial cell monolayers	Uncharacterized factor released by phagocyte	Golpon et al. (2004)
		Restoration of gut epithelium after infection or injury (*Drosophila*)	Release of IL-6-like cytokines	Jiang et al. (2009)
	Injured muscle:	Regeneration of mature multinucleated muscle fibers (mice)	Apoptotic recognition and signaling via BAI1	Hochreiter-Hufford et al. (2013)
	Atherosclerotic plaques:	Reduction in size and repair (mice)	CXCL12-dependent mechanism via uptake of microRNA-126-containing apoptotic bodies from apoptotic endothelial cells	Zernecke et al. (2009)
	Organ regeneration/renewal:	*Hydra* head following decapitation; Zebrafish hematopoietic stem cells following irradiation: murine hematopoietic stem cells following irradiation; murine liver following partial hepatectomy; *Xenopus laevis* tail following amputation	Wnt and/or PGE_2 generated by apoptotic cells	Chera et al. (2009), Goessling et al. (2009), Li et al. (2010), Tseng et al. (2007), and Pérez-Garijo (2018)

Angiogenesis	Directed *in vivo*:	Subcutaneous tissue of nude mice	PGE_2 generated by apoptotic cells	Li et al. (2010)
Barrier function maintenance	Epithelium:	Canine kidney epithelium during apoptotic cell extrusion	S1P generated by apoptotic cells	Gu, Forostyan, Sabbadini, and Rosenblatt (2011)
Development/ morphogenesis	Movement and/or shaping of tissue:	Fusion of palatal shelves (mice); neural tube closure, non-essential accelerant role (mice); coordinated left/right rotations of genital plates to create genital and anal structures (*Drosophila*)	Unclear mechanisms	Cuervo and Covarrubias (2004), Yamaguchi et al. (2011), and Suzanne et al. (2010)
	Organogenesis:	Myocardial development and osteogenesis (mice); neural path-finding, dendritic pruning, and synaptic plasticity (rodents)	Caspase-3 and/or – 7-dependent	Cardona et al. (2015), Miura et al. (2004), Yamaguchi and Miura (2015), and Pérez-Garijo (2018)
	Myogenesis:	Fusion of myoblasts into multinucleated muscle fibers (mice)	Apoptotic recognition and signaling via BAI1/BAI3	Hochreiter-Hufford et al. (2013) and Hamoud, Tran, Croteau, Kania, and Côté (2014)
	Joint formation:	Folding of epithelium as initial step in joint formation (*Drosophila*)	Actin-myosin cables within apoptotic cells generate tissue tension in surrounding live cells	Monier et al. (2015)

[a]Abbreviations. AMPK, AMP-activated protein kinase; BAI, brain angiogenesis inhibitor; HUVEC, human umbilical vein endothelial cells; IL, interleukin; mTORC1, mammalian target of rapamycin complex 1; PGE_2, prostaglandin E_2; S1P, sphingosine-1 phosphate; TNF, tumor necrosis factor; VEGF, vascular endothelial growth factor.

4. A universally heard language

For dead cells to function in the expanded homeostatic role described above, the ability to recognize and respond to cellular corpses should be universal across all cell types, regardless of phagocytic capacity. Given the wide variety of soluble mediators expressed in association with apoptotic cell death (Hochreiter-Hufford & Ravichandran, 2013; Medina & Ravichandran, 2016; Peter et al., 2010), a cell would need to be virtually deaf to fail to respond to at least one of these mediators. Even if the effects of soluble mediators are ruled out, the ability to respond to apoptotic cells still appears for all intents and purposes to be ubiquitous. It is important, however, to differentiate a specific recognition and response to apoptotic cells from a response to phagocytosis in general (Vujicic, Feng, Antoni, Rauch, & Levine, 2016). For example, some signaling responses, such as activation of Akt in peritoneal and bone marrow-derived macrophages, do not reflect specific interactions with apoptotic cells, but rather are associated with phagocytosis in general (Reddy et al., 2002). This lack of specificity can be revealed through abrogation of a response by inhibition of phagocytosis, and its mimicking by a non-specific phagocytic stimulus such as latex beads (Vujicic et al., 2016). Some such experimental maneuver is requisite, before a response can be ascribed unequivocally to selective discrimination of apoptotic cells, as distinct from generalized engulfment. Unfortunately, such maneuvers have not always been consistently reported in the literature. That said, complete dissociation is not always possible, as some responses, while specific to apoptotic cells, are nonetheless dependent on phagocytosis, and specificity is achieved through differential intracellular trafficking and/or processing of ingested apoptotic material (Röszer, 2017; Krysko et al., 2006; Krysko, Brouckaert, Kalai, Vandenabeele, & D'Herde, 2003; Peng & Elkon, 2011). With these caveats in mind, we may review the evidence in favor of the universality of responsiveness to apoptotic cells.

4.1 Phagocytic clearance and anti-inflammation

Phagocytic cells involved in the clearance of cellular corpses have traditionally been divided into professional and non-professional subsets. Professional phagocytes include immature dendritic cells and macrophages. Non-professional phagocytes include fibroblasts and epithelial cells, among others, and are characterized by a reduced phagocytic capacity, delayed uptake, and prolonged digestion of apoptotic material (Devitt & Marshall, 2011;

Parnaik, Raff, & Scholes, 2000; Wood et al., 2000). Macrophages themselves may be either tissue-resident or recruited from monocytes circulating within the bloodstream. These two pools of macrophages represent distinct populations of divergent origin. Tissue-resident macrophages derive for the most part from a common embryonic source (Gomez Perdiguero et al., 2015; Hoeffel et al., 2015); their names, however, vary depending on the organ of permanent residence. Examples include alveolar macrophages in the lung (Morimoto et al., 2001), Kupffer cells in the liver (Falasca, Bergamini, Serafino, Balabaud, & Dini, 1996), mesangial cells in the kidney (Savill et al., 1992), microglia in the brain (Parnaik et al., 2000), osteoclasts in bone (Harre et al., 2012), and resident peritoneal macrophages (Uderhardt et al., 2012). In situations of augmented cell death, monocytes are recruited from the bloodstream into affected tissues, where they differentiate into macrophages. Resident and recruited macrophages do not necessarily subserve the same function. Depending on the tissue, resident and recruited macrophages may differ not only in the receptors they use to recognize and ingest apoptotic cells, but also in their phagocytic capacity, and even in their very response to apoptotic cells (Freeman, Casasnovas, Umetsu, & DeKruyff, 2010; Ipseiz et al., 2014; Uderhardt et al., 2012).

While the ability of non-professional phagocytes to ingest apoptotic cells has long been recognized, the critical contribution of these cells to homeostasis has not. Early reports of the ability of non-professional phagocytes to ingest apoptotic cells more or less covered the gamut of exotic cell types and locations throughout the body: mammary epithelial cells (Monks et al., 2005), lens epithelial cells (Parnaik et al., 2000), high endothelial venule cells in lymph nodes (Hess et al., 1997), sinusoidal endothelial cells in the liver (Dini & Carla, 1998), vascular smooth muscle cells (Bennett, Gibson, Schwartz, & Tait, 1995), brain astrocytes (Parnaik et al., 2000), foreskin fibroblasts (Hall, Savill, Henson, & Haslett, 1994), and early embryonic mesenchymal cells (Wood et al., 2000). With a few notable exceptions (for example, post-lactational involution of the mammary gland) (Monks, Smith-Steinhart, Kruk, Fadok, & Henson, 2008), recognition of the important role of non-professional phagocytes, particularly epithelial cells, in the clearance of dead cells has required genetic manipulation and demonstration of disease worsening, increased susceptibility, or delayed healing upon impairment of phagocytic capacity. For example, inducible deletion of Rac1, a small GTPase required for phagocytosis of apoptotic cells, from bronchial epithelial cells increased susceptibility to allergen-induced inflammation (Juncadella et al., 2013). Loss of expression of brain-specific angiogenesis inhibitor-1 (BAI1), a receptor for apoptotic cells on colonic epithelium, increased morbidity and mortality in a model of colitis, whereas transgenic overexpression of BAI1 ameliorated disease (Lee et al., 2016). Finally, genetic deletion of a domain critical to phagocytosis from kidney injury molecule 1/T cell Ig and mucin domain 1 (KIM-1/TIM-1), a receptor for apoptotic cells up-regulated by kidney proximal tubular epithelial cells in response to injury, resulted in greater functional impairment and mortality in two distinct models of acute kidney injury (Yang et al., 2015).

These same three studies also demonstrated that the consequences of epithelial cell interaction with apoptotic cells, at least in the organs examined, extend beyond mere phagocytosis, entailing also a reduction in inflammatory responses (Juncadella et al., 2013; Lee et al., 2016; Yang et al., 2015). As discussed above, suppression of inflammation was one of the earliest activities attributed to apoptotic cells (Fadok, Bratton, et al., 1998; Voll et al., 1997). Subsequent work supported the concept that anti-inflammation is a ubiquitous aspect of the response to apoptotic corpses (Birge & Ucker, 2008). In a systematic study across a broad spectrum of cell types, recognition of apoptotic cells, independent of later engulfment or release of soluble factors by responding cells, triggered a set of potent anti-inflammatory responses (Cvetanovic et al., 2006). Exposure of cells of epithelial, endothelial, neuronal, and lymphoid lineages (untransformed as well as transformed) to apoptotic cells (but not to necrotic or viable cells) led to a rapid and profound ablation of nuclear factor κB (NFκB)-dependent pro-inflammatory transcription and subsequent pro-inflammatory cytokine release (Cvetanovic et al., 2006). Notably, since these anti-inflammatory responses occurred without regard to species, tissue type, or suicidal stimulus of the apoptotic cell, they represent an unconventional aspect of tissue surveillance, in which the essential distinction is that between viable and effete cells, as opposed to the traditional immunological distinction between "self" and "foreign" (Birge & Ucker, 2008; Cvetanovic et al., 2006).

4.2 Extended homeostatic effects

When it comes to the ability of cells to respond to apoptotic corpses in a manner other than anti-inflammation, the evidence for ubiquity is not nearly so solid. To date, the complementary homeostatic effects resulting from the interaction between apoptotic corpses and resident cells have been explored most carefully in invertebrate models of development, morphogenesis, and/or tissue regeneration (so-called apoptosis-induced compensatory proliferation) (Bergmann & Steller, 2010; Fan & Bergmann, 2008; King & Newmark, 2012; Suzanne & Steller, 2013). While these studies have provided invaluable insight into the elegant and diverse mechanisms by which recognition of apoptotic cells drives such complex processes as *Hydra* head regeneration following amputation (Chera et al., 2009), or the formation of an epithelial "fold" preceding joint formation in *Drosophila* (Pérez-Garijo et al., 2013), these results are not immediately translatable to humans.

Still, several noteworthy examples of the beneficial responses of resident cells to apoptotic cells in the restoration of homeostasis among higher organisms have been reported. One of the more striking is the so-called phoenix rising pathway (Li et al., 2010). Apoptotic cells, irrespective of source, were shown to stimulate the *in vitro* proliferation of a variety of progenitor cells, including epidermal keratinocyte precursors and mesenchymal stem cells. The proliferative effect of apoptotic cells depends strongly on events occurring within the apoptotic cells themselves as part of the apoptotic program. These events included cleavage and activation of caspase-3, followed by caspase-3-mediated activation of calcium-independent phospholipase A_2 ($iPLA_2$) and the $iPLA_2$-mediated generation of prostaglandin E_2 (PGE_2). PGE_2 released from apoptotic corpses then acted upon viable cells. The *in vivo* importance of this pathway was shown by comparison of wild-type versus caspase-3-deficient or caspase-7-deficient mice subjected to excision wounds or partial hepatectomy. In caspase-deficient mice, excision wounds evinced delayed wound closure, impaired epidermal cell migration, retarded epithelial cell recovery, and decreased overall cell proliferation. Similarly, following partial hepatectomy, deficient mice had impaired recovery of pre-surgical liver weight and overall decreased cell proliferation. A second noteworthy example involved the uptake of apoptotic bodies by endothelial cells and vascular smooth cells within atherosclerotic plaques (Zernecke et al., 2009). In addition to suppression of inflammation, apoptotic bodies promoted the recruitment and proliferation of both Stem cells antigen-1 (Sca-1)$^+$ stem cells and CD31$^+$ endothelial cell precursors. Together, these and other studies suggest that the ubiquity of responsiveness to apoptotic cells is not limited to anti-inflammation, but covers the full range of cellular activities.

4.3 Hierarchical considerations

The factors determining which of the available phagocytes undertake the clearance of dead cells and/or respond to their message remains a largely open question. In some tissues, such as the testis and eye, specialized phagocytes, recently proposed as a third category of phagocytic cells (Arandjelovic & Ravichandran, 2015), assume primary responsibility. Both the testis and eye are characterized by a large and continuous burden of apoptotic cells or debris, yet are barred from recruiting monocytes from the bloodstream for reasons of anatomic barrier and immunologic privilege (Elliott et al., 2010; Kim et al., 2013; Nakanishi & Shiratsuchi, 2004; Nandrot et al., 2004). Sertoli cells in the testes and retinal pigment epithelial cells in the eye are specialized multifunctional cells that, in addition to their other nursing and synthetic functions, account almost exclusively for the everyday removal of apoptotic germ cells and the shed outer segments of retinal photorecptors, respectively.

In most other tissues, professional and non-professional phagocytes share the responsibility of clearance. Under normal conditions, the choice between resident professional and non-professional phagocytes likely depends as much on the relative proportion of each cell type present as on issues of access and proximity. When injury or infection or some other environmental stress increases the local load of dead cells, then monocytes or other inflammatory cells may be recruited from the bloodstream to aid in the cleanup. While it is tempting to describe the apportioning of dead cells among phagocytic cells as occurring on a sort of "first-come, first-served" basis, the ability of dead cells to actively recruit certain cell types (Elliott et al., 2017; Hochreiter-Hufford & Ravichandran, 2013; Peter et al., 2010), while repelling others (Bournazou et al., 2009; Janssen et al., 2008; Voss, Tian, Murakami, Coligan, & Krzewski, 2015), suggests that, even in the case of non-migratory tissue-fixed cells, apoptotic corpses may retain a say in which cells ultimately dispose of them.

5. Paradigm

As used in this chapter, communication is defined as the act of conveying information from one cell to another through the use of mutually understood signals and rules. Like any communication system, the sending and receipt of a message involves a number of discrete steps (Fig. 1). The first step involves the assembling of the information to be contained in the message. Information includes any of the following: the reason for the cell's death (natural causes versus environmental stress); the type and severity of the environmental stress; the desired means of the dead cell's disposal; and instructions to living cells in the vicinity for restoration of a new homeostatic steady state, if possible. For communication to be effective, the information

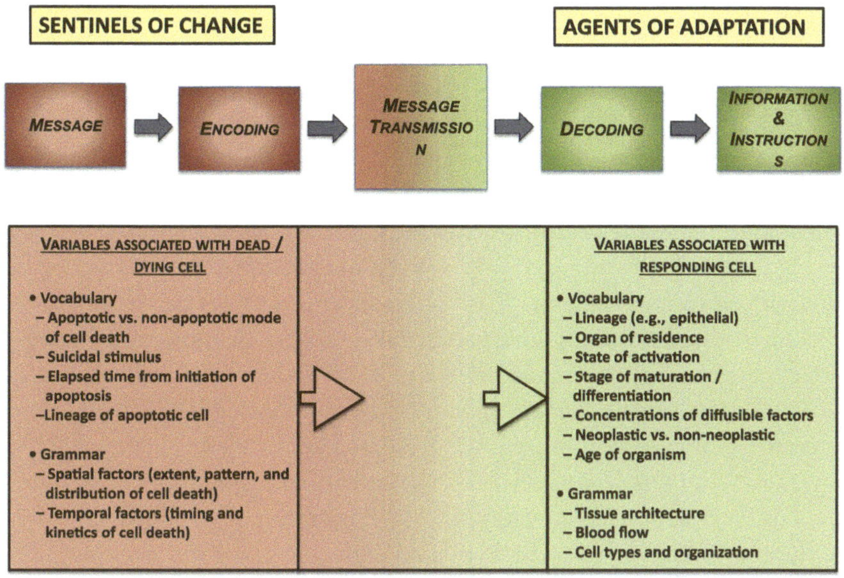

Fig. 1 Model of communication between apoptotic and live cells. Depicted are the discrete steps involved in the transmission of a message from dead or dying cells to live cells in the vicinity. The precision and robustness of the information contained within that message arises from the multiple variables available at each of the three most critical steps: encoding, transmission, and decoding of the message. Examples of these variables are listed for the encoding and decoding steps. Examples of the various modes of transmission are given in Fig. 2. Variables associated with an individual cell comprise the vocabulary. A grammar arises from the rules governing the integration of individual vocabulary elements into a coherent message. The resulting communication network permits apoptotic cells to act as both sentinels of local environmental change and agents of adaptation.

must be encoded in a format capable of transmission. Such encoding entails ultimately the alteration and/or rearrangement of the components and molecules of the dying cell, their synthesis or degradation, their cleavage or chemical modification, their release from the cell or translocation from one cellular compartment to another. Each of these events may contain information either singly or in aggregate, the arrangement of events being important not only in terms of spatial organization, but also temporal ordering.

Once the message is encoded, it is transmitted to live cells in the vicinity. Whereas neurons within the brain use electrical impulses, and most modern communication systems use electromagnetic waves, the medium by which dead cells transmit their message is biochemical. Typically this occurs either through direct physical contact between dead and live cells, via a variety of receptor-mediated interactions (Penberthy & Ravichandran, 2016), or at a distance via diffusible mediators (Peter et al., 2010). As primitive as this means of transmission may seem, it nonetheless permits enormous subtlety and robustness. Several examples provide a flavor. Instead of transmitting a signal, apoptotic cells may delete signals intended for live cells, through adsorption and sequestration of soluble mediators used by one live cell to recruit another (Ariel et al., 2006). Or the message may be enclosed within the dead cell or its fragments, like a letter inside an envelope, with the dead cell's surface ligands acting like an address, ensuring delivery of the message to the proper receptor-bearing recipient. Even in the case of enclosure, there is room for further layers of complexity. The message may be directly inside, in the form of cytosolic proteins, RNA, and lipids (Castellana et al., 2009; Lopatina, Gai, Deregibus, Kholia, & Camussi, 2016; Zernecke et al., 2009), or it may require further unwrapping, via digestion of the apoptotic cell or its fragments, with the byproducts of metabolic degradation, predominantly lipids, bearing the message (A-Gonzalez et al., 2009; Mukundan et al., 2009; Röszer, 2017; Roszer et al., 2011; Sarang et al., 2014).

After receipt of the message by nearby live cells, it must be decoded, its information extracted, and its instructions carried out. In one sense, this is a simple matter of the engagement of membrane receptors on responding cells—by soluble mediators and/or determinants on the surface of corpses and their fragments—or the activation of nuclear transcription factors by the metabolic products of digested corpses, as well as any of the other more subtle and elegant ways in which dead cells modulate intracellular signal transduction and gene expression. Interpretation of the dead cell's message and execution of its instructions result from integration of these various signaling and transcriptional events.

Specificity of effect arises in multiple ways, with the same message yielding potentially different effects in different cells. For example, many of the soluble mediators generated by apoptotic cells, such as sphingosine-1-phosphate (S1P) (Gude et al., 2008; Weigert et al., 2006) or various nucleotides including adenosine triphosphate (ATP) and uridine triphosphate (UTP) (Elliott et al., 2009), are recognized by multiple different receptors with diverse and divergent responses (Cekic & Linden, 2016; Weigert, Weis, & Brüne, 2009). Lineage specific differences in the expression pattern of these receptors, as well as concentration-dependent effects attributable to differences in receptor binding affinity, may determine not only which cells respond to any given message, but also the specific behavior of that response.

Even when only a single receptor is known to exist for a soluble mediator, the effects of that mediator may vary across cell types. The dimer of ribosomal protein S19 (RP S19), for example, which interacts with CD88 (the receptor for complement factor C5a), promotes the migration of monocytes, whereas it inhibits the migration of neutrophils (Revollo et al., 2005; Yamamoto, 2007). While in the case of RP S19 the opposing responses seem to be attributable to a C-terminal moiety, whose removal converts the effect of RP S19 on neutrophils from antagonist to agonist (Shrestha et al., 2003), differences in other cases may relate to cell-specific variations in the relative activity and/or nature of cross-talk among the assorted signal transduction pathways extant in any given cell.

6. A complex language

Any environmental perturbation or stress significant enough to require an adaptive response is typically accompanied by a change in the quantity and nature of cell death. The resulting dead cells serve a dual role. Not only do they signal the presence of an alteration in ambient conditions but they also serve to provide instructions to live cells in the vicinity on coping with the consequences of their altered environment. The instructions apply both to the short term (cleanup of the dead and mitigation of inflammation, if any) and to the long term (adaptation and a return to homeostatic steady state, albeit perhaps a new one, differing to a lesser or greater degree from the one before). While the essence of the message containing these instructions is biochemical, the precision and robustness of the information it contains derives from the multiplicity of variables available at each of the three critical steps within the process of communication (Fig. 1): encoding of the message by dead cells, transmission of the message, and receipt and decoding of the message by live recipient cells in the vicinity. The abundance and intricacy of these variables together comprise the vocabulary and grammar of the language by which dead cells achieve their voice.

Before examining in greater detail each of these three critical steps, it is worthwhile considering a few general principles. At an intuitive level, the information content of a dead cell is determined by the collection of its distinguishable characteristics. (A more precise definition involves the number of available values for each distinguishable characteristic, and the frequency with each of those values is chosen, but this is beyond the scope and needs of this chapter.) Characteristics, not necessarily distinguishable, may be discrete (for example, the mode of cell death, or the lineage of the dead cell) or continuous (for example, the elapsed time from application of a suicidal stimulus, or the diameter of a pore opened specifically in the surface membrane of a cell undergoing apoptosis). Characteristics may even vary with time (for example, the concentration of a released factor). Much of this potential information is extraneous, neither relevant nor interpretable. For information to be useful, the particular value (discrete or continuous) associated with the characteristic under consideration must be distinguishable by the intended recipient of that message. Moreover, the response of recipient cells to the available values for that characteristic (for example, necrotic versus apoptotic cell death) should be clear, specific, and reproducible, notwithstanding an unavoidable rate of error, inherent to the communication system itself, a concept embedded within the term "noise". Evolution has led to the selection of strategies that combat noise and enhance a message's signal-to-noise ratio; these will be discussed below. Importantly, a distinguishable characteristic need not in all cases yield distinct outcomes. For example, although macrophages recognize apoptotic and necrotic cells via distinct and non-competing receptors (Cocco & Ucker, 2001), exposure to either type of dead cell results in activation of Akt and enhanced survival (Reddy et al., 2002). In contrast, apoptotic and necrotic cells produce divergent responses with respect to ERK1/2 activity (Patel et al., 2006; Reddy et al., 2002), demonstrating that, at least for macrophages, the mode of cell death is not simply a distinguishable, but also a consequential, characteristic of cell death.

This last is a critical point, as a characteristic that is distinguishable and of consequence for one recipient cell may be irrelevant for another. A straightforward example would be two cell types, only one of which bears the receptor for a diffusible factor released by a dead cell. This fact carries a powerful implication: despite containing the same information, the delivered message by any given dead cell can vary according to the recipient. Thus, as depicted in Fig. 1, those variables associated with the dead or dying cell have greatest impact on the message's encoding, whereas those variables associated with the recipient cell have greatest impact on its decoding. To illustrate how each of these two sets of variables—dead cell-associated versus responding cell-associated—contributes to the combinatorial complexity of the language of dead cells, we will take the following approach. In discussing the encoding of the message, we will concentrate on the dead cell and those of its characteristics whose different values generate different outcomes in the same responding cell. A simple example is the one cited above, in which apoptotic and necrotic cells have opposite effects on ERK1/2 activity (Patel et al., 2006; Reddy et al., 2002). In discussing the decoding of the message, we will instead concentrate on the responding cell and those characteristics of the responding cell whose different values generate different outcomes for the same dead cell or message. As throughout this chapter, we will concern ourselves predominantly with the language of cells dying by apoptosis, the major form of physiological cell death (Nagata, 2018), and refer to other modes of cell death only as they highlight important aspects of the message transmitted by apoptotic cells.

6.1 Variables associated with the dead or dying cell

The two variables most utilized by dead or dying cells to encode their message, and also having domains with the greatest number of possible values, are: first, diffusible factors, passively or actively released during the process of death; and, second, newly expressed surface determinants or ligands, by which dead cells are recognized by professional and non-professional phagocytes. Other variables, though no less important, have more restricted domains of distinguishable values and include: mode of cell death; suicidal stimulus leading to initiation of apoptosis; elapsed time from induction of apoptosis; and lineage of the apoptotic cell. Most departures from homeostasis requiring an adaptive response lead to the death of huge numbers of cells, well in excess of the roughly one million cells per second dying naturally throughout the body (Arandjelovic & Ravichandran, 2015). Additional information may be extracted from the quantity and pattern of these dead cells, their rate of appearance and their arrangement, both spatially and temporally. While this last source of information has been systematically explored so far almost exclusively only in models of morphogenesis and organ regeneration (see Table 1 for references), such large-scale integration lies at the heart of what we mean by a grammar, in that it permits the synthesis of individual variables into a complex and coherent message.

In discussing the individual variables and respective values used by dead cells to encode their message, we will limit ourselves to general principles and illustrative examples. A comprehensive description of the domains for the two most prolific variables associated with apoptotic cell death is provided in Tables 2 and 3, which categorize the many diffusible factors released by apoptotic cells. Table 4 does the same for the many novel ligands and determinants they express on their surface. Because these variables play a meaningful role in both the encoding and the decoding of the dead cell's message, and because they are essential to the interaction between dead and responding cells, discussion of them will be deferred until the section on message transmission.

Table 2 Diffusible mediators generated by apoptotic cells.[a]

Mediator	Apoptotic cell of origin	Responders	Putative receptor	Other effects[b]	References
"FIND ME"					
Adenosine	W319L; extracellular conversion of ATP/AMP released from apoptotic cells through pannexin 1 channels	Plasmacytoid DC; neutrophils	AR (1, 2A, 2B, 3)	A1R: "find me" for plasmacytoid DC and neutrophils; A2AR: anti-inflammatory (most immune cells), ↑ Nr4a1-3 expression; A2BR: anti-inflammatory (macrophages and DC), ↑ IL-6 and VEGF; A3R: "don't find me" for neutrophils	Cekic and Linden (2016), Idzko, Ferrari, and Eltzschig (2014), Schnurr et al. (2004), Novitskiy et al. (2008), Yamaguchi, Maruyama, Urade, and Nagata (2014), Köröskényi et al. (2011), Duró, Pallai, Köröskényi, Sarang, and Szondy (2014), and Crean et al. (2015)
Annexin 1 fragment (amino acids 2–7)	Jurkat, MOLT4, PHA-treated THP-1, MDA-MB, Colo 800	Human PBM	?[c] ALX/FPR2	Sole biologic activity demonstrated for annexin 1 fragment (as opposed to intact protein) is "find me" for PBM	Blume et al. (2012), McArthur et al. (2015), and Sugimoto, Vago, Teixeira, and Sousa (2016)
ATP/UTP	Murine thymocytes, Jurkat, MCF-7_{casp3}	THP-1; many other cell types, though not directly established in case of apoptotic source of ATP/UTP (J773, HL60, rat mast cells, human neutrophils)	P2XR (10 subtypes) (ATP only) (EC_{50} ~0.5–1.0 μM, except P2X2R and P2X7R, ~10–100 μM)	Receptor-specific cell activation (T cells, mast cells, monocytes, macrophages, DC, iNK cells); ↑ phagocytosis (microglial cells, macrophages)	Corriden and Insel (2010), Idzko et al. (2014), Cekic and Linden (2016), Elliott et al. (2009), and Chekeni et al. (2010)
			P2YR (8 subtypes) (ATP and UTP) ($EC_{50} < 1.0$ μM)	Receptor-specific cell activation (monocytes, macrophages, DC, microglial cells, neutrophils); ↑ phagocytosis (microglial cells, macrophages); eNOS-dependent vasodilation; ↑ proliferation of multiple cell types	

CX3CL1/ fractalkine	Burkitt's lymphoma cells, germinal center B lymphocytes, CD19-depleted PBL, damaged neurons; NB4	MonoMac6, human MDM, murine BMDM, microglial cells	CX3CL1R	↑ phagocytosis via induction of MFG-E8; neural protection from glutamate toxicity	Segundo et al. (1999), Truman et al. (2008), Miksa, Amin, Wu, Ravikumar, and Wang (2007)), Noda et al. (2011), Fuller and Van Eldik (2008), Lee, Lee, Song, Lee, and Chang (2018), and Tsai et al. (2014)
EMAP-II	32D, MEF, Meth A	Monocytes, neutrophils, endothelial progenitor cells	Unknown (monocytes, neutrophils); CXCR3 (endothelial progenitors)	↑ tissue factor expression and TNF-α release by monocytes; ↑ release of myeloperoxidase by neutrophils; inhibition of neovascularization and tumor growth (via VEGFR1); pro-apoptotic for endothelial cells	Knies et al. (1998), Zhang and Schwarz (2002), Schwarz et al. (1999), Matschurat et al. (2003), Awasthi, Schwarz, Verma, Cappiello, and Schwarz (2009), and Hou et al. (2006)
HMGB1	Jurkat, HeLa, U937, Panc-1	DC	–	Essentially inactive when released during apoptosis, as oxidation of 3 cysteine residues within HMGB1 switches it from immunogenic to tolerogenic	Hou et al. (2006), Kazama et al. (2008), Urbonaviciute et al. (2009), Pisetsky (2014), and Scaffidi, Misteli, and Bianchi (2002)
LPC	HT29, COS7, L929	THP-1, MonoMac6, HMDM	G2A (? other receptor or process regulating G2A)	↑ endothelial expression of IL-8, MCP-1, RANTES; ↓ HMGB1 release	Lauber et al. (2003), Peter et al. (2008), Murugesan et al. (2003), Chen et al. (2005), Witte, Kabarowski, Xu, Le, and Zhu (2005)

Continued

Table 2 Diffusible mediators generated by apoptotic cells.[a]—cont'd

Mediator	Apoptotic cell of origin	Responders	Putative receptor	Other effects[b]	References
RP S19 dimer	HL60, NIH 3T3, AsPC-1, rheumatoid arthritis synovial tissue, aortic atheromatous plaque	Monocytes (but "don't find me" factor for neutrophils)	CD88 (C5a receptor)	↑ respiratory burst and anti-apoptotic (monocytes)	Yamamoto (2007), Nishiura, Shibuya, and Yamamoto (1998), and Revollo et al. (2005)
S1P	Jurkat, HEK293, MCF-7_{casp3}, RKO	RAW 264.7, THP-1, U937, human PBM, human/murine MDM	$S1P_{1\text{-}5}$	↑ COX2 protein via mRNA stabilization; ↓ NO via arginase II induction; angiogenesis; endothelial cell differentiation; ↑ VEGF and ↑ erythropoietin via HIF-1α transcriptional activation; barrier function maintenance during epithelial cell extrusion; inhibition of apoptosis	Gu et al. (2011), Herr et al. (2009), Johann et al. (2007), Barra, Kuhn, von Knethen, Weigert, and Brüne (2011), Weigert et al. (2006), Gude et al. (2008), Johann et al. (2008), Weichand et al. (2013), Bryan and Del Poeta (2018), Luo et al. (2016)
SAP-130	Murine thymocytes, 2B4	Neutrophils	Mincle	Indirect effect via SAP-130-induced secretion of CXCL1 and CXCL2 by Mincle-expressing macrophages (thymic, elicited peritoneal); ↑ release by macrophages of IL-6 and TNF-α; Mincle signals via association with ITAM-containing CD32 (FcγRII)	Yamasaki et al. (2008)

sIL-6R	Neutrophils	PBM	? CD130		Chalaris et al. (2007)
TSP-1	PBM, neutrophils	VSMC; viable neutrophils and PBM	Unknown (neutrophils, PMN); $\alpha_V\beta_3$ (VSMC)	Inhibition of LPS-induced up-regulation of MHC II and CD86 by DC	Mansfield, Boxer, and Suchard (1990), Lymn et al. (2002), Mansfield and Suchard (1994), Krispin et al. (2006), Peter et al. (2010)
TyrRS fragments	U937	Monocytes (C-terminal fragment); neutrophils (C- and N-terminal fragments); HUVEC (N-terminal fragment)	C-terminal fragment: unknown; N-terminal fragment: CXCR1	↑ tissue factor expression and TNF-α release by monocytes; ↑ release of myeloperoxidase by neutrophils; stimulation of angiogenesis (N-terminal fragment)	Park, Choi, and Kim (2010), Wakasugi and Schimmel (1999), and Wakasugi et al. (2002)
"DON'T FIND ME"					
Adenosine	W319L; extracellular conversion of ATP/AMP from apoptotic cells	Neutrophils (but "find me" for DC through A1R)	A3R	↓ superoxide production	Cekic and Linden (2016), Idzko et al. (2014), van der Hoeven, Wan, and Auchampach (2008), and Yamaguchi et al. (2014)
Lactoferrin	Burkitt's lymphoma cells, Jurkat, A549, human primary lymphocytes, MCF-7	Human neutrophils, human eosinophils; no effect on monocytes or macrophages	Multiple receptors, but unclear which one mediates effect	Interacts with and inhibits signaling by multiple innate immune receptors (TLR2, TLR4, TLR9, CD14); decreased expression by neutrophils of integrins and selectins	Bournazou, Mackenzie, Duffin, Rossi, and Gregory (2010), Bournazou et al. (2009), Kruzel, Zimecki, and Actor (2017), and Mayeur, Spahis, Pouliot, and Levy (2016)

Continued

Table 2 Diffusible mediators generated by apoptotic cells.[a]—cont'd

Mediator	Apoptotic cell of origin	Responders	Putative receptor	Other effects[b]	References
RP S19 dimer	HL60, NIH 3T3, AsPC-1, rheumatoid arthritis synovial tissue, aortic atheromatous plaque	Neutrophils (but "find me" for monocytes	CD88 (C5a receptor)	Pro-apoptotic (neutrophils)	Yamamoto (2007), Nishiura et al. (1998), and Revollo et al. (2005)

[a]Abbreviations. 2B4, T cell hybridoma cell line; 32D, murine myeloid precursor cell line; A549, human lung adenocarcinoma cell line; ALX/FRP2, lipoxin A4/formyl peptide receptor type 2; AR, adenosine receptor; AsPC-1, human pancreatic cancer cell line; ATP, adenosine triphosphate; BMDM, bone marrow-derived macrophages; COLO 800, human melanoma cell line; COS7, African green monkey kidney cell line; COX2, cyclo-oxygenase 2; DC, dendritic cells; EMAP-II, endothelial monocyte-activating peptide II; eNOS, endothelial nitric oxide synthase; GPCR, G-protein coupled receptor; HEK293, human embryonic kidney cell line; HeLa, human cervical carcinoma cell line; HIF-1α, hypoxia-inducible factor 1α; HL60, human promyelocytic leukemia cell line; HMGB1, high mobility group box 1; HT29, human colorectal adenocarcinoma cell line; HuT78, human T cell lymphoma cell line; IL, interleukin; iNK, invariant natural killer; J773, human chronic B cell leukemia cell line; Jurkat, human T cell line; L929, murine fibroblast cell line; LPC, lysophosphatidylcholine; MCF-7, human mammary cancer cell line; MCP-1, monocyte chemoattractant protein-1; MDA-MB, human breast carcinoma cell line; MDM, monocyte-derived macrophages; MEF, murine embryonic fibroblasts; Meth A, murine sarcoma cell line; MFG-E8, milk fat globule-EGF factor 8; Mincle, macrophage inducible Ca^{2+}-dependent lectin receptor; MOLT4, human acute lymphoblastic leukemia cell line; MonoMac6, human acute monocytic leukemia cell line; NB4, human acute promyelocytic leukemia cell line; NIH 3T3, murine fibroblast cell line; NO, nitric oxide; P2XR, purinergic receptor, X subtype (ligand-gated ionic channels); P2YR, purinergic receptor, Y subtype (GPCR); Panc-1, human pancreatic cancer cell line; PBL, peripheral blood lymphocytes; PBM, peripheral blood monocytes; PHA, phytohemagglutinin; RANTES, regulated on activation normal T expressed and secreted; RAW 264.7, murine macrophage cell line; RKO, human colorectal adenocarcinoma cell line; RP S19, ribosomal protein S19; S1P, sphingosine-1 phosphate; SAP-130, spliceosome-associated protein 130; sIL-6R, soluble IL-6 receptor; THP-1, human acute monocytic leukemia cell line; TLR, toll-like receptor; TSP-1, thrombospondin-1; TyrRS, human tyrosyl tRNA synthetase; U937, human lymphoblast cell line; UTP, uridine triphosphate; VEGF, vascular endothelial growth factor; VEGFR1, VEGF receptor 1; VSMC, vascular smooth muscle cells; WR19L, mouse lymphoma cell line.
[b]Additional effects of each mediator are mainly limited to those observed when the source of the mediator is release by apoptotic cells.
[c]indicates that the table entry is not yet confirmed, either tentatively identified or speculatively proposed.

Table 3 Factors affecting the signal and the noise associated with diffusible mediators from apoptotic cells[a].

Mediator	Sources of increased information capacity	Factors affecting fidelity of signal	References
"Find me"			
Adenosine	• Responder cell-specific differences in expression of AR • Responder cell-specific differences in coupling of AR to heterotrimeric G proteins • Multiple sources of plasma adenosine: (1) direct apoptotic cell-induced release by macrophages; (2) extracellular conversion via CD73 of AMP released from apoptotic cells to adenosine; (3) extracellular conversion of ATP to AMP (by CD39), followed by conversion of AMP to adenosine • Wide range of AR affinities leading to concentration-dependent effects • Adenosine and ATP have generally antagonistic effects (e.g., pro-inflammatory for ATP > adenosine, anti-inflammatory for adenosine > ATP) • ↑ A2AR, ↑ A2BR, and ↓ A3R expression levels by macrophages in response to apoptotic cell recognition and/or phagocytosis	• Release of ATP/AMP and subsequent generation of adenosine not exclusive to apoptotic cells (also released by necrotic, injured, and stressed cells) • Venous/arterial plasma adenosine levels ~20 nM, within affinity range of A1R (mediator of "find me" signaling), but below that of other AR • Up-regulated expression of AR not exclusive to apoptotic cell recognition/phagocytosis (also in response to hypoxia and multiple innate immune stimuli)	Cekic and Linden (2016), Idzko et al. (2014), Saito et al. (1999), Yamaguchi et al. (2014), Köröskényi et al. (2011), Eckle et al. (2014), Crean et al. (2015), Duró et al. (2014), and Cohen et al. (2013)

Continued

Table 3 Factors affecting the signal and the noise associated with diffusible mediators from apoptotic cells[a].—cont'd

Mediator	Sources of increased information capacity	Factors affecting fidelity of signal	References
Annexin 1 fragment (amino acids 2–7)	• Late cleavage (via ADAM10) and release after induction of apoptosis (?[b] only after loss of membrane integrity) • Cleavage and release restricted to certain cell types (cf. Table 1), and not seen in others (HL60, U937, THP-1) • Responder cell-specific differences in expression of putative receptor ALX/FPR2 • Intact cell-bound annexin 1 can also act as "eat me" signal • Anti-inflammatory and pro-resolving activities so far demonstrated only for intact annexin 1 (cell-bound and soluble)	• Release not exclusive to late apoptotic cells (intact annexin 1 also released by glucocorticoid-stimulated macrophages) • ? Competition between intact annexin 1 and fragment for receptor	Blume et al. (2012), McArthur et al. (2015), Sugimoto et al. (2016), and Maderna, Yona, Perretti, and Godson (2005)
ATP/UTP	• Responder cell-specific differences in expression of P2XR and P2YR • Responder cell-specific differences in coupling of P2YR to heterotrimeric G proteins • Extracellular degradation generates metabolites (e.g., ADP/UDP) that engage different P2YR • Extracellular degradation generates metabolites (e.g., adenosine) that engage separate class of receptors (adenosine receptors), often with antagonistic effects • Wide range of receptor affinities leading to concentration-dependent effects (e.g., pro-inflammatory at high concentrations versus anti-inflammatory at low concentrations)	• ATP present in plasma, though typically at levels below EC_{50} of most P2XR and P2YR • Release not exclusive to apoptotic cells (also released by necrotic, injured, and stressed cells) • Pannexin 1 channel, which is activated by caspase-3-mediated cleavage and mediates ATP/UTP release from apoptotic cells, is relatively non-selective, allowing release potentially of other molecules up to ~1 kDa	Corriden and Insel (2010), Idzko et al. (2014), Cekic and Linden (2016), Elliott et al. (2009), Chekeni et al. (2010), Cohen et al. (2013), and Gorman, Feigl, and Buffington (2007)

CX3CL1/fractalkine	• Responder cell-specific differences in expression of CX3CL1R • Several active forms, but only the 60-kDa microvesicular form released by apoptotic cells acts as "find me" signal	• ? Competition between 60 kDa microvesicular form and 90 kDa soluble/membrane-bound forms for CX3CL1R	Truman et al. (2008), Lee et al. (2018), and Tsai et al. (2014)
EMAP II	• Responder cell specific differences in expression of receptors (CXCR3, VEGFR1, unknown receptor on monocytes and neutrophils) • Delayed release from apoptotic cells (? only after loss of membrane integrity)	• Pro-inflammatory effect on monocytes and neutrophils not in keeping with anti-inflammatory effect of apoptotic cells in general • Precursor of EMAP-II (pro-EMAP-II or AIMP1/p43) secreted by intact non-apoptotic cells and itself biologically active (pro-inflammatory) • Processing and release of EMAP-II occurs also in response to hypoxia, but mechanism apparently distinct from that by apoptotic cells • Protease(s) responsible for cleavage of pro-EMAP II uncertain (caspase-7 versus proteasomal)	Knies et al. (1998), Behrensdorf, van de Craen, Knies, Vandenabeele, and Clauss (2000), Zhang and Schwarz (2002), Matschurat et al. (2003), Kwon et al. (2012), Bottoni et al. (2007), Park, Choi, and Kim (2010), and Hou et al. (2006)
HMGB1	• Rapid abundant release as DAMP by cells undergoing necrosis, necroptosis, and pyroptosis • Delayed limited release in biologically inactive form by apoptotic cells which have lost membrane integrity (loss of activity dependent on full oxidation of cysteine residues)	• Activity/inactivity critically dependent on redox state and other post-translational modifications	Bell, Jiang, Reich, and Pisetsky (2006), Kazama et al. (2008), Urbonaviciute et al. (2009), Pisetsky (2014), and Scaffidi et al. (2002)

Continued

Table 3 Factors affecting the signal and the noise associated with diffusible mediators from apoptotic cells[a].—cont'd

Mediator	Sources of increased information capacity	Factors affecting fidelity of signal	References
LPC	• Responder cell-specific differences in expression of G2A receptor • Responder cell-specific differences in coupling of G2A receptor to heterotrimeric G proteins ($G_{q/11}$, $G_{12/13}$, G_s, G_q) • Biased agonism (differences in signaling profiles elicited by different ligands through same receptor) • Acyl chain length and degree of saturation	• Existence of partial agonists (LSM) and antagonists (PAF, lyso-PAF) • Free versus bound LPC (albumin and lipoproteins) • Plasma levels of LPC exceed that in biologically active supernatants from apoptotic cells	Peter et al. (2008), Tomura, Mogi, Sato, and Okajima (2005), Wootten, Christopoulos, Marti-Solano, Babu, and Sexton (2018), and Riederer et al. (2010)
RP S19 dimers	• Only covalently linked dimer of RP S19 has biologic activity (monomer is inactive) • Cross-linkage occurs either intracellularly (via transglutaminase 2) or extracellularly (via factor XIIIa) • C-terminal moiety confers opposite activities on monocytes versus neutrophils despite same receptor • Very late release after induction of apoptosis (24–48 h) (? only after loss of membrane integrity)	–	Yamamoto (2007), Nishiura et al. (1998), Nishiura, Tanase, Sibuya, Nishimura, and Yamamoto (1999), Nishimura et al. (2001), and Shrestha et al. (2003)
S1P	• Responder cell-specific differences in expression of S1P receptors ($S1P_{1-5}$) • Apoptotic cell-specific use of SphK1 versus SphK2 to generate S1P • Responder cell-specific differences in coupling of $S1P_{1-5}$ to heterotrimeric G proteins ($G_{12/13}$, G_i, G_q) • Biased agonism • Intracellular versus extracellular generation of S1P	• Free versus bound S1P (albumin and lipoproteins) • Plasma levels of S1P exceed that in biologically active supernatants from apoptotic cells	Weigert et al. (2009), Weigert et al. (2010), Weigert et al. (2006), Gude et al. (2008), Bryan and Del Poeta (2018), and Wootten et al. (2018)

SAP-130	• Late release after induction of apoptosis (? only after loss of membrane integrity)	• Release not exclusive to apoptotic cells (also released by necrotic cells)	Yamasaki et al. (2008)
sIL-6R	• Shedding dependent on caspase-mediated activation of metalloproteinase ADAM17 • Binding of sIL-6R to circulating IL-6 creates agonistic sIL-6R/IL-6 complex	–	Chalaris et al. (2007)
TSP-1	• Biological activity seen not only with intact protein, but also with C-terminal fragment and N-terminal heparin-binding domain • Responder cell specific differences in response to intact protein and its fragments • Major role also as "eat me" signal • Actively synthesized and released by monocytes and neutrophils undergoing apoptosis	• Definitive role as "find me" signal remains to be established	Krispin et al. (2006), Mansfield and Suchard (1994), Mansfield et al. (1990), Lymn et al. (2002), and Peter et al. (2010)
TyrRS fragments	• Extracellular cleavage by neutrophil-released elastase into 2 active fragments (N- and C-terminal), with distinct receptors and activities • Intact TyrRS lacks activities of fragments • Responder cell specific differences in expression of receptors for different fragments • Delayed release from apoptotic cells (? only after loss of membrane integrity)	• C-terminal fragment has ~50% homology to EMAP-II and has similar activities • N-terminal fragment (pro-angiogenic) competes with CXCL10 (anti-angiogenic) for binding to CXCR1	Park, et al. (2010), Wakasugi and Schimmel (1999), and Wakasugi et al. (2002)

Continued

Table 3 Factors affecting the signal and the noise associated with diffusible mediators from apoptotic cells[a].—cont'd

Mediator	Sources of increased information capacity	Factors affecting fidelity of signal	References
"Don't find me"			
Lactoferrin	• Responder cell-specific differences in response • Existence of several isoforms with different activities • Multiple potential post-translational modifications affecting activity • Interaction with and modulation of the activity of multiple innate immune receptors	• Constitutively present at high concentrations in mucosal secretions • Present at low concentrations in blood circulation • Putative ligand for several receptors recognizing surface determinants on apoptotic cells (LRP1, CD14)	Bournazou et al. (2009), Bournazou et al. (2010), Kruzel et al. (2017), and Mayeur et al. (2016)

[a]Abbreviations not given in Table 2. ADAM10, a disintegrin and metalloprotease; AIMP1, aminoacyl-t-RNA-synthetase-interacting multifunctional protein 1; DAMP, danger-associated molecular pattern; LRP1, LDL-receptor-related protein 1; LSM, lysosphingomyelin; PAF, platelet activating factor; SphK, sphingosine kinase.

[b]? indicates that the table entry is not yet confirmed, either tentatively identified or speculatively proposed.

Table 4 Receptor-ligand pairs mediating apoptotic cell recognition.[a]

Receptor (alternate name)	Responding cells	Apoptotic ligand	Key features and consequences of recognition	References
Direct recognition (PtdSer)				
BAI1	PBM, MDM, bone marrow, spleen, colonic epithelium, skeletal muscle	PtdSer	• **"Eat me"** signaling via ELMO-DOCK180	Park et al. (2007), Das et al. (2014) Lee et al. (2016), and Hochreiter-Hufford et al. (2013)
BAI3	Skeletal muscle	PtdSer	• **"Eat me"** signaling via ELMO-DOCK180	Hamoud et al. (2014, 2018)
CD300A	Peritoneal macrophages, mast cells, MDM	PtdSer and/or PtdEth	• **"Don't eat me"** signaling via ITIM in cytoplasmic tail	Voss et al. (2015), Nakahashi-Oda et al. (2012), and Simhadri et al. (2012)
CD300B	J774.1, peritoneal macrophages	PtdSer	• **"Eat me"** signaling via ITAM in cytoplasmic tail	Voss et al. (2015), and Murakami et al. (2014)
CD300F	Macrophages, L929	PtdSer	• **"Eat me"** signaling in some cells: (ITAM > ITIM)	Voss et al. (2015), Tian et al. (2016) and Linjie Tian et al. (2014)
	DC, J774	PtdSer	• **"Don't eat me"** signaling in other cells (ITIM > ITAM)	
RAGE	Elicited peritoneal macrophages, alveolar macrophages, BMDM, pulmonary epithelial cells	PtdSer	• **"Eat me"** signaling • sRAGE (produced by alternative splicing and/or cleavage of membrane-associated RAGE) has same ligand-binding specificities, and may act as inhibitor of apoptotic uptake	Friggeri et al. (2011) and He et al. (2011)

Continued

Table 4 Receptor-ligand pairs mediating apoptotic cell recognition.[a]—cont'd

Receptor (alternate name)	Responding cells	Apoptotic ligand	Key features and consequences of recognition	References
Stabilin-1/-2	Sinusoidal endothelial cells (spleen, liver, lymph nodes); resident tissue macrophages; HMDM	PtdSer	• **"Eat me"** signaling via adaptor GULP • Stabilin-1/-2 similarly expressed in endothelial cells, but stabilin-1 expressed more broadly in tissue macrophages • Stabilin-2 constitutively expressed by HMDM, but stabilin-1 requires induction by IL-4 and dexamethasone	Park et al. (2008, 2009), Park, Kim, Kang, and Kim (2010), and Park et al., 2008
TIM-1 (KIM-1)	Kidney proximal tubular epithelium	PtdSer	• **"Eat me"** signaling via tyrosine phosphorylation site in cytoplasmic tail • TIM-1 up-regulated following injury or stress and mediates clearance of apoptotic cells • Co-stimulatory molecule on Th2 cells	Freeman et al. (2010), Kobayashi et al. (2007), Ichimura et al. (2008), and Yang et al. (2015)
TIM-3	Elicited (but *not* resident) peritoneal macrophages, some DC subsets	PtdSer	• **"Eat me"** signaling via tyrosine phosphorylation site in cytoplasmic tail • Inhibitory molecule on Th1 cells, inducing their apoptosis	Freeman et al. (2010)

TIM-4	Resident (but *not* elicited) peritoneal macrophages, some DC subsets	PtdSer	• **"Eat me"** recognition, but lacks signaling domain • May utilize integrins for **"eat me"** signaling	Freeman et al. (2010), Kobayashi et al. (2007), and Flannagan, Canton, Furuya, Glogauer, and Grinstein (2014)
TLT2	BMDM; elicited and resident peritoneal macrophages	PtdSer	• **"Eat me"** signaling • Expression induced by TLR3-mediated signaling	de Freitas et al. (2012)
Direct recognition (non-PtdSer ligands)				
CD14	HMDM; resident peritoneal macrophages; thymic and splenic macrophages	ICAM-3	• **"Eat me"** recognition, but lacks signaling domain • May utilize integrins for **"eat me"** signaling • CD14 recognizes ICAM-3 only on surface of apoptotic cells, not viable cells (? modified ICAM-3) • Microparticle-associated ICAM-3 also acts as "find me" signal	Moffatt, Devitt, Bell, Simmons, and Gregory (1999), Torr et al. (2012), and Devitt et al. (2004)
DD1α	Monocytes, macrophages, DC	DD1α	• **"Eat me"** signaling via homotypic interactions • DD1α expressed constitutively by phagocytes • p53-dependent expression by stressed cells • DD1α on apoptotic cells only (*not* live cells) triggers phagocytic removal • Immunosuppression of viable T cells by homotypic interaction with apoptotic cells	Yoon et al. (2015)

Continued

Table 4 Receptor-ligand pairs mediating apoptotic cell recognition.[a]—cont'd

Receptor (alternate name)	Responding cells	Apoptotic ligand	Key features and consequences of recognition	References
PECAM-1 (CD31)	Human MDM	CD31	• **"Don't eat me"** signaling via uncharacterized mechanism • Homophilic CD31-CD31 interaction disabled in apoptotic cells	Brown et al. (2002)
SIRPα (CD172a)	Human PBM, THP-1	CD47	• **"Don't eat me"** signaling via inhibition of non-muscle myosin IIa • Dispersion of CD47 out of lipid rafts decreases its avidity for SIRPα and facilitates apoptotic uptake	Lv et al. (2015) and Tsai and Discher (2008)
Indirect recognition (PtdSer-dependent bridging molecules)				
$\alpha_V\beta_{3/5}$	Elicited peritoneal macrophages; $CD68^+$ macrophages	PtdSer-bound MFG-E8	• **"Eat me"** signaling via FAK and ELMO-DOCK180 • MFG-E8 secreted by activated and glucocorticoid-treated macrophages • Cross-linking of MFG-E8 and β_3 by transglutaminase is critical to phagocytosis	Hanayama et al. (2004, 2002), Akakura et al. (2004), Albert, Kim, and Birge (2000), and Tóth et al. (2009)
	Macrophages (wound-associated)	PtdSer-bound CCN1	• **"Eat me"** signaling • CCN1 = matricellular protein mediating uptake of apoptotic neutrophils in cutaneous wounds	Jun, Kim, and Lau (2015)
	Elicited peritoneal macrophages	PtdSer-bound HMGB1	• **"Don't eat me"** signaling for $\alpha_V\beta_3$ only (not $\alpha_V\beta_5$) • Inhibits interaction between MFG-E8 and $\alpha_V\beta_3$	Liu et al. (2008) and Friggeri et al. (2011)

Tyro3	Myeloid cells; osteoclasts; Sertoli cells	PtdSer-bound Gas6 or protein S	• Innate immune suppression of myeloid cells via **"eat me"** signaling • **"Eat me"** signaling and bone resorption by osteoclasts	Rothlin, Carrera-Silva, Bosurgi, and Ghosh (2015), Graham, DeRyckere, Davies, and Earp (2014), Burstyn-Cohen (2017), Sandahl, Hunter, Strunk, Earp, and Cook (2010), D'Cruz et al. (2000), Caberoy, Alvarado, and Li (2012), Ji et al. (2013), Chung et al. (2013), Nakamura et al. (1998), Happonen et al. (2016), Caberoy, Alvarado, Bigcas, and Li (2012), Caberoy, Zhou, and Li (2010), and Guo et al. (2015)
Axl	Myeloid cells; endothelial cells (aortic, umbilical vein); Sertoli cells	PtdSer-bound Gas6	• Innate immune suppression of myeloid cells via **"eat me"** signaling • **"Eat me"** signaling for platelet microparticles by endothelial cells	
MerTK	Myeloid cells; mammary epithelial cells; RPE; Sertoli cells	PtdSer-bound Gas6, protein S, Tubby (microglia and astrocytes), galectin-3 (RPE), Tulp1 (RPE), Lyar (RPE)	• Innate immune suppression of myeloid cells via **"eat me"** signaling • "**Eat me**" signaling in post-partum mammary gland involution • **"Eat me"** signaling for uptake of shed photoreceptor outer segments by RPE • **"Eat me"** signaling in neurogenesis and synaptic remodeling of brain	
Indirect recognition (PtdSer-independent bridging molecules, or incompletely characterized)				
Calreticulin/LRP1	Professional phagocytes	Collectins (C1q, MBL, surfactant A, surfactant D)	• Recognition via calreticulin and **"eat me"** signaling via LRP1 and adaptor GULP • Collectins bind predominantly to late apoptotic cells • Collectins also mediate recognition of necrotic cells	Gaipl et al. (2001), Ogden et al. (2001), Trouw, Blom, and Gasque (2008) Vandivier et al. (2002), and Janssen et al. (2008)
CD11b/CD18, CD11c/CD18	THP-1	iC3b3	• **"Eat me"** signaling	Takizawa, Tsuji, and Nagasawa (1996)

Continued

Table 4 Receptor-ligand pairs mediating apoptotic cell recognition.[a]—cont'd

Receptor (alternate name)	Responding cells	Apoptotic ligand	Key features and consequences of recognition	References
CD16 (FcγRIII), CD64 (FcγRI)	Human MDM	PtdEth-bound SAP	• **"Eat me"** signaling	Mold, Baca, and Du Clos (2002), Familian et al. (2001)
CD16 (FcγRIII), CD64 (FcγRI)	Human MDM	PtdEth-bound CRP	• **"Eat me"** signaling	Gershov, Kim, Brot, and Elkon (2000), Mold et al. (2002)
LRP1 (CD91, α2-macroglobulin receptor)	Professional phagocytes	Calreticulin	• **"Eat me"** signaling via adaptor GULP • Calreticulin exposure on apoptotic cell surface determines immunogenicity of apoptotic cancer cells	Su et al. (2002), Obeid et al. (2007), and Gardai et al. (2005)
MEGF10	Astrocytes, glial precursors	C1q	• **"Eat me"** signaling for apoptotic neurons via ITAM in cytoplasmic tail and adaptor GULP	Iram et al. (2016) and Scheib, Sullivan, and Carter (2012)
SIRPα (CD172a)	Alveolar macrophages	Collectins (surfactant A, surfactant D)	• **"Don't eat me"** signaling via inhibition of Rac • **"Don't eat me"** signaling via SIRPα is dominant over opsonizing effect of surfactant A/D binding to apoptotic cells	Vandivier et al. (2002), and Janssen et al. (2008)
TREM2	Microglia	TREM2-L (β-amyloid oligomers)	• **"Eat me"** signaling for apoptotic neurons by microglia	Hsieh et al. (2009) and Zhao et al. (2018)

Scavenger receptors				
CD36	Elicited peritoneal macrophages	OxPtdSer	• **"<u>Eat me</u>"** signaling • OxPtdCho recognized to lesser degree • Non-oxidized PtdSer not recognized	Greenberg et al. (2006)
CD36/$\alpha_V\beta_3$	HMDM	Apoptotic cell-bound TSP-1	• **"<u>Eat me</u>"** signaling • TSP-1 contacts both CD36 and $\alpha_V\beta_3$	Savill, Hogg, Ren, and Haslett (1992) and Fadok, Warner, Bratton, and Henson (1998)
CD68/macrosialin	Elicited peritoneal macrophages	? OxPtdSer, OxLDL-like	• **"<u>Eat me</u>"** signaling	Ramprasad et al. (1995)
LOX-1	Bovine aortic endothelial cells	? PtdSer, OxPtdSer, OxLDL-like	• **"<u>Eat me</u>"** signaling	Oka et al. (1998), and Murphy et al. (2006)
SCARA1	Thymic macrophages	? OxPtdSer, OxLDL-like	• **"<u>Eat me</u>"** signaling	Platt, Suzuki, Kodama, and Gordon (2000)
SCARB1	Sertoli cells, thymic nurse cells	? OxPtdSer, OxLDL-like	• **"<u>Eat me</u>"** signaling for apoptotic spermatocytes	Shiratsuchi, Kawasaki, Ikemoto, Arai, and Nakanishi (1999) and Imachi et al. (2000)
SCARF1	CD8α^+ DC; macrophages, endothelial cells	PtdSer-bound C1q	• **"<u>Eat me</u>"** signaling	Berwin, Delneste, Lovingood, Post, and Pizzo (2004), and Ramirez-Ortiz et al. (2013)

Continued

Table 4 Receptor-ligand pairs mediating apoptotic cell recognition.[a]—cont'd

Receptor (alternate name)	Responding cells	Apoptotic ligand	Key features and consequences of recognition	References
Incompletely characterized receptors-ligand pairs				
CD93 (C1qR1)	Elicited peritoneal macrophages	??	• **"Eat me"** signaling • Effect on peritoneal macrophages seen *in vivo* but not *in vitro* • Ligand appears *not* to be C1q	Norsworthy et al. (2004)
Jedi-1	Glial precursors	??	• **"Eat me"** signaling for apoptotic neurons via ITAM in cytoplasmic tail	Wu et al. (2009) and Scheib et al. (2012)
??	THP-1, elicited peritoneal macrophages	PtdSer-bound annexin 1	• Caspase-dependent expression predominantly on late apoptotic cells, *but* **"eat me"** signaling seen only for early apoptotic cells (intact membrane) • Anti-inflammatory signaling, even in absence of phagocytosis	Blume et al. (2009), Arur et al. (2003), Fan, Krahling, Smith, Williamson, and Schlegel (2004)
??	DC (human monocyte-derived)	PTX3	• **"Don't eat me"** signaling	Rovere et al. (2000)

[a]Abbreviations not given in Table 2. BAI, brain angiogenesis inhibitor; C1qR1, complement factor C1q receptor 1; CCN1, cysteine-rich angiogenic protein 61; DD1α, death domain 1α; CRP, C-reactive protein; DOCK180, dedicator of cytokinesis 180; ELMO, engulfment and cell motility; FAK, focal adhesion kinase; GULP, PTB-containing engulfment adaptor; iC3b, inactivated complement factor 3b; ICAM-3, intercellular adhesion molecule 3; ITAM, immuno-tyrosine activating motif; ITIM, immuno-tyrosine inhibitory motif; KIM-1, kidney injury molecule 1; LOX-1, lectin-like oxidized low-density lipoprotein receptor 1; LRP1, LDL-receptor-related protein 1; Lyar, Ly-1 antibody reactive clone; MBL, mannose binding lectin; MEGF10, multiple EGF-domains 10; OxLDL, oxidized low density lipoprotein; OxPtdCho, oxidized phosphatidylcholine; OxPtdSer, oxidized PtdSer; PECAM-1, platelet endothelial cell adhesion molecule 1; PtdEth, phosphatidylethanolamine; PtdSer, phosphatidylserine; PTX3, pentraxin 3; RAGE, receptor for advanced glycation end products; RPE, retinal pigment epithelium; SAP, serum amyloid P; SCARA1, scavenger receptor type A family member 1; SCARB1, scavenger receptor type B family member 1; SCARF1, scavenger receptor type F family member 1; SIRPα, signal regulatory protein α; sRAGE, soluble RAGE; TIM, T cell immunoglobulin mucin protein; TLR3, Toll-like receptor 3; TLT2, TREM-like protein 2; TREM, triggering receptor expressed on myeloid cells; Tulp1, tubby-like protein 1.

6.1.1 Mode of cell death

The mode of cell death as a distinguishable characteristic of dead cells has so far been established definitively only for apoptotic versus necrotic corpses. Not only macrophages (Cocco & Ucker, 2001), but also kidney proximal tubular epithelial cells (PTEC) (Patel et al., 2010), possess distinct non-competing receptors for cells undergoing these two modes of death. Moreover, for both macrophages and PTEC, separate recognition is linked to contrasting signaling outcomes (Cocco & Ucker, 2001; Cvetanovic et al., 2006; Patel et al., 2006; Reddy et al., 2002). In addition to the aforementioned differences in ERK1/2 activity, observed in both macrophages and PTEC, there are multiple other kinases, with contrasting signaling events linked to recognition of the mode of cell death and affecting proliferation, survival, and metabolism (Patel et al., 2012, 2010, 2006, 2015; Reddy et al., 2002). When both apoptotic and necrotic cells were added to bone marrow-derived macrophages, the effect of apoptotic cells was dominant over that of necrotic cells with respect to the activity of ERK1/2 and several other kinases (Patel et al., 2006). Dominance persisted even when apoptotic cells were present in reduced numbers, as much as eightfold fewer (Patel et al., 2006). A similar dominance of apoptotic over necrotic cells was observed with respect to secretion of inflammatory cytokines by macrophages (Cocco & Ucker, 2001). Notably, the signaling events induced by late apoptotic cells (lacking membrane integrity) were identical to and just as potent as those induced by early apoptotic cells (retained membrane integrity) (Patel et al., 2006). Thus, at least with regard to these specific signaling events, apoptotic cells are functionally equivalent throughout the cell death process, irrespective of membrane integrity. While the loss of membrane integrity can and does affect the information content of apoptotic cells, especially with regard to the late release of certain diffusible factors (discussed below), the consistent recognition of an apoptotic cell as apoptotic, without an obscuring or overriding of its identity by leaked intracellular contents, is essential to the stability and unambiguousness of its message.

Several caveats should be noted. First, the inclusion of necrotic cells has not been a consistent control in studies examining the effect of apoptotic cells. When necrotic cells have been included, mostly in studies of professional phagocytes, they either produced a contrasting effect or failed to reproduce the effect of apoptotic cells (Chung et al., 2007; Johann et al., 2007; Kim, Elkon, & Ma, 2004; Luo et al., 2016; Mukundan et al., 2009; Obeid et al., 2007). Second, the stimuli typically used to induce necrosis in nearly all studies, such as heating or freeze-thawing or exposure to detergents or harsh chemicals, are extremely artificial and bear little relation to the inducers of necrosis *in vivo*, most commonly severe ischemia or infection. Since, by definition, necrosis represents a catastrophic disruption of the cell, one might argue that the stimulus of that unregulated destruction seems hardly to matter. Nonetheless, it remains uncertain whether the artificial means used to induce necrosis *in vitro* reflect what occurs *in vivo*. Third, as recently defined by the Nomenclature Committee on Cell Death 2018 (Galluzzi et al., 2018), there are as many as 10 or more modes of cell death, with distinct biochemical and molecular programs. While considerable knowledge exists regarding the pathways involved in the regulation and execution of these modes of cell death, there is a relative paucity of knowledge regarding the specificity of the outcomes they produce in responding cells. For example, necroptotic cells resemble apoptotic cells in their externalization of phosphatidylserine (PtdSer), albeit brought about in a caspase-independent manner (Kaczmarek et al., 2013; Martinez et al., 2011; Zargarian et al., 2017). Even though PtdSer exposure drives recognition and phagocytosis, the intracellular trafficking and processing of ingested material from apoptotic and necroptotic cells appear to differ (Krysko et al., 2006, 2003; Peng & Elkon, 2011; Röszer, 2017). In addition, necroptotic, but not apoptotic, cells induced release of the chemokine CCL2 by phagocytes (Zargarian et al., 2017). It is unclear whether necroptotic and apoptotic cells are recognized by distinct receptors, or whether their recognition occurs via the same PtdSer-dependent receptor(s), and differences in outcome arise from other signals specific to the necroptotic cell. Necroptotic cells may even engage the receptors for both apoptotic and necrotic cells, since necroptosis combines features of the two modes of cell death. Addressing these issues will be crucial not only in the case of necroptosis but also each of the other more exotic forms of cell death, especially since different modes of cell death likely co-exist in the same tissue (González-Polo et al., 2005; Günther et al., 2013; Stevens et al., 2013; Xiao, Wei, Xia, Rothman, & Yu, 2002).

6.1.2 Suicidal stimuli initiating apoptosis

As previously discussed, anti-inflammation appears to be a ubiquitous aspect of the response to apoptotic corpses, independent of species, tissue type, or suicidal stimulus of the apoptotic cell (Birge & Ucker, 2008; Cvetanovic et al., 2006). The ability of responding cells to distinguish apoptotic cells on the basis of the suicide-inducing stimulus has so far been observed only for outcomes other than anti-inflammation. In one of the first studies to explore this issue, increased surface exposure of calreticulin proved to be a critical event differentiating immunogenic from non-immunogenic cancer cell death (Obeid et al., 2007). Increased surface exposure of calreticulin by a mouse colon cancer cell line occurred when death was induced by ultraviolet irradiation or several anthracycline chemotherapeutic agents, but not in response to a wide assortment of other suicidal stimuli (Obeid et al., 2007). Calreticulin may govern immunogenicity by shifting the receptor responsible for recognition and uptake of the apoptotic cell to CD91 on professional phagocytes (see Table 4) (Gardai et al., 2005; Obeid et al., 2007; Su et al., 2002). Along similar lines, the suicidal stimulus was found to correlate with the efficiency of phagocytic uptake (fraction of apoptotic cells not taken up in a given amount of time) by professional and non-professional phagocytic cell lines (Wiegand, Corbach, Prescott, Savill, & Spruce, 2001). Receptor-mediated recognition of apoptotic cells by viable PTEC leads to a synchronized mass suicide of nearly all PTEC (denoted "death-induced-death"), but only for certain suicidal stimuli (Patel et al., 2012). Finally, the choice of sphingosine kinase isoform used to catalyze the production of S1P, a "find me" factor released by apoptotic cells (Tables 2 and 3), may depend on the suicidal stimulus, although it is uncertain whether this constitutes a distinguishing characteristic (Gude et al., 2008; Weigert et al., 2006).

6.1.3 Elapsed time from apoptotic initiation

As apoptosis is a highly regulated form of cell death, it is hardly surprising that its morphologic, biochemical, and molecular features evolve over time. Among the earliest events, occurring within 2–4 h, are blebbing of the cell surface membrane and the caspase-dependent opening of pannexin I channels, permitting release of ATP and UTP (Elliott & Ravichandran, 2016). Exposure of PtdSer is a somewhat later event, as is release of other "find me" factors like lysophosphatidylcholine (LPC) (Elliott & Ravichandran, 2016). The lifespan of apoptotic cells can be divided into two clearly defined stages, marked by the unregulated rupture of the cell membrane. The later of these two stages is commonly referred to as "late apoptosis" or "secondary necrosis". Most of the apoptotic cell-associated characteristics shown to be distinguishable in a time-dependent manner by responding cells involve later events in the apoptotic program, occurring around or after the transition to compromised membrane integrity. These include: the release of various "find me" factors such as endothelial monocyte-activating peptide II (EMAP II) (Knies et al., 1998), human tyrosyl tRNA synthetase (TyrRS) fragments (Park, Choi, & Kim, 2010; Wakasugi & Schimmel, 1999), and RP S19 dimers

(Yamamoto, 2007); the increased surface expression of the anti-inflammatory protein annexin I (Blume et al., 2009) and an inactive form of the chemokine receptor CCR5 (Ariel et al., 2006); and the altered surface binding of various acute phase reactants, complement proteins, and regulators of either (Mold et al., 2002; Trouw et al., 2008). The purpose of many of these changes seems to be to enhance the phagocytosis of apoptotic cells that have somehow eluded clearance as well as to counteract their inflammatory potential, since several proteins also show parallel alterations in their binding to necrotic cells (Trouw et al., 2008). Of these later characteristics, the one whose message leads to the most rigorously established outcome is the increased expression of inactive CCR5, which helps to terminate recruitment of inflammatory cells by sequestering several chemokines (Ariel et al., 2006). Fewer examples exist of time-dependent characteristics in early apoptotic cells before the loss of membrane integrity. In one study, although apoptotic cells were recognized immediately by both professional and non-professional phagocytes, phagocytosis by non-professional phagocytes was delayed and required time-dependent changes of the apoptotic cell (Parnaik et al., 2000) This was confirmed by deferred addition of apoptotic cells to non-professional phagocytes. A time- and stimulus-dependent change in the response to apoptotic cells was also observed in the "death-induced-death" response by kidney PTEC (Patel et al., 2012). This was tested by fixing apoptotic cells at various times after addition of a suicidal stimulus. Actinomycin D-induced apoptotic cells were almost immediately potent, but lost potency over time. Staurosporine-induced apoptotic cells displayed an opposite dependency, increasing in potency over time. Unfortunately, in neither of these two studies was the identity of the characteristic displaying time-dependent behavior identified.

6.1.4 Lineage of the apoptotic cell

The identity of the apoptotic cell appears to contribute little to the information content of its message. Compared to variables associated with the death process itself, variables associated with the dying cell—for example, its lineage, organ of residence, stage of differentiation, etc.—apparently are of minor importance. Recipient cells seem largely ignorant of the apoptotic cell's identity. The only distinguishable characteristics, unique to the apoptotic cell, which recipient cells seem able to identify are the diffusible factors apoptotic cells release or the surface determinants and ligands they express, and these relate less to the content of the message than to which cells the message is delivered. For example, cleavage of the anti-inflammatory protein annexin I by the metalloproteinase ADAM10, restricted largely to cells of lymphocytic lineage, generates a "find me" signal for peripheral blood monocytes (Blume et al., 2012). Part of the apparent lack of distinguishability of an apoptotic cell's identity may relate to the use of inflammatory status as a measure of the effect of apoptotic cells, since, as previously discussed, anti-inflammation appears to be a universal component of the message delivered by all apoptotic cells (Birge & Ucker, 2008; Cvetanovic et al., 2006). Just as the ability to distinguish apoptotic cells on the basis of suicidal stimulus has so far been described only for outcomes other than anti-inflammation, the impact of cellular identity on an apoptotic cell's message may require a more careful examination of its extended homeostatic effects. At least in the case of the "death-induced-death" response of kidney PTEC, however, no differences were found with respect to the species or tissue type of the apoptotic cells (Patel et al., 2012).

6.2 Variables associated with the responding cell

The number of variables that impact the decoding of an apoptotic cell's message greatly exceeds that which impact its encoding. This is not simply the result of selective and specific delivery of the apoptotic cell's message to particular responding cells. Numerous examples exist in which the same message produces divergent outcomes depending upon the characteristics of the recipient cell. Relevant variables include: lineage of the responding cell, its organ of residence, its state of activation, its stage of maturation or differentiation, the concentrations of the secreted factors to which it is responding, whether or not it is neoplastic, and the age of the organism in which it resides. The response of a recipient cell may even vary over time following its initial interaction with an apoptotic cell, not simply as expected in its magnitude, but in its direction, whether stimulatory or inhibitory. For example, exposure of bone marrow- and monocyte-derived macrophages to apoptotic cells led to an early enhancement of lipopolysaccharide (LPS)-induced release of tumor necrosis factor (TNF)-α for the first 2–4 h, followed by later inhibition (Lucas, Stuart, Savill, & Lacy-Hulbert, 2003). It is likely that this enhancement reflects release of already synthesized TNF-α, rather than new transcription or translation of protein. No other cytokine or chemokine showed a similarly biphasic response. As in our discussion of the variables used by dead cells to encode their message, we will limit ourselves here to general principles and illustrative examples.

6.2.1 Lineage of the responding cell

As previously discussed, a fairly clear separation exists between professional and non-professional phagocytes in terms of the efficiency and kinetics of uptake, intracellular trafficking, and processing of apoptotic cells (Devitt & Marshall, 2011; Parnaik et al., 2000; Wood et al., 2000). Even among professional phagocytes, there may be differences in intracellular trafficking, as that observed between macrophages and dendritic cells, which is mediated by differential activation of the small GTPase protein Rho (Erwig et al., 2006). We have already touched upon one of the more noteworthy example of the importance of lineage, in which apoptotic cells activated Akt and promoted the survival of macrophages, whereas they inhibited Akt and induced the apoptotic death of kidney PTEC (Patel et al., 2012, 2010; Reddy et al., 2002). Not only were the survival responses opposite for cells of these two lineages, but the mechanisms producing the contrasting responses also differed. Signaling for macrophages was linked to phagocytosis, independent of the material being engulfed—apoptotic, necrotic, and inert debris gave equivalent outcomes—whereas the response of PTEC was specific to apoptotic cell recognition, and necrotic cells actually activated Akt and promoted survival (Patel et al., 2012, 2010; Reddy et al., 2002). Even recognition by the same receptor does not guarantee a uniform response in cells of different lineage. Thus, recognition of apoptotic cells by the cell surface receptor CD300f mediated increased phagocytic uptake in a fibroblast cell line but decreased phagocytic uptake in a macrophage cell line (Voss et al., 2015; Linjie Tian et al., 2014).

6.2.2 Organ of residence of the responding cell

Among cells of the same lineage, the response to apoptotic cells may still differ according to the responding cell's organ of residence. For example, as compared to resident peritoneal macrophages, the phagocytic uptake of apoptotic cells by alveolar macrophages is markedly slower. This has been variably attributed to diminished activation of the protein kinase C (PKC) isoform, PKC βII (Todt et al., 2002) or to engagement of signal regulatory protein α (SIRPα; CD172a) on the surface of alveolar macrophages by surfactants A and D (Janssen et al., 2008; Vandivier et al., 2002). Exposure to apoptotic cells stimulated phosphorylation of ERK1/2 in alveolar and resident peritoneal macrophages (Park, Choi, Cho, Henson, & Kang, 2011), but inhibited ERK1/2 phosphorylation in bone marrow-derived and elicited peritoneal macrophages (Reddy et al., 2002). Epithelial cells too may display differences in response. Like kidney PTEC cell lines, epithelial cell lines from the ovary respond to apoptotic cells with inhibition of proliferation and "death-induced-death." In contrast, epithelial cell lines from the uterus and breast show enhanced proliferation and survival (Patel et al., 2012, 2010).

6.2.3 State of activation of the responding cell

Naive neutrophils, freshly isolated from peripheral blood, have a limited capacity to phagocytose and respond to apoptotic cells, in large part attributable to apoptotic release of "don't find me" factors specific for neutrophils, such as lactoferrin (Bournazou et al., 2010), RP S19 dimer (Yamamoto, 2007), and adenosine (van der Hoeven et al., 2008). Activation of neutrophils with LPS, interferon (IFN)-γ, and/or granulocyte-macrophage colony-stimulating factor (GM-CSF) enabled neutrophils to recognize and respond to apoptotic cells, with resultant anti-inflammatory activity (Esmann et al., 2010). Recognition and uptake of apoptotic cells was found to depend on a heat-labile serum factor, most likely a complement component such as C1q (Ramirez-Ortiz et al., 2013; Trouw et al., 2008), iC3b3 (Takizawa et al., 1996), or C5a ((Yamamoto, 2007).

6.2.4 Stage of maturation or differentiation of the responding cell

Tissue-resident macrophages and recruited monocyte-derived macrophages represent separate populations of distinct embryological origin (Gomez Perdiguero et al., 2015; Hoeffel et al., 2015). Although circulating monocytes are often recruited from the bloodstream to aid tissue-resident macrophages in the cleanup of dead cells, the interaction of these two types of macrophages is not always cooperative, and may even be antagonistic (Uderhardt et al., 2012). During thioglycollate-induced peritonitis, for example, resident macrophages comprise only about 3% of the macrophage population, yet they ingest many more apoptotic cells than do recruited macrophages (Uderhardt et al., 2012). The mechanism by which resident peritoneal macrophages achieve this highlights a number of differences in the responses of these two macrophage populations. Unlike recruited macrophages, resident peritoneal macrophages express the enzyme, 12/15-lipoxygenase (12/15-LO), which specifically oxidizes the unsaturated fatty acid side chains of phosphatidylethanolamine (PtdEth), leading to the enrichment of resident peritoneal macrophage plasma membranes with PtdEth oxidation products (Uderhardt et al., 2012). Uptake of apoptotic cells by resident versus elicited peritoneal macrophages occurs via distinct receptors (Freeman et al., 2010; Ipseiz et al., 2014). A bridging molecule, milk fat globule-EGF factor 8 (MFG-E8), essential for the uptake of apoptotic cells by elicited macrophages, but not resident macrophages, preferentially binds to 12/15-LO-modified PtdEth on resident peritoneal macrophages (Uderhardt et al., 2012). Sequestration of MFG-E8 diminishes apoptotic cell uptake by elicited macrophages, enabling resident macrophages, despite their greatly lower frequency, to ingest a disproportionate number of apoptotic cells.

Differential responsiveness also occurs within the individual macrophage populations, especially among recruited monocytes as they mature into macrophages. For example, TNF-α inhibits the ability of mature monocyte-derived macrophages to ingest apoptotic cells, whereas it stimulates the ability of immature monocytes (McPhillips et al., 2007). In addition, signaling through the MerTK receptor for apoptotic cells plays a central role in macrophage polarization and alters the distribution of macrophage subsets, increasing the frequency of anti-inflammatory (so-called M2c) macrophages (Zizzo, Hilliard, Monestier, & Cohen, 2012).

6.2.5 Concentration-dependent responses to diffusible factors released by apoptotic cells

The concentrations of diffusible factors released by apoptotic cells will typically decrease with increased distance from the cell and elapsed time from the moment of release. While spatial variation is essential to chemotaxis, as phagocytes travel up an increasing gradient of "find me" signals, spatial and temporal variation can also lead to differences in cellular response, especially if recipient cells bear several receptors for the diffusible factor, with different affinities and linked to different outcomes. Both ATP/UTP, which is released from apoptotic cells through caspase-dependent opening of pannexin 1 channels (Chekeni et al., 2010; Elliott et al., 2009), and adenosine, which is released by apoptotic cells and/or extracellularly generated from ATP by phagocytes and lymphocytes (Cekic & Linden, 2016; Köröskényi et al., 2011), fulfill these criteria. ATP (but not UTP) is recognized by 10 distinct X-subtype purinergic receptors (P2XR), and ATP and UTP are jointly recognized by 8 distinct Y-subtype purinergic receptors (P2YR), with EC_{50} spanning a log range of 2–3 units (Cekic & Linden, 2016; Corriden & Insel, 2010; Idzko et al., 2014). These receptors are variably expressed by cells of immunological lineage with widely varying responses, from pro- to anti-inflammatory, in addition to chemotactic and other activities (Cekic & Linden, 2016; Corriden & Insel, 2010; Idzko et al., 2014). Similarly, adenosine is recognized by four distinct receptors, also with varying affinities and responses

(Cekic & Linden, 2016; Crean et al., 2015; Duró et al., 2014; Idzko et al., 2014; Köröskényi et al., 2011; Novitskiy et al., 2008; Schnurr et al., 2004; Yamaguchi et al., 2014).

An additional layer of complexity arises from the extracellular conversion of ATP to adenosine, and provides an excellent example of the grammatical interrelationship among individual elements of the apoptotic cell's message. A recent review provides an excellent summary of this interrelationship as it plays out over time (Cekic & Linden, 2016). Acute tissue injury is characterized by the release of ATP/UTP from apoptotic, as well as necrotic and stressed cells. The unregulated release of ATP/UTP by necrotic cells greatly exceeds the regulated release by apoptotic cells. At early times (minutes to hours), the high concentrations of ATP/UTP exert a pro-inflammatory effect through both P2XR and P2YR. Chemotaxis of professional phagocytes and lymphocytes is mediated by P2Y2R (Elliott et al., 2009). Of the purinergic receptors, P2X7R has the lowest affinity for ATP, and is only activated by the extremely high concentrations of ATP ($\sim$100 μM) seen in cases of severe injury or infection. Such concentrations are not achievable by the regulated opening of pannexin 1 channels in apoptotic cells (Qu et al., 2011). Activated P2X7R forms large pores that permit the passage of molecules as large as 900 kDa (versus 1 kDa for pannexin 1 channels), including ATP itself (Cekic & Linden, 2016). Extracellular ectonucleotidases, up-regulated on the surface of lymphocytes and macrophages (recruited via P2Y2R) (Cekic & Linden, 2016; Cohen et al., 2013), rapidly degrade ATP to ADP and AMP. In turn, AMP is converted to adenosine by CD73 (5-nucleotidase), also expressed on the surface of recruited lymphocytes and macrophages (Cekic & Linden, 2016). During the ensuing hours to days, as ATP concentrations decline and adenosine concentrations rise, the balance is shifted from pro-inflammation to anti-inflammation, with decreased engagement of pro-inflammatory P2XR and P2YR, and increased engagement of anti-inflammatory adenosine receptors, A2RA and A2RB (Cekic & Linden, 2016). Apoptotic cells up-regulate expression of A2RA and A2RB by macrophages (Köröskényi et al., 2011), thereby increasing the potency of adenosine in limiting inflammation and promoting healing. Apoptotic cells also down-regulate expression of A3R, whose engagement by adenosine attenuates A2AR-dependent signaling (Duró et al., 2014). A2AR mediates anti-inflammation in part via induced expression of the nuclear receptor NR4a, which directly opposes NFκB-mediated effects (Crean et al., 2015; Yamaguchi et al., 2014). While the role of A2RB is less clear, its engagement skews the differentiation of dendritic cells to a pro-angiogenic, anti-inflammatory, and tolerogenic phenotype, characterized by the production of high levels of vascular endothelial growth factor (VEGF), interleukin (IL)-10, transforming growth factor (TGF)-β, cyclooxygenase 2 (COX2), and indoleamine 2,3-dioxygenase (Novitskiy et al., 2008).

6.2.6 Neoplastic versus non-neoplastic responding cells

The increased rate of cell death typically seen in nearly all tumors, including that associated with specific therapy, may enable neoplastic cells to subvert the normal homeostatic processes associated with apoptosis, especially those related to proliferation and survival, with the unfortunate consequence of facilitating tumorigenesis (Gregory et al., 2016; Ucker & Levine, 2018). It is not necessary that neoplastic cells respond differently to individual apoptotic cells. Rather, the prominence and persistence of apoptotic cells locally, coupled with unique features of the tumorous microenvironment, can enhance the mitogenic and pro-survival effects of apoptotic cells (Donato et al., 2014; Huang et al., 2011; Lauber et al., 2011; Mao et al., 2013), and even promote the selective outgrowth of tumor cells more susceptible to these effects (Feng et al., 2018). Although less definitively established, some evidence exists that neoplastic cells may also respond abnormally to apoptotic cells (Graham et al., 2014). Aberrant decoding of the apoptotic cell's message has been looked at the most so far with respect to the Tyro-Axl-MerTK (TAM) family of receptors (see Table 4). In contrast to inhibition of NFκB-dependent inflammatory signaling, which has emerged as an ubiquitous aspect of the response to apoptotic cells, signaling through one or more of the TAM receptors in a wide variety of cancer cell lines has been linked to activation of NFκB and enhancement of tumor cell survival, proliferation, invasiveness, dedifferentiation, and overall progression (Asiedu et al., 2014; Lee et al., 2014; Paccez et al., 2013; Shiozawa, Pedersen, & Taichman, 2010; Tai, Shieh, Lee, Shiah, & Wu, 2008). Lacking, however, in all of these studies is a direct connection between these events and TAM-dependent recognition of apoptotic cells.

6.2.7 Age of the organism in which the responding cell resides

The anti-inflammatory response induced by apoptotic cells in macrophages (both resident and recruited) has been shown to alter with the age of the mouse from which the macrophages are derived. The change does not affect the magnitude of the initial anti-inflammatory response, but rather the duration of the response and the synchronicity with which it is terminated (Pattabiraman, Palasiewicz, Galvin, & Ucker, 2017; Pattabiraman, Palasiewicz, & Ucker, 2016). In macrophages from older mice, not only did suppression of LPS-stimulated expression of TNF-α persist for a longer duration (~twofold), but loss of suppression at a single cell level displayed a greater variability, spread out over a much broader range of times (~sixfold) (Pattabiraman et al., 2017, 2016). These effects of age could potentially lead to a blurring of the apoptotic cell's message, through a reduction in its sharpness and coherence, as well as to a diminution in the nimbleness with which tissues respond to subsequent environmental perturbations, through a delay and impairment of homeostatic recovery.

6.3 Variables associated with transmission of the message

The diversity, precision, and nuance with which apoptotic cells transmit their message are astonishing on many levels. Not only does the message contain an abundance of comprehensive information, with specific instructions to recipient cells regarding virtually all of their vital functions (see Table 1), but also, with remarkably few exceptions, the message is delivered precisely to the cells for which it is intended. Among its intended recipients, as discussed above, the very same message may be variously interpreted by different recipients, depending on a broad range of recipient-associated variables, yielding discrete sets of instructions, each tailored specifically for the particular recipient cell at hand and its designated role in homeostatic recovery.

The two most extensively studied and diversified modes of transmission are: first, diffusible mediators (soluble or microvesicle-associated), which are released by apoptotic cells and engage their corresponding receptors on recipient cells; and, second, direct physical interaction between apoptotic and responding cells, via receptor-mediated recognition of newly expressed determinants or ligands on the apoptotic cell surface. In addition to these two modes of delivery, a number of other creative and sophisticated ways to transmit the apoptotic cell's message are evident (Fig. 2).

6.3.1 Diffusible mediators

An extremely important function of the assorted mediators released by apoptotic cells is to attract phagocytes for their disposal. Although adjacent cells do not need to migrate to an apoptotic cell, their number and their capacity, especially if they are non-professional phagocytes, may be insufficient to handle the load of dead cells. In this case, the recruitment of professional

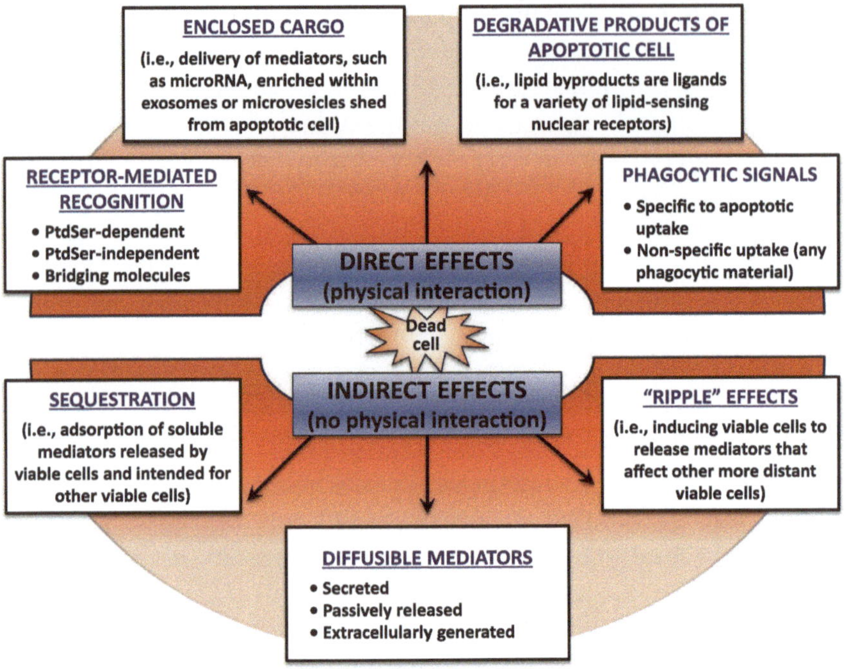

Fig. 2 Modes of transmission used by apoptotic cells to deliver a message to live recipient cells. Depicted are several of the best characterized mechanisms for transmission of information from apoptotic cells to nearby live responding cells. These are broadly divided into those which require direct physical interaction between the apoptotic and responding cells, and those in which the apoptotic cell exerts its effects indirectly, either from afar or through an induced intermediate.

phagocytes is required, and so-called "find me" signals direct the migration of macrophages and other cells up a concentration gradient toward sites in need of cleanup. Tables 2 and 3 summarize the properties of the >10 "find me" factors so far described. There are also three "don't find me" factors, which act to antagonize the recruitment of cells, mainly neutrophils. One of these factors, RP S19, acts as a "find me" signal for peripheral blood monocytes and as a "don't find me" signal for neutrophils (Nishiura et al., 1998; Revollo et al., 2005; Yamamoto, 2007).

Specific details regarding each of these factors are provided in Tables 2 and 3. We discuss here a number of important principles and generalizations. First, the effects of these factors are not limited to chemotaxis. While directed migration was the effect by which these factors were most commonly sought and identified, their range of activities is exceptionally diverse, and not at all limited to the modulation of inflammatory and immunogenic responsiveness. For example, S1P has a number of extended homeostatic effects, including maintenance of epithelial barrier function during extrusion of apoptotic cells (Gu et al., 2011), protection of macrophages from several suicidal stimuli (Weigert et al., 2006), and increased protein expression by macrophages of COX2 (Johann et al., 2008) and erythropoietin (Luo et al., 2016). Second, release of these factors is not specific in all cases to apoptotic cells. ATP/UTP is also released by stressed and sublethally injured cells, as well as by necrotic cells, although in the latter case, in markedly increased quantities as compared to apoptotic cells (Cekic & Linden, 2016; Corriden & Insel, 2010; Idzko et al., 2014). Moreover, since extracellular ATP is rapidly converted to adenosine (Cekic & Linden, 2016), necrotic cells also contribute large amounts of this mediator. Third, the time following initiation of apoptosis at which these factors are released varies considerably. While most are actively released by early apoptotic cells, some such as EMAP II (Knies et al., 1998), TyrRS fragments (Patel et al., 2010; Wakasugi & Schimmel, 1999), and RP S19 dimers (Yamamoto, 2007) are only released by late apoptotic cells, near or following their loss of membrane integrity. Fourth, several of these factors are released in more than one form, with specificity ensured by restricting "find me" or other activities to a single form. For example, the chemokine CX3CL1/fractalkine can be released in either a microparticle-associated 60-kDa form or a soluble 90-kDa form. While both retain biological activity, only the microparticle-associated form has "find me" activity (Lee et al., 2018; Truman et al., 2008; Tsai et al., 2014). High mobility group box 1 (HMGB1), a pro-inflammatory danger-associated molecular pattern (DAMP) molecule that is rapidly and abundantly released by cells undergoing necrosis, necroptosis, and pyroptosis (Scaffidi et al., 2002), is also released in smaller amounts by cells undergoing apoptosis, but is inactivated in apoptotic cells by oxidation of critical cysteine residues (Bell et al., 2006; Kazama et al., 2008; Urbonaviciute et al., 2009). Finally, establishing the *in vivo* role of these factors is problematic. Many factors, being non-proteins, are not amenable to the generation of transgenic or knockout mice. Rather, mice with genetic deletion of their receptors or the enzymes responsible for their production or release must be used. Even when a suitable knockout exists, one can still not be sure whether its phenotypic features are attributable to impaired phagocyte recruitment or to one of the factor's non-"find me" activities (Medina & Ravichandran, 2016).

6.3.1.1 Sources of increased information capacity associated with diffusible mediators

All of the diffusible mediators released by apoptotic cells have activities beyond chemotaxis, expanding the content of the apoptotic cells' message, and thereby increasing its information capacity. Two basic strategies are available. Differential processing and/or chemical or other modifications of a mediator, if distinguishable, can create distinct versions targeting specific responder cells, or even different receptors. Alternatively, a single receptor may be coupled to different downstream intermediates, leading to potentially different signal transduction events in different cells. We provide examples for each of these strategies.

We have previously discussed the example of CX3CL1/fractalkine, which can exist in at least four different forms, with differing biological activity: 90-kDa cell-bound, 60-kDa microparticle-associated, 90-kDa soluble, and 55-kDa soluble forms (Lee et al., 2018; Truman et al., 2008; Tsai et al., 2014). Other examples abound. The acyl side chains of LPC can vary in length and saturation, modifying its affinity for its GPCR receptor (Riederer et al., 2010). RP S19 is biologically active only as a covalently linked dimer; cross-linkage can occur either intracellularly or extracellularly, with perhaps different effects (Yamamoto, 2007). Similarly, there are two isoforms of sphingosine kinase, the enzyme responsible for formation of S1P; generation of S1P, like that of RP S19 dimers, can occur both inside and outside cells (Gude et al., 2008; Weigert et al., 2009, 2010, 2006). Finally, TyrRS is cleaved extracellularly by neutrophil-derived elastase into two fragments, each with distinct receptors and activities (Park, Choi, & Kim, 2010).

The second strategy focuses on the receptors engaging these mediators. One of the simplest ways to diversify a mediator's message is through the provision of multiple receptors, each expressed by a distinct set of responder cells and generating receptor-specific outcomes. This is true of adenosine (Cekic & Linden, 2016), ATP/UTP (Cekic & Linden, 2016), EMAP II (Awasthi et al., 2009; Hou et al., 2006), S1P (Bryan & Del Poeta, 2018), and possibly lactoferrin (Kruzel et al., 2017; Mayeur et al., 2016). As previously discussed for ATP/UTP and adenosine, a cell may contain several different receptors of varying affinity for any one single mediator (Cekic & Linden, 2016; Idzko et al., 2014); the mediator's effect is determined in a concentration-dependent manner by the degree to which it engages each receptor. The relative abundance of each receptor type may also be modified in response to extracellular cues, including recognition or uptake of apoptotic cells, as seen in the case of adenosine (Duró et al., 2014; Köröskényi et al., 2011). In this way, a single cell's response to a mediator may vary widely over time as the concentration of the mediator and/or the number and configuration of its receptors changes.

In addition, as the receptors for many diffusible mediators are G-protein coupled receptors (GPCR), differential responsiveness may also be achieved by variable coupling of the GPCR to different heterotrimeric G proteins, both inhibitory and excitatory (Wootten et al., 2018). Additional flexibility in the signals generated by GPCR engagement derives from a relatively new concept known as biased agonism, in which ligands acting at the same GPCR elicit distinct cellular signaling profiles by preferentially stabilizing different active conformational states of the receptor (Wootten et al., 2018). This may be especially relevant for a mediator like LPC, with demonstrated variability in its side chain composition (Riederer et al., 2010).

6.3.1.2 Sources of increased noise associated with diffusible mediators

"Noise" may be broadly defined as anything that interferes with the transmission or interpretation of information. To one degree or another, all forms of communication are subject to noise, and that between apoptotic and responder cells is no exception. The sources of noise are of varying disruptiveness, but all have the potential to reduce the fidelity of the apoptotic cell's message.

First, factors whose release or generation is not exclusive to apoptosis may not reliably convey the mode of cell death. In the case of ATP/UTP and adenosine (Cekic & Linden, 2016; Corriden & Insel, 2010; Idzko et al., 2014), the lack of specificity is seeming only, since the requisite information is largely contained in the combined concentrations of ATP/UTP and adenosine. Non-apoptotic modes of cell death, such as necrosis, necroptosis, and pyroptosis, result in the release of far greater quantities of ATP/UTP (Qu et al., 2011), easily enabling their distinction. Even when concentrations of ATP/UTP decline into a range associated with apoptotic death, by virtue of diffusive distance or conversion to adenosine, the disproportionately increased concentrations of adenosine converted from the previous surfeit of ATP/UTP will offer the telltale clue, revealing the non-apoptotic mode of death.

A second source of noise is the fact that circulating levels *in vivo* for several of these factors lie within the affinity range of their receptors. For example, plasma levels of both LPC (Tomura et al., 2005) and S1P (Gude et al., 2008; Weigert et al., 2009) exceed that found in biologically active supernatants from apoptotic cells. Both LPC and S1P are heavily bound to plasma albumin and lipoproteins (Tomura et al., 2005; Weigert et al., 2009), so part of the discrepancy may be attributed to differences in free versus bound forms, since plasma levels reflect both. The relative potencies of free versus bound forms are not known and may also help to account for the discrepancy. Adenosine and ATP, too, circulate in the plasma (Gorman et al., 2007; Saito et al., 1999). Concentrations of ATP are typically well below the EC_{50} of most P2XR and P2YR. Concentrations of adenosine are within the affinity range of only A1R, which mediates "find me" activity for neutrophils and plasmacytoid dendritic cells (Schnurr et al., 2004).

Other potential sources of noise, or interfering signals, include the following. Circulating plasma contains several partial agonists of LPC (for example, lysosphingomyelin) and even antagonists (for example, platelet activating factor [PAF] and lyso-PAF) (Lauber et al., 2003). In some cases, hypoxia can mimic or modulate the effects of mediators, such as release of EMAP II (Matschurat et al., 2003) or up-regulated expression of A2AR and A2BR (Eckle et al., 2014), although it is not clear whether these effects are brought about directly by hypoxia, or indirectly through induced apoptotic cell death, or both. All in all, as should be clear from this discussion, a high degree of fidelity pertains to the apoptotic cell's message, with disruption by noise kept to a surprising minimum.

6.3.2 Receptor-ligand interactions

More than 40 distinct receptors are known to mediate the recognition and/or phagocytosis of apoptotic cells (Table 4). As for diffusible mediators, we will restrict our discussion to general principles and illustrative examples. These receptors may grouped into the following broad categories: those which directly recognize externalized PtdSer on the surface of apoptotic cells; those which directly recognize a non-PtdSer ligand or determinant; those which indirectly recognize apoptotic cells via a PtdSer-dependent bridging molecule; those which indirectly recognize apoptotic cells through a PtdSer-independent bridging molecule; scavenger receptors recognizing oxidized or otherwise altered surface lipids; and, finally, those which are as yet incompletely characterized.

First, and most obviously, tissue- and cell-specific expression of receptors and/or bridging molecules ensures precise and accurate delivery of a message. Many receptors are broadly expressed across the range of professional phagocytes, and play a predominant role in the clearance of apoptotic debris and maintenance of homeostasis, whereas others have a more restricted distribution, for example, triggering receptor expressed on myeloid cells (TREM)-like protein 2 (TREM2) (Hsieh et al., 2009; Zhao et al., 2018), multiple EGF-domains 10 (MEGF10) (Iram et al., 2016; Scheib et al., 2012), and Jedi (Scheib et al., 2012; Wu et al., 2009) in the brain, or scavenger receptor B1 (SCARB1) in the testes (Imachi et al., 2000; Shiratsuchi et al., 1999). Second, not all receptors are capable of downstream signaling or initiating phagocytosis. Some, like TIM-4 (Flannagan et al., 2014; Freeman et al., 2010; Kobayashi et al., 2007) and CD14 (Devitt et al., 2004; Moffatt et al., 1999; Torr et al., 2012), lack cytoplasmic tails, and therefore can only recognize and tether apoptotic cells; they require accessory molecules, such as integrins, for internalization. Third, just as with "find me" and "don't find me" factors, some receptors provide "don't eat me" signals and prevent uptake of apoptotic cells. Examples include CD300A (Nakahashi-Oda et al., 2012; Simhadri et al., 2012; Voss et al., 2015), CD31 (Brown et al., 2002), and SIRPα (CD172a) (Janssen et al., 2008; Lv et al., 2015; Tsai & Discher, 2008; Vandivier et al., 2002). Fourth, many cells express multiple receptors, with the result that combinatorial "mix-and-match" of receptors and ligands dramatically increases the range and complexity of downstream achievable signaling profiles. Fifth, individual molecules may serve a dual purpose as both ligands and receptors. For example, when expressed on the apoptotic cell surface, calreticulin acts as a ligand for LDL receptor-related protein 1 (LRP1; CD91) (Gardai et al., 2005; Obeid et al., 2007; Su et al., 2002), but when expressed by phagocytes it acts as a co-receptor for collectins like C1q and surfactants A and D bound to apoptotic cells (Gaipl et al., 2001; Ogden et al., 2001; Trouw et al., 2008) (Janssen et al., 2008; Vandivier et al., 2002). Finally, unlike diffusible mediators, most receptors and bridging molecules are proteins and therefore capable of genetic deletion, enabling their *in vivo* role to be assessed more accurately. The diverse phenotypes of these knockout mice are consistent with the expanded homeostatic role for apoptotic cells put forth in this chapter (Arandjelovic & Ravichandran, 2015). For example, genetic deletion of BAI1 and BAI3 revealed an important role for apoptotic cells in both myogenesis and repair of injured muscle (Hamoud et al., 2014; Hochreiter-Hufford et al., 2013).

A last important issue relating to receptor-ligand interactions is the mechanism by which receptors like CD300 (Voss et al., 2015) and TIM (Freeman et al., 2010) directly recognize PtdSer (Birge et al., 2016). While externalization of PtdSer on the outer aspect of the cell membrane represents one of the cardinal features marking a cell as apoptotic, many viable cells also express PtdSer on their surface, including monocytes and activated B cells (Birge et al., 2016; Dillon, Mancini, Rosen, & Schlissel, 2000). Since both professional and non-professional phagocytes easily discriminate apoptotic from PtdSer-bearing live cells, additional distinguishable characteristics must come into play. For example, while there are at least two scramblases, which collapse the asymmetric distribution of phospholipids between inner and outer aspects of the cell membrane, only one (Xkr8) contributes to apoptosis, and its constitutive activation leads to recognition and engulfment of cells (Birge et al., 2016). In marked contrast, in cells containing a constitutively active form of the other scramblase (TREM16F), no engulfment occurs, unless apoptosis is triggered, for example, by anti-Fas antibody (Segawa, Suzuki, & Nagata, 2011). Opsonization of PtdSer may require a critical concentration or clustered arrangement of PtdSer (Ishii et al., 2005; Menke, Gerke, & Steinem, 2005), or modification of PtdSer or its side chains may be necessary. For example, some bridging molecules like Gas6 and MFG-E8 may bind with higher affinity to oxidized PtdSer (Tyurin et al., 2014), and some receptors, like CD36, specifically recognize oxidized PtdSer (Greenberg et al., 2006).

6.3.3 Additional mechanisms of message transmission

Several other mechanisms exist for transmission of the apoptotic cell's message, for each of which we will provide a remarkable example. It should be noted that, in transmitting its message, an apoptotic cell can, and often will, utilize more than one of these mechanisms. The mechanisms can be broadly divided into those which require direct physical interaction between the apoptotic and responding cells, and those in which the apoptotic cell exerts its effects indirectly, either from afar or through an induced intermediate.

Within the first category requiring direct physical interaction, besides receptor-mediated recognition and signaling, message transmission may depend, for example, on activation of elements of the phagocytic machinery. This can occur in a non-specific manner, irrespective of the nature of the phagocytic material (apoptotic, necrotic, or latex beads), as in the case of activation of Akt and suppression of apoptosis in macrophages (Reddy et al., 2002). Transmission may ultimately depend on phagocytosis, yet still be specific to apoptotic cells. For example, release of adenosine and increased expression of A2AR was shown to be dependent on phagocytosis of apoptotic corpses, as demonstrated by loss of these effects upon inhibition of phagocytosis with cytochalasin D (Köröskényi et al., 2011). In contrast, cytochalasin D had no effect on apoptotic cell-induced decreased expression of A3R, indicating its dependence on recognition alone, without a need for phagocytosis (Duró et al., 2014).

Alternatively, the message may be enclosed within the apoptotic cell or, more commonly, within exosomes or microvesicles that are shed by the apoptotic cell. Receipt of the message is brought about not by receptor-mediated recognition of the exosomes or microvesicles themselves, but rather by uptake of their hidden cargo, in the form of cytosolic proteins, RNA, or lipids (Castellana et al., 2009; Lopatina et al., 2016; Zernecke et al., 2009). During atherosclerosis, for example, exosomes or microvesicles from apoptotic endothelial cells, enriched with microRNA (miRNA)-126, are taken up by other endothelial and vascular smooth muscle cells. The miRNA-126 destabilizes the expression of a number of proteins, including an inhibitor of signaling through the chemokine receptor, CXCR4, which binds CXCL12 (also known as stromal cell-derived factor-1 [SDF-1]). Released signaling through CXCL12 initiates a positive feedback loop leading to increased production of CXCL12, followed by recruitment of stem cell antigen (Sca)-1-positive progenitor cells, and their subsequent differentiation into $CD31^+$ endothelial cells. Notably, apoptotic bodies from the atherosclerotic plaques of patients undergoing carotid endarterectomy were shown to reduce plaque area and macrophage content in a murine model of accelerated atherosclerosis.

The final mechanism requiring direct interaction resembles the preceding one in that uptake of a cargo is involved, but there is an essential difference. In this case, engulfment is not enough, and the cargo must be processed further to reveal its message. Also, in a sense, the entire apoptotic cell is the cargo, not just what it contains. As described in a recent comprehensive review (Röszer, 2017), the intracellular degradation of apoptotic cells leads to the efflux from phagosomes into the cytoplasm of breakdown products, mostly cholesterol and other lipids. Although as yet not definitively established, the extruded lipids (or their metabolic products) include the ligands for a variety of lipid-sensing nuclear receptors, including PPARδ (Mukundan et al., 2009), PPARγ (Roszer et al., 2011), retinoic acid receptor (RAR) (Sarang et al., 2014), liver X receptor (LXR) α/β (A-Gonzalez et al., 2009), and retinoid X receptor α (RXRα) (Roszer et al., 2011). Upon ligand binding, heterodimers of these lipid-sensing nuclear receptors (commonly including RXRα) translocate to the nucleus, where they augment the transcription of genes involved in apoptotic cell recognition, internalization, and signaling, as well as genes for other nuclear receptors (Röszer, 2017).

There are also other modes of indirect transmission. In addition to the release of diffusible mediators, apoptotic cells may interfere with message transmission between live cells, in effect deleting another cell's message by sequestering it and preventing it from reaching its intended cell (Ariel et al., 2006). Late apoptotic neutrophils and activated T cells have increased surface expression of the chemokine receptor CCR5, which binds CCL3, CCL4, and CCL5. The CCR5 is not newly synthesized, but externalized in a caspase-dependent manner. Importantly, externalized CCR5 on apoptotic corpses is non-functional, mediating no chemotactic activity, even though it binds CCL3, CCL4, and CCL5 with high avidity. In a murine peritonitis model, genetic deletion of CCR5 led to persistent and higher levels of CCL3 and CCL5, suggesting that late apoptotic CCR5-positive cells help to terminate chemokine signaling during the resolution of inflammation.

Finally, any of the above mechanisms may prove the first in a series of cascading effects, each involving the transcriptional expression or repression of multiple genes, including cytokines, chemokines, and other soluble mediators. The apoptotic cell's message is ultimately transmitted indirectly through the altered profile of expressed genes. A series of cascading indirect effects, beginning with release of S1P, has been shown in elicited peritoneal macrophages (Luo et al., 2016). Engagement of S1P with one of its receptors, S1PR1, triggers a transcriptionally-mediated increase in expression of the transcription factor, hypoxia-inducible factor 1α (HIF-1α). HIF-1α then mediates a transcriptional increase in expression of erythropoietin. Autocrine and/or paracrine interaction of erythropoietin with its receptor activates the transcription factor, CCAAT/enhancer binding protein β (C/EBPβ), which mediates transcription of the erythropoietin receptor and the transcription factor peroxisome proliferator activated receptor γ (PPARγ). In the final step, PPARγ mediates the expression of a number of proteins important to the recognition, phagocytosis, and effects of apoptotic corpses, including MerTK, Gas6, MFG-E8, and CD36.

As should be clear from the discussion so far, a sizeable portion of the instructions contained in the apoptotic cell's message, aside from "find me" and "eat me" signaling, is dedicated to the modulation of transcriptional activity. The range of transcription factors whose activity is either stimulated or repressed is broad, as is the panel of genes whose expression they regulate. Many of the activities ascribed to apoptotic cells can be traced back to the altered transcriptional profile they elicit in responding cells. Table 5 summarizes the major transcription factors and corresponding genes that apoptotic cells have been shown to modulate.

Table 5 Transcriptional factors modulated by apoptotic cells.[a]

Transcription factor	Responding cells	Mode of information transmission	Modulated genes	References
C/EBPβ	Elicited peritoneal macrophages	Erythropoietin-mediated, via autocrine/paracrine mechanism leading to HIF1α-dependent up-regulation of erythropoietin	Erythropoietin receptor, PPARγ	Luo et al. (2016)
CREB	RAW 264.7	Soluble mediator (S1P via $S1P_2$)	↑ Arginase II	Barra et al. (2011)
GC-BP (zinc finger nuclear factor)	RAW 264.7, murine BMDM	PS-dependent recognition of apoptotic cells, independent of phagocytosis	↓ IL-12 p35, IL-12 p40	Kim et al. (2004)
HIF-1α	RAW 264.7, HMDM, MMDM	Soluble mediators (S1P via $S1P_{1/3}$ and TGF-β)	↑ VEGF	Herr et al. (2009)
	Elicited peritoneal macrophages, BMDM	S1P via $S1P_1$	↑ Erythropoietin	Luo et al. (2016)
LXRα/-β/RXR	Elicited peritoneal macrophages	Intracellular digestion of engulfed apoptotic cells generates ligands for LXR (e.g., cholesterol, oxycholesterol)	↑ MerTK, ABCA1, ABCG1, PLTP, SCL2A4	Röszer (2017), A-Gonzalez et al. (2009)
	Resident peritoneal macrophages		↑ RALDH, Tg2	Garabuczi et al. (2013)
	BMDM, HMDM		↑ MerTK, ABCA1, SREBP1c, RARα, LXRα	Sarang et al. (2014) and Rébé et al. (2009)
NFAT	RAW 264.7	Soluble mediator (uncharacterized)	↑ HIF-1α	Herr et al. (2009)

Continued

Table 5 Transcriptional factors modulated by apoptotic cells.[a]—cont'd

Transcription factor	Responding cells	Mode of information transmission	Modulated genes	References
NFκB	Universal among nearly all cells	Recognition-dependent, immediate-early effect, independent of phagocytosis, protein synthesis, or secreted factors; unimpaired degradation of IκB	↓ NFκB transcriptional activity	Birge and Ucker (2008), Cvetanovic and Ucker (2004), and Cvetanovic et al. (2006)
	DC	MerTK-dependent inhibition of IκB kinase, delayed effect requiring new protein synthesis		Sen et al. (2007)
	RAW264.7, splenic macrophages	Trans-repression (sumoylation of PPARγ prevents LPS-induced induced NCoR removal from κB binding sites)		Jennewein et al. (2008)
Nr4a1/2	Resident peritoneal macrophages	PtdSer-dependent immediate early transcription via activation of p38 and ERK1/2	↓ IL-12 p40, inhibition of LPS-mediated transactivation of NFκB	Ipseiz et al. (2014), and Crean et al. (2015)
Pbx-1b/Prep-1 (Hox nuclear factors)	RAW 264.7, HMDM	PS-dependent recognition of apoptotic cells, independent of phagocytosis	↑ IL-10	Chung et al. (2007)
PPARδ/RXRα	Elicited peritoneal macrophages; BMDM	Intracellular digestion of engulfed apoptotic cells generates ligands for PPARδ (e.g., unsaturated FA, carbaprostacyclin, VLDL components)	↑ Axl, MerTK, C1q, Gas6, thrombospondin 1, MFG-E8	Röszer (2017), Grygiel-Górniak (2014), and Mukundan et al. (2009)
	Resident peritoneal macrophages		↑ RALDH, Tg2	Garabuczi et al. (2013)

PPARγ/RXRα	Elicited peritoneal macrophages	Intracellular digestion of engulfed apoptotic cells generates ligands for PPARγ (e.g., unsaturated FA, PGJ2)	↑ CD36, Axl, MerTK, Tg2, C1q, CD64 (FcγRI), Gas6, MFG-E8	Röszer (2017), Grygiel-Górniak (2014), Roszer et al. (2011), and Luo et al. (2016)
	Resident peritoneal macrophages		↑ RALDH, Tg2	Garabuczi et al. (2013)
RAR/RXRα	BMDM	Intracellular digestion of engulfed apoptotic cells generates ligand for RAR (? non-classical retinoid) through LXR-mediated up-regulation of RALDH	↑ Tg2, CD14, C1q, Stabilin2, TIM4	Röszer (2017) and Sarang et al. (2014)
	Resident peritoneal macrophages		↑ RALDH, Tg2	Garabuczi et al. (2013)
RXRα (dimers with LXR or PPARγ)	Elicited peritoneal macrophages	Intracellular digestion of engulfed apoptotic cells generates ligands for RXRα (e.g., 9cRA≫ATRA)	↑ CD36, Axl, MerTK, Tg2, C1q, CD64 (FcγRI)	Röszer (2017), A-Gonzalez et al. (2009), Roszer et al. (2011)
	Resident peritoneal macrophages		↑ RALDH, Tg2	Garabuczi et al. (2013)
SREBP1c	BMDM	Intracellular digestion of engulfed apoptotic cells leads to LXR/RXP-mediated transcription of SREBP1c	↑ RALDH	Sarang et al. (2014)
STAT1	Murine DC	Axl- and MerTK-dependent recognition of apoptotic cells	↑ SOCS1, SOCS3	Rothlin, Ghosh, Zuniga, Oldstone, and Lemke (2007)

[a]Abbreviations not given in Table 2. 9cRA, 9-*cis* retinoic acid; ABCA1, ATP-binding cassette transporter A1; ABCG1, ATP-binding cassette transporter G1; ATRA, all-*trans* retinoic acid; C/EBPβ, CCAAT/enhancer binding protein β; CREB, cAMP response element-binding protein; FA, fatty acids; FcγR, immunoglobulin G Fc receptor; GC-BP, GC-binding protein; Hox, homeobox; IκB, inhibitor of κB; IL-10, interleukin 10; IL-12, interleukin 12; LPS, lipopolysaccharide; LXR, liver X receptor; NCoR, nuclear receptor co-repressor; NFAT, nuclear factor of activated T cell; NFκB, nuclear factor κB; Nr4a1/2, nuclear receptor subfamily 4 group A member 1/2; Pbx-1b, pre-B cell leukemia transcription factor-1b; PGJ2, prostaglandin J2; PLTP, phospholipid transfer protein; PPAR, peroxisome proliferator activated receptor; Prep-1, Pbx-regulating protein-1; PS, phosphatidylserine; RALDH, retinaldehyde dehydrogenase; RAR, retinoic acid receptor; RXR, retinoid X receptor; SCL2A4, glucose transporter type 4; SOCS, suppressor of cytokine secretion; SREBP1c, serum response element binding protein 1c; STAT, signal transducer and activator of transcription; Tg2, transglutaminase 2; TGF-β, transforming growth factor β; TIM, T cell immunoglobulin mucin protein; VEGF, vascular endothelial growth factor; VLDL, very low density lipoprotein.

7. Inbuit mechanisms that improve the signal-to-noise ratio

The error rate, or probability that a message will be incorrectly encoded or decoded, is critically determined by the signal-to-noise ratio. The lower this ratio, the more difficult it is to distinguish specific signal from random noise. There are at least four mechanisms, two population-based and two single cell-based, by which the signal-to-noise ratio is increased and the error rate reduced, thereby ensuring fidelity of the apoptotic message. The two population-based mechanisms are a consequence of the very large numbers of both apoptotic cells, which transmit the message, and responder cells, which receive the message. Errors in the encoding of the apoptotic cell's message become of negligible consequence with the millions of apoptotic cells seen in most non-trivial stresses or injuries. A few rogue corpses, bearing a false message, can easily be distinguished from the overwhelming excess of apoptotic corpses carrying the true message. The ineffectiveness of false messages is enhanced by the fact that their generation is random. As opposed to the uniformity of corpses bearing a true message, rogue corpses will differ among themselves in the content of their false messages, further mitigating the likelihood of an aberrant response. Similarly, the very large number of responding cells of any one lineage or type works to enhance fidelity. The consequence of a few responders incorrectly decoding a message and setting forth on some aberrant activity will again be negligible compared to the uniform activity of the vast majority of phagocytes adhering to the message's instructions.

The two cell-based mechanisms derive from similar probabilistic principles. Just as in experimental investigation, repeated measurements improve signal detection and reduce the likelihood of error. In situations where dead cells are in excess, a single responding cell can engage multiple apoptotic corpses. Each engagement is like a single measurement. Moreover, if a responding cell possesses several different apoptotic receptors with overlapping downstream signaling profiles, then simultaneous engagement of these different receptors can act as a crosscheck of the apoptotic message. The second cell-based mechanism has its greatest impact when the number of apoptotic corpses is limiting, and not every phagocyte is able to interact with a corpse. In this case, it is better to concentrate receptor engagement to a reduced number of responders, rather than chancing an even distribution, with its attendant risk that receptor engagement will fail to exceed the threshold for activation of intracellular signaling in all but a very few responders. Several positive feedback loops exist to ensure that concentration of this sort in fact happens. Recognition and/or phagocytosis of apoptotic cells leads to activation of the small G protein and cytoskeletal regulator, Rac1 (Tosello-Trampont, Brugnera, & Ravichandran, 2001), and up-regulation of the mitochondrial protein, Ucp2 (Park, Han, et al., 2011). Both of these events in turn increase the phagocytic capacity of responder cells (Park, Han, et al., 2011; Tosello-Trampont et al., 2001), thus ensuring that effective intracellular signaling is triggered by phagocytic ingestion in at least a subpopulation of responding cells.

8. An intricate grammar

So far, in discussing the language of dead cells, we have dealt only with the effect of single characteristics and variables. Individually, each of these can influence the encoding, transmission, and/or decoding of the message from dead to responding cells. In nature, of course, these variables are rarely altered in isolation. In any situation of augmented cell death, many or all of them vary simultaneously, and their combination and interaction become as important a consideration as their individual impacts. Additional death-associated variables, not previously considered, enter the picture too, such as the extent, pattern, spatial distribution, timing, and kinetics of cell death. The architecture of the tissue in which cell death has occurred needs also to be taken into account, its barriers and partitions, its blood flow, its cell types and their organization, its ease of cellular entry and egress, all of which help to determine which cells will be recruited or banished, and the sequence and rapidity with which they arrive or depart. For the most part, the rules governing such large-scale integration are as yet unknown, a black box encompassing what is meant by a grammar. That such a grammar exists, and that it is characterized by a remarkable power, precision, and intricacy is revealed by the range, consistency, and power of the outcomes it determines, not only inflammatory and immunologic responsiveness, but also the expanded homeostatic effects described in Table 1, the most dramatic and far-reaching of which occur within the realm of development and organogenesis.

Previously, in describing the effects of the soluble mediators ATP/UTP and adenosine, we have given a flavor for how the individual apoptotic signals may be integrated into a coherent and comprehensive message. Despite the fact that ATP/UTP is released from both apoptotic and necrotic cells, the response of tissues somehow remains tailored to the correct mode of cell death. Even when levels of ATP/UTP decline to those most characteristic of apoptotic death, the extracellular conversion of ATP to adenosine, coupled with the ability to monitor the concentrations of both ATP/UTP and adenosine through multiple receptors of varying affinity, may allow responding cells not only to determine the mode of death, but also to guide the tissue successfully from acute injury through the various phase of inflammation to ultimate healing and restoration of a new homeostatic state.

We conclude with two further examples, the first from developmental morphogenesis and the second from vertebrate wound healing. Flexible structures called joints divide the legs of *Drosophila* into multiple segments. Apoptosis plays a critical role in the formation of these joints (Manjón, Sánchez-Herrero, & Suzanne, 2007; Monier et al., 2015). Induction of apoptosis is dependent on signaling by decapentaplegic (Dpp), a member of the bone morphogenic protein (BMP) family. BMP family members are also responsible for digital sculpting in vertebrates, but the mechanisms of apoptosis induction differ dramatically. In vertebrates, it is the absolute activity of BMP's that triggers cell death, whereas in *Drosophila* it is a discontinuity in the level of BMP signaling. This comes about in the following way. Dpp, which is expressed in the most distal portion of each segment-to-be, only diffuses proximally, where it accumulates at the proximal end of that segment-to-be. The discontinuity between high Dpp signaling at the proximal end of one segment-to-be and low Dpp signaling at the distal end of the adjacent segment-to-be triggers apoptotic cell death. Extrusion of the apoptotic cell requires the formation of a ring of actin and myosin, whose contraction squeezes the dying cell out of the epithelium. The synergistic interaction of multiple apoptotic cells aligned at the boundary between segments-to-be leads to a concentration of myosin, increased tension, and eventual constriction, with subsequent tissue folding and joint formation.

In wound healing, a network of interacting signals may work to coordinate events. PGE_2 plays a central role in the "phoenix rising" pathway, which promotes wound healing and liver regeneration (Li et al., 2010). The production of PGE_2 depends on release of arachidonic acid, and its conversion to PGE_2 by a series of enzymes including COX2. Release of arachidonic acid itself depends on cleavage of phosphatidylcholine by calcium-independent $iPLA_2$ (Li et al., 2010), the other product of this cleavage being the diffusible mediator LPC (Lauber et al., 2003). Moreover, S1P, also released by apoptotic cells, increases protein expression of COX2 by stabilizing its mRNA (Johann et al., 2008). Thus, while PGE_2 is promoting the survival and proliferation of epidermal keratinocyte progenitors and mesenchymal stem cells, LPC and S1P are recruiting professional phagocytes to clear away the dead. While it remains unclear whether these individual events cooperate in the manner described, the potential of such a coordinated interaction among these factors and events is nonetheless revealing of the powerful grammar by which apoptotic cells can convey their message and accomplish their ends.

9. Conclusions

Long regarded as mere waste, apoptotic cells are now regarded as key participants in multiple aspects of homeostasis. We have argued that the capacity of apoptotic cells to play an active role in homeostatic recovery and maintenance rests almost entirely on the transmission of information in the form of biochemical signals from cells undergoing apoptosis to live responding cells in the vicinity. Examples of the types of information transmitted by apoptotic cells include: the reason for its death, whether natural or induced by environmental stress; the type and severity of the environmental stress; the desired means of the dead cell's disposal, including various "find me" and "eat me" signals; and instructions to living cells in the vicinity for restoration of a new homeostatic steady state, if its achievement is possible. The scope and combinatorial complexity of the apoptotic cell's message is revealed by the number of variables affecting separately each of the three critical steps in communication: encoding of the message by apoptotic cells, transmission of the message, and receipt and decoding of the message by live recipient cells. A huge body of work over the last 10–20 years has identified many of the individual factors and variables that determine the content of the apoptotic cell's "message." The resulting communication network, centered on dead cells, allows tissues to both monitor and regulate their environment. While there are almost certainly more factors and variables to be discovered, the greatest challenge ahead lies in deciphering the manner in which these manifold variables combine and interact to bring about such a broad range of exquisitely fine-tuned and far-reaching effects, from the sculpting of tissues during development to the regeneration of severely damaged organs.

References

A-Gonzalez, N., Bensinger, S. J., Hong, C., Beceiro, S., Bradley, M. N., Zelcer, N., et al. (2009). Apoptotic cells promote their own clearance and immune tolerance through activation of the nuclear receptor LXR. *Immunity*, *31*(2), 245–258. https://doi.org/10.1016/j.immuni.2009.06.018.

Akakura, S., Singh, S., Spataro, M., Akakura, R., Kim, J.-I., Albert, M. L., et al. (2004). The opsonin MFG-E8 is a ligand for the alphavbeta5 integrin and triggers DOCK180-dependent Rac1 activation for the phagocytosis of apoptotic cells. *Experimental Cell Research*, *292*(2), 403–416.

Albert, M. L., Kim, J. I., & Birge, R. B. (2000). Alphavbeta5 integrin recruits the CrkII-Dock180-rac1 complex for phagocytosis of apoptotic cells. *Nature Cell Biology*, *2*(12), 899–905. https://doi.org/10.1038/35046549.

Arandjelovic, S., & Ravichandran, K. S. (2015). Phagocytosis of apoptotic cells in homeostasis. *Nature Immunology*, *16*(9), 907–917. https://doi.org/10.1038/ni.3253.

Ariel, A., Fredman, G., Sun, Y.-P., Kantarci, A., Van Dyke, T. E., Luster, A. D., et al. (2006). Apoptotic neutrophils and T cells sequester chemokines during immune response resolution through modulation of CCR5 expression. *Nature Immunology*, 7(11), 1209–1216. https://doi.org/10.1038/ni1392.

Arur, S., Uche, U. E., Rezaul, K., Fong, M., Scranton, V., Cowan, A. E., et al. (2003). Annexin I is an endogenous ligand that mediates apoptotic cell engulfment. *Developmental Cell*, *4*(4), 587–598.

Asiedu, M. K., Beauchamp-Perez, F. D., Ingle, J. N., Behrens, M. D., Radisky, D. C., & Knutson, K. L. (2014). AXL induces epithelial-to-mesenchymal transition and regulates the function of breast cancer stem cells. *Oncogene*, *33*(10), 1316–1324. https://doi.org/10.1038/onc.2013.57.

Awasthi, N., Schwarz, M. A., Verma, V., Cappiello, C., & Schwarz, R. E. (2009). Endothelial monocyte activating polypeptide II interferes with VEGF-induced proangiogenic signaling. *Laboratory Investigation; A Journal of Technical Methods and Pathology*, *89*(1), 38–46. https://doi.org/10.1038/labinvest.2008.106.

Barra, V., Kuhn, A.-M., von Knethen, A., Weigert, A., & Brüne, B. (2011). Apoptotic cell-derived factors induce arginase II expression in murine macrophages by activating ERK5/CREB. *Cellular and Molecular Life Sciences: CMLS*, *68*(10), 1815–1827. https://doi.org/10.1007/s00018-010-0537-x.

Behrensdorf, H. A., van de Craen, M., Knies, U. E., Vandenabeele, P., & Clauss, M. (2000). The endothelial monocyte-activating polypeptide II (EMAP II) is a substrate for caspase-7. *FEBS Letters*, *466*(1), 143–147.

Bell, C. W., Jiang, W., Reich, C. F., & Pisetsky, D. S. (2006). The extracellular release of HMGB1 during apoptotic cell death. *American Journal of Physiology. Cell Physiology*, *291*(6), C1318–C1325. https://doi.org/10.1152/ajpcell.00616.2005.

Bennett, M. R., Gibson, D. F., Schwartz, S. M., & Tait, J. F. (1995). Binding and phagocytosis of apoptotic vascular smooth muscle cells is mediated in part by exposure of phosphatidylserine. *Circulation Research*, 77(6), 1136–1142.

Bergmann, A., & Steller, H. (2010). Apoptosis, stem cells, and tissue regeneration. *Science Signaling*, *3*(145), re8. https://doi.org/10.1126/scisignal.3145re8.

Berwin, B., Delneste, Y., Lovingood, R. V., Post, S. R., & Pizzo, S. V. (2004). SREC-I, a type F scavenger receptor, is an endocytic receptor for calreticulin. *The Journal of Biological Chemistry*, *279*(49), 51250–51257. https://doi.org/10.1074/jbc.M406202200.

Birge, R. B., Boeltz, S., Kumar, S., Carlson, J., Wanderley, J., Calianese, D., et al. (2016). Phosphatidylserine is a global immunosuppressive signal in efferocytosis, infectious disease, and cancer. *Cell Death and Differentiation*, *23*(6), 962–978. https://doi.org/10.1038/cdd.2016.11.

Birge, R. B., & Ucker, D. S. (2008). Innate apoptotic immunity: The calming touch of death. *Cell Death and Differentiation*, *15*(7), 1096–1102. https://doi.org/10.1038/cdd.2008.58.

Blume, K. E., Soeroes, S., Keppeler, H., Stevanovic, S., Kretschmer, D., Rautenberg, M., et al. (2012). Cleavage of annexin A1 by ADAM10 during secondary necrosis generates a monocytic "find-me" signal. *Journal of Immunology (Baltimore, Md.: 1950)*, *188*(1), 135–145. https://doi.org/10.4049/jimmunol.1004073.

Blume, K. E., Soeroes, S., Waibel, M., Keppeler, H., Wesselborg, S., Herrmann, M., et al. (2009). Cell surface externalization of annexin A1 as a failsafe mechanism preventing inflammatory responses during secondary necrosis. *Journal of Immunology (Baltimore, Md.: 1950)*, *183*(12), 8138–8147. https://doi.org/10.4049/jimmunol.0902250.

Bottoni, A., Vignali, C., Piccin, D., Tagliati, F., Luchin, A., Zatelli, M. C., et al. (2007). Proteasomes and RARS modulate AIMP1/EMAP II secretion in human cancer cell lines. *Journal of Cellular Physiology*, *212*(2), 293–297. https://doi.org/10.1002/jcp.21083.

Bournazou, I., Mackenzie, K. J., Duffin, R., Rossi, A. G., & Gregory, C. D. (2010). Inhibition of eosinophil migration by lactoferrin. *Immunology and Cell Biology*, *88*(2), 220–223. https://doi.org/10.1038/icb.2009.86.

Bournazou, I., Pound, J. D., Duffin, R., Bournazos, S., Melville, L. A., Brown, S. B., et al. (2009). Apoptotic human cells inhibit migration of granulocytes via release of lactoferrin. *The Journal of Clinical Investigation*, *119*(1), 20–32. https://doi.org/10.1172/JCI36226.

Brown, S., Heinisch, I., Ross, E., Shaw, K., Buckley, C. D., & Savill, J. (2002). Apoptosis disables CD31-mediated cell detachment from phagocytes promoting binding and engulfment. *Nature*, *418*(6894), 200–203. https://doi.org/10.1038/nature00811.

Bryan, A. M., & Del Poeta, M. (2018). Sphingosine-1-phosphate receptors and innate immunity. *Cellular Microbiology*, *20*(5) e12836. https://doi.org/10.1111/cmi.12836.

Burstyn-Cohen, T. (2017). TAM receptor signaling in development. *The International Journal of Developmental Biology*, *61*(3-4–5), 215–224. https://doi.org/10.1387/ijdb.160285tb.

Byun, J. S., & Gardner, K. (2013). Wounds that will not heal: Pervasive cellular reprogramming in cancer. *The American Journal of Pathology*, *182*(4), 1055–1064. https://doi.org/10.1016/j.ajpath.2013.01.009.

Caberoy, N. B., Alvarado, G., Bigcas, J.-L., & Li, W. (2012). Galectin-3 is a new MerTK-specific eat-me signal. *Journal of Cellular Physiology*, *227*(2), 401–407. https://doi.org/10.1002/jcp.22955.

Caberoy, N. B., Alvarado, G., & Li, W. (2012). Tubby regulates microglial phagocytosis through MerTK. *Journal of Neuroimmunology*, *252*(1–2), 40–48. https://doi.org/10.1016/j.jneuroim.2012.07.009.

Caberoy, N. B., Zhou, Y., & Li, W. (2010). Tubby and tubby-like protein 1 are new MerTK ligands for phagocytosis. *The EMBO Journal*, *29*(23), 3898–3910. https://doi.org/10.1038/emboj.2010.265.

Cannon, W. B., & Cannon, C. J. (1967). *The wisdom of the body: How the human body reacts to disturbance and danger and maintains the stability essential to life (Rev. and enl. ed. Renewed)*. New York: Norton.

Cardona, M., López, J. A., Serafín, A., Rongvaux, A., Inserte, J., García-Dorado, D., et al. (2015). Executioner caspase-3 and 7 deficiency reduces myocyte number in the developing mouse heart. *PloS One*, *10*(6) e0131411. https://doi.org/10.1371/journal.pone.0131411.

Castellana, D., Zobairi, F., Martinez, M. C., Panaro, M. A., Mitolo, V., Freyssinet, J.-M., et al. (2009). Membrane microvesicles as actors in the establishment of a favorable prostatic tumoral niche: A role for activated fibroblasts and CX3CL1-CX3CR1 axis. *Cancer Research*, *69*(3), 785–793. https://doi.org/10.1158/0008-5472.CAN-08-1946.

Cekic, C., & Linden, J. (2016). Purinergic regulation of the immune system. *Nature Reviews Immunology*, *16*(3), 177–192. https://doi.org/10.1038/nri.2016.4.

Chalaris, A., Rabe, B., Paliga, K., Lange, H., Laskay, T., Fielding, C. A., et al. (2007). Apoptosis is a natural stimulus of IL6R shedding and contributes to the proinflammatory trans-signaling function of neutrophils. *Blood*, *110*(6), 1748–1755. https://doi.org/10.1182/blood-2007-01-067918.

Chekeni, F. B., Elliott, M. R., Sandilos, J. K., Walk, S. F., Kinchen, J. M., Lazarowski, E. R., et al. (2010). Pannexin 1 channels mediate "find-me" signal release and membrane permeability during apoptosis. *Nature*, *467*(7317), 863–867. https://doi.org/10.1038/nature09413.

Chen, G., Li, J., Qiang, X., Czura, C. J., Ochani, M., Ochani, K., et al. (2005). Suppression of HMGB1 release by stearoyl lysophosphatidylcholine:an additional mechanism for its therapeutic effects in experimental sepsis. *Journal of Lipid Research, 46*(4), 623–627. https://doi.org/10.1194/jlr.C400018-JLR200.

Chera, S., Ghila, L., Dobretz, K., Wenger, Y., Bauer, C., Buzgariu, W., et al. (2009). Apoptotic cells provide an unexpected source of Wnt3 signaling to drive hydra head regeneration. *Developmental Cell, 17*(2), 279–289. https://doi.org/10.1016/j.devcel.2009.07.014.

Chung, W.-S., Clarke, L. E., Wang, G. X., Stafford, B. K., Sher, A., Chakraborty, C., et al. (2013). Astrocytes mediate synapse elimination through MEGF10 and MERTK pathways. *Nature, 504*(7480), 394–400. https://doi.org/10.1038/nature12776.

Chung, E. Y., Liu, J., Homma, Y., Zhang, Y., Brendolan, A., Saggese, M., et al. (2007). Interleukin-10 expression in macrophages during phagocytosis of apoptotic cells is mediated by homeodomain proteins Pbx1 and Prep-1. *Immunity, 27*(6), 952–964. https://doi.org/10.1016/j.immuni.2007.11.014.

Cocco, R. E., & Ucker, D. S. (2001). Distinct modes of macrophage recognition for apoptotic and necrotic cells are not specified exclusively by phosphatidylserine exposure. *Molecular Biology of the Cell, 12*(4), 919–930. https://doi.org/10.1091/mbc.12.4.919.

Cohen, H. B., Briggs, K. T., Marino, J. P., Ravid, K., Robson, S. C., & Mosser, D. M. (2013). TLR stimulation initiates a CD39-based autoregulatory mechanism that limits macrophage inflammatory responses. *Blood, 122*(11), 1935–1945. https://doi.org/10.1182/blood-2013-04-496216.

Corriden, R., & Insel, P. A. (2010). Basal release of ATP: An autocrine-paracrine mechanism for cell regulation. *Science Signaling, 3*(104), re1. https://doi.org/10.1126/scisignal.3104re1.

Crean, D., Cummins, E. P., Bahar, B., Mohan, H., McMorrow, J. P., & Murphy, E. P. (2015). Adenosine modulates NR4A orphan nuclear receptors to attenuate hyperinflammatory responses in monocytic cells. *Journal of Immunology (Baltimore, Md.: 1950), 195*(4), 1436–1448. https://doi.org/10.4049/jimmunol.1402039.

Cuervo, R., & Covarrubias, L. (2004). Death is the major fate of medial edge epithelial cells and the cause of basal lamina degradation during palatogenesis. *Development (Cambridge, England), 131*(1), 15–24. https://doi.org/10.1242/dev.00907.

Cvetanovic, M., Mitchell, J. E., Patel, V., Avner, B. S., Su, Y., van der Saag, P. T., et al. (2006). Specific recognition of apoptotic cells reveals a ubiquitous and unconventional innate immunity. *The Journal of Biological Chemistry, 281*(29), 20055–20067. https://doi.org/10.1074/jbc.M603920200.

Cvetanovic, M., & Ucker, D. S. (2004). Innate immune discrimination of apoptotic cells: Repression of proinflammatory macrophage transcription is coupled directly to specific recognition. *Journal of Immunology (Baltimore, Md.: 1950), 172*(2), 880–889.

D'Cruz, P. M., Yasumura, D., Weir, J., Matthes, M. T., Abderrahim, H., LaVail, M. M., et al. (2000). Mutation of the receptor tyrosine kinase gene Mertk in the retinal dystrophic RCS rat. *Human Molecular Genetics, 9*(4), 645–651.

Das, S., Sarkar, A., Ryan, K. A., Fox, S., Berger, A. H., Juncadella, I. J., et al. (2014). Brain angiogenesis inhibitor 1 is expressed by gastric phagocytes during infection with Helicobacter pylori and mediates the recognition and engulfment of human apoptotic gastric epithelial cells. *FASEB Journal: Official Publication of the Federation of American Societies for Experimental Biology, 28*(5), 2214–2224. https://doi.org/10.1096/fj.13-243238.

de Freitas, A., Banerjee, S., Xie, N., Cui, H., Davis, K. I., Friggeri, A., et al. (2012). Identification of TLT2 as an engulfment receptor for apoptotic cells. *Journal of Immunology (Baltimore, Md.: 1950), 188*(12), 6381–6388. https://doi.org/10.4049/jimmunol.1200020.

Devitt, A., & Marshall, L. J. (2011). The innate immune system and the clearance of apoptotic cells. *Journal of Leukocyte Biology, 90*(3), 447–457. https://doi.org/10.1189/jlb.0211095.

Devitt, A., Parker, K. G., Ogden, C. A., Oldreive, C., Clay, M. F., Melville, L. A., et al. (2004). Persistence of apoptotic cells without autoimmune disease or inflammation in CD14-/- mice. *The Journal of Cell Biology, 167*(6), 1161–1170. https://doi.org/10.1083/jcb.200410057.

Dillon, S. R., Mancini, M., Rosen, A., & Schlissel, M. S. (2000). Annexin V binds to viable B cells and colocalizes with a marker of lipid rafts upon B cell receptor activation. *Journal of Immunology (Baltimore, Md.: 1950), 164*(3), 1322–1332.

Dini, L., & Carla, E. C. (1998). Hepatic sinusoidal endothelium heterogeneity with respect to the recognition of apoptotic cells. *Experimental Cell Research, 240*(2), 388–393.

Donato, A. L., Huang, Q., Liu, X., Li, F., Zimmerman, M. A., & Li, C.-Y. (2014). Caspase 3 promotes surviving melanoma tumor cell growth after cytotoxic therapy. *The Journal of Investigative Dermatology, 134*(6), 1686–1692. https://doi.org/10.1038/jid.2014.18.

Drappa, J., Vaishnaw, A. K., Sullivan, K. E., Chu, J. L., & Elkon, K. B. (1996). Fas gene mutations in the Canale-Smith syndrome, an inherited lymphoproliferative disorder associated with autoimmunity. *The New England Journal of Medicine, 335*(22), 1643–1649. https://doi.org/10.1056/NEJM199611283352204.

Duró, E., Pallai, A., Köröskényi, K., Sarang, Z., & Szondy, Z. (2014). Adenosine A3 receptors negatively regulate the engulfment-dependent apoptotic cell suppression of inflammation. *Immunology Letters, 162*(2 Pt B), 292–301. https://doi.org/10.1016/j.imlet.2014.06.014.

Dvorak, H. F. (1986). Tumors: Wounds that do not heal. Similarities between tumor stroma generation and wound healing. *The New England Journal of Medicine, 315*(26), 1650–1659. https://doi.org/10.1056/NEJM198612253152606.

Eckle, T., Kewley, E. M., Brodsky, K. S., Tak, E., Bonney, S., Gobel, M., et al. (2014). Identification of hypoxia-inducible factor HIF-1A as transcriptional regulator of the A2B adenosine receptor during acute lung injury. *Journal of Immunology (Baltimore, Md.: 1950), 192*(3), 1249–1256. https://doi.org/10.4049/jimmunol.1100593.

Elliott, M. R., Chekeni, F. B., Trampont, P. C., Lazarowski, E. R., Kadl, A., Walk, S. F., et al. (2009). Nucleotides released by apoptotic cells act as a find-me signal to promote phagocytic clearance. *Nature, 461*(7261), 282–286. https://doi.org/10.1038/nature08296.

Elliott, M. R., Koster, K. M., & Murphy, P. S. (2017). Efferocytosis signaling in the regulation of macrophage inflammatory responses. *Journal of Immunology (Baltimore, Md.: 1950), 198*(4), 1387–1394. https://doi.org/10.4049/jimmunol.1601520.

Elliott, M. R., & Ravichandran, K. S. (2016). The dynamics of apoptotic cell clearance. *Developmental Cell, 38*(2), 147–160. https://doi.org/10.1016/j.devcel.2016.06.029.

Elliott, M. R., Zheng, S., Park, D., Woodson, R. I., Reardon, M. A., Juncadella, I. J., et al. (2010). Unexpected requirement for ELMO1 in clearance of apoptotic germ cells in vivo. *Nature, 467*(7313), 333–337. https://doi.org/10.1038/nature09356.

Erwig, L.-P., McPhilips, K. A., Wynes, M. W., Ivetic, A., Ridley, A. J., & Henson, P. M. (2006). Differential regulation of phagosome maturation in macrophages and dendritic cells mediated by Rho GTPases and ezrin-radixin-moesin (ERM) proteins. *Proceedings of the National Academy of Sciences of the United States of America, 103*(34), 12825–12830. https://doi.org/10.1073/pnas.0605331103.

Esmann, L., Idel, C., Sarkar, A., Hellberg, L., Behnen, M., Möller, S., et al. (2010). Phagocytosis of apoptotic cells by neutrophil granulocytes: Diminished proinflammatory neutrophil functions in the presence of apoptotic cells. *Journal of Immunology (Baltimore, Md.: 1950), 184*(1), 391–400. https://doi.org/10.4049/jimmunol.0900564.

Evan, G. I., & Vousden, K. H. (2001). Proliferation, cell cycle and apoptosis in cancer. *Nature, 411*(6835), 342–348. https://doi.org/10.1038/35077213.

Fadok, V. A., Bratton, D. L., Konowal, A., Freed, P. W., Westcott, J. Y., & Henson, P. M. (1998). Macrophages that have ingested apoptotic cells in vitro inhibit proinflammatory cytokine production through autocrine/paracrine mechanisms involving TGF-beta, PGE2, and PAF. *The Journal of Clinical Investigation*, *101*(4), 890–898. https://doi.org/10.1172/JCI1112.

Fadok, V. A., Warner, M. L., Bratton, D. L., & Henson, P. M. (1998). CD36 is required for phagocytosis of apoptotic cells by human macrophages that use either a phosphatidylserine receptor or the vitronectin receptor (alpha v beta 3). *Journal of Immunology (Baltimore, Md.: 1950)*, *161*(11), 6250–6257.

Falasca, L., Bergamini, A., Serafino, A., Balabaud, C., & Dini, L. (1996). Human Kupffer cell recognition and phagocytosis of apoptotic peripheral blood lymphocytes. *Experimental Cell Research*, *224*(1), 152–162. https://doi.org/10.1006/excr.1996.0123.

Familian, A., Zwart, B., Huisman, H. G., Rensink, I., Roem, D., Hordijk, P. L., et al. (2001). Chromatin-independent binding of serum amyloid P component to apoptotic cells. *Journal of Immunology (Baltimore, Md.: 1950)*, *167*(2), 647–654.

Fan, Y., & Bergmann, A. (2008). Apoptosis-induced compensatory proliferation. The cell is dead. Long live the cell!. *Trends in Cell Biology*, *18*(10), 467–473. https://doi.org/10.1016/j.tcb.2008.08.001.

Fan, X., Krahling, S., Smith, D., Williamson, P., & Schlegel, R. A. (2004). Macrophage surface expression of annexins I and II in the phagocytosis of apoptotic lymphocytes. *Molecular Biology of the Cell*, *15*(6), 2863–2872. https://doi.org/10.1091/mbc.e03-09-0670.

Feng, L., Vujicic, S., Dietrich, M. E., Litbarg, N., Setty, S., Antoni, A., et al. (2018). Repeated exposure of epithelial cells to apoptotic cells induces the specific selection of an adaptive phenotype: Implications for tumorigenesis. *The Journal of Biological Chemistry*, *293*(26), 10245–10263. https://doi.org/10.1074/jbc.RA117.001290.

Flannagan, R. S., Canton, J., Furuya, W., Glogauer, M., & Grinstein, S. (2014). The phosphatidylserine receptor TIM4 utilizes integrins as coreceptors to effect phagocytosis. *Molecular Biology of the Cell*, *25*(9), 1511–1522. https://doi.org/10.1091/mbc.E13-04-0212.

Freeman, G. J., Casasnovas, J. M., Umetsu, D. T., & DeKruyff, R. H. (2010). TIM genes: A family of cell surface phosphatidylserine receptors that regulate innate and adaptive immunity. *Immunological Reviews*, *235*(1), 172–189. https://doi.org/10.1111/j.0105-2896.2010.00903.x.

Friggeri, A., Banerjee, S., Biswas, S., de Freitas, A., Liu, G., Bierhaus, A., et al. (2011). Participation of the receptor for advanced glycation end products in efferocytosis. *Journal of Immunology (Baltimore, Md.: 1950)*, *186*(11), 6191–6198. https://doi.org/10.4049/jimmunol.1004134.

Fuller, A. D., & Van Eldik, L. J. (2008). MFG-E8 regulates microglial phagocytosis of apoptotic neurons. *Journal of Neuroimmune Pharmacology: The Official Journal of the Society on NeuroImmune Pharmacology*, *3*(4), 246–256. https://doi.org/10.1007/s11481-008-9118-2.

Gaipl, U. S., Kuenkele, S., Voll, R. E., Beyer, T. D., Kolowos, W., Heyder, P., et al. (2001). Complement binding is an early feature of necrotic and a rather late event during apoptotic cell death. *Cell Death and Differentiation*, *8*(4), 327–334. https://doi.org/10.1038/sj.cdd.4400826.

Galluzzi, L., Vitale, I., Aaronson, S. A., Abrams, J. M., Adam, D., Agostinis, P., et al. (2018). Molecular mechanisms of cell death: Recommendations of the nomenclature committee on cell death 2018. *Cell Death and Differentiation*, *25*(3), 486–541. https://doi.org/10.1038/s41418-017-0012-4.

Garabuczi, É., Kiss, B., Felszeghy, S., Tsay, G. J., Fésüs, L., & Szondy, Z. (2013). Retinoids produced by macrophages engulfing apoptotic cells contribute to the appearance of transglutaminase 2 in apoptotic thymocytes. *Amino Acids*, *44*(1), 235–244. https://doi.org/10.1007/s00726-011-1119-4.

Gardai, S. J., McPhillips, K. A., Frasch, S. C., Janssen, W. J., Starefeldt, A., Murphy-Ullrich, J. E., et al. (2005). Cell-surface calreticulin initiates clearance of viable or apoptotic cells through trans-activation of LRP on the phagocyte. *Cell*, *123*(2), 321–334. https://doi.org/10.1016/j.cell.2005.08.032.

Gershov, D., Kim, S., Brot, N., & Elkon, K. B. (2000). C-Reactive protein binds to apoptotic cells, protects the cells from assembly of the terminal complement components, and sustains an antiinflammatory innate immune response: Implications for systemic autoimmunity. *The Journal of Experimental Medicine*, *192*(9), 1353–1364.

Goessling, W., North, T. E., Loewer, S., Lord, A. M., Lee, S., Stoick-Cooper, C. L., et al. (2009). Genetic interaction of PGE2 and Wnt signaling regulates developmental specification of stem cells and regeneration. *Cell*, *136*(6), 1136–1147. https://doi.org/10.1016/j.cell.2009.01.015.

Golpon, H. A., Fadok, V. A., Taraseviciene-Stewart, L., Scerbavicius, R., Sauer, C., Welte, T., et al. (2004). Life after corpse engulfment: Phagocytosis of apoptotic cells leads to VEGF secretion and cell growth. *FASEB Journal: Official Publication of the Federation of American Societies for Experimental Biology*, *18*(14), 1716–1718. https://doi.org/10.1096/fj.04-1853fje.

Gomez Perdiguero, E., Klapproth, K., Schulz, C., Busch, K., Azzoni, E., Crozet, L., et al. (2015). Tissue-resident macrophages originate from yolk-sac-derived erythro-myeloid progenitors. *Nature*, *518*(7540), 547–551. https://doi.org/10.1038/nature13989.

González-Polo, R.-A., Boya, P., Pauleau, A.-L., Jalil, A., Larochette, N., Souquère, S., et al. (2005). The apoptosis/autophagy paradox: Autophagic vacuolization before apoptotic death. *Journal of Cell Science*, *118*(Pt 14), 3091–3102. https://doi.org/10.1242/jcs.02447.

Gorman, M. W., Feigl, E. O., & Buffington, C. W. (2007). Human plasma ATP concentration. *Clinical Chemistry*, *53*(2), 318–325. https://doi.org/10.1373/clinchem.2006.076364.

Graham, D. K., DeRyckere, D., Davies, K. D., & Earp, H. S. (2014). The TAM family: Phosphatidylserine sensing receptor tyrosine kinases gone awry in cancer. *Nature Reviews Cancer*, *14*(12), 769–785. https://doi.org/10.1038/nrc3847.

Green, D. R., Galluzzi, L., & Kroemer, G. (2014). Cell biology. Metabolic control of cell death. *Science (New York, N.Y.)*, *345*(6203), 1250256. https://doi.org/10.1126/science.1250256.

Green, D. R., Oguin, T. H., & Martinez, J. (2016). The clearance of dying cells: Table for two. *Cell Death and Differentiation*, *23*(6), 915–926. https://doi.org/10.1038/cdd.2015.172.

Greenberg, M. E., Sun, M., Zhang, R., Febbraio, M., Silverstein, R., & Hazen, S. L. (2006). Oxidized phosphatidylserine-CD36 interactions play an essential role in macrophage-dependent phagocytosis of apoptotic cells. *The Journal of Experimental Medicine*, *203*(12), 2613–2625. https://doi.org/10.1084/jem.20060370.

Gregory, C. D., Ford, C. A., & Voss, J. J. L. P. (2016). Microenvironmental effects of cell death in malignant disease. *Advances in Experimental Medicine and Biology*, *930*, 51–88. https://doi.org/10.1007/978-3-319-39406-0_3.

Grygiel-Górniak, B. (2014). Peroxisome proliferator-activated receptors and their ligands: Nutritional and clinical implications—A review. *Nutrition Journal*, *13*, 17. https://doi.org/10.1186/1475-2891-13-17.

Gu, Y., Forostyan, T., Sabbadini, R., & Rosenblatt, J. (2011). Epithelial cell extrusion requires the sphingosine-1-phosphate receptor 2 pathway. *The Journal of Cell Biology*, *193*(4), 667–676. https://doi.org/10.1083/jcb.201010075.

Gude, D. R., Alvarez, S. E., Paugh, S. W., Mitra, P., Yu, J., Griffiths, R., et al. (2008). Apoptosis induces expression of sphingosine kinase 1 to release sphingosine-1-phosphate as a "come-and-get-me" signal. *FASEB Journal: Official Publication of the Federation of American Societies for Experimental Biology*, *22*(8), 2629–2638. https://doi.org/10.1096/fj.08-107169.

Günther, C., Neumann, H., Neurath, M. F., & Becker, C. (2013). Apoptosis, necrosis and necroptosis: Cell death regulation in the intestinal epithelium. *Gut, 62*(7), 1062–1071. https://doi.org/10.1136/gutjnl-2011-301364.

Guo, F., Ding, Y., Caberoy, N. B., Alvarado, G., Liu, R., Shen, C., et al. (2015). Lyar is a new ligand for retinal pigment epithelial phagocytosis. *Journal of Cellular Biochemistry, 116*(10), 2177–2187. https://doi.org/10.1002/jcb.25089.

Hall, S. E., Savill, J. S., Henson, P. M., & Haslett, C. (1994). Apoptotic neutrophils are phagocytosed by fibroblasts with participation of the fibroblast vitronectin receptor and involvement of a mannose/fucose-specific lectin. *Journal of Immunology (Baltimore, Md.: 1950), 153*(7), 3218–3227.

Hamoud, N., Tran, V., Aimi, T., Kakegawa, W., Lahaie, S., Thibault, M.-P., et al. (2018). Spatiotemporal regulation of the GPCR activity of BAI3 by C1qL4 and Stabilin-2 controls myoblast fusion. *Nature Communications, 9*(1), 4470. https://doi.org/10.1038/s41467-018-06897-5.

Hamoud, N., Tran, V., Croteau, L.-P., Kania, A., & Côté, J.-F. (2014). G-protein coupled receptor BAI3 promotes myoblast fusion in vertebrates. *Proceedings of the National Academy of Sciences of the United States of America, 111*(10), 3745–3750. https://doi.org/10.1073/pnas.1313886111.

Hanayama, R., Tanaka, M., Miwa, K., Shinohara, A., Iwamatsu, A., & Nagata, S. (2002). Identification of a factor that links apoptotic cells to phagocytes. *Nature, 417*(6885), 182–187. https://doi.org/10.1038/417182a.

Hanayama, R., Tanaka, M., Miyasaka, K., Aozasa, K., Koike, M., Uchiyama, Y., et al. (2004). Autoimmune disease and impaired uptake of apoptotic cells in MFG-E8-deficient mice. *Science (New York, NY), 304*(5674), 1147–1150. https://doi.org/10.1126/science.1094359.

Happonen, K. E., Tran, S., Mörgelin, M., Prince, R., Calzavarini, S., Angelillo-Scherrer, A., et al. (2016). The Gas6-Axl protein interaction mediates endothelial uptake of platelet microparticles. *The Journal of Biological Chemistry, 291*(20), 10586–10601. https://doi.org/10.1074/jbc.M115.699058.

Harre, U., Keppeler, H., Ipseiz, N., Derer, A., Poller, K., Aigner, M., et al. (2012). Moonlighting osteoclasts as undertakers of apoptotic cells. *Autoimmunity, 45*(8), 612–619. https://doi.org/10.3109/08916934.2012.719950.

He, M., Kubo, H., Morimoto, K., Fujino, N., Suzuki, T., Takahasi, T., et al. (2011). Receptor for advanced glycation end products binds to phosphatidylserine and assists in the clearance of apoptotic cells. *EMBO Reports, 12*(4), 358–364. https://doi.org/10.1038/embor.2011.28.

Herr, B., Zhou, J., Werno, C., Menrad, H., Namgaladze, D., Weigert, A., et al. (2009). The supernatant of apoptotic cells causes transcriptional activation of hypoxia-inducible factor-1alpha in macrophages via sphingosine-1-phosphate and transforming growth factor-beta. *Blood, 114*(10), 2140–2148. https://doi.org/10.1182/blood-2009-01-201889.

Hess, K. L., Tudor, K. S., Johnson, J. D., Osati-Ashtiani, F., Askew, D. S., & Cook-Mills, J. M. (1997). Human and murine high endothelial venule cells phagocytose apoptotic leukocytes. *Experimental Cell Research, 236*(2), 404–411. https://doi.org/10.1006/excr.1997.3745.

Hochreiter-Hufford, A. E., Lee, C. S., Kinchen, J. M., Sokolowski, J. D., Arandjelovic, S., Call, J. A., et al. (2013). Phosphatidylserine receptor BAI1 and apoptotic cells as new promoters of myoblast fusion. *Nature, 497*(7448), 263–267. https://doi.org/10.1038/nature12135.

Hochreiter-Hufford, A., & Ravichandran, K. S. (2013). Clearing the dead: Apoptotic cell sensing, recognition, engulfment, and digestion. *Cold Spring Harbor Perspectives in Biology, 5*(1), a008748. https://doi.org/10.1101/cshperspect.a008748.

Hoeffel, G., Chen, J., Lavin, Y., Low, D., Almeida, F. F., See, P., et al. (2015). C-Myb(+) erythro-myeloid progenitor-derived fetal monocytes give rise to adult tissue-resident macrophages. *Immunity, 42*(4), 665–678. https://doi.org/10.1016/j.immuni.2015.03.011.

Hou, Y., Plett, P. A., Ingram, D. A., Rajashekhar, G., Orschell, C. M., Yoder, M. C., et al. (2006). Endothelial-monocyte-activating polypeptide II induces migration of endothelial progenitor cells via the chemokine receptor CXCR3. *Experimental Hematology, 34*(8), 1125–1132. https://doi.org/10.1016/j.exphem.2006.05.021.

Hsieh, C. L., Koike, M., Spusta, S. C., Niemi, E. C., Yenari, M., Nakamura, M. C., et al. (2009). A role for TREM2 ligands in the phagocytosis of apoptotic neuronal cells by microglia. *Journal of Neurochemistry, 109*(4), 1144–1156. https://doi.org/10.1111/j.1471-4159.2009.06042.x.

Huang, Q., Li, F., Liu, X., Li, W., Shi, W., Liu, F.-F., et al. (2011). Caspase 3-mediated stimulation of tumor cell repopulation during cancer radiotherapy. *Nature Medicine, 17*(7), 860–866. https://doi.org/10.1038/nm.2385.

Ichimura, T., Asseldonk, E. J. P. V., Humphreys, B. D., Gunaratnam, L., Duffield, J. S., & Bonventre, J. V. (2008). Kidney injury molecule-1 is a phosphatidylserine receptor that confers a phagocytic phenotype on epithelial cells. *The Journal of Clinical Investigation, 118*(5), 1657–1668. https://doi.org/10.1172/JCI34487.

Idzko, M., Ferrari, D., & Eltzschig, H. K. (2014). Nucleotide signalling during inflammation. *Nature, 509*(7500), 310–317. https://doi.org/10.1038/nature13085.

Imachi, H., Murao, K., Hiramine, C., Sayo, Y., Sato, M., Hosokawa, H., et al. (2000). Human scavenger receptor B1 is involved in recognition of apoptotic thymocytes by thymic nurse cells. *Laboratory Investigation; A Journal of Technical Methods and Pathology, 80*(2), 263–270.

Ipseiz, N., Uderhardt, S., Scholtysek, C., Steffen, M., Schabbauer, G., Bozec, A., et al. (2014). The nuclear receptor Nr4a1 mediates anti-inflammatory effects of apoptotic cells. *Journal of Immunology (Baltimore, Md.: 1950), 192*(10), 4852–4858. https://doi.org/10.4049/jimmunol.1303377.

Iram, T., Ramirez-Ortiz, Z., Byrne, M. H., Coleman, U. A., Kingery, N. D., Means, T. K., et al. (2016). Megf10 is a receptor for C1Q that mediates clearance of apoptotic cells by astrocytes. *The Journal of Neuroscience: The Official Journal of the Society for Neuroscience, 36*(19), 5185–5192. https://doi.org/10.1523/JNEUROSCI.3850-15.2016.

Ishii, H., Mori, T., Shiratsuchi, A., Nakai, Y., Shimada, Y., Ohno-Iwashita, Y., et al. (2005). Distinct localization of lipid rafts and externalized phosphatidylserine at the surface of apoptotic cells. *Biochemical and Biophysical Research Communications, 327*(1), 94–99. https://doi.org/10.1016/j.bbrc.2004.11.135.

Janssen, W. J., McPhillips, K. A., Dickinson, M. G., Linderman, D. J., Morimoto, K., Xiao, Y. Q., et al. (2008). Surfactant proteins A and D suppress alveolar macrophage phagocytosis via interaction with SIRP alpha. *American Journal of Respiratory and Critical Care Medicine, 178*(2), 158–167. https://doi.org/10.1164/rccm.200711-1661OC.

Jennewein, C., Kuhn, A.-M., Schmidt, M. V., Meilladec-Jullig, V., von Knethen, A., Gonzalez, F. J., et al. (2008). Sumoylation of peroxisome proliferator-activated receptor gamma by apoptotic cells prevents lipopolysaccharide-induced NCoR removal from kappaB binding sites mediating transrepression of proinflammatory cytokines. *Journal of Immunology (Baltimore, Md.: 1950), 181*(8), 5646–5652.

Ji, R., Tian, S., Lu, H. J., Lu, Q., Zheng, Y., Wang, X., et al. (2013). TAM receptors affect adult brain neurogenesis by negative regulation of microglial cell activation. *Journal of Immunology (Baltimore, Md.: 1950), 191*(12), 6165–6177. https://doi.org/10.4049/jimmunol.1302229.

Jiang, H., Patel, P. H., Kohlmaier, A., Grenley, M. O., McEwen, D. G., & Edgar, B. A. (2009). Cytokine/Jak/Stat signaling mediates regeneration and homeostasis in the Drosophila midgut. *Cell, 137*(7), 1343–1355. https://doi.org/10.1016/j.cell.2009.05.014.

Johann, A. M., Barra, V., Kuhn, A.-M., Weigert, A., von Knethen, A., & Brüne, B. (2007). Apoptotic cells induce arginase II in macrophages, thereby attenuating NO production. *FASEB Journal: Official Publication of the Federation of American Societies for Experimental Biology*, *21*(11), 2704–2712. https://doi.org/10.1096/fj.06-7815com.

Johann, A. M., Weigert, A., Eberhardt, W., Kuhn, A.-M., Barra, V., von Knethen, A., et al. (2008). Apoptotic cell-derived sphingosine-1-phosphate promotes HuR-dependent cyclooxygenase-2 mRNA stabilization and protein expression. *Journal of Immunology (Baltimore, Md.: 1950)*, *180*(2), 1239–1248.

Johnston, L. A. (2009). Competitive interactions between cells: Death, growth, and geography. *Science (New York, N.Y.)*, *324*(5935), 1679–1682. https://doi.org/10.1126/science.1163862.

Jorgensen, I., & Miao, E. A. (2015). Pyroptotic cell death defends against intracellular pathogens. *Immunological Reviews*, *265*(1), 130–142. https://doi.org/10.1111/imr.12287.

Jun, J.-I., Kim, K.-H., & Lau, L. F. (2015). The matricellular protein CCN1 mediates neutrophil efferocytosis in cutaneous wound healing. *Nature Communications*, *6*, 7386. https://doi.org/10.1038/ncomms8386.

Juncadella, I. J., Kadl, A., Sharma, A. K., Shim, Y. M., Hochreiter-Hufford, A., Borish, L., et al. (2013). Apoptotic cell clearance by bronchial epithelial cells critically influences airway inflammation. *Nature*, *493*(7433), 547–551. https://doi.org/10.1038/nature11714.

Kaczmarek, A., Vandenabeele, P., & Krysko, D. V. (2013). Necroptosis: The release of damage-associated molecular patterns and its physiological relevance. *Immunity*, *38*(2), 209–223. https://doi.org/10.1016/j.immuni.2013.02.003.

Kazama, H., Ricci, J.-E., Herndon, J. M., Hoppe, G., Green, D. R., & Ferguson, T. A. (2008). Induction of immunological tolerance by apoptotic cells requires caspase-dependent oxidation of high-mobility group box-1 protein. *Immunity*, *29*(1), 21–32. https://doi.org/10.1016/j.immuni.2008.05.013.

Kim, S., Elkon, K. B., & Ma, X. (2004). Transcriptional suppression of interleukin-12 gene expression following phagocytosis of apoptotic cells. *Immunity*, *21*(5), 643–653. https://doi.org/10.1016/j.immuni.2004.09.009.

Kim, J.-Y., Zhao, H., Martinez, J., Doggett, T. A., Kolesnikov, A. V., Tang, P. H., et al. (2013). Noncanonical autophagy promotes the visual cycle. *Cell*, *154*(2), 365–376. https://doi.org/10.1016/j.cell.2013.06.012.

King, R. S., & Newmark, P. A. (2012). The cell biology of regeneration. *The Journal of Cell Biology*, *196*(5), 553–562. https://doi.org/10.1083/jcb.201105099.

Knies, U. E., Behrensdorf, H. A., Mitchell, C. A., Deutsch, U., Risau, W., Drexler, H. C., et al. (1998). Regulation of endothelial monocyte-activating polypeptide II release by apoptosis. *Proceedings of the National Academy of Sciences of the United States of America*, *95*(21), 12322–12327.

Kobayashi, N., Karisola, P., Peña-Cruz, V., Dorfman, D. M., Jinushi, M., Umetsu, S. E., et al. (2007). TIM-1 and TIM-4 glycoproteins bind phosphatidylserine and mediate uptake of apoptotic cells. *Immunity*, *27*(6), 927–940. https://doi.org/10.1016/j.immuni.2007.11.011.

Köröskényi, K., Duró, E., Pallai, A., Sarang, Z., Kloor, D., Ucker, D. S., et al. (2011). Involvement of adenosine A2A receptors in engulfment-dependent apoptotic cell suppression of inflammation. *Journal of Immunology (Baltimore, Md.: 1950)*, *186*(12), 7144–7155. https://doi.org/10.4049/jimmunol.1002284.

Krispin, A., Bledi, Y., Atallah, M., Trahtemberg, U., Verbovetski, I., Nahari, E., et al. (2006). Apoptotic cell thrombospondin-1 and heparin-binding domain lead to dendritic-cell phagocytic and tolerizing states. *Blood*, *108*(10), 3580–3589. https://doi.org/10.1182/blood-2006-03-013334.

Kruzel, M. L., Zimecki, M., & Actor, J. K. (2017). Lactoferrin in a context of inflammation-induced pathology. *Frontiers in Immunology*, *8*, 1438. https://doi.org/10.3389/fimmu.2017.01438.

Krysko, D. V., Brouckaert, G., Kalai, M., Vandenabeele, P., & D'Herde, K. (2003). Mechanisms of internalization of apoptotic and necrotic L929 cells by a macrophage cell line studied by electron microscopy. *Journal of Morphology*, *258*(3), 336–345. https://doi.org/10.1002/jmor.10161.

Krysko, D. V., Denecker, G., Festjens, N., Gabriels, S., Parthoens, E., D'Herde, K., et al. (2006). Macrophages use different internalization mechanisms to clear apoptotic and necrotic cells. *Cell Death and Differentiation*, *13*(12), 2011–2022. https://doi.org/10.1038/sj.cdd.4401900.

Kwon, H.-S., Park, M. C., Kim, D. G., Cho, K., Park, Y. W., Han, J. M., et al. (2012). Identification of CD23 as a functional receptor for the proinflammatory cytokine AIMP1/p43. *Journal of Cell Science*, *125*(Pt 19), 4620–4629. https://doi.org/10.1242/jcs.108209.

Lauber, K., Bohn, E., Kröber, S. M., Xiao, Y., Blumenthal, S. G., Lindemann, R. K., et al. (2003). Apoptotic cells induce migration of phagocytes via caspase-3-mediated release of a lipid attraction signal. *Cell*, *113*(6), 717–730.

Lauber, K., Munoz, L. E., Berens, C., Jendrossek, V., Belka, C., & Herrmann, M. (2011). Apoptosis induction and tumor cell repopulation: The yin and yang of radiotherapy. *Radiation Oncology (London, England)*, *6*, 176. https://doi.org/10.1186/1748-717X-6-176.

Lee, M., Lee, Y., Song, J., Lee, J., & Chang, S.-Y. (2018). Tissue-specific role of CX3CR1 expressing immune cells and their relationships with human disease. *Immune Network*, *18*(1) e5. https://doi.org/10.4110/in.2018.18.e5.

Lee, C.-H., Liu, S.-Y., Chou, K.-C., Yeh, C.-T., Shiah, S.-G., Huang, R.-Y., et al. (2014). Tumor-associated macrophages promote oral cancer progression through activation of the Axl signaling pathway. *Annals of Surgical Oncology*, *21*(3), 1031–1037. https://doi.org/10.1245/s10434-013-3400-0.

Lee, C. S., Penberthy, K. K., Wheeler, K. M., Juncadella, I. J., Vandenabeele, P., Lysiak, J. J., et al. (2016). Boosting apoptotic cell clearance by colonic epithelial cells attenuates inflammation in vivo. *Immunity*, *44*(4), 807–820. https://doi.org/10.1016/j.immuni.2016.02.005.

Li, F., Huang, Q., Chen, J., Peng, Y., Roop, D. R., Bedford, J. S., et al. (2010). Apoptotic cells activate the "phoenix rising" pathway to promote wound healing and tissue regeneration. *Science Signaling*, *3*(110), ra13. https://doi.org/10.1126/scisignal.2000634.

Li, M., Lindblad, J. L., Perez, E., Bergmann, A., & Fan, Y. (2016). Autophagy-independent function of Atg1 for apoptosis-induced compensatory proliferation. *BMC Biology*, *14*, 70. https://doi.org/10.1186/s12915-016-0293-y.

Linkermann, A., Himmerkus, N., Rölver, L., Keyser, K. A., Steen, P., Bräsen, J.-H., et al. (2011). Renal tubular Fas ligand mediates fratricide in cisplatin-induced acute kidney failure. *Kidney International*, *79*(2), 169–178. https://doi.org/10.1038/ki.2010.317.

Liu, G., Wang, J., Park, Y.-J., Tsuruta, Y., Lorne, E. F., Zhao, X., et al. (2008). High mobility group protein-1 inhibits phagocytosis of apoptotic neutrophils through binding to phosphatidylserine. *Journal of Immunology (Baltimore, Md.: 1950)*, *181*(6), 4240–4246.

Lockshin, R. A., & Zakeri, Z. (2001). Programmed cell death and apoptosis: Origins of the theory. *Nature Reviews. Molecular Cell Biology*, *2*(7), 545–550. https://doi.org/10.1038/35080097.

Lopatina, T., Gai, C., Deregibus, M. C., Kholia, S., & Camussi, G. (2016). Cross talk between cancer and mesenchymal stem cells through extracellular vesicles carrying nucleic acids. *Frontiers in Oncology*, *6*, 125. https://doi.org/10.3389/fonc.2016.00125.

Lucas, M., Stuart, L. M., Savill, J., & Lacy-Hulbert, A. (2003). Apoptotic cells and innate immune stimuli combine to regulate macrophage cytokine secretion. *Journal of Immunology (Baltimore, Md.: 1950)*, *171*(5), 2610–2615.

Luo, B., Gan, W., Liu, Z., Shen, Z., Wang, J., Shi, R., et al. (2016). Erythropoeitin signaling in macrophages promotes dying cell clearance and immune tolerance. *Immunity*, *44*(2), 287–302. https://doi.org/10.1016/j.immuni.2016.01.002.

Lv, Z., Bian, Z., Shi, L., Niu, S., Ha, B., Tremblay, A., et al. (2015). Loss of cell surface CD47 clustering formation and binding avidity to SIRPα facilitate apoptotic cell clearance by macrophages. *Journal of Immunology (Baltimore, Md.: 1950)*, *195*(2), 661–671. https://doi.org/10.4049/jimmunol.1401719.

Lymn, J. S., Patel, M. K., Clunn, G. F., Rao, S. J., Gallagher, K. L., & Hughes, A. D. (2002). Thrombospondin-1 differentially induces chemotaxis and DNA synthesis of human venous smooth muscle cells at the receptor-binding level. *Journal of Cell Science*, *115*(Pt 22), 4353–4360.

Maderna, P., Yona, S., Perretti, M., & Godson, C. (2005). Modulation of phagocytosis of apoptotic neutrophils by supernatant from dexamethasone-treated macrophages and annexin-derived peptide Ac(2-26). *Journal of Immunology (Baltimore, Md.: 1950)*, *174*(6), 3727–3733.

Manjón, C., Sánchez-Herrero, E., & Suzanne, M. (2007). Sharp boundaries of Dpp signalling trigger local cell death required for Drosophila leg morphogenesis. *Nature Cell Biology*, *9*(1), 57–63. https://doi.org/10.1038/ncb1518.

Mansfield, P. J., Boxer, L. A., & Suchard, S. J. (1990). Thrombospondin stimulates motility of human neutrophils. *The Journal of Cell Biology*, *111*(6 Pt 2), 3077–3086.

Mansfield, P. J., & Suchard, S. J. (1994). Thrombospondin promotes chemotaxis and haptotaxis of human peripheral blood monocytes. *Journal of Immunology (Baltimore, Md.: 1950)*, *153*(9), 4219–4229.

Mao, P., Smith, L., Xie, W., & Wang, M. (2013). Dying endothelial cells stimulate proliferation of malignant glioma cells via a caspase 3-mediated pathway. *Oncology Letters*, *5*(5), 1615–1620. https://doi.org/10.3892/ol.2013.1223.

Marinari, E., Mehonic, A., Curran, S., Gale, J., Duke, T., & Baum, B. (2012). Live-cell delamination counterbalances epithelial growth to limit tissue overcrowding. *Nature*, *484*(7395), 542–545. https://doi.org/10.1038/nature10984.

Martinez, J., Almendinger, J., Oberst, A., Ness, R., Dillon, C. P., Fitzgerald, P., et al. (2011). Microtubule-associated protein 1 light chain 3 alpha (LC3)-associated phagocytosis is required for the efficient clearance of dead cells. *Proceedings of the National Academy of Sciences of the United States of America*, *108*(42), 17396–17401. https://doi.org/10.1073/pnas.1113421108.

Matschurat, S., Knies, U. E., Person, V., Fink, L., Stoelcker, B., Ebenebe, C., et al. (2003). Regulation of EMAP II by hypoxia. *The American Journal of Pathology*, *162*(1), 93–103. https://doi.org/10.1016/S0002-9440(10)63801-1.

Matsusaka, T., Sandgren, E., Shintani, A., Kon, V., Pastan, I., Fogo, A. B., et al. (2011). Podocyte injury damages other podocytes. *Journal of the American Society of Nephrology: JASN*, *22*(7), 1275–1285. https://doi.org/10.1681/ASN.2010090963.

Mayeur, S., Spahis, S., Pouliot, Y., & Levy, E. (2016). Lactoferrin, a pleiotropic protein in health and disease. *Antioxidants & Redox Signaling*, *24*(14), 813–836. https://doi.org/10.1089/ars.2015.6458.

McArthur, S., Gobbetti, T., Kusters, D. H. M., Reutelingsperger, C. P., Flower, R. J., & Perretti, M. (2015). Definition of a novel pathway centered on lysophosphatidic acid to recruit monocytes during the resolution phase of tissue inflammation. *Journal of Immunology (Baltimore, Md.: 1950)*, *195*(3), 1139–1151. https://doi.org/10.4049/jimmunol.1500733.

McPhillips, K., Janssen, W. J., Ghosh, M., Byrne, A., Gardai, S., Remigio, L., et al. (2007). TNF-alpha inhibits macrophage clearance of apoptotic cells via cytosolic phospholipase A2 and oxidant-dependent mechanisms. *Journal of Immunology (Baltimore, Md.: 1950)*, *178*(12), 8117–8126.

Medina, C. B., & Ravichandran, K. S. (2016). Do not let death do us part: "Find-me" signals in communication between dying cells and the phagocytes. *Cell Death and Differentiation*, *23*(6), 979–989. https://doi.org/10.1038/cdd.2016.13.
Menke, M., Gerke, V., & Steinem, C. (2005). Phosphatidylserine membrane domain clustering induced by annexin A2/S100A10 heterotetramer. *Biochemistry*, *44*(46), 15296–15303. https://doi.org/10.1021/bi051585i.
Miksa, M., Amin, D., Wu, R., Ravikumar, T. S., & Wang, P. (2007). Fractalkine-induced MFG-E8 leads to enhanced apoptotic cell clearance by macrophages. *Molecular Medicine (Cambridge, Mass.)*, *13*(11–12), 553–560. https://doi.org/10.2119/2007-00019.Miksa.
Miura, M., Chen, X.-D., Allen, M. R., Bi, Y., Gronthos, S., Seo, B.-M., et al. (2004). A crucial role of caspase-3 in osteogenic differentiation of bone marrow stromal stem cells. *The Journal of Clinical Investigation*, *114*(12), 1704–1713. https://doi.org/10.1172/JCI20427.
Moffatt, O. D., Devitt, A., Bell, E. D., Simmons, D. L., & Gregory, C. D. (1999). Macrophage recognition of ICAM-3 on apoptotic leukocytes. *Journal of Immunology (Baltimore, Md.: 1950)*, *162*(11), 6800–6810.
Mold, C., Baca, R., & Du Clos, T. W. (2002). Serum amyloid P component and C-reactive protein opsonize apoptotic cells for phagocytosis through Fcgamma receptors. *Journal of Autoimmunity*, *19*(3), 147–154.
Monier, B., Gettings, M., Gay, G., Mangeat, T., Schott, S., Guarner, A., et al. (2015). Apico-basal forces exerted by apoptotic cells drive epithelium folding. *Nature*, *518*(7538), 245–248. https://doi.org/10.1038/nature14152.
Monks, J., Rosner, D., Geske, F. J., Lehman, L., Hanson, L., Neville, M. C., et al. (2005). Epithelial cells as phagocytes: Apoptotic epithelial cells are engulfed by mammary alveolar epithelial cells and repress inflammatory mediator release. *Cell Death and Differentiation*, *12*(2), 107–114. https://doi.org/10.1038/sj.cdd.4401517.
Monks, J., Smith-Steinhart, C., Kruk, E. R., Fadok, V. A., & Henson, P. M. (2008). Epithelial cells remove apoptotic epithelial cells during post-lactation involution of the mouse mammary gland. *Biology of Reproduction*, *78*(4), 586–594. https://doi.org/10.1095/biolreprod.107.065045.
Morimoto, K., Amano, H., Sonoda, F., Baba, M., Senba, M., Yoshimine, H., et al. (2001). Alveolar macrophages that phagocytose apoptotic neutrophils produce hepatocyte growth factor during bacterial pneumonia in mice. *American Journal of Respiratory Cell and Molecular Biology*, *24*(5), 608–615. https://doi.org/10.1165/ajrcmb.24.5.4292.
Mukundan, L., Odegaard, J. I., Morel, C. R., Heredia, J. E., Mwangi, J. W., Ricardo-Gonzalez, R. R., et al. (2009). PPAR-delta senses and orchestrates clearance of apoptotic cells to promote tolerance. *Nature Medicine*, *15*(11), 1266–1272. https://doi.org/10.1038/nm.2048.
Murakami, Y., Tian, L., Voss, O. H., Margulies, D. H., Krzewski, K., & Coligan, J. E. (2014). CD300b regulates the phagocytosis of apoptotic cells via phosphatidylserine recognition. *Cell Death and Differentiation*, *21*(11), 1746–1757. https://doi.org/10.1038/cdd.2014.86.
Murphy, J. E., Tacon, D., Tedbury, P. R., Hadden, J. M., Knowling, S., Sawamura, T., et al. (2006). LOX-1 scavenger receptor mediates calcium-dependent recognition of phosphatidylserine and apoptotic cells. *The Biochemical Journal*, *393*(Pt 1), 107–115. https://doi.org/10.1042/BJ20051166.
Murugesan, G., Sandhya Rani, M. R., Gerber, C. E., Mukhopadhyay, C., Ransohoff, R. M., Chisolm, G. M., et al. (2003). Lysophosphatidylcholine regulates human microvascular endothelial cell expression of chemokines. *Journal of Molecular and Cellular Cardiology*, *35*(11), 1375–1384.
Nagata, S. (2018). Apoptosis and clearance of apoptotic cells. *Annual Review of Immunology*, *36*, 489–517. https://doi.org/10.1146/annurev-immunol-042617-053010.

Nakahashi-Oda, C., Tahara-Hanaoka, S., Shoji, M., Okoshi, Y., Nakano-Yokomizo, T., Ohkohchi, N., et al. (2012). Apoptotic cells suppress mast cell inflammatory responses via the CD300a immunoreceptor. *The Journal of Experimental Medicine, 209*(8), 1493–1503. https://doi.org/10.1084/jem.20120096.

Nakamura, Y. S., Hakeda, Y., Takakura, N., Kameda, T., Hamaguchi, I., Miyamoto, T., et al. (1998). Tyro 3 receptor tyrosine kinase and its ligand, Gas6, stimulate the function of osteoclasts. *Stem Cells (Dayton, Ohio), 16*(3), 229–238. https://doi.org/10.1002/stem.160229.

Nakanishi, Y., & Shiratsuchi, A. (2004). Phagocytic removal of apoptotic spermatogenic cells by Sertoli cells: Mechanisms and consequences. *Biological & Pharmaceutical Bulletin, 27*(1), 13–16.

Nandrot, E. F., Kim, Y., Brodie, S. E., Huang, X., Sheppard, D., & Finnemann, S. C. (2004). Loss of synchronized retinal phagocytosis and age-related blindness in mice lacking alphavbeta5 integrin. *The Journal of Experimental Medicine, 200*(12), 1539–1545. https://doi.org/10.1084/jem.20041447.

Nishimura, T., Horino, K., Nishiura, H., Shibuya, Y., Hiraoka, T., Tanase, S., et al. (2001). Apoptotic cells of an epithelial cell line, AsPC-1, release monocyte chemotactic S19 ribosomal protein dimer. *Journal of Biochemistry, 129*(3), 445–454.

Nishiura, H., Shibuya, Y., & Yamamoto, T. (1998). S19 ribosomal protein cross-linked dimer causes monocyte-predominant infiltration by means of molecular mimicry to complement C5a. *Laboratory Investigation; a Journal of Technical Methods and Pathology, 78*(12), 1615–1623.

Nishiura, H., Tanase, S., Sibuya, Y., Nishimura, T., & Yamamoto, T. (1999). Determination of the cross-linked residues in homo-dimerization of S19 ribosomal protein concomitant with exhibition of monocyte chemotactic activity. *Laboratory Investigation; A Journal of Technical Methods and Pathology, 79*(8), 915–923.

Noda, M., Doi, Y., Liang, J., Kawanokuchi, J., Sonobe, Y., Takeuchi, H., et al. (2011). Fractalkine attenuates excito-neurotoxicity via microglial clearance of damaged neurons and antioxidant enzyme heme oxygenase-1 expression. *The Journal of Biological Chemistry, 286*(3), 2308–2319. https://doi.org/10.1074/jbc.M110.169839.

Norsworthy, P. J., Fossati-Jimack, L., Cortes-Hernandez, J., Taylor, P. R., Bygrave, A. E., Thompson, R. D., et al. (2004). Murine CD93 (C1qRp) contributes to the removal of apoptotic cells in vivo but is not required for C1q-mediated enhancement of phagocytosis. *Journal of Immunology (Baltimore, Md.: 1950), 172*(6), 3406–3414.

Novitskiy, S. V., Ryzhov, S., Zaynagetdinov, R., Goldstein, A. E., Huang, Y., Tikhomirov, O. Y., et al. (2008). Adenosine receptors in regulation of dendritic cell differentiation and function. *Blood, 112*(5), 1822–1831. https://doi.org/10.1182/blood-2008-02-136325.

Obeid, M., Tesniere, A., Ghiringhelli, F., Fimia, G. M., Apetoh, L., Perfettini, J.-L., et al. (2007). Calreticulin exposure dictates the immunogenicity of cancer cell death. *Nature Medicine, 13*(1), 54–61. https://doi.org/10.1038/nm1523.

Ogden, C. A., deCathelineau, A., Hoffmann, P. R., Bratton, D., Ghebrehiwet, B., Fadok, V. A., et al. (2001). C1q and mannose binding lectin engagement of cell surface calreticulin and CD91 initiates macropinocytosis and uptake of apoptotic cells. *The Journal of Experimental Medicine, 194*(6), 781–795.

Oka, K., Sawamura, T., Kikuta, K., Itokawa, S., Kume, N., Kita, T., et al. (1998). Lectin-like oxidized low-density lipoprotein receptor 1 mediates phagocytosis of aged/apoptotic cells in endothelial cells. *Proceedings of the National Academy of Sciences of the United States of America, 95*(16), 9535–9540.

Paccez, J. D., Vasques, G. J., Correa, R. G., Vasconcellos, J. F., Duncan, K., Gu, X., et al. (2013). The receptor tyrosine kinase Axl is an essential regulator of prostate cancer

proliferation and tumor growth and represents a new therapeutic target. *Oncogene, 32*(6), 689–698. https://doi.org/10.1038/onc.2012.89.
Park, H.-J., Choi, Y.-H., Cho, Y. J., Henson, P. M., & Kang, J. L. (2011). RhoA-mediated signaling up-regulates hepatocyte growth factor gene and protein expression in response to apoptotic cells. *Journal of Leukocyte Biology, 89*(3), 399–411. https://doi.org/10.1189/jlb.0710414.
Park, S. G., Choi, E.-C., & Kim, S. (2010). Aminoacyl-tRNA synthetase-interacting multifunctional proteins (AIMPs): A triad for cellular homeostasis. *IUBMB Life, 62*(4), 296–302. https://doi.org/10.1002/iub.324.
Park, D., Han, C. Z., Elliott, M. R., Kinchen, J. M., Trampont, P. C., Das, S., et al. (2011). Continued clearance of apoptotic cells critically depends on the phagocyte Ucp2 protein. *Nature, 477*(7363), 220–224. https://doi.org/10.1038/nature10340.
Park, S.-Y., Jung, M.-Y., Kim, H.-J., Lee, S.-J., Kim, S.-Y., Lee, B.-H., et al. (2008). Rapid cell corpse clearance by stabilin-2, a membrane phosphatidylserine receptor. *Cell Death and Differentiation, 15*(1), 192–201. https://doi.org/10.1038/sj.cdd.4402242.
Park, S.-Y., Jung, M.-Y., Lee, S.-J., Kang, K.-B., Gratchev, A., Riabov, V., et al. (2009). Stabilin-1 mediates phosphatidylserine-dependent clearance of cell corpses in alternatively activated macrophages. *Journal of Cell Science, 122*(Pt 18), 3365–3373. https://doi.org/10.1242/jcs.049569.
Park, S.-Y., Kang, K.-B., Thapa, N., Kim, S.-Y., Lee, S.-J., & Kim, I.-S. (2008). Requirement of adaptor protein GULP during stabilin-2-mediated cell corpse engulfment. *The Journal of Biological Chemistry, 283*(16), 10593–10600. https://doi.org/10.1074/jbc.M709105200.
Park, S.-Y., Kim, S.-Y., Kang, K.-B., & Kim, I.-S. (2010). Adaptor protein GULP is involved in stabilin-1-mediated phagocytosis. *Biochemical and Biophysical Research Communications, 398*(3), 467–472. https://doi.org/10.1016/j.bbrc.2010.06.101.
Park, D., Tosello-Trampont, A.-C., Elliott, M. R., Lu, M., Haney, L. B., Ma, Z., et al. (2007). BAI1 is an engulfment receptor for apoptotic cells upstream of the ELMO/Dock180/Rac module. *Nature, 450*(7168), 430–434. https://doi.org/10.1038/nature06329.
Parnaik, R., Raff, M. C., & Scholes, J. (2000). Differences between the clearance of apoptotic cells by professional and non-professional phagocytes. *Current Biology: CB, 10*(14), 857–860.
Patel, V. A., Feng, L., Lee, D. J., Massenburg, D., Pattabiraman, G., Antoni, A., et al. (2012). Recognition-dependent signaling events in response to apoptotic targets inhibit epithelial cell viability by multiple mechanisms: Implications for non-immune tissue homeostasis. *The Journal of Biological Chemistry, 287*(17), 13761–13777. https://doi.org/10.1074/jbc.M112.350843.
Patel, V. A., Lee, D. J., Feng, L., Antoni, A., Lieberthal, W., Schwartz, J. H., et al. (2010). Recognition of apoptotic cells by epithelial cells: Conserved versus tissue-specific signaling responses. *Journal of Biological Chemistry, 285*(3), 1829–1840. https://doi.org/10.1074/jbc.M109.018440.
Patel, V. A., Longacre, A., Hsiao, K., Fan, H., Meng, F., Mitchell, J. E., et al. (2006). Apoptotic cells, at all stages of the death process, trigger characteristic signaling events that are divergent from and dominant over those triggered by necrotic cells: Implications for the delayed clearance model of autoimmunity. *The Journal of Biological Chemistry, 281*(8), 4663–4670. https://doi.org/10.1074/jbc.M508342200.
Patel, V. A., Massenburg, D., Vujicic, S., Feng, L., Tang, M., Litbarg, N., et al. (2015). Apoptotic cells activate AMP-activated protein kinase (AMPK) and inhibit epithelial cell growth without change in intracellular energy stores. *The Journal of Biological Chemistry, 290*(37), 22352–22369. https://doi.org/10.1074/jbc.M115.667345.

Pattabiraman, G., Palasiewicz, K., Galvin, J. P., & Ucker, D. S. (2017). Aging-associated dysregulation of homeostatic immune response termination (and not initiation). *Aging Cell, 16*(3), 585–593. https://doi.org/10.1111/acel.12589.
Pattabiraman, G., Palasiewicz, K., & Ucker, D. S. (2016). Toll-like receptor function of murine macrophages, probed by cytokine induction, is biphasic and is not impaired globally with age. *Mechanisms of Ageing and Development, 157*, 44–59. https://doi.org/10.1016/j.mad.2016.07.008.
Penberthy, K. K., & Ravichandran, K. S. (2016). Apoptotic cell recognition receptors and scavenger receptors. *Immunological Reviews, 269*(1), 44–59. https://doi.org/10.1111/imr.12376.
Peng, Y., & Elkon, K. B. (2011). Autoimmunity in MFG-E8-deficient mice is associated with altered trafficking and enhanced cross-presentation of apoptotic cell antigens. *The Journal of Clinical Investigation, 121*(6), 2221–2241. https://doi.org/10.1172/JCI43254.
Pérez-Garijo, A. (2018). When dying is not the end: Apoptotic caspases as drivers of proliferation. *Seminars in Cell & Developmental Biology, 82*, 86–95. https://doi.org/10.1016/j.semcdb.2017.11.036.
Pérez-Garijo, A., Fuchs, Y., & Steller, H. (2013). Apoptotic cells can induce non-autonomous apoptosis through the TNF pathway. *ELife, 2* e01004. https://doi.org/10.7554/eLife.01004.
Peter, C., Waibel, M., Radu, C. G., Yang, L. V., Witte, O. N., Schulze-Osthoff, K., et al. (2008). Migration to apoptotic "find-me" signals is mediated via the phagocyte receptor G2A. *The Journal of Biological Chemistry, 283*(9), 5296–5305. https://doi.org/10.1074/jbc.M706586200.
Peter, C., Wesselborg, S., Herrmann, M., & Lauber, K. (2010). Dangerous attraction: Phagocyte recruitment and danger signals of apoptotic and necrotic cells. *Apoptosis: An International Journal on Programmed Cell Death, 15*(9), 1007–1028. https://doi.org/10.1007/s10495-010-0472-1.
Pisetsky, D. S. (2014). The expression of HMGB1 on microparticles released during cell activation and cell death in vitro and in vivo. *Molecular Medicine (Cambridge, Mass), 20*, 158–163. https://doi.org/10.2119/molmed.2014.00014.
Platt, N., Suzuki, H., Kodama, T., & Gordon, S. (2000). Apoptotic thymocyte clearance in scavenger receptor class A-deficient mice is apparently normal. *Journal of Immunology (Baltimore, Md.: 1950), 164*(9), 4861–4867.
Qu, Y., Misaghi, S., Newton, K., Gilmour, L. L., Louie, S., Cupp, J. E., et al. (2011). Pannexin-1 is required for ATP release during apoptosis but not for inflammasome activation. *Journal of Immunology (Baltimore, Md.: 1950), 186*(11), 6553–6561. https://doi.org/10.4049/jimmunol.1100478.
Ramirez-Ortiz, Z. G., Pendergraft, W. F., Prasad, A., Byrne, M. H., Iram, T., Blanchette, C. J., et al. (2013). The scavenger receptor SCARF1 mediates the clearance of apoptotic cells and prevents autoimmunity. *Nature Immunology, 14*(9), 917–926. https://doi.org/10.1038/ni.2670.
Ramprasad, M. P., Fischer, W., Witztum, J. L., Sambrano, G. R., Quehenberger, O., & Steinberg, D. (1995). The 94- to 97-kDa mouse macrophage membrane protein that recognizes oxidized low density lipoprotein and phosphatidylserine-rich liposomes is identical to macrosialin, the mouse homologue of human CD68. *Proceedings of the National Academy of Sciences of the United States of America, 92*(21), 9580–9584.
Rathmell, J. C., & Thompson, C. B. (2002). Pathways of apoptosis in lymphocyte development, homeostasis, and disease. *Cell, 109*(Suppl), S97–107.
Rébé, C., Raveneau, M., Chevriaux, A., Lakomy, D., Sberna, A.-L., Costa, A., et al. (2009). Induction of transglutaminase 2 by a liver X receptor/retinoic acid receptor alpha pathway increases the clearance of apoptotic cells by human macrophages. *Circulation Research, 105*(4), 393–401. https://doi.org/10.1161/CIRCRESAHA.109.201855.

Reddy, S. M., Hsiao, K.-H. K., Abernethy, V. E., Fan, H., Longacre, A., Lieberthal, W., et al. (2002). Phagocytosis of apoptotic cells by macrophages induces novel signaling events leading to cytokine-independent survival and inhibition of proliferation: Activation of Akt and inhibition of extracellular signal-regulated kinases 1 and 2. *Journal of Immunology (Baltimore, Md.: 1950)*, *169*(2), 702–713.

Revollo, I., Nishiura, H., Shibuya, Y., Oda, Y., Nishino, N., & Yamamoto, T. (2005). Agonist and antagonist dual effect of the cross-linked S19 ribosomal protein dimer in the C5a receptor-mediated respiratory burst reaction of phagocytic leukocytes. *Inflammation Research: Official Journal of the European Histamine Research Society ... [et Al.]*, *54*(2), 82–90. https://doi.org/10.1007/s00011-004-1327-4.

Riederer, M., Ojala, P. J., Hrzenjak, A., Graier, W. F., Malli, R., Tritscher, M., et al. (2010). Acyl chain-dependent effect of lysophosphatidylcholine on endothelial prostacyclin production. *Journal of Lipid Research*, *51*(10), 2957–2966. https://doi.org/10.1194/jlr.M006536.

Röszer, T. (2017). Transcriptional control of apoptotic cell clearance by macrophage nuclear receptors. *Apoptosis: An International Journal on Programmed Cell Death*, *22*(2), 284–294. https://doi.org/10.1007/s10495-016-1310-x.

Roszer, T., Menéndez-Gutiérrez, M. P., Lefterova, M. I., Alameda, D., Núñez, V., Lazar, M. A., et al. (2011). Autoimmune kidney disease and impaired engulfment of apoptotic cells in mice with macrophage peroxisome proliferator-activated receptor gamma or retinoid X receptor alpha deficiency. *Journal of Immunology (Baltimore, Md.: 1950)*, *186*(1), 621–631. https://doi.org/10.4049/jimmunol.1002230.

Rothlin, C. V., Carrera-Silva, E. A., Bosurgi, L., & Ghosh, S. (2015). TAM receptor signaling in immune homeostasis. *Annual Review of Immunology*, *33*, 355–391. https://doi.org/10.1146/annurev-immunol-032414-112103.

Rothlin, C. V., Ghosh, S., Zuniga, E. I., Oldstone, M. B. A., & Lemke, G. (2007). TAM receptors are pleiotropic inhibitors of the innate immune response. *Cell*, *131*(6), 1124–1136. https://doi.org/10.1016/j.cell.2007.10.034.

Rovere, P., Peri, G., Fazzini, F., Bottazzi, B., Doni, A., Bondanza, A., et al. (2000). The long pentraxin PTX3 binds to apoptotic cells and regulates their clearance by antigen-presenting dendritic cells. *Blood*, *96*(13), 4300–4306.

Saito, H., Nishimura, M., Shinano, H., Makita, H., Tsujino, I., Shibuya, E., et al. (1999). Plasma concentration of adenosine during normoxia and moderate hypoxia in humans. *American Journal of Respiratory and Critical Care Medicine*, *159*(3), 1014–1018. https://doi.org/10.1164/ajrccm.159.3.9803100.

Sandahl, M., Hunter, D. M., Strunk, K. E., Earp, H. S., & Cook, R. S. (2010). Epithelial cell-directed efferocytosis in the post-partum mammary gland is necessary for tissue homeostasis and future lactation. *BMC Developmental Biology*, *10*, 122. https://doi.org/10.1186/1471-213X-10-122.

Sarang, Z., Joós, G., Garabuczi, É., Rühl, R., Gregory, C. D., & Szondy, Z. (2014). Macrophages engulfing apoptotic cells produce nonclassical retinoids to enhance their phagocytic capacity. *Journal of Immunology (Baltimore, Md.: 1950)*, *192*(12), 5730–5738. https://doi.org/10.4049/jimmunol.1400284.

Savill, J., Hogg, N., Ren, Y., & Haslett, C. (1992). Thrombospondin cooperates with CD36 and the vitronectin receptor in macrophage recognition of neutrophils undergoing apoptosis. *The Journal of Clinical Investigation*, *90*(4), 1513–1522. https://doi.org/10.1172/JCI116019.

Savill, J., Smith, J., Sarraf, C., Ren, Y., Abbott, F., & Rees, A. (1992). Glomerular mesangial cells and inflammatory macrophages ingest neutrophils undergoing apoptosis. *Kidney International*, *42*(4), 924–936.

Scaffidi, P., Misteli, T., & Bianchi, M. E. (2002). Release of chromatin protein HMGB1 by necrotic cells triggers inflammation. *Nature*, *418*(6894), 191–195. https://doi.org/10.1038/nature00858.

Schäfer, M., & Werner, S. (2008). Cancer as an overhealing wound: An old hypothesis revisited. *Nature Reviews Molecular Cell Biology*, *9*(8), 628–638. https://doi.org/10.1038/nrm2455.
Scheib, J. L., Sullivan, C. S., & Carter, B. D. (2012). Jedi-1 and MEGF10 signal engulfment of apoptotic neurons through the tyrosine kinase Syk. *The Journal of Neuroscience: The Official Journal of the Society for Neuroscience*, *32*(38), 13022–13031. https://doi.org/10.1523/JNEUROSCI.6350-11.2012.
Schnurr, M., Toy, T., Shin, A., Hartmann, G., Rothenfusser, S., Soellner, J., et al. (2004). Role of adenosine receptors in regulating chemotaxis and cytokine production of plasmacytoid dendritic cells. *Blood*, *103*(4), 1391–1397. https://doi.org/10.1182/blood-2003-06-1959.
Schwarz, M. A., Kandel, J., Brett, J., Li, J., Hayward, J., Schwarz, R. E., et al. (1999). Endothelial-monocyte activating polypeptide II, a novel antitumor cytokine that suppresses primary and metastatic tumor growth and induces apoptosis in growing endothelial cells. *The Journal of Experimental Medicine*, *190*(3), 341–354.
Segawa, K., Suzuki, J., & Nagata, S. (2011). Constitutive exposure of phosphatidylserine on viable cells. *Proceedings of the National Academy of Sciences of the United States of America*, *108*(48), 19246–19251. https://doi.org/10.1073/pnas.1114799108.
Segundo, C., Medina, F., Rodríguez, C., Martínez-Palencia, R., Leyva-Cobián, F., & Brieva, J. A. (1999). Surface molecule loss and bleb formation by human germinal center B cells undergoing apoptosis: Role of apoptotic blebs in monocyte chemotaxis. *Blood*, *94*(3), 1012–1020.
Sen, P., Wallet, M. A., Yi, Z., Huang, Y., Henderson, M., Mathews, C. E., et al. (2007). Apoptotic cells induce Mer tyrosine kinase-dependent blockade of NF-kappaB activation in dendritic cells. *Blood*, *109*(2), 653–660. https://doi.org/10.1182/blood-2006-04-017368.
Shiozawa, Y., Pedersen, E. A., & Taichman, R. S. (2010). GAS6/Mer axis regulates the homing and survival of the E2A/PBX1-positive B-cell precursor acute lymphoblastic leukemia in the bone marrow niche. *Experimental Hematology*, *38*(2), 132–140. https://doi.org/10.1016/j.exphem.2009.11.002.
Shiratsuchi, A., Kawasaki, Y., Ikemoto, M., Arai, H., & Nakanishi, Y. (1999). Role of class B scavenger receptor type I in phagocytosis of apoptotic rat spermatogenic cells by Sertoli cells. *The Journal of Biological Chemistry*, *274*(9), 5901–5908.
Shrestha, A., Shiokawa, M., Nishimura, T., Nishiura, H., Tanaka, Y., Nishino, N., et al. (2003). Switch moiety in agonist/antagonist dual effect of S19 ribosomal protein dimer on leukocyte chemotactic C5a receptor. *The American Journal of Pathology*, *162*(4), 1381–1388. https://doi.org/10.1016/S0002-9440(10)63934-X.
Simhadri, V. R., Andersen, J. F., Calvo, E., Choi, S.-C., Coligan, J. E., & Borrego, F. (2012). Human CD300a binds to phosphatidylethanolamine and phosphatidylserine, and modulates the phagocytosis of dead cells. *Blood*, *119*(12), 2799–2809. https://doi.org/10.1182/blood-2011-08-372425.
Stevens, J. B., Abdallah, B. Y., Liu, G., Horne, S. D., Bremer, S. W., Ye, K. J., et al. (2013). Heterogeneity of cell death. *Cytogenetic and Genome Research*, *139*(3), 164–173. https://doi.org/10.1159/000348679.
Stockwell, B. R., Friedmann Angeli, J. P., Bayir, H., Bush, A. I., Conrad, M., Dixon, S. J., et al. (2017). Ferroptosis: A regulated cell death nexus linking metabolism, redox biology, and disease. *Cell*, *171*(2), 273–285. https://doi.org/10.1016/j.cell.2017.09.021.
Su, H. P., Nakada-Tsukui, K., Tosello-Trampont, A.-C., Li, Y., Bu, G., Henson, P. M., et al. (2002). Interaction of CED-6/GULP, an adapter protein involved in engulfment of apoptotic cells with CED-1 and CD91/low density lipoprotein receptor-related protein (LRP). *The Journal of Biological Chemistry*, *277*(14), 11772–11779. https://doi.org/10.1074/jbc.M109336200.

Sugimoto, M. A., Vago, J. P., Teixeira, M. M., & Sousa, L. P. (2016). Annexin A1 and the resolution of inflammation: Modulation of neutrophil recruitment, apoptosis, and clearance. *Journal of Immunology Research*, *2016*, 8239258. https://doi.org/10.1155/2016/8239258.

Surh, C. D., & Sprent, J. (1994). T-cell apoptosis detected in situ during positive and negative selection in the thymus. *Nature*, *372*(6501), 100–103. https://doi.org/10.1038/372100a0.

Suzanne, M., Petzoldt, A. G., Spéder, P., Coutelis, J.-B., Steller, H., & Noselli, S. (2010). Coupling of apoptosis and L/R patterning controls stepwise organ looping. *Current Biology: CB*, *20*(19), 1773–1778. https://doi.org/10.1016/j.cub.2010.08.056.

Suzanne, M., & Steller, H. (2013). Shaping organisms with apoptosis. *Cell Death and Differentiation*, *20*(5), 669–675. https://doi.org/10.1038/cdd.2013.11.

Tai, K.-Y., Shieh, Y.-S., Lee, C.-S., Shiah, S.-G., & Wu, C.-W. (2008). Axl promotes cell invasion by inducing MMP-9 activity through activation of NF-kappaB and Brg-1. *Oncogene*, *27*(29), 4044–4055. https://doi.org/10.1038/onc.2008.57.

Takizawa, F., Tsuji, S., & Nagasawa, S. (1996). Enhancement of macrophage phagocytosis upon iC3b deposition on apoptotic cells. *FEBS Letters*, *397*(2–3), 269–272.

Tian, L., Choi, S.-C., Lee, H.-N., Murakami, Y., Qi, C.-F., Sengottuvelu, M., et al. (2016). Enhanced efferocytosis by dendritic cells underlies memory T-cell expansion and susceptibility to autoimmune disease in CD300f-deficient mice. *Cell Death and Differentiation*, *23*(6), 1086–1096. https://doi.org/10.1038/cdd.2015.161.

Tian, L., Choi, S.-C., Murakami, Y., Allen, J., Morse, H. C., Qi, C.-F., et al. (2014). p85α recruitment by the CD300f phosphatidylserine receptor mediates apoptotic cell clearance required for autoimmunity suppression. *Nature Communications*, *5*, 3146. https://doi.org/10.1038/ncomms4146.

Todt, J. C., Hu, B., Punturieri, A., Sonstein, J., Polak, T., & Curtis, J. L. (2002). Activation of protein kinase C beta II by the stereo-specific phosphatidylserine receptor is required for phagocytosis of apoptotic thymocytes by resident murine tissue macrophages. *The Journal of Biological Chemistry*, *277*(39), 35906–35914. https://doi.org/10.1074/jbc.M202967200.

Tomura, H., Mogi, C., Sato, K., & Okajima, F. (2005). Proton-sensing and lysolipid-sensitive G-protein-coupled receptors: A novel type of multi-functional receptors. *Cellular Signalling*, *17*(12), 1466–1476. https://doi.org/10.1016/j.cellsig.2005.06.002.

Torr, E. E., Gardner, D. H., Thomas, L., Goodall, D. M., Bielemeier, A., Willetts, R., et al. (2012). Apoptotic cell-derived ICAM-3 promotes both macrophage chemoattraction to and tethering of apoptotic cells. *Cell Death and Differentiation*, *19*(4), 671–679. https://doi.org/10.1038/cdd.2011.167.

Tosello-Trampont, A. C., Brugnera, E., & Ravichandran, K. S. (2001). Evidence for a conserved role for CRKII and Rac in engulfment of apoptotic cells. *The Journal of Biological Chemistry*, *276*(17), 13797–13802. https://doi.org/10.1074/jbc.M011238200.

Tóth, B., Garabuczi, E., Sarang, Z., Vereb, G., Vámosi, G., Aeschlimann, D., et al. (2009). Transglutaminase 2 is needed for the formation of an efficient phagocyte portal in macrophages engulfing apoptotic cells. *Journal of Immunology (Baltimore, Md.: 1950)*, *182*(4), 2084–2092. https://doi.org/10.4049/jimmunol.0803444.

Trouw, L. A., Blom, A. M., & Gasque, P. (2008). Role of complement and complement regulators in the removal of apoptotic cells. *Molecular Immunology*, *45*(5), 1199–1207. https://doi.org/10.1016/j.molimm.2007.09.008.

Truman, L. A., Ford, C. A., Pasikowska, M., Pound, J. D., Wilkinson, S. J., Dumitriu, I. E., et al. (2008). CX3CL1/fractalkine is released from apoptotic lymphocytes to stimulate macrophage chemotaxis. *Blood*, *112*(13), 5026–5036. https://doi.org/10.1182/blood-2008-06-162404.

Tsai, R. K., & Discher, D. E. (2008). Inhibition of "self" engulfment through deactivation of myosin-II at the phagocytic synapse between human cells. *The Journal of Cell Biology, 180*(5), 989–1003. https://doi.org/10.1083/jcb.200708043.

Tsai, W.-H., Shih, C.-H., Feng, S.-Y., Li, I.-T., Chang, S.-C., Lin, Y.-C., et al. (2014). CX3CL1(+) microparticles mediate the chemoattraction of alveolar macrophages toward apoptotic acute promyelocytic leukemic cells. *Cellular Physiology and Biochemistry: International Journal of Experimental Cellular Physiology, Biochemistry, and Pharmacology, 33*(3), 594–604. https://doi.org/10.1159/000358637.

Tseng, A.-S., Adams, D. S., Qiu, D., Koustubhan, P., & Levin, M. (2007). Apoptosis is required during early stages of tail regeneration in Xenopus laevis. *Developmental Biology, 301*(1), 62–69. https://doi.org/10.1016/j.ydbio.2006.10.048.

Tyurin, V. A., Balasubramanian, K., Winnica, D., Tyurina, Y. Y., Vikulina, A. S., He, R. R., et al. (2014). Oxidatively modified phosphatidylserines on the surface of apoptotic cells are essential phagocytic "eat-me" signals: Cleavage and inhibition of phagocytosis by Lp-PLA2. *Cell Death and Differentiation, 21*(5), 825–835. https://doi.org/10.1038/cdd.2014.1.

Ucker, D. S., & Levine, J. S. (2018). Exploitation of apoptotic regulation in cancer. *Frontiers in Immunology, 9*, 241. https://doi.org/10.3389/fimmu.2018.00241.

Uderhardt, S., Herrmann, M., Oskolkova, O. V., Aschermann, S., Bicker, W., Ipseiz, N., et al. (2012). 12/15-lipoxygenase orchestrates the clearance of apoptotic cells and maintains immunologic tolerance. *Immunity, 36*(5), 834–846. https://doi.org/10.1016/j.immuni.2012.03.010.

Urbonaviciute, V., Meister, S., Fürnrohr, B. G., Frey, B., Gückel, E., Schett, G., et al. (2009). Oxidation of the alarmin high-mobility group box 1 protein (HMGB1) during apoptosis. *Autoimmunity, 42*(4), 305–307.

van der Hoeven, D., Wan, T. C., & Auchampach, J. A. (2008). Activation of the A(3) adenosine receptor suppresses superoxide production and chemotaxis of mouse bone marrow neutrophils. *Molecular Pharmacology, 74*(3), 685–696. https://doi.org/10.1124/mol.108.048066.

Vandivier, R. W., Ogden, C. A., Fadok, V. A., Hoffmann, P. R., Brown, K. K., Botto, M., et al. (2002). Role of surfactant proteins A, D, and C1q in the clearance of apoptotic cells in vivo and in vitro: Calreticulin and CD91 as a common collectin receptor complex. *Journal of Immunology (Baltimore, Md.: 1950), 169*(7), 3978–3986.

Voll, R. E., Herrmann, M., Roth, E. A., Stach, C., Kalden, J. R., & Girkontaite, I. (1997). Immunosuppressive effects of apoptotic cells. *Nature, 390*(6658), 350–351. https://doi.org/10.1038/37022.

Voss, O. H., Tian, L., Murakami, Y., Coligan, J. E., & Krzewski, K. (2015). Emerging role of CD300 receptors in regulating myeloid cell efferocytosis. *Molecular & Cellular Oncology, 2*(4) e964625. https://doi.org/10.4161/23723548.2014.964625.

Vujicic, S., Feng, L., Antoni, A., Rauch, J., & Levine, J. S. (2016). Identification of intracellular signaling events induced in viable cells by interaction with neighboring cells undergoing apoptotic cell death. *Journal of Visualized Experiments: JoVE, 118*. https://doi.org/10.3791/54980.

Wakasugi, K., & Schimmel, P. (1999). Two distinct cytokines released from a human aminoacyl-tRNA synthetase. *Science (New York, N.Y.), 284*(5411), 147–151.

Wakasugi, K., Slike, B. M., Hood, J., Ewalt, K. L., Cheresh, D. A., & Schimmel, P. (2002). Induction of angiogenesis by a fragment of human tyrosyl-tRNA synthetase. *The Journal of Biological Chemistry, 277*(23), 20124–20126. https://doi.org/10.1074/jbc.C200126200.

Wallach, D., Kang, T.-B., & Kovalenko, A. (2014). Concepts of tissue injury and cell death in inflammation: A historical perspective. *Nature Reviews. Immunology, 14*(1), 51–59. https://doi.org/10.1038/nri3561.

Weichand, B., Weis, N., Weigert, A., Grossmann, N., Levkau, B., & Brüne, B. (2013). Apoptotic cells enhance sphingosine-1-phosphate receptor 1 dependent macrophage migration. *European Journal of Immunology*, *43*(12), 3306–3313. https://doi.org/10.1002/eji.201343441.

Weigert, A., Cremer, S., Schmidt, M. V., von Knethen, A., Angioni, C., Geisslinger, G., et al. (2010). Cleavage of sphingosine kinase 2 by caspase-1 provokes its release from apoptotic cells. *Blood*, *115*(17), 3531–3540. https://doi.org/10.1182/blood-2009-10-243444.

Weigert, A., Johann, A. M., von Knethen, A., Schmidt, H., Geisslinger, G., & Brüne, B. (2006). Apoptotic cells promote macrophage survival by releasing the antiapoptotic mediator sphingosine-1-phosphate. *Blood*, *108*(5), 1635–1642. https://doi.org/10.1182/blood-2006-04-014852.

Weigert, A., Weis, N., & Brüne, B. (2009). Regulation of macrophage function by sphingosine-1-phosphate. *Immunobiology*, *214*(9–10), 748–760. https://doi.org/10.1016/j.imbio.2009.06.003.

Wiegand, U. K., Corbach, S., Prescott, A. R., Savill, J., & Spruce, B. A. (2001). The trigger to cell death determines the efficiency with which dying cells are cleared by neighbours. *Cell Death and Differentiation*, *8*(7), 734–746. https://doi.org/10.1038/sj.cdd.4400867.

Willenborg, S., & Eming, S. A. (2018). Cellular networks in wound healing. *Science (New York, N.Y.)*, *362*(6417), 891–892. https://doi.org/10.1126/science.aav5542.

Witte, O. N., Kabarowski, J. H., Xu, Y., Le, L. Q., & Zhu, K. (2005). Retraction. *Science (New York, N.Y.)*, *307*(5707), 206. https://doi.org/10.1126/science.307.5707.206b.

Wood, W., Turmaine, M., Weber, R., Camp, V., Maki, R. A., McKercher, S. R., et al. (2000). Mesenchymal cells engulf and clear apoptotic footplate cells in macrophageless PU.1 null mouse embryos. *Development (Cambridge, England)*, *127*(24), 5245–5252.

Wootten, D., Christopoulos, A., Marti-Solano, M., Babu, M. M., & Sexton, P. M. (2018). Mechanisms of signalling and biased agonism in G protein-coupled receptors. *Nature Reviews. Molecular Cell Biology*, *19*(10), 638–653. https://doi.org/10.1038/s41580-018-0049-3.

Wu, H.-H., Bellmunt, E., Scheib, J. L., Venegas, V., Burkert, C., Reichardt, L. F., et al. (2009). Glial precursors clear sensory neuron corpses during development via Jedi-1, an engulfment receptor. *Nature Neuroscience*, *12*(12), 1534–1541. https://doi.org/10.1038/nn.2446.

Xiao, A. Y., Wei, L., Xia, S., Rothman, S., & Yu, S. P. (2002). Ionic mechanism of ouabain-induced concurrent apoptosis and necrosis in individual cultured cortical neurons. *The Journal of Neuroscience: The Official Journal of the Society for Neuroscience*, *22*(4), 1350–1362.

Yamaguchi, H., Maruyama, T., Urade, Y., & Nagata, S. (2014). Immunosuppression via adenosine receptor activation by adenosine monophosphate released from apoptotic cells. *ELife*, *3* e02172. https://doi.org/10.7554/eLife.02172.

Yamaguchi, Y., & Miura, M. (2015). Programmed cell death and caspase functions during neural development. *Current Topics in Developmental Biology*, *114*, 159–184. https://doi.org/10.1016/bs.ctdb.2015.07.016.

Yamaguchi, Y., Shinotsuka, N., Nonomura, K., Takemoto, K., Kuida, K., Yosida, H., et al. (2011). Live imaging of apoptosis in a novel transgenic mouse highlights its role in neural tube closure. *The Journal of Cell Biology*, *195*(6), 1047–1060. https://doi.org/10.1083/jcb.201104057.

Yamamoto, T. (2007). Roles of the ribosomal protein S19 dimer and the C5a receptor in pathophysiological functions of phagocytic leukocytes. *Pathology International*, *57*(1), 1–11. https://doi.org/10.1111/j.1440-1827.2007.02049.x.

Yamasaki, S., Ishikawa, E., Sakuma, M., Hara, H., Ogata, K., & Saito, T. (2008). Mincle is an ITAM-coupled activating receptor that senses damaged cells. *Nature Immunology*, *9*(10), 1179–1188. https://doi.org/10.1038/ni.1651.

Yang, L., Brooks, C. R., Xiao, S., Sabbisetti, V., Yeung, M. Y., Hsiao, L.-L., et al. (2015). KIM-1-mediated phagocytosis reduces acute injury to the kidney. *The Journal of Clinical Investigation, 125*(4), 1620–1636. https://doi.org/10.1172/JCI75417.

Yoon, K. W., Byun, S., Kwon, E., Hwang, S.-Y., Chu, K., Hiraki, M., et al. (2015). Control of signaling-mediated clearance of apoptotic cells by the tumor suppressor p53. *Science (New York, N.Y.), 349*(6247), 1261669. https://doi.org/10.1126/science.1261669.

Zargarian, S., Shlomovitz, I., Erlich, Z., Hourizadeh, A., Ofir-Birin, Y., Croker, B. A., et al. (2017). Phosphatidylserine externalization, "necroptotic bodies" release, and phagocytosis during necroptosis. *PLoS Biology, 15*(6), e2002711. https://doi.org/10.1371/journal.pbio.2002711.

Zernecke, A., Bidzhekov, K., Noels, H., Shagdarsuren, E., Gan, L., Denecke, B., et al. (2009). Delivery of microRNA-126 by apoptotic bodies induces CXCL12-dependent vascular protection. *Science Signaling, 2*(100) ra81. https://doi.org/10.1126/scisignal.2000610.

Zhang, F. R., & Schwarz, M. A. (2002). Pro-EMAP II is not primarily cleaved by caspase-3 and -7. *American Journal of Physiology. Lung Cellular and Molecular Physiology, 282*(6), L1239–L1244. https://doi.org/10.1152/ajplung.00141.2001.

Zhao, Y., Wu, X., Li, X., Jiang, L.-L., Gui, X., Liu, Y., et al. (2018). TREM2 is a receptor for β-amyloid that mediates microglial function. *Neuron, 97*(5), 1023–1031.e7. https://doi.org/10.1016/j.neuron.2018.01.031.

Zitvogel, L., Kepp, O., & Kroemer, G. (2010). Decoding cell death signals in inflammation and immunity. *Cell, 140*(6), 798–804. https://doi.org/10.1016/j.cell.2010.02.015.

Zizzo, G., Hilliard, B. A., Monestier, M., & Cohen, P. L. (2012). Efficient clearance of early apoptotic cells by human macrophages requires M2c polarization and MerTK induction. *Journal of Immunology (Baltimore, Md.: 1950), 189*(7), 3508–3520. https://doi.org/10.4049/jimmunol.1200662.

CHAPTER TWO

Nucleobindins and encoded peptides: From cell signaling to physiology

Adelaine Kwun-Wai Leung[a], Naresh Ramesh[a], Christine Vogel[b], Suraj Unniappan[a,*]
[a]Department of Veterinary Biomedical Sciences, Western College of Veterinary Medicine, Saskatoon, SK, Canada
[b]Department of Biology, New York University, New York, NY, United States
*Corresponding author: e-mail address: suraj.unniappan@usask.ca

Contents

Abstract

Nucleobindins (NUCBs) are DNA and calcium binding, secreted proteins with various signaling functions. Two NUCBs, nucleobindin-1 (NUCB1) and nucleobindin-2 (NUCB2), were discovered during the 1990s. These two peptides are shown to have diverse functions, including the regulation of inflammation and bone formation, among others. In 2006, Oh-I and colleagues discovered that three peptides encoded within the

Advances in Protein Chemistry and Structural Biology, Volume 116
ISSN 1876-1623
https://doi.org/10.1016/bs.apcsb.2019.02.001

NUCB2 could be processed by prohormone convertases. These peptides were named nesfatin-1, 2 and 3, mainly due to the satiety and fat influencing properties of nesfatin-1. However, it was found that nesfatin-2 and -3 have no such effects. Nesfatin-1, especially its mid-segment, is very highly conserved across vertebrates. Although the receptor(s) that mediate nesfatin-1 effects are currently unknown, it is now considered an endogenous peptide with multiple functions, affecting central and peripheral tissues to regulate metabolism, reproduction, endocrine and other functions. We recently identified a nesfatin-1-like peptide (NLP) encoded within the NUCB1. Like nesfatin-1, NLP suppressed feed intake in mice and fish, and stimulated insulin secretion from pancreatic beta cells. There is considerable evidence available to indicate that nucleobindins and its encoded peptides are multifunctional regulators of cell biology and whole animal physiology. This review aims to briefly discuss the structure, distribution, functions and mechanism of action nucleobindins and encoded peptides.

1. Introduction

Nucleobindins (NUCB) are DNA and calcium binding proteins encoded by two unlinked genes. Nucleobindin-1 is a 55 kDa multi-domain protein identified first in a culture supernatant of B lymphocyte cell line identified from mice prone to systemic lupus erythematosus, an autoimmune disorder. The first paper that reported nucleobindin indicated that it has a potential role in autoimmunity and apoptosis (Miura, Titani, Kurosawa, & Kanai, 1992). It is also called CALNUC owing to its Ca^{2+} and DNA-binding ability. NUCB2 belongs to the EF-hand family of calcium-binding proteins (Lin et al., 1998; Valencia, Cotten, Duan, & Liu, 2008). This protein shares a structural motif of calcium binding domain known as the EF-hand (Moncrief, Kretsinger, & Goodman, 1990; Nakayama, Moncrief, & Kretsinger, 1992). The name EF-hand was coined by Kretsinger and Nockolds (1973) to visually describe the calcium-binding motif they observed in parvalbumin, a calcium-binding protein (Celio & Heizmann, 1981; Henrotte, 1952). Several proteins with EF-hand motif have been extensively characterized and were shown to undergo several processes including splicing, translocation, fusion and gene replication. Phylogenetic analyses were conducted to determine the primary ancestor of these proteins. This helped to understand the origin of the proteins and how it changes during evolution. One protein that was recently demonstrated to have rooted phylogeny from the EF-hand domain is nucleobindin-2/NEFA (Nucleic acid/DNA binding/EF-hand/Acidic amino acid rich region) (Miura et al., 1992).

Phylogenetic and sequence analyses have shown that through a series of genetic modifications, such as duplication and fusion, NEFA arose from the CTER EF-hand domains (Miura et al., 1992). This common domain led to the formation of four-domain precursor ancestor. With the ultimate steps of domain deletion, duplication and fusion, two sequences were fused together leading to NEFA-N and NEFA-C (Miura et al., 1992). A signal peptide is present at the N-terminal region of NEFA-N, and two functional EF-hand calcium-binding domains present at the N-terminal region of NEFA-C. With the subsequent divergence of this precursor protein sequence with other domains such as heptad repeated domain, a putative hydrophilic DNA-binding region and a leucine zipper region resulted in the genetic evolution of NEFA. Duplication of the NEFA gene led to the formation of nucleobindin-2 and NEFA-like proteins (Miura et al., 1992).

The discovery of nucleobindin was a result of research conducted in lupus-prone MRL/l mice. These mice are known to secrete large amounts of antibodies targeted against single-stranded and double-stranded DNA (Nakayama et al., 1992). A cell line (KML 1–7) that produced a soluble protein that induced the formation of antibodies against single- and double-stranded DNA in vivo and in vitro was established from the MRL/l mice (Kanai et al., 1990; Kubota, Akatsuka, & Kanai, 1986). From this cell line, a 55 kDa protein was purified (Kanai & Tanuma, 1992) and the gene encoding this protein was later cloned (Barnikol-Watanabe et al., 1994). Even though this protein is secreted and consists of a signal peptide, its capability to bind to DNA resulted in the name, nucleobindin.

2. Structural characteristics of the NUCB protein family

NUCB is a highly conserved protein family with five distinct sequence regions: (1) an N-terminal signal peptide (SP); (2) a putative neuropeptide region (NP); (3) a basic region that can bind DNA in human (DB) (Miura et al., 1992); (4) an EF-hand domain that consists of two EF-hand motifs (EF); and (5) a coiled-coil region that is a "leucine zipper" domain with a heptad Leu repeat (CC) (Fig. 1). Yeast is the simplest organism that has an annotated NUCB1 gene which mainly contains the EF-hand domain. Both *C. elegans* and *Drosophila* have a single annotated NUCB1 copy that contains all five conserved regions (Fig. 1). Coelacanth, a rare order of fish, appears to be the earliest species to have two NUCB paralogs. However, only one homolog contains all domains, while the other copy is shorter. The sequence regions may be lost during divergence of the two duplicates

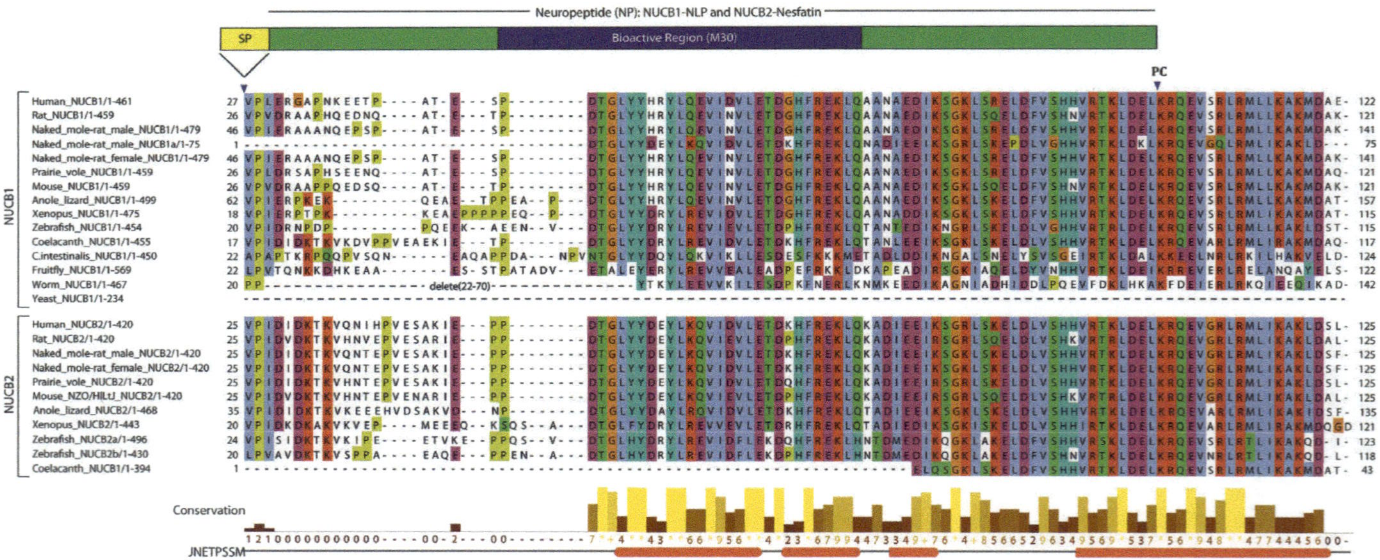

Fig. 1 Annotated sequence alignment of the NUCB protein family across select eukaryotes. NUCB is a protein family containing four regions of highly conserved sequence regions. The similarity patterns suggest that gene duplication occurred before the appearance of mammals. NUCB1 and NUCB2 sequences are grouped with their respective orthologs. The alignment focuses on human, rat, mouse, naked mole-rats (model system for aging), prairie vole (model system for social behavior), anole lizard (the first reptile which has two full copies of NUCB), zebra fish, coelacanth (the first order that has two copies of NUCB but one copy is incomplete), and *Ciona intestinalis* (a sea squirt invertebrate that is the closest relative to mammals). The sequence is color coded by the default Clustal X color scheme (blue: hydrophobic, red: positive charge, magenta: negative charge; green: polar; pink: cysteines; orange: gly; yellow: pro; cyan: aromatic; white: unconserved). Each residue in the alignment is assigned a color based on a threshold profile. Under the sequences are the annotation for alignment conservation and secondary structure prediction. Conservation is a numerical index reflecting the conservation of physcio-chemical properties in the alignment (* = conserved; + = all physico-chemical properties are conserved; numerical value indicates degrees of conservation from low to high).

(Continued)

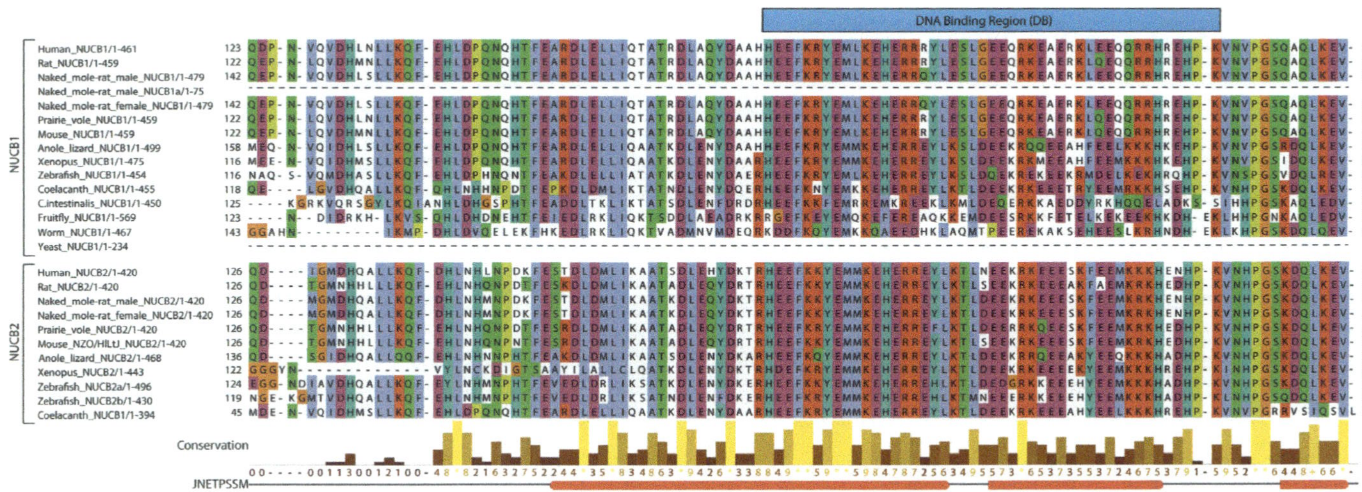

Fig. 1—Cont'd Stretches of sequences for some species with little conservation (dash line) are not included in the calculated conservation score. Secondary structure annotation either originates from experimental structure if available (e.g., the EF-hand domain) or from secondary structure prediction algorithm. We only show annotation from position-specific-scoring matrices (PSSM) profile. Shown above the sequence alignment are annotations of the four conserved sequence domains. All the species contain a signal peptide, except for yeast, the third NUCB copy of naked mole-rat, and the second NUCB copy of coelacanth (SP-yellow; sequence excluded). The neuropeptide domain (NP-green) is present in all species, except for yeast. The second NUCB copy of coelacanth contains a partial sequence of NP while the *C. elegans* sequence shows little conservation except for the central bioactive region (M30-blue). "PC" above an inverted triangle represents the putative cleavage site of proconvertase. The DNA-binding domain (DB-cyan) is present in all species, except for yeast and the third copy of naked mole-rat (DB-cyan). The EF-hand domain contains two EF-hand motifs and is present in all species (EF-purple). The second NUCB copy of Xenopus is missing the calcium binding loop in the first EF-hand motif; therefore, we excluded the sequence. Residues important for calcium ion coordination are highlighted in black boxes. The caspase cleavage site is highlighted in a red box. The secondary structure annotation for the EF-hand domain is from the NMR structure of human NUCB1 (PDB:1SNL). The coiled-coil domain is the least conserved region (CC-orange). The simpler species (yeast, *C. elegans*, drosophila) show limited similarity and are excluded from the alignment.

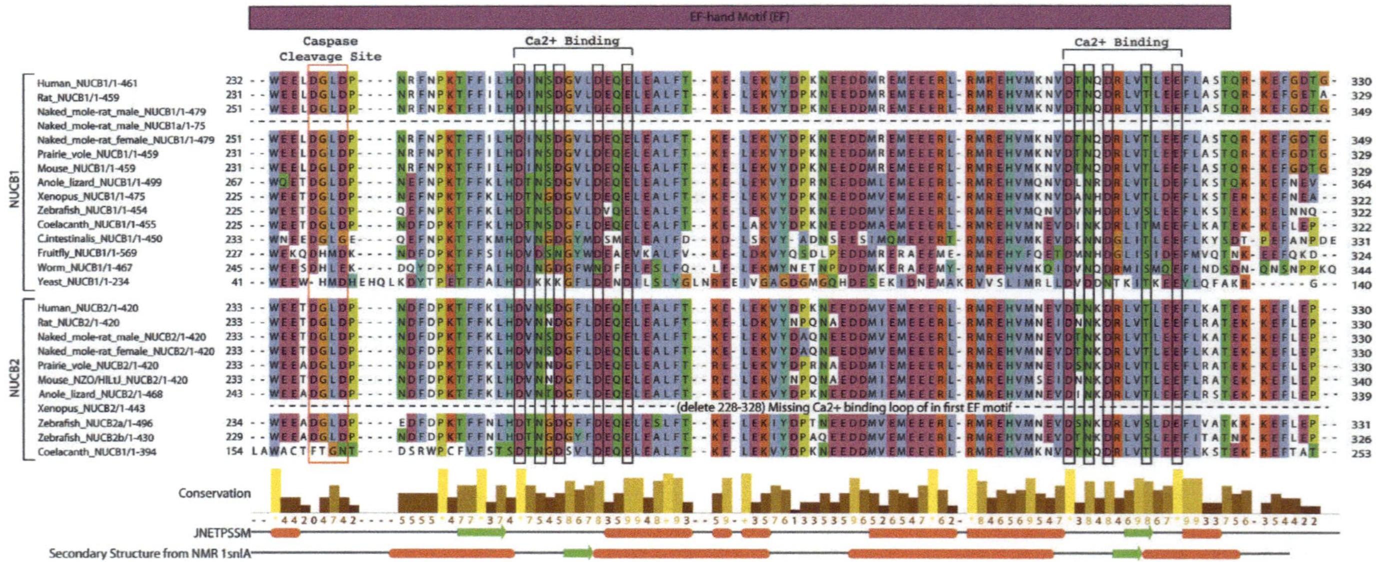
EF-hand Motif (EF)
Caspase Cleavage Site
Ca2+ Binding
Ca2+ Binding
NUCB1
Human_NUCB1/1-461
Rat_NUCB1/1-459
Naked_mole-rat_male_NUCB1/1-479
Naked_mole-rat_male_NUCB1a/1-75
Naked_mole-rat_female_NUCB1/1-479
Prairie_vole_NUCB1/1-459
Mouse_NUCB1/1-459
Anole_lizard_NUCB1/1-499
Xenopus_NUCB1/1-475
Zebrafish_NUCB1/1-454
Coelacanth_NUCB1/1-455
C.intestinalis_NUCB1/1-450
Fruitfly_NUCB1/1-569
Worm_NUCB1/1-467
Yeast_NUCB1/1-234
NUCB2
Human_NUCB2/1-420
Rat_NUCB2/1-420
Naked_mole-rat_male_NUCB2/1-420
Naked_mole-rat_female_NUCB2/1-420
Prairie_vole_NUCB2/1-420
Mouse_NZO/HILtJ_NUCB2/1-420
Anole_lizard_NUCB2/1-468
Xenopus_NUCB2/1-443
Zebrafish_NUCB2a/1-496
Zebrafish_NUCB2b/1-430
Coelacanth_NUCB1/1-394
(delete 228-328) Missing Ca2+ binding loop of in first EF motif
Conservation
JNETPSSM
Secondary Structure from NMR 1snlA

Fig. 1—Cont'd

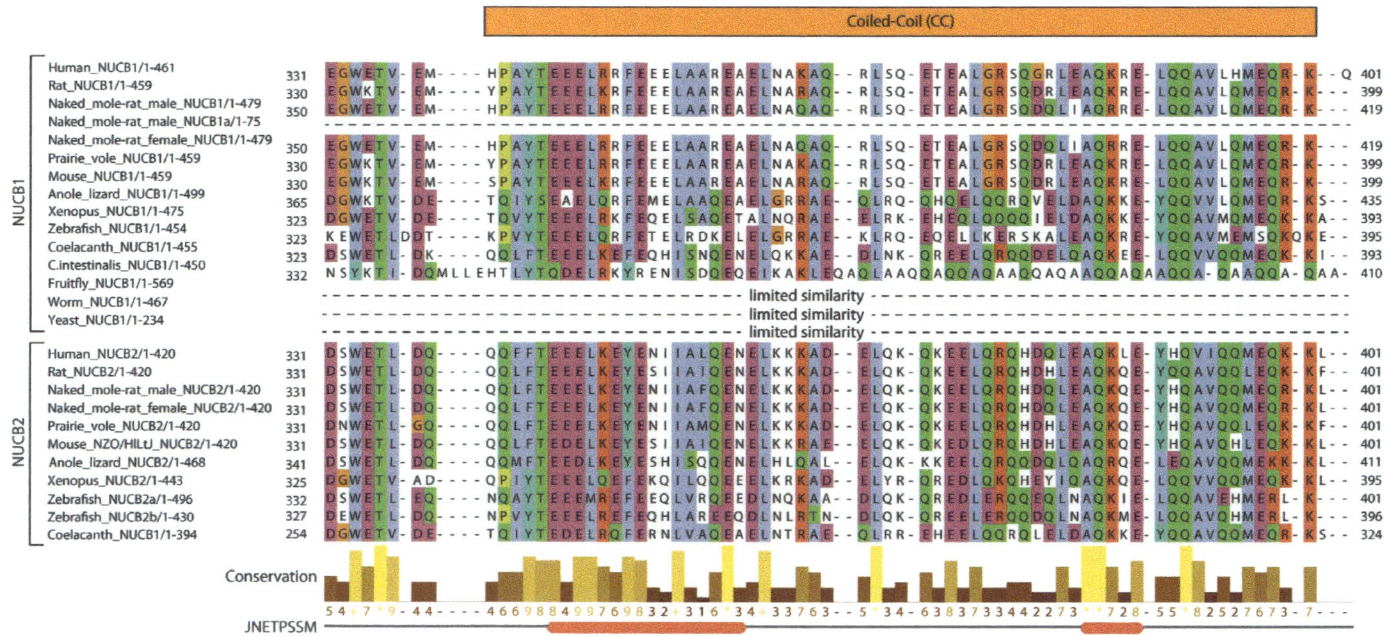
Coiled-Coil (CC)
NUCB1
Human_NUCB1/1-461
Rat_NUCB1/1-459
Naked_mole-rat_male_NUCB1/1-479
Naked_mole-rat_male_NUCB1a/1-75
Naked_mole-rat_female_NUCB1/1-479
Prairie_vole_NUCB1/1-459
Mouse_NUCB1/1-459
Anole_lizard_NUCB1/1-499
Xenopus_NUCB1/1-475
Zebrafish_NUCB1/1-454
Coelacanth_NUCB1/1-455
C.intestinalis_NUCB1/1-450
Fruitfly_NUCB1/1-569
Worm_NUCB1/1-467
Yeast_NUCB1/1-234
limited similarity
limited similarity
limited similarity
NUCB2
Human_NUCB2/1-420
Rat_NUCB2/1-420
Naked_mole-rat_male_NUCB2/1-420
Naked_mole-rat_female_NUCB2/1-420
Prairie_vole_NUCB2/1-420
Mouse_NZO/HILtJ_NUCB2/1-420
Anole_lizard_NUCB2/1-468
Xenopus_NUCB2/1-443
Zebrafish_NUCB2a/1-496
Zebrafish_NUCB2b/1-430
Coelacanth_NUCB1/1-394
Conservation
JNETPSSM

Fig. 1—Cont'd

or may be due to incorrect gene prediction. By comparison, the Anole lizard from the reptiles and bird branch has NUCB genes with all five regions. Some species in caniforms, cetaceans and even-toed ungulates have three NUCB paralogs, but only two copies contain all five regions. Additional gene duplication may have occurred in these organisms. In tetraploid fish, for example, a whole genome duplication produced at least two copies of the NUCB2 gene (NUCB2A and NUCB2B) in goldfish (*Carassius auratus*).

3. Biochemical and structural properties of NUCB

The human genome has two NUCB paralogs. NUCB1 appears to be the more ancient copy, as it is more similar to NUCB orthologs in invertebrates than NUCB2. In addition, it is more ubiquitously expressed across tissues than NUCB2, as discussed below.

In general, paralogs tend to diverge rapidly during evolution to account for functional redundancy due to the duplicate. However, the sequence conservation between NUCB1 and NUCB2, in particular in key regions, suggests that the two paralogs still have similar functions. Unfortunately, most of the experimental results discussed below were obtained for NUCB1, leaving the conservation with NUCB2 subject to speculation.

The secondary structures of NUCB1 and NUCB2 observed from the NMR structure and predicted by JPred are shown in Fig. 1. The putative bioactive peptides are proposed to be cleaved by prohormone convertases that recognize the KR cleavage site (Fig. 1, blue triangle labeled "PC," and Figs. 3 and 4). The peptides nesfatin-1 and nesfatin-1-like peptide (NLP) from NUCB2 and NUCB1, respectively, are discussed in detailed later in this chapter (Gonzalez, Kerbel, Chun, & Unniappan, 2010; Gonzalez, Mohan, & Unniappan, 2012; Mohan & Unniappan, 2013; Oh-I et al., 2006; Ramesh, Mohan, & Unniappan, 2015; Sundarrajan et al., 2016). Within the nesfatin-1 region is a highly conserved sequence, which constitutes the bioactive segment of the neuropeptides (M30) (Fig. 1B—light blue box). The DNA binding property of NUCB was first shown in an in vitro gel shift assay in human NUCB1 and required both DNA-binding and coiled-coil regions (Miura et al., 1992). The EF-hand domain contains two EF-hand motifs, each can bind a calcium ion (Fig. 1).

The NUCB1 protein appears to be monomeric, as experiments with recombinant protein have shown (Kapoor et al., 2010). Dynamic light scattering shows that the shape of the protein deviates from a sphere.

The asymmetrical shape is mainly due to the "leucine zipper" at the C-terminus. A truncation construct without the "leucine zipper" remains monomeric and has a hydrodynamic radius value that resembles closer to a spherical object. Interestingly, CD spectroscopic analysis of the same NUCB1 construct shows that the protein has a smooth concentration-dependent transition of secondary structure content, switching from highly helical at low concentration (~24 μM or less) to predominantly β-sheet at high concentration (~40–60 μM) (Kapoor et al., 2010). This property suggests that NUCB1 undergoes concentration-dependent conformational changes, which can affect its function by inhibiting or facilitating certain protein-protein interactions. Since NUCB2 shares high sequence homology with NUCB1, it is expected to have similar shape and concentration-dependent conformational flexibility.

4. Structure of the EF-hand pair in NUCB1

The EF-hand is a diverse motif class consisting of 30 amino acids that fold into a helix-loop-helix structure. Its structure resembles a right hand fist, with the index finger and thumb extended. The index finger represents the N-terminal "E" helix and the thumb represents the C-terminal "F" helix. The loop between the helices (at the palm) comprises ~12 conserved residues that can coordinate a Ca^{2+} or calcium ion. This general structure predictions have been confirmed for the human NUCB1 proteins, i.e., residues 228–326 fold into an EF-hand pair (EF1 and EF2 in Fig. 2) (de Alba & Tjandra, 2004). The chemical shifts upon the addition of calcium indicate that the backbone carbonyls of Gly-258 from EF1 and Arg-310 from EF2 are involved in calcium binding. Both of these residues belong to the sixth residue of the binding loop, which is most frequently a glycine (Gifford, Walsh, & Vogel, 2007). The two peptides have different calcium affinities: EF2 can still bind calcium at low affinity with a K_d of 73.5 μM, compared to a K_d of 6.3 μM for EF1 (Kapoor et al., 2010). Fig. 1 shows with black boxes additional amino acid side chains that are predicted to bind calcium (Lewit-Bentley & Rety, 2000). The conformation of the EF-hand is open or closed based on the orientation of the two helices relative to each other. In the NMR structure, both EF-hands are in the open conformation, which is frequently the case for calcium-bound EF-hand structures. However, NUCB1's EF-hands appear to be less compact and more flexible compared to a typical EF-hand domain (de Alba & Tjandra, 2004).

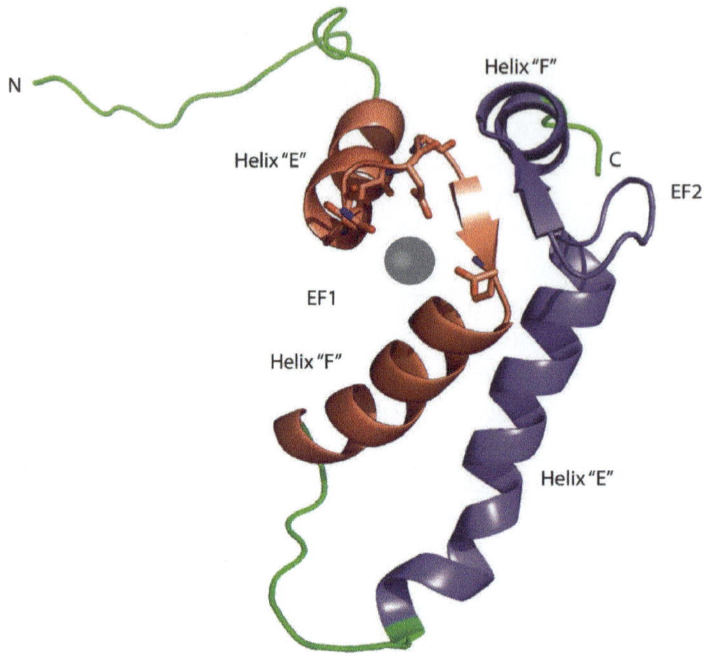

Fig. 2 NMR Structure of the EF-hand domain of Human NUCB1 (residue 228–326) (PDB:1SNL). EF1 (residues 244–272; pink), EF2 (residues 297–325; light purple). Residues involved in calcium ion coordination are represented as a stick model (red = oxygen, blue = nitrogen). The gray sphere represents the location of the calcium ion binding site in the EF1 motif. Note that there is no direct evidence of calcium coordination in the EF-hand domain in the NMR study, as calcium ion is magnetically inactive (de Alba & Tjandra, 2004). The EF-hand domain of human NUCB2 is expected to have similar structure.

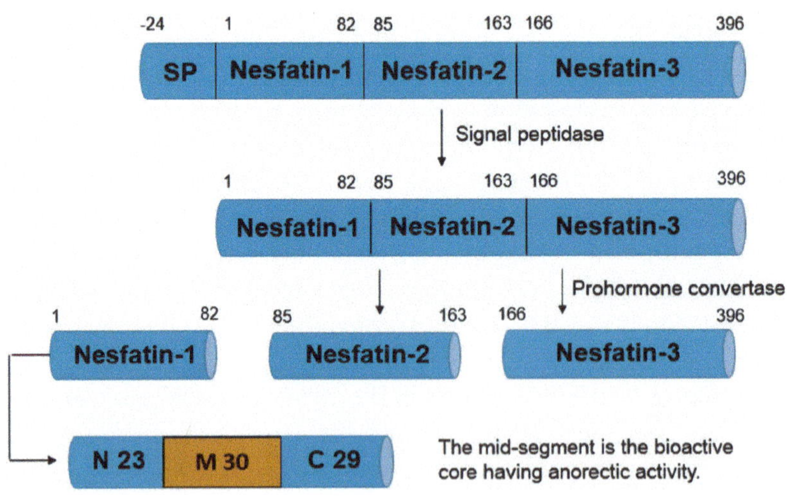

Fig. 3 A schematic of nesfatin-1 processing from its precursor NUCB2. The pre-prohormone is cleaved by signal peptidase, yielding the prohormone. This sequence is then post-translationally cleaved by prohormone convertases PC 1/3 and PC 2 resulting in three distinct fragments including nesfatin-1 (82 amino acids). The mid segment of nesfastin-1 (30 amino acids) is critical for its anorexigenic action.

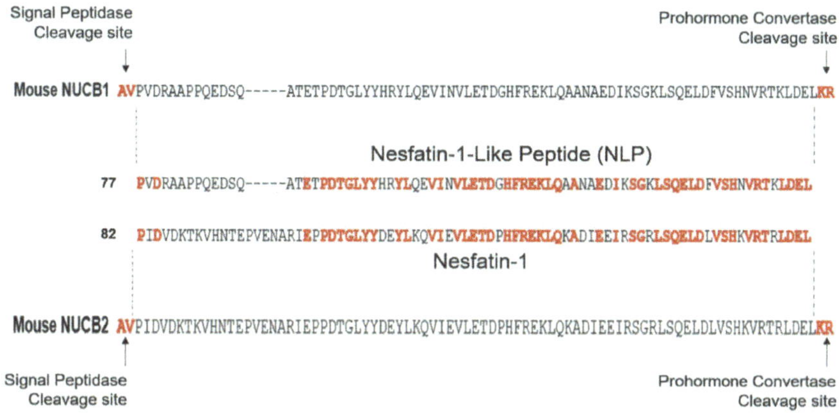

Fig. 4 Comparison of mouse NUCB1 and NUCB2 amino acid sequences. The scheme shows full-length amino acid sequences of nesfatin-1 and NLP. Identical amino acids are shown in red.

5. Calcium induces conformational changes in NUCB1

Multiple biophysical studies in human NUCB1 have shown that the addition of calcium induces large conformational changes likely impacting the function of the protein (de Alba & Tjandra, 2004; Kapoor et al., 2010; Miura, Kurosawa, & Kanai, 1994). Recombinant human NUCB1 without the signal peptide increases in helical content from 29.5% to 34.5% in the presence of calcium (Miura et al., 1994). Without calcium, the EF-hands are essentially unstructured as NMR analysis of a construct showed (de Alba & Tjandra, 2004). Moreover, the conformational change involves the whole EF-hand domain rather than just the 12-residue binding loop. More recently, the effect of calcium on the conformation of the EF-hand domain was investigated by tryptophan fluorescence quenching experiment. There are two Trp residues on NUCB1 (residues 232 and 333 before and after the EF-hand domain, respectively). Changes in Trp fluorescence measurement with and without quenchers suggest that calcium binding causes the reorientation of the Trp residues from a solvent-exposed environment to a more hydrophobic or well-packed environment. Finally, calcium binding also increases the thermal stability of NUCB1, which indicates a transition to a more ordered three-dimensional structure (Kapoor et al., 2010). In sum, the EF-hands appear to have crucial roles in NUCB structure and function. Based on the findings above, it is tempting to hypothesize that the more compact calcium-bound conformation will create binding surfaces for some interacting partners while blocking access to others. Similarly, the more flexible calcium-free conformation will inhibit or enhance the binding of a different set of interacting partners. Therefore, the EF-hands are the major contributor to the multiple physiological roles of NUCB1 and NUCB2.

6. DNA binding property of NUCB1

NUCB1 was first reported as a soluble factor secreted from a cell line developed from a mouse model of lupus (Kanai et al., 1990; Kanai, Katagiri, Mori, & Kubota, 1986). Both human and mouse NUCB1 bind DNA in the basic region of the protein (Fig. 1B—in blue box). Binding DNA requires the coiled-coil region near the C-terminus (Miura et al., 1992) (Fig. 1B). One of the hallmarks of lupus disease is the presence of autoantibodies to DNA in patient serum. Conversely, NUCB1 promotes the production of antibodies against DNA in a lupus mouse model, tempting speculation of a possible role of the protein in the disease. However, the DNA binding property of NUCB1 remains elusive. DNA-binding domains together with leucine zippers are typical components of transcription factors, but NUCB1 has structural properties that are inconsistent with a role as transcription factor. For example, while NUCB1's leucine zipper is responsible for NUCB1's dimerization (Kapoor et al., 2010), the EF-hands domain intervenes with the basic region and the leucine zipper, rendering it dysfunctional. NUCB1 might either function as a non-canonical transcription factor or serve as a competitive inhibitor of other transcription factor functions. It may, under certain conditions, sequester DNA and therefore promote the production of anti-DNA antibodies without DNA binding being part of its normal role.

7. Post-translational modification

Post-translational modifications (PTMs) of proteins, such as phosphorylation, acetylation, SUMOylation, are highly diverse and affect protein function, localization, and stability. The experimentally observed modification sites in NUCB1 and NUCB2 are very different, as reported by the PhosphoSitePlus database (Hornbeck et al., 2015). Table 1 lists the modifications that were identified by more than two large-scale proteomic studies and are highly conserved across species (Hornbeck et al., 2015). Except for two phosphoserine sites (NUCB1: S82p, S86p and NUCB2: S85p, S89p) that are common in both proteins, all others are unique to each protein.

Table 1 Sites of post-translational modifications (PTMs) in human NUCB1 and NUCB2.

NUCB1	NUCB2	Conservation (excluding partial sequences)[a]
–	S42p	In signal peptide, no conservation
Y50	Y53p	8
S82p	S85p	8
S86p	S89p	8
A121	S124p	0
T148p	K149	7
Y168	Y169p	9
K177 ac	K178	+
Y179p	Y180	+
Y190p	Y191	7
S224p	S225	+
S256	S257p	7
G332	S332p	4
S369p	Q369	6
T372p	K372	6
L389	Y389p, Y389-ga	7

[a]Numerical index reflecting conservation of physico-chemical properties. See Fig. 1 legend for a detailed description.
Conserved and more than two records using ONLY proteomic discovery mass spec
Green, conserved PTM
Yellow, conserved residue but not PTM
No color, PTM not conserved across the two paralogs.

The two phosphorylation sites are located in the neuropeptide region, suggesting that they may be recognized by the same receptor. While the exact modification sites are different between the proteins, both NUCB1 and NUCB2 have several phosphorylation events in the highly conserved EF-hand region (Table 1, Fig. 1). Again, these modification events may be regulatory and affect the structure and function of the EF-hand and the entire protein.

8. Subcellular distribution

Both NUCB1 and NUCB2 have been detected in the Golgi apparatus in various cell lines (Lin et al., 1998; Nesselhut et al., 2001). From an in vitro study of NUCB2, the Golgi retention signal was first identified to be in the N-terminal region, starting from the NP domain (NUCB1-NLP, NUCB2-Nesfatin-1) until the DNA-binding domain (residues 25–170 in rat) (Nesselhut et al., 2001). In a pulse-chase experiment using various cell lines, NUCB1 was found to be synthesized in the endoplasmic reticulum, transported to the Golgi apparatus for post-translational modification, and secreted to the medium via the constitutive pathway (Lavoie, Meerloo, Lin, & Farquhar, 2002). Subsequently, by expressing various mutants of NUCB1 in HT1080 cells, a single Pro residue +2 from the signal peptide was identified as the endoplasmic reticulum export signal (Tsukumo, Tsukahara, Saito, Tsuruo, & Tomida, 2009). When this residue was mutated to Ala, NUCB1 was never exported to the Golgi apparatus and secreted.

The in vivo subcellular distribution of NUCB1 and NUCB2 is likely variable and cell-type specific. In the CNS, NUCB1 is expressed only in neurons but not glial or ependymal cells. Within neurons, NUCB1 is detected in the soma and dendrites but not along the axons or the presynaptic terminals or the nucleus (Tulke et al., 2016). On the other hand, nuclear expression of NUCB1 was reported in odontoblasts (Somogyi, Petersson, Sugars, Hultenby, & Wendel, 2004) and gastric adenocarcinomas (Wang et al., 1994).

9. NUCB protein-protein interactions that mediate different physiological functions

Physical interactions with other proteins can provide important insights into a protein's function and regulation. Human NUCB1 and NUCB2 form unique protein-protein interaction networks as inferred from several lines of evidence. A large multi-evidence interaction database lists 42 and 31 unique physical interactors identified for NUCB1 and NUCB2, respectively (Chatr-Aryamontri et al., 2017). NUCB2's interactors are not enriched in specific functions. In contrast, NUCB1's interactors are enriched for proteins involved in cAMP-mediated GPCR signaling, regulation of cell proliferation and cell death—suggesting that NUCB1 is also involved in these processes. Perhaps its DNA-binding function discussed above is related to interaction with nucleotide derivatives such as cAMP and others, rather than being strictly DNA. Below, we will discuss some of these interactors where specific interactions sites have been determined.

9.1 NUCB1/COXs and NUCB2/ART-1: Functions in inflammation response

Cyclooxygenases (COXs) also called prostaglandin-endoperoxide synthase (PTGS) catalyze the conversion of arachidonate to prostanoids. Therefore, they are mediators of inflammatory response and targets for the non-steroidal anti-inflammatory drugs. COX-1 and COX-2 interact with a fragment of NUCB1 that includes the N-terminal signal peptide and the NLP sequences (1−123) (Ballif, Mincek, Barratt, Wilson, & Simmons, 1996). COX-1 is a constitutively expressed enzyme, whereas COX-2 expression is induced by stimuli (Kujubu, Fletcher, Varnum, Lim, & Herschman, 1991; O'Banion, Sadowski, Winn, & Young, 1991; Smith & Marnett, 1991; Xie, Chipman, Robertson, Erikson, & Simmons, 1991). Intriguingly, the COX substrate arachidonate can also bind to the N-terminal region of NUCB1 (Table 2) (Niphakis et al., 2015). How can NUCB1 bind to COX and its substrate at an overlapping region? COX are intraluminal bound proteins at the ER. The calcium concentration in the ER is in the millimolar range (Bygrave & Benedetti, 1996). Since the K_m of calcium for both EF-hands for NUCB1 are in the micromolar range (6.3 and 73.5 μM) (Kapoor et al., 2010), it is expected that NUCB1 in the ER lumen would be in the calcium-bound conformation as observed in the NMR structure (Fig. 2). While it is unclear whether NUCB1/COX interactions require calcium, the binding of its substrate arachidonate does (Niphakis et al., 2015). Therefore, it is possible that the calcium-bound NUCB1 conformation facilitates the binding of arachidonate, thus inhibiting the activity of COX. When the local calcium concentration drops to the range that induces a conformational change of NUCB1, arachidonate will be dislodged and perhaps COX can then bind NUCB1. The NUCB1/COX interaction enhances the COX catalytic activity (Leclerc et al., 2008) and leads to the retention of NUCB1 in the ER (Ballif et al., 1996). Once in a binary complex with COX, NUCB1 may be locked into a conformation that makes it unresponsive to calcium. While there is no information on NUCB2 interactions with COXs, NUCB2 is also implicated in inflammatory response by its involvement in TNF/TNFR1 signaling. TNF is a cytokine responsible for systemic inflammation and TNFR1 is one of its surface receptors. Upon TNF binding, TNF+TNFR1 is translocated to a lipid raft, where several other proteins associate to activate the NF-κB signaling pathway. Dysregulation in TNF activity is implicated in a variety of human diseases including Alzheimer's disease, cancer, and inflammatory bowel disease. Extracellular release of TNFR1 is one mechanism that TNF activity can be modulated. NUCB2 interacts with ART-1, which is an integral membrane protein that associates with TNFR1 and promotes its release to the extracellular compartment. Therefore, by interacting with different partners, NUCB1 and NUCB2 regulate inflammatory response in distinct mechanisms.

9.2 NUCB1/Gαi interaction: Functions in G protein signaling

As experimental evidence suggests, the calcium-dependent conformational changes in NUCB1 discussed above trigger interactions with the Gα subunit, which is a component of the heterotrimeric αβγ G protein complex that functions with G protein coupled receptors (GPCRs) in signaling. When the signaling system is inactive, the Gα subunit bound with a GDP is associated with the βγ subunit and anchored to the GPCR. Upon activation of the GPCR by its ligand, the GDP on the Gα subunit is exchanged with a GTP. The nucleotide exchange causes a conformational change in the Gα subunit that leads to its dissociation from the GPCR and the βγ subunits, thus activating a cascade of downstream effectors. The signaling event continues until the GTP on the Gα subunit is hydrolyzed back to GDP. Then the GDP-bound Gα subunit is re-associated with the βγ subunit and the GPCR. In the calcium-free conformation, the EF-hands domain of NUCB1 interacts with the C-terminal helix (α5) of a Gα protein subtype Gαi and inhibits the nucleotide exchange step (GDP→GTP), thus effectively inactivating the signaling pathway (Table 2) (Kapoor et al., 2010; Lin, Fischer, Weiss, & Farquhar, 2000). Therefore, the calcium-dependent conformational changes of NUCB1 affect its ability to bind Gα and in turn modulate the activity level of this ubiquitous signaling pathway.

9.3 NUCB2/Necdin: Functions in energy metabolism

The physiological consequence of how NUCB1 (and likely NUCB2) interacts with the Gα subunit is demonstrated in the NUCB2/necdin interaction involved in energy metabolism. Necdin was first identified as a gene expressed exclusively in post-mitotic neurons, but not in proliferating progenitors (Aizawa, Maruyama, Kondo, & Yoshikawa, 1992; Maruyama, Usami, Aizawa, & Yoshikawa, 1991). Necdin was discovered to interact with NUCB2 and inhibit its secretion and consequently prolong its cytoplasmic retention (Taniguchi et al., 2000). Necdin itself is a multifunctional protein with many protein partners (Table 2). Therefore, NUCB2/necdin interaction can affect other functions of necdin via its other binding partners.

The *necdin* gene is one of the deleted genes found in Prader-Willi syndrome (PWS). Uncontrolled appetite, obesity, and abnormal fat distribution

Table 2 Protein-protein interactions discussed in this review.

Protein 1	Protein 2	Experiment that detected PPI	Physiological Consequence of the Interaction	Reference
NUCB1 (264–301) regions between two EF-hands	Gαi (α5 helix)	Y2H	Inhibits GDP-GTP exchange	Kapoor et al. (2010)
NUCB2 (214–358) EF-hand domain	Necdin (83–292)	Co-IP and Y2H	Inhibits NUCB2 secretion	Taniguchi et al. (2000)
Necdin (83–292) WH domain	E2F1 (383–430) C-terminal transactivation domain	Co-IP and Y2H	Represses E2F1 transcriptional activity	Kuwako, Taniura, and Yoshikawa (2004)
Necdin (83–292) WH domain	Gαo (214–354 include α5 helix)	Co-IP and Y2H	Increases necdin/E2F1 interaction	Ju, Lee, Kang, Kim, and Ghil (2014)
Necdin (83–292)	Leptin receptor intracellular tail (E1041-L1092)	Mammalian protein-protein interaction trap	Leptin receptor trafficking	Wijesuriya et al. (2017)
NUCB1 (1–123)	COX1 domain 1 and 2 that contains Met381-Gly498	Co-IP and Y2H	Inhibit NUCB1 secretion, increases COX-2 mediated formation of PGE2	Ballif et al. (1996)
NUCB2	ARTS-1	Co-IP and Y2H	Extracellular release of TNFR1	Islam et al. (2006)

The table lists the sites, experimental evidence, and likely physiological role for selected physical protein-protein interactions. Y2H, yeast two hybrid; Co-IP, co-immunoprecipitation.

Source: BioGrid (Chatr-Aryamontri et al., 2017).

are consistent clinical presentations of PWS patients, suggesting a dysfunctional/dysregulated endocrine system that results in energy imbalance. Through its interaction with the leptin receptor and the E2F1 transcription factor, necdin is implicated in the regulation of leptin receptor trafficking and the induction of adipocyte differentiation, both of which play important roles in maintaining energy homeostasis. Leptin is an anti-hunger hormone. Its effect is dependent on the availability of its receptor (LepR) that activates the aneroxigenic proopiomelanocortin neurons in the hypothalamus. Necdin interacts with the intracellular C-terminus of LepR and plays a role in regulating its internalization, processing, and stability through the ubiquitination system (Wijesuriya et al., 2017) (Table 2).

Necdin interacts with the C-terminal transactivation domain of the transcription factor E2F1 (Kuwako et al., 2004). The interaction represses the transcriptional activity of E2F1, which leads to cellular differentiation (Pamuklar, Chen, Muehlbauer, Spagnoli, & Torquati, 2013). Necdin also interacts with the C-terminus of the G protein subunit Gαo (Ju et al., 2014). Gαo enhances the ability of necdin to repress the activity of E2F1 that results in cellular differentiation (e.g., adipocyte) (Ju et al., 2014). The mechanism of how Gαo, necdin, and E2F1 function to promote adipocyte differentiation is unclear. The EF-hands domain of NUCB2 interacts with the middle region of necdin (residues 83–292); this same region of necdin interacts with E2F1 (Table 2). Furthermore, necdin interacts with 214–354 of Gαo, which encompasses the equivalent α5 helix on Gαi that was shown to interact with the region between the two EF-hands on NUCB1 (Kapoor et al., 2010). In summary, NUCB2/necdin, necdin/E2F1, NUCB1/Gα, and Gα/necdin have all been observed. Although it is unclear whether these binary complexes are exclusive or they are part of a bigger complex, it does suggest an interplay between G protein signaling, NUCB secretion, and cellular differentiation.

9.4 Other physiological functions of NUCB

Both NUCB1 and NUCB2 are substrates for caspases, which are proteolytic enzymes implicated in apoptosis and inflammation. The cleavage occurs in conserved sites DXXD between the NP and EF-domains (Fig. 1—red box) (Valencia et al., 2008). Therefore, upon caspase cleavage, the functional domains of NUCB1 and NUCB2 will be separated and may perform unique functions during apoptosis or inflammation. NUCB1 has many more documented physiological functions compared to NUCB2. For example,

NUCB1 has been linked to Alzheimer's disease by its interaction with the C-terminus of the β-amyloid precursor protein (APP). The level of NUCB1 is inversely related to the amount of endogenous APP. Interestingly, NUCB1 is significantly reduced in the brains of patients with Alzheimer's disease (Lin et al., 2007). In addition, NUCB1 has been implicated in osteoblast differentiation (Xue et al., 2016), endoplasmic reticulum stress (Tsukumo et al., 2007) and lipid metabolism (Niphakis et al., 2015).

10. NUCB encoded peptides

Early investigation of the NUCB family focused on the full length protein until Oh-I and colleagues discovered the satiety property of nesfatin-1 encoded in the NUCB2. Subsequently, the equivalent peptide of NUCB1 was also shown to have similar activity. Henceforth, we discuss the current status of the physiological functions of these two bioactive peptides encoded in NUCBs.

10.1 Nesfatin-1

NUCB2 structure was discussed earlier in this review. It has been shown that a secretory protein NEFA (Nuclear EF-hand acidic) or NUCB2, which has a calcium-binding domain (EF-domain) and DNA-binding domain, is present in the hypothalamic appetite regulatory regions. Since previous clinical data on type 2 diabetic patients indicated that peroxisome proliferator activated receptor γ (PPARγ) agonist, troglitazone, might modify satiety (Oh-I et al., 2006), Mori and colleagues tried to find a new protein regulated by PPARγ, which modulates feeding behavior. They identified a gene present both in the brain and adipocyte cell clones and found that it encodes NEFA/NUCB2, a secreted protein of unknown function (Oh-I et al., 2006). Intracerebroventricular (ICV) injection of NEFA/NUCB2 suppressed food intake in rats. Therefore, they referred to this protein as nesfatin (NEFA/nucleobindin-2-Encoded Satiety- and Fat-INfluencing protein). In silico analysis of NUCB2 sequence found potential cleavage sites for prohormone convertases (both 1/3 and 2) at Lys83-Arg84 site and Arg164-Arg165, respectively. This could give rise to three possible processed fragments: nesfatin-1 (residues 1–82), nesfatin-2 (residues 85–163) and nesfatin-3 (residues 166–396). The expression of nesfatin-1, the N-terminal fragment of NUCB2, in hypothalamic PVN and cerebrospinal fluid was decreased by fasting in rats. ICV injection of nesfatin-1 suppressed food intake dose-dependently and an anti-nesfastin-1 antibody promoted food intake. In contrast, ICV injection of synthetic peptide corresponding to nesfatin-2 and -3 did not affect food intake. Therefore, nesfatin-1 is a naturally occurring 82-amino acid anorectic peptide encoded by nucleobindin-2 (NUCB2) (Oh-I et al., 2006). The 396-amino acid precursor protein NUCB2 is highly conserved among rodents, humans and non-mammals, indicative of its importance (Oh-I et al., 2006). The active site of nesfatin-1 is in the mid-segment of 30 amino acids (23–53 amino acid, Fig. 3), as the injections of C-terminal (1 – 23) and N-terminal regions (54–82) had no anorectic effect in mammals (Oh-I et al., 2006). The C-terminal region of the mid-segment contains an amino acid sequence H-F-R identical to that of α-melanin stimulating hormones (MSH) this sequence has the characteristic motif of melanocortin receptor 3 (MC3R) and (MC4R) recognition site (Cai et al., 2005; Wilson, Todorovic, Proneth, & Haskell-Luevano, 2006). The central region of the mid-segment contains the amino acid sequence LKQVIDV that shows 42.9% identity and 85.7% similarity to an amino acid sequence in human AgRP. This sequence includes the site in AgRP involved in increased energy expenditure (Oh-I et al., 2006).

The details of tissue and cell specific expression of nesfatin-1 in the brain and periphery are provided in Table 3. Nesfatin-1 is colocalized with a large number of regulatory proteins in the brain (Brailoiu et al., 2007; Foo et al., 2008; Fort et al., 2008; Kohno et al., 2008). Nesfatin-1 expression in the hypothalamus suggests a role for endogenous NUCB2/nesfatin-1 in the regulation of energy balance. In agreement with this, both central and peripheral acute administration of nesfatin-1 reduces food intake (Atsuchi et al., 2010; Goebel, Stengel, Wang, & Taché, 2011; Könczöl et al., 2012; Maejima et al., 2009; Oh-I et al., 2006; Stengel et al., 2009; Yosten & Samson, 2010) and adipose tissue weight (Oh-I et al., 2006). Nesfatin-1 also reduced the dark phase-specific food intake in mice (Stengel et al., 2009). Nesfatin-1 reduces gastroduodenal motility in mice, and the lowering of transit time might be one of the modes by which nesfatin-1 reduces feeding (Atsuchi et al., 2010). Nesfatin-1 was found abundant in the adipose tissue, and the expression augmented with an increase in adiposity due to high-fat (HF) diet feeding, and decreased after food deprivation (Ramanjaneya et al., 2010). Further, it was found that nesfatin-1 synthesis and secretion increased when 3T3L1 preadipocytes differentiated to mature adipocytes (Ramanjaneya et al., 2010). NUCB2 overexpression was shown to reduce 3T3L1 differentiation (Ramanjaneya et al., 2010). Nesfatin-1 levels were higher in obese individuals, and the decreased ratio of nesfatin-1/NUCB2 in the cerebrospinal fluid and the plasma provides evidence for nesfatin-1 resistance in human obesity (Tan, Hallschmid, Kern, Lehnert, & Randeva, 2011). Nesfatin-1 polymorphisms

Table 3 NUCB1 and NUCB2 tissue expression in non-human vertebrates.

Species	Tissues	Regions or cell types	*NUCB1* (Gene/ mRNA)	NUCB1 (Protein)	*NUCB2* (Gene/ mRNA)	NUCB2 (Protein)	References
	Brain		✓or × or –	✓or × or –	✓or × or –	✓or × or –	
Rat	Olfactory bulb		✓	✓	–	–	NUCB1: Tulke et al. (2016) NUCB2: Foo, Brismar, and Broberger (2008), Oh-I et al. (2006), and Goebel, Stengel, Wang, Lambrecht, and Tache (2009)
		Cells of olfactory nucleus	✓	✓	–	–	
		Granular cells	✓	✓	–	–	
		Anterior commissure	×	×	–	–	
	Cerebral cortex		✓	✓	–	–	
		Pre-frontal cortex	✓	✓	–	–	
		Motor cortex	✓	✓	–	–	
		Visual cortex	✓	✓	–	–	
		Corpus callosum	×	×	–	–	
		Dentate gyrus	×	×	–	–	
		Piriform cortex	–	–	✓	✓	
		Insular cortex	–	–	✓	✓	
	Ventricles						
		Ependymal cells	×	×	–	–	
		Cells of choroid plexus	✓	✓	–	–	

Continued

Table 3 NUCB1 and NUCB2 tissue expression in non-human vertebrates.—cont'd

Species	Tissues	Regions or cell types	*NUCB1* (Gene/ mRNA)	NUCB1 (Protein)	*NUCB2* (Gene/ mRNA)	NUCB2 (Protein)	References
	Brain		**✓or × or –**	**✓or × or –**	**✓or × or –**	**✓or × or –**	
	Hypothalamus						
		Arcuate nucleus (ARC)	✓	✓	✓	✓	
		Supraoptic nucleus (SON)	✓	✓	✓	✓	
		Lateral hypothalamic area (LHA)	✓	✓	✓	✓	
	Thalamus						
		Parafascicular nucleus	–	–	✓	✓	
		Edinger-Westphal nucleus	–	–	✓	✓	
		Locus Coeruleus	–	–	✓	✓	
	Cerebellum						
		Purkinje cells	✓	✓	✓	✓	
		Golgi cells	✓	✓	–	–	
	Hippocampus	CA3 region	✓	✓	–	–	
	Pituitary						
		Anterior lobe	–	–	✓	✓	
		Posterior lobe	–	–	×	–	

Brain stem					
	Raphe nuclei	✓	✓	✓	✓
	Pontine nuclei	✓	✓	–	–
Medulla Oblongata	Inferior olive	✓	✓	–	–
Spinal cord					
	Motor neurons in anterior horn	✓	✓	✓	✓
	Posterior horn	✓	✓	✓	✓
	Gray commissure	✓	✓	–	–
Dorsal root ganglion					
	Primary sensory neurons	✓	✓	–	–
	Nerve fibers	×	×	–	–
	Satellite cells	×	×	–	–

Continued

Table 3 NUCB1 and NUCB2 tissue expression in non-human vertebrates.—cont'd

Species	Tissues	Regions or cell types	*NUCB1* (Gene/mRNA)	NUCB1 (Protein)	*NUCB2* (Gene/mRNA)	NUCB2 (Protein)	References
	Brain		**✓or × or –**	**✓or × or –**	**✓or × or –**	**✓or × or –**	
Mouse	Olfactory bulb						
		Cells of olfactory nucleus	–	–	–	✓	
		Granular cells	–	–	–	✓	
		Anterior Commissure	–	–	–	✓	
	Cerebral cortex						
		Piriform cortex	–	–	–	✓	
		Insular cortex	–	–	–	✓	
		Cingulate cortex	–	–	–	✓	
		Somatomotor cortex	–	–	–	✓	
	Hypothalamus						
		Arcuate nucleus (ARC)	–	–	–	✓	
		Supraoptic nucleus (SON)	–	–	–	✓	
		Lateral hypothalamic area (LHA)	–	–	–	✓	

Thalamus					
	Dorsal Raphe nuclei	–	–	–	✓
	Edinger-Westphal nucleus	–	–	–	✓
	Locus Coeruleus	–	–	–	✓
Cerebellum	Purkinje cells	–	–	–	✓
Medulla oblongata					
	Inferior olive	–	–	–	✓
	Nucleus gigantocellularis	–	–	–	✓
	Area postrema	–	–	–	✓
	Nucleus of solitary tract	–	–	–	✓
	dorsal motor nucleus of vagus nerve	–	–	–	✓
	Nucleus ambiguus	–	–	–	✓

Continued

Table 3 NUCB1 and NUCB2 tissue expression in non-human vertebrates.—cont'd

Species	Tissues	Regions or cell types	*NUCB1* (Gene/mRNA)	NUCB1 (Protein)	*NUCB2* (Gene/mRNA)	NUCB2 (Protein)	References
	Brain		✓or × or –	✓or × or –	✓or × or –	✓or × or –	
Goldfish	Olfactory bulb		✓	–	✓	–	NUCB1: Sundarrajan et al. (2016) NUCB2: Gonzalez et al. (2010) and Gonzalez et al. (2012)
	Hypothalamus		✓	–	✓	–	
		Nucleus anterior tuberis (NAT)	–	–	–	✓	
		Nucleus lateralis tuberis (NLT)	–	–	–	✓	
	Telencephalon		✓	–	✓	–	
	Midbrain		✓	✓	✓	–	
	Hindbrain		✓	✓	✓	–	
	Pituitary		✓	✓	✓	✓	
Zebrafish*	Brain		–	–	✓	–	*Two isoforms of NUCB2: NUCB2A and NUCB2B (Hatef, Shajan, & Unniappan, 2015)
Frog	Olfactory system						Senejani, Gaupale, Unniappan, and Bhargava (2014)
		Medial diverticulum epithelium	–	–	–	✓	
		Bowman's glands	–	–	–	✓	
	Telencephalon					✓	
	Diencephalon	Preoptic area	–	–	–	✓	

Ventricles					
	Suprachiasmatic nucleus	–	–	–	✓
	Nucleus infundibularis ventralis	–	–	–	✓
	Nucleus infundibularis dorsalis	–	–	–	✓
Hypothalamus					
	Lateral hypothalamic area (LHA)	–	–	–	✓
	Ventromedial thalamic nucleus	–	–	–	✓
Pituitary					
	Neural lobe	–	–	–	✓
	Intermediate lobe	–	–	–	✓
	Distal lobe	–	–	–	✓

Continued

Table 3 NUCB1 and NUCB2 tissue expression in non-human vertebrates.—cont'd

Species	Tissues	Regions or cell types	*NUCB1* (Gene/ mRNA)	NUCB1 (Protein)	*NUCB2* (Gene/ mRNA)	NUCB2 (Protein)	References
	Peripheral Tissues		✓or × or –	✓or × or –	✓or × or –	✓or × or –	
Rat	Pancreas	Beta cells of islet	–	–	✓	✓	Zhang et al. (2010), Ramanjaneya et al. (2015), Osaki et al. (2012), García-Galiano et al. (2012), Angelone et al. (2012), and Xu, Zhang, Li, Lao, and Wang (2017)
	Stomach	Oxyntic mucosa (X/A cells)	–	–	✓	✓	
	Duodenum	Submucosal brunner's glands	–	–	✓	✓	
	Colon		–	–	✓	✓	
	Thyroid		–	–	–	–	
	Parathyroid		–	–	–	–	
	Liver		–	–	–	–	
	Lung		–	–	–	–	
	Heart	Cardiac extracts	–	–	✓	✓	
	Adrenal	All zones of adrenal cortex	–	–	✓	✓	
	Adipose tissue	Visceral, subcutaneous and interscapular brown adipose tissue	–	–	✓	✓	

	Kidney		–	–	–	–	
	Testes	Interstitial mature Leydig cells	–	–	✓	✓	
	Ovaries		–	–	✓	✓	
Mouse	Pancreas	All five cell types of islets	✓	✓	✓	✓	NUCB1: Williams, Tulke, Ilegems, Berggren, and Broberger (2014) NUCB2: Chung, Jung, Kim, Kim, and Yang (2013), Kim et al. (2014), Zhang et al. (2010), and García-Galiano et al. (2012)
	Stomach		✓	✓	✓	✓	
	Duodenum	L-, I- and K-cells	✓	✓	✓	✓	
	Colon		✓	✓	✓	✓	
	Thyroid	Follicular cells	✓	✓	–	–	
	Parathyroid	Principal cells	✓	✓	–	–	
	Liver		–	–	✓	✓	
	Lung		–	–	✓	✓	
	Heart		–	–	✓	✓	

Continued

Table 3 NUCB1 and NUCB2 tissue expression in non-human vertebrates.—cont'd

Species	Tissues	Regions or cell types	*NUCB1* (Gene/mRNA)	NUCB1 (Protein)	*NUCB2* (Gene/mRNA)	NUCB2 (Protein)	References
	Peripheral Tissues		✓or × or –	✓or × or –	✓or × or –	✓or × or –	
	Adrenal	Y1 mouse adrenal tumor cells	✓	✓	✓	✓	
	Adipose tissue		–	–			
	Kidney		–	–	✓	✓	
	Testes	Seminiferous tubules (NUCB1), interstitial mature Leydig cells (NUCB2)	✓	✓	✓	✓	
	Ovaries	Cells of stratum granulosum, follicular cells	✓	✓	✓	✓	
Human	Adrenals	H295R human adrenocortical cells	–	–	✓	✓	NUCB1: Tsukumo et al. (2007) NUCB2: Ramanjaneya et al. (2015), García-Galiano et al. (2012), andTakagi et al. (2016)
	Testes	Interstitial mature Leydig cells	–	–	✓	✓	
	Uterus	Endometrial carcinoma tissues	–	–	✓	–	
	Connective tissue	Fibrosarcoma cell line HT-1080	✓	✓	–	–	

Goldfish	Gut	✓	✓	–	–	NUCB1: Sundarrajan et al. (2016) NUCB2: Gonzalez et al. (2010) and Bertucci, Blanco, Canosa, and Unniappan (2017)
	Testes	✓	✓	✓	–	
	Ovary	✓	✓	✓	–	
	Whole brain	✓	✓	✓	✓	
	Pituitary	✓	✓	–	–	
	Olfactory bulb	✓	–	✓	–	
	Hypothalamus	✓	–	✓	–	
	Telencephalon	✓	–	✓	–	
	Gall bladder	✓	–	✓	–	
	Adipose tissue		–	✓	–	
	Midbrain	✓	–	✓	–	
	Hindbrain	✓	–	✓	–	
	Eye	✓	–	✓	–	
	Kidney	–	–	✓	–	
	Midgut	✓	✓	✓	–	
	Rectum	✓	–	✓	–	
	Heart	✓	–	✓	–	
	Skin	✓	–	✓	–	
	J-loop	✓	–	✓	–	

Continued

Table 3 NUCB1 and NUCB2 tissue expression in non-human vertebrates.—cont'd

Species	Tissues	Regions or cell types	*NUCB1* (Gene/ mRNA)	NUCB1 (Protein)	*NUCB2* (Gene/ mRNA)	NUCB2 (Protein)	References
	Peripheral Tissues		✓or × or –	✓or × or –	✓or × or –	✓or × or –	
	Muscle		✓	–	✓	–	
	Gill		✓	–	✓	–	
	Intestine		–	–	✓	✓	
	Hepatopancreas		–	–	✓	✓	
Zebrafish	Liver		–	–	✓	–	Hatef, Shajan, and Unniappan (2014)
	Gut	Mucosal layer cells	–	–	✓	–	
	J-loop		–	–	✓	–	
	Heart		–	–	✓	–	
	Eye		–	–	✓	–	
	Muscle		–	–	✓	–	
	Gill		–	–	✓	–	
	Ovary		–	–	✓	–	
	Testes		–	–	✓	–	
	Skin		–	–	✓	–	

✓, Presence.
×, Absence.
–, Yet to be explored.

were reported in human diabetes and obesity (Zegers, Beckers, Mertens, Van Gaal, & Van Hul, 2011). Nesfatin-1 inhibits gastric acid secretion via a vagal mediated mechanism in rats (Xia et al., 2012), increases insulin sensitivity when administered into the brain of HF diet fed rats (Yang et al., 2012), and protects against cardiac ischemia-reperfusion injury (Angelone et al., 2012). Hypoglycemia activated nesfatinergic neurons in the hypothalamus-brainstem neural circuitry of rats, and it was found that these glucopenia responsive neurons innervate the stomach and the pancreas (Bonnet et al., 2012). Central administration of nesfatin-1 induces testicular expression of 3 beta-hydroxysteroid dehydrogenase (3β-HSD), 17β-hydroxysteroid dehydrogenase (17β-HSD), and cytochrome P450 cleavage (P450scc) in pubertal rats, but attenuates its expression and serum follicle stimulating hormone (FSH), luteinizing hormone (LH), and testosterone (T) concentrations in adult rats (Gao et al., 2016). Nesfatin-1 stimulates T secretion from testis in vitro (Kujubu et al., 1991) and rat testicular explants ex vivo (García-Galiano et al., 2012). Hypothalamic NUCB2/nesfatin-1 expression in female rats increased during puberty, but 48 hour fasting eliminated this increase (García-Galiano et al., 2010). Central administration of nesfatin-1 caused a significant increase in gonadotropin secretion in female rats at puberty, but this increase was abolished when the rats were fasted. These two studies suggest sexually dimorphic expression and action of the nesfatinergic system in rats, and suggest a role for this peptide in integrating metabolism and reproduction. There are numerous other tissue- and species-specific functions of nesfatin-1. A detailed discussion of all of these functions is beyond the focus of this review. The readers are encouraged to consult several recent reviews from our group and others for more information on the diverse biological actions of nesfatin-1 (Dore, Levata, Lehnert, & Schulz, 2017; Imbrogno, Angelone, & Cerra, 2015; Ramesh, Gawli, Pasupulleti, & Unniappan, 2017; Schalla & Stengel, 2018; Stengel, 2015; Weibert, Hofmann, & Stengel, 2019). Collectively, the literature available to date strongly supports important physiological roles for nesfatin-1 in mammals and other vertebrates.

The identity of nesfatin-1 receptor currently remains unknown. However, recent observations suggest that neuronal effects of nesfatin-1 are mediated via a G-protein coupled receptor (GPCR) (Brailoiu et al., 2007). Preliminary results suggest that GPCR3, 6 and/or 12 are potential receptors for nesfatin-1 (Ishida et al., 2012; Osei-Hyiaman, Sophie-Dreher, Nishimura, & Encinas, 2011). Some studies have investigated intracellular signaling events of nesfatin-1 in dorsal root ganglion neurons, pancreatic β cells and cardiomyocytes (Iwasaki et al., 2009; Ozcan, Gok,

Kacar, Serhatlioglu, & Kelestimur, 2016; Ying et al., 2015). In most cell types, nesfatin-1 causes calcium influx. Nesfatin-1 causes an increase in intracellular calcium concentrations in cultured hypothalamic neurons of rats, an effect that was attenuated by pertussis toxin, a calcium channel blocker. Nesfatin-1 could interact with GPCR causing an increase in intracellular calcium concentration mediated via either the L- or N-type calcium channels (Ozcan et al., 2016). In the vagal afferent ganglion of mice, nesfatin-1 stimulates an influx of calcium ions through N-type calcium channels present in these neurons (Ozcan et al., 2016). Nesfatin-1 increases intracellular calcium concentration by activating protein kinase C (PKC) in cultured neurons of dorsal root ganglion (Ying et al., 2015). In insulin secreting β cells of pancreatic islets, nesfatin-1 stimulates influx of calcium via L-type channels (Nakata, Manaka, Yamamoto, Mori, & Yada, 2011). Recent efforts have focused on localizing nesfatin-1 receptor using autoradiography. Upon injection of ^{125}I labeled nesfatin-1, its binding was abundantly detected in the gastric mucosa of corpus and antrum, duodenum, jejunum and ileum. Besides these tissues, labeling was also detected in the endocrine pancreas and adrenal glands. In the brain, ^{125}I labeled nesfatin-1 was detected in the paraventricular nucleus (PVN), arcuate nucleus (ARC), area postrema (AP), dorsal motor nucleus of the vagus (DMNV) and cerebellum (Prinz et al., 2016).

10.2 Nesfatin-1-like peptide (NLP)

NUCB1 structure and distribution were discussed previously. In silico analysis of the NUCB1 sequence (aligned using ClustalW and eBioX software) showed that it has a 24-amino acid signal peptide sequence followed by a signal peptidase cleavage site between position 26 and 27 (Fig. 4). This is followed by a region that has high similarity to nesfatin-1 region of NUCB2 (76.6% identity) containing a mid-segment of 30 amino acids. Analysis of NUCB1 sequence using Prop 1.0 server and NeuroPred™ tools revealed potential proprotein convertase cleavage site at Lysine-Arginine (KR, highlighted in red above), forming a 77-amino acid peptide. Owing to its high similarity with nesfatin-1, we named the peptide as Nesfatin-1-Like Peptide (NLP) (Fig. 4) (Gonzalez et al., 2012; Ramesh et al., 2015; Sundarrajan et al., 2016).

The endogenous expression profile of NUCB1 was evaluated extensively, showing that NUCB1 is present in pancreatic islets as well as in other endocrine tissues including stomach, intestine, adrenal gland, pituitary, ovary and testis (Table 3). Immunofluorescence staining performed on formalin fixed sections showed NUCB1 to be localized in the endocrine pancreas with no signals in the surrounding exocrine acinar cells or pancreatic ducts. NUCB1 immunoreactivity is distributed in pancreatic islets and it co-localizes only with insulin in β cells. Subcellular localization of NUCB1 using organelle specific protein stains showed NUCB1 to be co-distributed with Golgi apparatus protein giantin. NUCB1-IR was found all through the GI tract, but not in all endocrine cell types. In the stomach, NUCB1 signal was concentrated in the fundic glands. NUCB1-IR was found in the duodenum, jejunum and colon, but it is primarily present in the duodenal enterocytes. Strong IR was observed in deep intestinal glands wherein cells immunopositive for somatostatin and ghrelin showed NUCB1-IR. NUCB1-IR was also observed in the principal cells of parathyroid, thyroid follicles, adrenal medulla and zona glomerulosa of adrenal cortex, seminiferous tubules in the testes, and follicular cells of the ovary. NUCB1-IR was detected in the CNS and virtually in all anterior pituitary cells, including cortocotrophes, somatotrophes, thyrotrophes, lactotrophes and gonadotrophes (Williams et al., 2014). These studies indicate the role of NUCB1 in multiple cellular processes. Given the cytoplasmic localization of this protein, and that it could be secreted points toward a potential endocrine function for NUCB1 and/or encoded peptides.

NLP was first reported as an insulinotropic peptide (Ramesh et al., 2015). Treatment of MIN6 cells with synthetic NLP dose-dependently stimulates insulin secretion. This effect is ablated upon treatment with a peptide sequence with the amino acids scrambled. The study highlighted that NLP is biologically active and that the intact 77-amino acid sequence (with the M30 region) is required for its action (Kanai & Tanuma, 1992). We recently assessed the role of NLP as a potential metabolic factor in rats. Consistent with our hypothesis, single IP injection of NLP decreases food intake and increases ambulatory movement when compared to saline-injected controls. No changes in energy expenditure were observed between control and treatment groups (Gawli, Ramesh, & Unniappan, 2017). We then carried out a similar study, evaluating the effects of long-term NLP infusion in rats. Subsequently, continuous 7-day subcutaneous infusion of NLP using osmotic mini-pumps causes reduction of dark phase food and water intake. This was accompanied by increased respiratory exchange ratio (RER) and energy expenditure, measured using the comprehensive lab animal monitoring system (CLAMS). While NLP infusion causes a decrease in whole-body fat oxidation, no changes in body weight were observed between the control and treatment groups (Gawli et al., 2017). NLP infusion also causes a twofold increase in cholecystokinin (CCK) expression in the duodenum (Gawli et al., 2017). NLP was also reported to be a bioactive peptide in goldfish. IP injection of NLP in goldfish causes a potent inhibition of food intake (Sundarrajan et al., 2016). While NLP injection downregulates the expression of genes encoding orexigenic hormones *preproghrelin* and *orexin-A*, it upregulates anorexigenic cocaine and amphetamine regulated transcript (CART) in goldfish brain (Sundarrajan et al., 2016). Collectively, the NLP literature highlights its role as an insulinotropic peptide that modulates energy metabolism in mammals and non-mammals alike. The mechanism of nesfatin-1 action remains unknown. Considering the presence of a bioactive core region very similar to the one in nesfatin-1 within the NLP, it is possible that NLP also elicits its cellular action via a GPCR mediated mechanism.

11. Perspectives and considerations for future research

The field of research that commenced with the discovery of nucleobindins has now grown into a much bigger portfolio that includes nesfatin-1 and NLP. The discovery of nesfatin-1 and NLP as novel endogenous bioactive peptides led to the generation of remarkable knowledge on biological importance of NUCBs. It is clear that the precursors and encoded peptides function to regulate a large number of physiological processes. However, several issues also arose with the emergence of new peptides. Given that the NUCBs are also secreted proteins makes it difficult to determine whether some of the functions attributed to nesfatin-1 and NLP could also be carried out by the precursor peptides. The lack of an antibody or an assay that distinguishes the precursor and smaller peptides within it makes it difficult to distinguish between them within circulation. The highly conserved bioactive core regions of both nesfatin-1 and NLP suggest conserved biological actions for these peptides. Whether nesfatin-1 compensates for NLP when one of them is absent or downregulated, and vice versa will be interesting to explore. While many endocrine-like actions for nesfatin-1 and NLP have been reported, the receptors for these two peptides remain unknown. Putative receptors and signaling mechanisms should also be the focus of future research. NUCBs are considered ubiquitous proteins. However, it is evident that not all cells express these proteins, and there is a very clear cell-specific NUCB expression pattern within each tissue. Cell specific differences in biological actions were also found. Some species-specific variations in NUCB expression and the biological effects were also observed. Possible gender or sex-based differences in nesfatin-1 or NLP actions are anticipated, but currently remain unknown.

Acknowledgments

The nesfatin-1 and NLP research in Suraj Unniappan's laboratory is funded by the Natural Sciences and Engineering Research Council (NSERC), Canadian Institutes of Health Research (CIHR), Canada Foundation for Innovation (CFI), and the Saskatchewan Health Research Foundation (SHRF). A.K.-W.L. is funded by NSERC, CFI and SHRF. C.V. is funded by the National Institutes of Health (R35 GM127089-02). N.R. was funded by a Dean's Scholarship, and the Teacher-Scholar Award from the University of Saskatchewan.

References

Aizawa, T., Maruyama, K., Kondo, H., & Yoshikawa, K. (1992). Expression of necdin, an embryonal carcinoma-derived nuclear protein, in developing mouse brain. *Brain Research. Developmental Brain Research*, *68*(2), 265–274.

Angelone, T., Filice, E., Pasqua, T., Amodio, N., Galluccio, M., Montesanti, G., et al. (2012). Nesfatin-1 as a novel cardiac peptide: Identification, functional characterization, and protection against ischemia/reperfusion injury. *Cellular and Molecular Life Sciences*, *70*, 495–509. https://doi.org/10.1007/s00018-012-1138-7.

Atsuchi, K., Asakawa, A., Ushikai, M., Ataka, K., Tsai, M., Koyama, K., et al. (2010). Centrally administered nesfatin-1 inhibits feeding behaviour and gastroduodenal motility in mice. *Neuroreport*, *21*, 1008–1011.

Ballif, B. A., Mincek, N. V., Barratt, J. T., Wilson, M. L., & Simmons, D. L. (1996). Interaction of cyclooxygenases with an apoptosis- and autoimmunity-associated protein. *Proceedings of the National Academy of Sciences of the United States of America*, *93*(11), 5544–5549.

Barnikol-Watanabe, S., Gross, N. A., Gotz, H., Henkel, T., Karabinos, A., Kratzin, H., et al. (1994). Human protein NEFA, a novel DNA binding/EF-hand/leucine zipper protein. Molecular cloning and sequence analysis of the cDNA, isolation and characterization of the protein. *Biological Chemistry Hoppe-Seyler*, *375*, 497–512.

Bertucci, J. I., Blanco, A. M., Canosa, L. F., & Unniappan, S. (2017). Glucose, amino acids and fatty acids directly regulate ghrelin and NUCB2/nesfatin-1 in the intestine and hepatopancreas of goldfish (Carassius auratus) in vitro. *Comparative Biochemistry and Physiology. Part A, Molecular & Integrative Physiology*, *206*, 24–35.

Bonnet, M., Djelloul, M., Tillement, V., Tardivel, C., Mounien, L., Trouslard, J., et al. (2012). Central NUCB2/nesfatin-1-expressing neurones belong to the hypothalamus-brainstem circuitry activated by hypoglycaemia. *Journal of Neuroendocrinology*, *25*, 1–13. https://doi.org/10.1111/j.1365-2826.2012.02375.x.

Brailoiu, G. C., Dun, S. L., Brailoiu, E., Inan, S., Yang, J., Chang, J. K., et al. (2007). Nesfatin-1: Distribution and interaction with a G protein-coupled receptor in the rat brain. *Endocrinology*, *148*, 5088–5094.

Bygrave, F. L., & Benedetti, A. (1996). What is the concentration of calcium ions in the endoplasmic reticulum? *Cell Calcium*, *19*(6), 547–551.

Cai, M., Mayorov, A. V., Ying, J., Stankova, M., Trivedi, D., Cabello, C., et al. (2005). Design of novel melanotropin agonists and antagonists with high potency and selectivity for human melanocortin receptors. *Peptides*, *26*, 1481–1485.

Celio, M. R., & Heizmann, C. W. (1981). Calcium-binding protein parvalbumin as a neuronal marker. *Nature*, *293*, 300–302.

Chatr-Aryamontri, A., Oughtred, R., Boucher, L., Rust, J., Chang, C., Kolas, N. K., et al. (2017). The BioGRID interaction database: 2017 update. *Nucleic Acids Research*, *45*(D1), D369–D379.

Chung, Y., Jung, E., Kim, H., Kim, J., & Yang, H. (2013). Expression of nesfatin-1/NUCB2 in fetal, neonatal and adult mice. *Development & Reproduction*, *17*, 461–467.

de Alba, E., & Tjandra, N. (2004). Structural studies on the Ca2+-binding domain of human nucleobindin (calnuc). *Biochemistry*, *43*(31), 10039–10049.

Dore, R., Levata, L., Lehnert, H., & Schulz, C. (2017). Nesfatin-1: Functions and physiology of a novel regulatory peptide. *The Journal of Endocrinology*, *232*(1), R45–R65.

Foo, K., Brismar, H., & Broberger, C. (2008). Distribution and neuropeptide coexistence of nucleobindin-2 mRNA/nesfatin-like immunoreactivity in the rat CNS. *Neuroscience*, *156*, 563–579.

Fort, P., Salvert, D., Hanriot, L., Jego, S., Shimizu, H., Hashimoto, K., et al. (2008). The satiety molecule nesfatin-1 is co-expressed with melanin concentrating hormone in tuberal hypothalamic neurons of the rat. *Neuroscience*, *155*, 174–181.

Gao, X., Zhang, K., Song, M., Li, X., Luo, L., Tian, Y., et al. (2016). Role of Nesfatin-1 in the reproductive axis of male rat. *Scientific Reports*, *6*, 32877.

García-Galiano, D., Navarro, V. M., Roa, J., Ruiz-Pino, F., Sánchez-Garrido, M. A., Pineda, R., et al. (2010). The anorexigenic neuropeptide, nesfatin-1, is indispensable for normal puberty onset in the female rat. *The Journal of Neuroscience*, *30*(23), 7783–7792.

García-Galiano, D., Pineda, R., Ilhan, T., Castellano, J. M., Ruiz-Pino, F., Sánchez-Garrido, M. A., et al. (2012). Cellular distribution, regulated expression, and functional role of the anorexigenic peptide, NUCB2/nesfatin-1, in the testis. *Endocrinology*, *153*(4), 1959–1971.

Gawli, Ramesh, N., & Unniappan, S. (2017). Nesfatin-1-like peptide is a novel metabolic factor that suppresses feeding, and regulates whole-body energy homeostasis in male Wistar rats. *PLoS One*, *12*, e0178329.

Gifford, J. L., Walsh, M. P., & Vogel, H. J. (2007). Structures and metal-ion-binding properties of the Ca2+-binding helix-loop-helix EF-hand motifs. *The Biochemical Journal*, *405*(2), 199–221.

Goebel, M., Stengel, A., Wang, L., Lambrecht, N. W., & Tache, Y. (2009). Nesfatin-1 immunoreactivity in rat brain and spinal cord autonomic nuclei. *Neuroscience Letters*, *452*, 241–246.

Goebel, M., Stengel, A., Wang, L., & Taché, Y. (2011). Central nesfatin-1 reduces the nocturnal food intake in mice by reducing meal size and increasing inter-meal intervals. *Peptides*, *32*, 36–43.

Gonzalez, R., Kerbel, B., Chun, A., & Unniappan, S. (2010). Molecular, cellular and physiological evidences for the anorexigenic actions of nesfatin-1 in goldfish. *PLoS One*, *5*, e15201.

Gonzalez, R., Mohan, H., & Unniappan, S. (2012). Nucleobindins: Bioactive precursor proteins encoding putative endocrine factors? *General and Comparative Endocrinology*, *76*, 341–346.

Hatef, A., Shajan, S., & Unniappan, S. (2014). Nutrient status modulates the expression of nesfatin-1 encoding nucleobindin 2A and 2B mRNAs in zebrafish gut, liver and brain. *General and Comparative Endocrinology*, *215*, 51–60.
Henrotte, J. G. (1952). A crystalline constituent from myogen of carp muscles. *Nature*, *169*, 968–969.
Hornbeck, P. V., Zhang, B., Murray, B., Kornhauser, J. M., Latham, V., & Skrzypek, E. (2015). PhosphoSitePlus, 2014: Mutations, PTMs and recalibrations. *Nucleic Acids Research*, *43*(Database issue), D512–D520.
Imbrogno, S., Angelone, T., & Cerra, M. C. (2015). Nesfatin-1 and the cardiovascular system: Central and pheripheral actions and cardioprotection. *Current Drug Targets*, *16*(8), 877–883.
Ishida, E., Hashimoto, K., Shimizu, H., Okada, S., Satoh, T., Kato, I., et al. (2012). Nesfatin-1 induces the phosphorylation levels of cAMP response element-binding protein for intracellular signaling in a neural cell line. *PLoS One*, 7, e50918.
Islam, A., Adamik, B., Hawari, F. I., Ma, G., Rouhani, F. N., Zhang, J., et al. (2006). Extracellular TNFR1 release requires the calcium-dependent formation of a nucleobindin 2-ARTS-1 complex. *The Journal of Biological Chemistry*, *281*(10), 6860–6873.
Iwasaki, Y., Nakabayashi, H., Kakei, M., Shimizu, H., Mori, M., & Yada, T. (2009). Nesfatin-1 evokes Ca2+ signaling in isolated vagal afferent neurons via Ca2+ influx through N-type channels. *Biochemical and Biophysical Research Communications*, *390*, 958–962.
Ju, H., Lee, S., Kang, S., Kim, S. S., & Ghil, S. (2014). The alpha subunit of Go modulates cell proliferation and differentiation through interactions with Necdin. *Cell Communication and Signaling: CCS*, *12*, 39.
Kanai, Y., Isonishi, S., Katagiri, T., Koizumi, T., Miura, K., & Kurosawa, Y. (1990). Early induction of anti-double-stranded DNA antibodies in lupus-prone MRL mice inoculated with Ly-24+ cells cloned from lymph node cells of an MRL/Mp-lpr/lpr mouse: Possible effect of putative cytokines produced by cloned Ly-24+ cells. *Immunology Letters*, *24*, 49–55.
Kanai, Y., Katagiri, T., Mori, S., & Kubota, T. (1986). An established MRL/Mp-lpr/lpr cell line with null cell properties produces a B cell differentiation factor(s) that promotes anti-single-stranded DNA antibody production in MRL spleen cell culture. *International Archives of Allergy and Applied Immunology*, *81*(1), 92–94.
Kanai, Y., & Tanuma, S. (1992). Purification of a novel B cell growth and differentiation factor associated with lupus syndrome. *Immunology Letters*, *32*, 43–48.
Kapoor, N., Gupta, R., Menon, S. T., Folta-Stogniew, E., Raleigh, D. P., & Sakmar, T. P. (2010). Nucleobindin 1 is a calcium-regulated guanine nucleotide dissociation inhibitor of G{alpha}i1. *The Journal of Biological Chemistry*, *285*(41), 31647–31660.
Kim, J., Chung, Y., Kim, H., Im, E., Lee, H., & Yang, H. (2014). The tissue distribution of nesfatin-1/NUCB2 in mouse. *Development & Reproduction*, *18*, 301–309.
Kohno, D., Nakata, M., Maejima, Y., Shimizu, H., Sedbazar, U., Yoshida, N., et al. (2008). Nesfatin-1 neurons in paraventricular and supraoptic nuclei of the rat hypothalamus coexpress oxytocin and vasopressin and are activated by refeeding. *Endocrinology*, *149*, 1295–1301.
Könczöl, K., Pintér, O., Ferenczi, S., Varga, J., Kovács, K., Palkovits, M., et al. (2012). Nesfatin-1 exerts long-term effect on food intake and body temperature. *International Journal of Obesity*, *36*, 1514–1521. https://doi.org/10.1038/ijo.2012.2.
Kretsinger, R. H., & Nockolds, C. E. (1973). Carp muscle calcium-binding protein. II. Structure determination and general description. *The Journal of Biological Chemistry*, *248*, 3313–3326.
Kubota, T., Akatsuka, T., & Kanai, Y. (1986). A monoclonal anti-double stranded DNA antibody from an autoimmune MRL/Mp-lpr/lpr mouse: Specificity and idiotype in serum immunoglobulins. *Immunology Letters*, *14*, 53–58.

Kujubu, D. A., Fletcher, B. S., Varnum, B. C., Lim, R. W., & Herschman, H. R. (1991). TIS10, a phorbol ester tumor promoter-inducible mRNA from Swiss 3T3 cells, encodes a novel prostaglandin synthase/cyclooxygenase homologue. *The Journal of Biological Chemistry, 266*(20), 12866–12872.

Kuwako, K., Taniura, H., & Yoshikawa, K. (2004). Necdin-related MAGE proteins differentially interact with the E2F1 transcription factor and the p75 neurotrophin receptor. *The Journal of Biological Chemistry, 279*(3), 1703–1712.

Lavoie, C., Meerloo, T., Lin, P., & Farquhar, M. G. (2002). Calnuc, an EF-hand Ca(2+)-binding protein, is stored and processed in the Golgi and secreted by the constitutive-like pathway in AtT20 cells. *Molecular Endocrinology, 16*(11), 2462–2474.

Leclerc, P., Biarc, J., St-Onge, M., Gilbert, C., Dussault, A. A., Laflamme, C., et al. (2008). Nucleobindin co-localizes and associates with cyclooxygenase (COX)-2 in human neutrophils. *PLoS One, 3*(5), e2229.

Lewit-Bentley, A., & Rety, S. (2000). EF-hand calcium-binding proteins. *Current Opinion in Structural Biology, 10*(6), 637–643.

Lin, P., Fischer, T., Weiss, T., & Farquhar, M. G. (2000). Calnuc, an EF-hand Ca(2+) binding protein, specifically interacts with the C-terminal alpha5-helix of G(alpha)i3. *Proceedings of the National Academy of Sciences of the United States of America, 97*(2), 674–679.

Lin, P., Le-Niculescu, H., Hofmeister, R., McCaffery, J. M., Jin, M., Hennemann, H., et al. (1998). The mammalian calcium-binding protein, nucleobindin (CALNUC), is a Golgi resident protein. *The Journal of Cell Biology, 141*, 1515–1527.

Lin, P., Li, F., Zhang, Y. W., Huang, H., Tong, G., Farquhar, M. G., et al. (2007). Calnuc binds to Alzheimer's beta-amyloid precursor protein and affects its biogenesis. *Journal of Neurochemistry, 100*(6), 1505–1514.

Maejima, Y., Sedbazar, U., Suyama, S., Kohno, D., Onaka, T., Takano, E., et al. (2009). Nesfatin-1-regulated oxytocinergic signaling in the paraventricular nucleus causes anorexia through a leptin-independent melanocortin pathway. *Cell Metabolism, 10*, 355–365.

Maruyama, K., Usami, M., Aizawa, T., & Yoshikawa, K. (1991). A novel brain-specific mRNA encoding nuclear protein (necdin) expressed in neurally differentiated embryonal carcinoma cells. *Biochemical and Biophysical Research Communications, 178*(1), 291–296.

Miura, K., Kurosawa, Y., & Kanai, Y. (1994). Calcium-binding activity of nucleobindin mediated by an EF hand moiety. *Biochemical and Biophysical Research Communications, 199*(3), 1388–1393.

Miura, Titani, K., Kurosawa, Y., & Kanai, Y. (1992). Molecular cloning of nucleobindin, a novel DNA-binding protein that contains both a signal peptide and a leucine zipper structure. *Biochemical and Biophysical Research Communications, 187*, 375–380.

Mohan, H., & Unniappan, S. (2013). Phylogenetic aspects of nesfatin-1. *Current Pharmaceutical Design, 19*(39), 6929–6934.

Moncrief, N. D., Kretsinger, R. H., & Goodman, M. (1990). Evolution of EF-hand calcium-modulated proteins. I. Relationships based on amino acid sequences. *Journal of Molecular Evolution, 30*, 522–562.

Nakata, M., Manaka, K., Yamamoto, S., Mori, M., & Yada, T. (2011). Nesfatin-1 enhances glucoseinduced insulin secretion by promoting Ca(2+) influx through L-type channels in mouse islet beta cells. *Endocrine Journal, 58*, 305–313.

Nakayama, S., Moncrief, N. D., & Kretsinger, R. H. (1992). Evolution of EF-hand calcium-modulated proteins. II. Domains of several subfamilies have diverse evolutionary histories. *Journal of Molecular Evolution, 34*, 416–448.

Nesselhut, J., Jurgan, U., Onken, E., Gotz, H., Barnikol, H. U., Hirschfeld, G., et al. (2001). Golgi retention of human protein NEFA is mediated by its N-terminal Leu/Ile-rich region. *FEBS Letters, 509*(3), 469–475.

Niphakis, M. J., Lum, K. M., Cognetta, A. B., 3rd, Correia, B. E., Ichu, T. A., Olucha, J., et al. (2015). A global map of lipid-binding proteins and their ligandability in cells. *Cell, 161*(7), 1668–1680.

O'Banion, M. K., Sadowski, H. B., Winn, V., & Young, D. A. (1991). A serum- and glucocorticoid-regulated 4-kilobase mRNA encodes a cyclooxygenase-related protein. *The Journal of Biological Chemistry, 266*(34), 23261–23267.

Oh-I, S., Shimizu, H., Satoh, T., Okada, S., Adachi, S., Inoue, K., et al. (2006). Identification of nesfatin-1 as a satiety molecule in the hypothalamus. *Nature, 443*, 709–712.

Osaki, A., Shimizu, H., Ishizuka, N., Suzuki, Y., Mori, M., & Inoue, S. (2012). Enhanced expression of nesfatin/nucleobindin-2 in white adipose tissue of ventromedial hypothalamus-lesioned rats. *Neuroscience Letters, 521*, 46–51.

Osei-Hyiaman, D., Sophie-Dreher, L., Nishimura, S., & Encinas, J. (2011). Fasting co-supresses Nesfatin-1 and GPR12 in Mouse Hypothalamic Appetite Center: Implications for energy metabolism. *Endocrine Reviews, 32*, 2–300.

Ozcan, M., Gok, Z. B., Kacar, E., Serhatlioglu, I., & Kelestimur, H. (2016). Nesfatin-1 increases intracellular calcium concentration by protein kinase C activation in cultured rat dorsal root ganglion neurons. *Neuroscience Letters, 619*, 177–181.

Pamuklar, Z. N., Chen, J., Muehlbauer, M., Spagnoli, A., & Torquati, A. (2013). Necdin-E2F4 interaction provides insulin-sensitizing effect after weight loss induced by gastric bypass surgery. *Surgery for Obesity and Related Diseases, 9*(1), 94–99.

Prinz, P., Goebel-Stengel, M., Teuffel, P., Rose, M., Klapp, B. F., & Stengel, A. (2016). Peripheral and central localization of the nesfatin-1 receptor using autoradiography in rats. *Biochemical and Biophysical Research Communications, 470*, 521–527.

Ramanjaneya, M., Chen, J., Brown, J. E., Tripathi, G., Hallschmid, M., Patel, S., et al. (2010). Identification of nesfatin-1 in human and murine adipose tissue: A novel depot-specific adipokine with increased levels in obesity. *Endocrinology, 15*, 3169–3180.

Ramanjaneya, M., Tan, B. K., Rucinski, M., Kawan, M., Hu, J., Kaur, J., et al. (2015). Nesfatin-1 inhibits proliferation and enhances apoptosis of human adrenocortical H295R cells. *Journal of Endocrinology, 226*, 1–11.

Ramesh, N., Gawli, K., Pasupulleti, V. K., & Unniappan, S. (2017). Metabolic and cardiovascular actions of Nesfatin-1: Implications in health and disease. *Current Pharmaceutical Design, 23*(10), 1453–1464.

Ramesh, N., Mohan, H., & Unniappan, S. (2015). Nucleobindin-1 encodes a nesfatin-1-like peptide that stimulates insulin secretion. *General and Comparative Endocrinology, 216*, 182–189.

Schalla, M. A., & Stengel, A. (2018). Current understanding of the role of nesfatin-1. *Journal of the Endocrine Society, 2*(10), 1188–1206.

Senejani, A. G., Gaupale, T. C., Unniappan, S., & Bhargava, S. (2014). Nesfatin-1/nucleobindin-2 like immunoreactivity in the olfactory system, brain and pituitary of frog, Microhyla ornata. *General and Comparative Endocrinology, 202*, 8–14.

Smith, W. L., & Marnett, L. J. (1991). Prostaglandin endoperoxide synthase: Structure and catalysis. *Biochimica et Biophysica Acta, 1083*(1), 1–17.

Somogyi, E., Petersson, U., Sugars, R. V., Hultenby, K., & Wendel, M. (2004). Nucleobindin—A Ca2+-binding protein present in the cells and mineralized tissues of the tooth. *Calcified Tissue International, 74*(4), 366–376.

Stengel, A. (2015). Nesfatin-1—More than a food intake regulatory peptide. *Peptides, 72*, 175–183.

Stengel, A., Goebel, M., Wang, L., Rivier, J., Kobelt, P., Mönnikes, H., et al. (2009). Central nesfatin-1 reduces dark-phase food intake and gastric emptying in rats: Differential role of corticotropin-releasing factor2 receptor. *Endocrinology, 150*, 4911–4919.

Sundarrajan, L., Blanco, A. M., Bertucci, J. I., Ramesh, N., Canosa, L. F., & Unniappan, S. (2016). Nesfatin-1-like peptide encoded in nucleobindin-1 in goldfish is a novel anorexigen modulated by sex steroids, macronutrients and daily rhythm. *Nature Scientific Reports*, *6*, 28377.

Takagi, K., Miki, Y., Tanaka, S., Hashimoto, C., Watanabe, M., Sasano, H., et al. (2016). Nucleobindin 2 (NUCB2) in human endometrial carcinoma: A potent prognostic factor associated with cell proliferation and migration. *Endocrine Journal*, *63*, 287–299.

Tan, B. K., Hallschmid, M., Kern, W., Lehnert, H., & Randeva, H. S. (2011). Decreased cerebrospinal fluid/plasma ratio of the novel satiety molecule, nesfatin-1/NUCB-2, in obese humans: Evidence of nesfatin-1/NUCB-2 resistance and implications for obesity treatment. *The Journal of Clinical Endocrinology and Metabolism*, *96*, E669–E673.

Taniguchi, N., Taniura, H., Niinobe, M., Takayama, C., Tominaga-Yoshino, K., Ogura, A., et al. (2000). The postmitotic growth suppressor necdin interacts with a calcium-binding protein (NEFA) in neuronal cytoplasm. *The Journal of Biological Chemistry*, *275*(41), 31674–31681.

Tsukumo, Y., Tomida, A., Kitahara, O., Nakamura, Y., Asada, S., Mori, K., et al. (2007). Nucleobindin 1 controls the unfolded protein response by inhibiting ATF6 activation. *The Journal of Biological Chemistry*, *282*(40), 29264–29272.

Tsukumo, Y., Tsukahara, S., Saito, S., Tsuruo, T., & Tomida, A. (2009). A novel endoplasmic reticulum export signal: Proline at the +2-position from the signal peptide cleavage site. *The Journal of Biological Chemistry*, *284*(40), 27500–27510.

Tulke, S., Williams, P., Hellysaz, A., Ilegems, E., Wendel, M., & Broberger, C. (2016). Nucleobindin 1 (NUCB1) is a Golgi-resident marker of neurons. *Neuroscience*, *314*, 179–188.

Valencia, C. A., Cotten, S. W., Duan, J., & Liu, R. (2008). Modulation of nucleobindin-1 and nucleobindin-2 by caspases. *FEBS Letters*, *582*, 286–290.

Wang, S. N., Miyauchi, M., Koshikawa, N., Maruyama, K., Kubota, T., Miura, K., et al. (1994). Antigen expression associated with lymph node metastasis in gastric adenocarcinomas. *Pathology International*, *44*(12), 844–849.

Weibert, E., Hofmann, T., & Stengel, A. (2019). Role of nesfatin-1 in anxiety, depression and the response to stress. *Psychoneuroendocrinology*, *100*, 58–66.

Wijesuriya, T. M., De Ceuninck, L., Masschaele, D., Sanderson, M. R., Carias, K. V., Tavernier, J., et al. (2017). The Prader-Willi syndrome proteins MAGEL2 and necdin regulate leptin receptor cell surface abundance through ubiquitination pathways. *Human Molecular Genetics*, *26*(21), 4215–4230.

Williams, P., Tulke, S., Ilegems, E., Berggren, P. O., & Broberger, C. (2014). Expression of nucleobindin 1 (NUCB1) in pancreatic islets and other endocrine tissues. *Cell and Tissue Research*, *358*, 331–342.

Wilson, K. R., Todorovic, A., Proneth, B., & Haskell-Luevano, C. (2006). Overview of endogenous and synthetic melanocortin peptides. Cellular and Molecular Biology (Noisy-le-Grand, France). *52*, 3–20.

Xia, Z. F., Fritze, D. M., Li, J. Y., Chai, B., Zhang, C., Zhang, W., et al. (2012). Nesfatin-1 inhibits gastric acid secretion via a central vagal mechanism in rats. *American Journal of Physiology. Gastrointestinal and Liver Physiology*, *303*, G570–G577.

Xie, W. L., Chipman, J. G., Robertson, D. L., Erikson, R. L., & Simmons, D. L. (1991). Expression of a mitogen-responsive gene encoding prostaglandin synthase is regulated by mRNA splicing. *Proceedings of the National Academy of Sciences of the United States of America*, *88*(7), 2692–2696.

Xu, Y., Zhang, H., Li, Q., Lao, K., & Wang, Y. (2017). The role of nesfatin-1 expression in letrozole-induced polycystic ovaries in the rat. *Gynecological Endocrinology: The Official Journal of the International Society of Gynecological Endocrinology*, *33*, 438–441.

Xue, F., Wu, Y., Zhao, X., Zhao, T., Meng, Y., Zhao, Z., et al. (2016). CHIP mediates down-regulation of nucleobindin-1 in preosteoblast cell line models. *Cellular Signalling*, *28*(8), 1058–1065.

Yang, M., Zhang, Z., Wang, C., Li, K., Li, S., Boden, G., et al. (2012). Nesfatin-1 action in the brain increases insulin sensitivity through Akt/AMPK/TORC2 pathway in diet-induced insulin resistance. *Diabetes*, *61*, 1959–1968.

Ying, J., Zhang, Y., Gong, S., Chang, Z., Zhou, X., Li, H., et al. (2015). Nesfatin-1 suppresses cardiac L-type Ca(2)(+) channels through melanocortin type 4 receptor and the novel protein kinase C theta isoform pathway. *Cellular Physiology and Biochemistry*, *36*, 555–568.

Yosten, G. L., & Samson, W. K. (2010). The anorexigenic and hypertensive effects of nesfatin-1 are reversed by pretreatment with an oxytocin receptor antagonist. *American Journal of Physiology. Regulatory, Integrative and Comparative Physiology*, *298*, R1642–R1647.

Zegers, D., Beckers, S., Mertens, I. L., Van Gaal, L. F., & Van Hul, W. (2011). Association between polymorphisms of the Nesfatin gene, NUCB2, and obesity in men. *Molecular Genetics and Metabolism*, *103*, 282–286.

Zhang, A. Q., Li, X. L., Jiang, C. Y., Lin, L., Shi, R. H., Chen, J. D., et al. (2010). Expression of nesfatin-1/NUCB2 in rodent digestive system. *World Journal of Gastroenterology*, *16*, 1735–1741.

CHAPTER THREE

Estrogen receptor signaling mechanisms

Nathalie Fuentes[a], Patricia Silveyra[a,b,*]
[a]Department of Pediatrics, The Pennsylvania State University College of Medicine, Hershey, PA, United States
[b]The University of North Carolina at Chapel Hill, School of Nursing, Chapel Hill, NC, United States
*Corresponding author: e-mail address: patry@email.unc.edu

Contents

Abstract

The primary female sex hormones, estrogens, are responsible for the control of functions of the female reproductive system, as well as the development of secondary sexual characteristics that appear during puberty and sexual maturity. Estrogens exert their actions by binding to specific receptors, the estrogen receptors (ERs), which in turn activate transcriptional processes and/or signaling events that result in the control of gene expression. These actions can be mediated by direct binding of estrogen receptor complexes to specific sequences in gene promoters (genomic effects), or by mechanisms that do not involve direct binding to DNA (non-genomic effects). Whether acting via direct nuclear effects, indirect non-nuclear actions, or a combination of both, the effects of estrogens on gene expression are controlled by highly regulated complex mechanisms. In this chapter, we summarize the knowledge gained in the past 60years since the discovery of the estrogen receptors on the mechanisms governing

Advances in Protein Chemistry and Structural Biology, Volume 116
ISSN 1876-1623
https://doi.org/10.1016/bs.apcsb.2019.01.001

estrogen-mediated gene expression. We provide an overview of estrogen biosynthesis, and we describe the main mechanisms by which the female sex hormone controls gene transcription in different tissues and cell types. Specifically, we address the molecular events governing regulation of gene expression via the nuclear estrogen receptors (ERα, and ERβ) and the membrane estrogen receptor (GPER1). We also describe mechanisms of cross-talk between signaling cascades activated by both nuclear and membrane estrogen receptors. Finally, we discuss natural compounds that are able to target specific estrogen receptors and their implications for human health and medical therapeutics.

1. Estrogens: Definition and history

The term "estrogens" refers to a group of female hormones, including estrone, estradiol, estriol, and estretrol (Fig. 1). Chemically, estrogens belong to the family of organic compounds known as steroids. As such, their core structure is composed of 17 carbon-carbon bonds arranged as four fused rings (three cyclohexane rings and a cyclopentane ring). All four estrogens contain 18 carbons ($C_{18}H_{24}O_2$) and are collectively known as C18 steroids. They consist of one benzene ring, a phenolic hydroxyl group, and a ketone group (estrone), or one (17β-estradiol), two (estriol), or three (estretrol) hydroxyl groups.

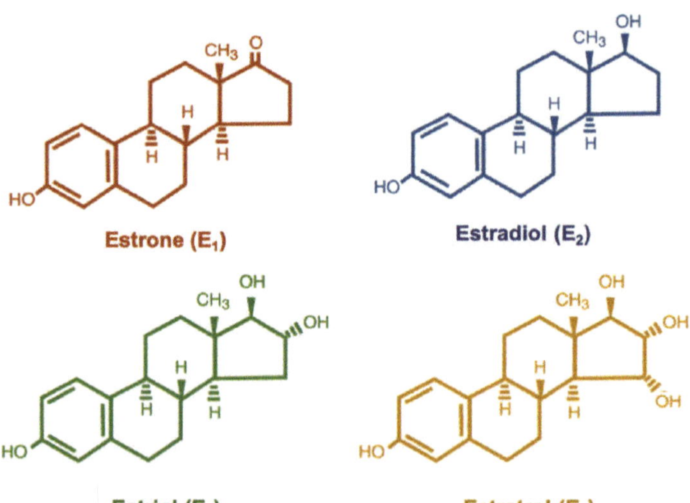

Fig. 1 Chemical structures of endogenous estrogens. Estrone (E1; orange), estradiol (E2; blue), estriol (E3; green) and estretrol (E4; yellow).

Estrogens are primarily synthesized in the ovaries, but also in the adrenal glands and adipose tissue. They were discovered in the early 1900s, when ovarian extracts ("liquor folliculi") from cattle and hogs were injected in rodents, and found to be effective in inducing sexual activity or "estrus" (Allen & Doisy, 1983). It was later determined that the hormone was produced by mature ovarian follicles, and that it was likely common to all female animals. The term estrogen derives from the Greek words *oistros* (frenzy, in heat) and *gennan* (to produce). As mentioned above, estrogens are a group of C18 hormones with similar chemical structures and function (Fig. 1). In addition, all four estrogens are able to bind to both nuclear and membrane estrogen receptors, with different affinity and strength of the response (Watson, Jeng, & Kochukov, 2008). However, the word estrogen is commonly used to refer to estradiol (or 17β-estradiol), due to its physiological relevance and predominance during reproductive years. While females produce all estrogens throughout life, the hormones 16-hydroxyestradiol (estriol) and 15α-hydroxyestriol (estretrol) are predominantly found during pregnancy, and estrone is usually found at higher levels during menopause (Samavat & Kurzer, 2015).

Estradiol, the predominant circulating estrogen in humans, it is mainly secreted by the granulosa cells of the ovarian follicles, and the corpora lutea. On the other hand, estretrol is synthesized exclusively by the fetal liver and reaches maternal circulation through the placenta (Coelingh Bennink, Holinka, Visser, & Coelingh Bennink, 2008; Holinka, Diczfalusy, & Coelingh Bennink, 2008). Estrone, which is produced by aromatization of androstenedione in extraglandular tissues, can be reversibly transformed to estradiol by the enzyme 17β-hydroxysteroid dehydrogenase in peripheral tissues (Bulun, Zeitoun, Sasano, & Simpson, 1999; RYAN, 1959).

2. Estrogen biosynthesis

The main substrate for steroid hormone biosynthesis is dietary cholesterol, specifically low-density lipoprotein (LDL)-cholesterol (Carr, MacDonald, & Simpson, 1982). Through a process called steroidogenesis, cholesterol is converted to the 21-carbon (pregnanes, progestogens), 19-carbon (androstanes), and 18-carbon (estranes) steroid hormones in gonads, adrenal cortex, and adipose tissue (Miller, 2017). The main site of estrogen synthesis is the ovaries, and specifically the granulosa cells (Fig. 2).

The first step in the biosynthesis of steroid hormones is the translocation of cholesterol into the inner mitochondrial membrane, a process regulated

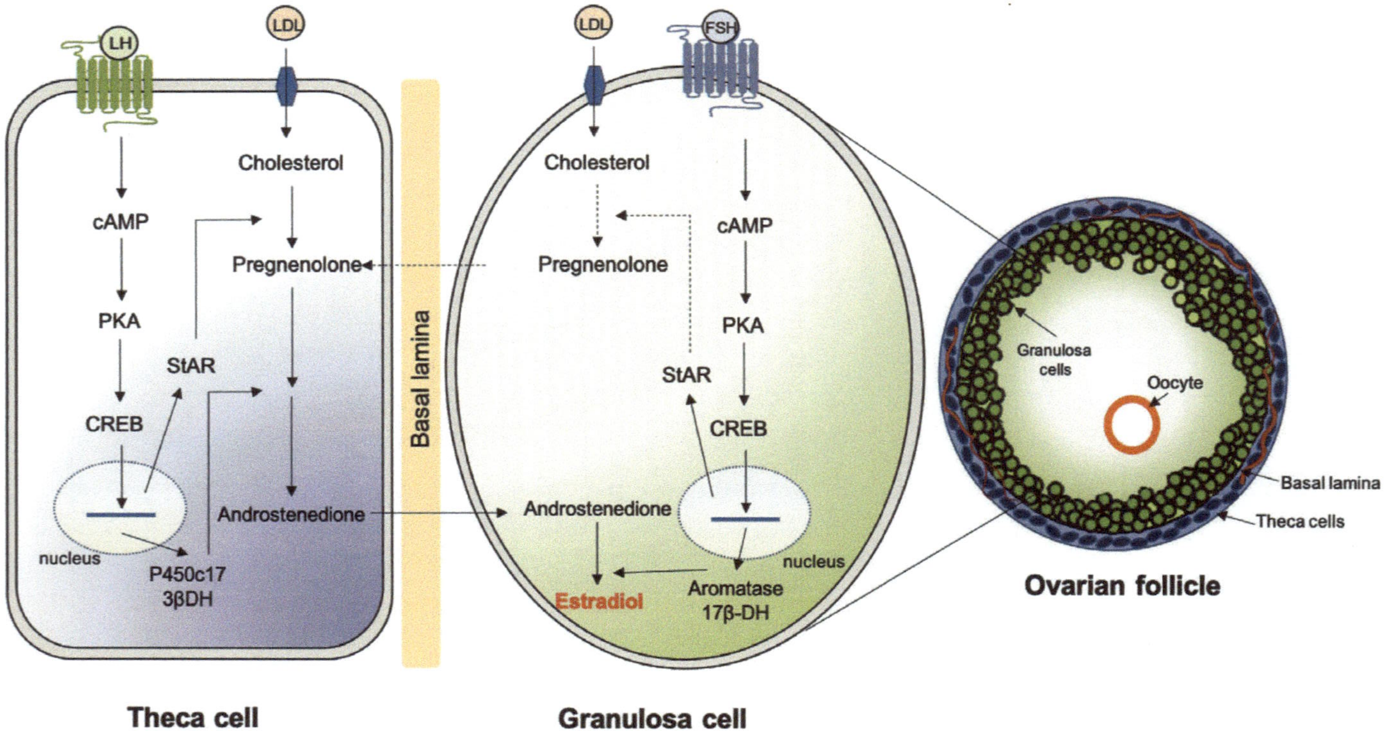

Fig. 2 Association of theca and granulosa cell in estrogen synthesis. The luteinizing hormone (LH) induces the production of androgens in theca cells. The follicle-stimulating hormone (FSH) stimulates granulosa cells via aromatization of androgens to estrogens and by using cholesterol to produce pregnenolone. The process occurs in the ovarian follicle, which is composed of granulosa cells, oocyte, basal lamina and theca cells. CREB, cyclic AMP response element binding protein; PKA, protein kinase A; LDL, low density lipoproteins; cAMP, cyclic adenosine monophosphate; StAR, steroid acute regulatory protein; P450c17, 17α-hydroxylase/lyase; 17βHSD, 17β-hydroxysteroid dehydrogenase.

by the steroidogenic acute regulatory protein STARD1 (also known as StAR), which is believed to act as a shuttle enzyme (Miller & Strauss, 1999). This is the rate-limiting step of steroidogenesis in all tissues. The expression of StAR is controlled by a mechanism involving binding of luteinizing hormone (LH) to its G protein-coupled receptor in the theca cells of the ovary and stimulation of adenylate cyclase, which catalyzes the production of cyclic adenosine monophosphate (cAMP) from adenosine triphosphate (ATP). The cAMP produced activates protein kinase A, which catalyzes phosphorylation of cAMP response element binding protein (CREB) leading to activation of transcription of StAR and other factors associated with steroid hormone production (Fig. 2). At the inner mitochondrial membrane, cholesterol is converted to pregnenolone by the enzyme P450scc, or cholesterol side-chain cleavage enzyme, encoded by the CYP11A1 gene (Belfiore, Hawkins, Wiltbank, & Niswender, 1994). Pregnenolone then acts as a precursor for all steroid hormones (Fig. 3), and can diffuse between adjacent granulosa and theca cells of the ovary. The synthesis continues with the conversion of pregnenolone to androstenedione by the enzymes CYP17A1 (steroid 17-α-hydroxylase/17,20-lyase) and 3β-HSD (3β-hydroxysteroid dehydrogenase/Δ^{5-4} isomerase), via dehydroepiandrosterone (DHEA). Androstenedione can be either converted

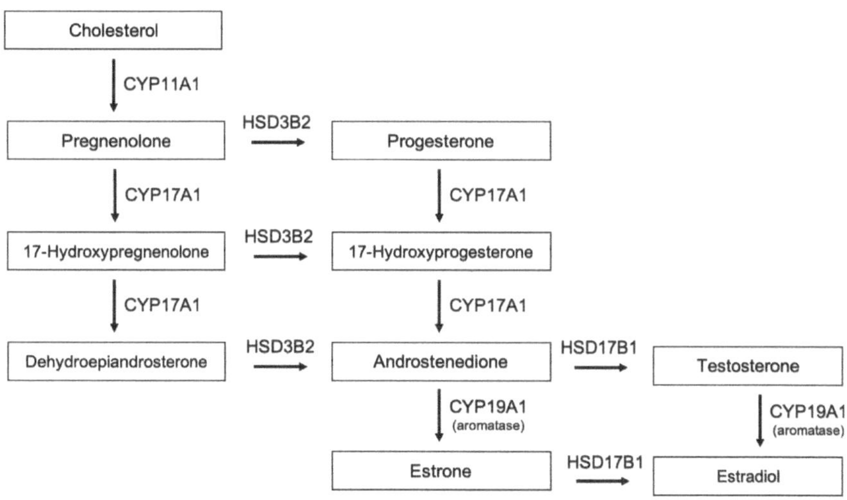

Fig. 3 Estrogen biosynthesis pathway. The estrogen biosynthetic pathway involves the conversion of cholesterol to progestogens, androgens and finally estrogens. The conversion of androgen to estrone (E1) and estradiol (E2) catalyzed by aromatase is the final step for synthesis of estrogen.

to other androgens, such as testosterone and dihydrotestosterone, or diffuse to the granulosa cells through the basal lamina (Fig. 2). At the granulosa cells, androstenedione is converted to estrone by the enzyme CYP19A1 (also known as aromatase). Estrone is then converted to estradiol by the enzyme 17β-HSD (17β-hydroxysteroid dehydrogenase). In the granulosa cells, the expression of both aromatase and 17β-HSD is controlled by follicle stimulating hormone (FSH) stimulation. Interestingly, testosterone can be metabolized to estradiol and estrone by the action of aromatase in peripheral tissues, including adipose cells and bone (Simpson et al., 2002). Males also produce local estrogen by aromatization in cells of the reproductive tract, including Sertoli cells, Leydig cells, and mature spermatocytes. Overall, estrogens are normally produced by the ovaries and in smaller amounts by other tissues such as the liver, pancreas, adrenal glands, adipose tissue, and breast (Barakat, Oakley, Kim, Jin, & Ko, 2016). In specific physiological conditions, such as pregnancy, estrogen is also synthesized by the placenta. However, the biosynthesis of estrogen in non-gonadal sites follows rather unusual mechanisms, since these tissues are not able to generate C19 steroids from cholesterol. In these tissues, estrogen production is largely dependent on C19 steroids transported from other tissues and conversion by local CYP19A1 aromatase (Labrie et al., 1998; Nelson & Bulun, 2001).

Estradiol, the predominant circulating estrogen in humans, it is mainly secreted by the granulosa cells of the ovarian follicles, and the corpora lutea, by the mechanisms indicated above. On the other hand, estretrol is synthesized exclusively during pregnancy by the fetal liver and reaches maternal circulation through the placenta (Coelingh Bennink et al., 2008; Holinka et al., 2008). Estriol, which is also primarily synthesized during pregnancy, is almost exclusively produced by the placenta. To produce estriol, dietary cholesterol is converted to pregnenolone and progesterone in the placenta, and these steroids are further metabolized to DHEA and DHEA-sulfate (DHEA-S) in the fetal adrenal glands. DHEA-S is later hydroxylated to 16α-OH-DHEA-S in the fetal liver by the action of the CYP3A7 enzyme, and transported back to the placenta where it is converted to 16α-OH-DHEA by the steroid sulfatase. The enzyme 3β-HSD1 converts 16α-OH-DHEA into 16α-OH-androstenedione, which is later aromatized to 16α-OH-estrone. In the final step, 16α-OH-estrone is converted to estriol by the 17β-HSD enzyme, and secreted into maternal circulation (ITTRICH & NEUMANN, 1963; WILSON, ERIKSSON, & DICZFALUSY, 1964). In non-pregnant women, estriol is produced mainly in the liver by 16α-hydroxylation of estradiol and estrone by CYP enzymes (Samavat & Kurzer, 2015; Tsuchiya, Nakajima, & Yokoi, 2005). Finally, estrone is mainly produced during menopause by aromatization of androstenedione in extra-glandular tissues, where it can act locally as a paracrine or intracrine factor (Simpson, 2003). Estrone can also be transformed to estradiol by the enzyme 17β-hydroxysteroid dehydrogenase in peripheral tissues, including adipose and breast tissue, vascular endothelium, smooth muscle cells, brain tissue, and bone cells, where it is metabolized or enters the circulation in small quantities (Bulun et al., 1999; RYAN, 1959; Simpson, 2003).

3. Estrogen metabolism

Physiologically, the metabolic conversion of estrogens allows their excretion from the body via urine, feces, and/or bile, along with the production of estrogen analogs, which have been shown to present antiproliferative effects (Tsuchiya et al., 2005). In target cells, there are different pathways capable of metabolizing estradiol and estrone. Members of the cytochrome P450 superfamily of enzymes (CYP1A1, CYP1B1, and CYP1A2) catalyze hydroxylation of estrone and estradiol at positions C2, C4 and C16. Due to the high expression of these enzymes in the liver, a large proportion of estrogen metabolism occurs in this tissue, although CYP1B1 is also expressed in target tissues such as mammary gland, uterus, kidney, brain, and pituitary gland, where estradiol and estrone can also be metabolized. Estradiol hydroxylation is followed by conversions to 2-hydroxyestrone, 4-hydroxyestrone, 2-hydroxyestradiol, 4-hydroxyestradiol, and 16α-hydroxyestrone, which are also known as catechol estrogens, due to their presence of pharmacological properties of both catecholamines and estrogens. The hydroxylation of estradiol or 16α-hydroxyestrone forms estriol. In addition, catechol estrogens can be methylated via the catechol-O-methyltransferase (COMT) enzyme to methoxy estrogens (Samavat & Kurzer, 2015). These compounds have gained significant attention due to their little estrogenic effects, antiproliferative properties, and ability to control estrogen synthesis (Purohit & Reed, 2002; Purohit et al., 2006). Moreover, catechol estrogens can also be conjugated by estrogen sulfotransferases and UDP-glucuronyltransferases (Cheng et al., 1998; Garbacz, Jiang, & Xie, 2017). In a conjugation reaction, hormones become water soluble and excreted from the body (Lakhani, Venitz, Figg, & Sparreboom, 2003).

4. Physiological functions of estrogens

Estrogens are sex steroid hormones, and as such display a broad spectrum of physiological functions. These include regulation of the menstrual cycle and reproduction, bone density, brain function, cholesterol mobilization, development of breast tissue and sexual organs, and control of inflammation (Liang & Shang, 2013). While estrogens play diverse roles in normal male and female physiology, in certain physiological situations they can play similar roles in both sexes (Simpson et al., 2005). In females, estrogens are responsible for primary and secondary sexual characteristics. Estradiol promotes epithelial cell proliferation in the uterine endometrium and mammary glands starting in puberty (Gruber, Tschugguel, Schneeberger, & Huber, 2002; Koos, 2011; Simpson et al., 2005). During pregnancy, estrogens produced by the placenta help prepare the mammary gland for milk production (Voogt, 1978). On the other hand, lower levels of estrogens produced in men are essential for functions including sperm maturation, erectile function and maintenance of a healthy libido (Schulster, Bernie, & Ramasamy, 2016). It is important to mention here that all the estrogenic physiological functions previously described are mediated by estrogen receptors, which we describe in the next sections.

5. The estrogen receptors: History and discovery

In 1958, Elwood Jensen discovered the estrogen receptor, the first receptor ever encountered for any hormone, by showing that reproductive female tissues were able to uptake estrogen from the circulation by binding to proteins. He later demonstrated that estrogen-bound receptors were able to migrate to the nucleus, where they could stimulate gene transcription (Jensen et al., 1967; Jensen et al., 1968). More than 20 years later, the first human estrogen receptor (known today as ERα) was cloned using RNA from the human breast cancer cell line MCF-7 (Green et al., 1986; Greene et al., 1986). Similarly, the second estrogen receptor (known today as ERβ) was described 10 years later by the research team lead by Dr. Jan-Ake Gustafsson (Kuiper, Enmark, Pelto-Huikko, Nilsson, & Gustafsson, 1996). Gustafsson's lab discovered that a newly identified protein that was mainly expressed in the secretory epithelial cells of the prostate and in the granulosa cells of the ovary, shared a high degree of homology with the ERα (DNA-binding domain, 95%; ligand-binding domain, 55%). As a result of these similarities, the team suggested for the protein be named ERβ.

More recently, a new type of estrogen binding protein was discovered in target cells: The G Protein-Coupled Estrogen Receptor GPER1, or membrane estrogen receptor. Unlike the nuclear estrogen receptors ERα and ERβ, which were isolated by traditional biochemical approaches, GPER1 was identified by molecular cloning methods (Filardo & Thomas, 2012). Almost two decades ago, several research laboratories had reported the isolation of a G Protein-Coupled Receptor homolog, which was ascribed the orphan term GPR30 (Carmeci, Thompson, Ring, Francke, & Weigel, 1997; Feng & Gregor, 1997; Kvingedal & Smeland, 1997; O'Dowd et al., 1998; Owman, Blay, Nilsson, & Lolait, 1996; Takada, Kato, Kondo, Korenaga, & Ando, 1997). It was assumed that the ligand for GPR30 was a hormone or chemotactic peptide due to its structural similarities to the receptors for angiotensin II and other peptides such as such as interleukin-8, monocyte chemotactic proteins, and complement factors (Filardo & Thomas, 2012). However, after screening of multiple chemotactic peptides and factors, no molecules with binding affinities to GPR30 were found, the receptor continued to be classified as orphan (Feng & Gregor, 1997). However, in the year 2000, a research team was able to show that fast estrogen-mediated activation of extracellular signal-regulated kinases (ERKs) was dependent on GPR30 (Filardo, Quinn, Bland, & Frackelton, 2000). Five years later, this and other groups were able to demonstrate direct binding of 17β-estradiol to GPR30 in GPR30-transfected cells and breast cancer cell lines (Revankar, Cimino, Sklar, Arterburn, & Prossnitz, 2005; Thomas, Pang, Filardo, & Dong, 2005). Finally, in 2007 GPR30 was officially named G protein-coupled estrogen receptor 1 (also known as GPER or GPER1), and its role in mediating fast responses to estrogens and overall physiological and pathological processes has been studied extensively in human and animal models (Boonyaratanakornkit & Edwards, 2007; Filardo et al., 2007; Molina, Figueroa, Bhoola, & Ehrenfeld, 2017; Prossnitz & Barton, 2014; Sharma & Prossnitz, 2016).

6. Structural properties of estrogen receptors

The full-length size of ERα is 595 amino acids and 67 kDa. ERβ is 530 amino acids in length and 59 kDa. The main difference between the two proteins is that ERβ has a shorter amino terminal domain than ERα (Fig. 4).

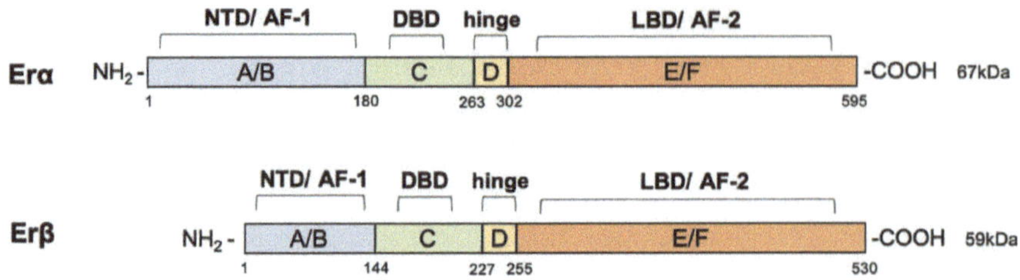

Fig. 4 Structural organization of estrogen receptors. Structural domains of estrogen receptor α (ERα) (595 aa) and ERβ (530 aa) are labeled A-F. Both receptors have six different structural and functional domains: N- terminal (NTD, A/B domains, AF-1), DNA binding domain (DBD, C domain), the hinge (D domain), the C-terminal region containing the ligand binding domain (LBD, E/F domain, AF-2).

As members of the nuclear hormone receptors superfamily of transcription regulators, the structures of the estrogen receptors ERα and ERβ are composed of various functional domains and have several structural regions in common (Schwabe & Teichmann, 2004). The principal functional domains are termed A/B, C, D, and E/F, and are present in both receptor full-length structures (Fig. 4). The A/B region represents the amino-terminal domain (NTD), which is involved in gene transcription transactivation, and contains a zinc-finger that mediates binding to target sequences. The C region corresponds to the DNA binding domain (DBD), which contributes to estrogen receptor dimerization and binding to specific sequences in the chromatin. These canonical sequences known collectively as estrogen response elements (ERE) (Scheidereit et al., 1986; Truss & Beato, 1993). The D domain is a hinge region that connects the C and E domains, and is able to bind to chaperone proteins. This region also contains the nuclear localization signal, that is unmasked upon estrogen binding, allowing for the receptor-ligand complexes to translocate to the nucleus. In the carboxy-terminal E/F region, also known as the ligand binding domain, contains the estrogen binding area, along with binding sites for coactivators and corepressors. Finally, two additional regulators of the estrogen receptor transcriptional activity known as activation function (AF) domains AF1 and AF2, are located within the NTD and DBD, respectively (Kumar et al., 2011). The mechanisms of transcriptional regulation mediated by these receptors appear to involve a synergistic effect of AF1 and AF2 (Tora et al., 1989). Contrarily to AF2, AF1 does not require binding to hormones or steroids to be activated (Kumar et al., 2011).

In humans, the ERα is encoded by the gene ESR1, located on chromosome 6, locus 6q25.1 (Gosden, Middleton, & Rout, 1986). In addition to

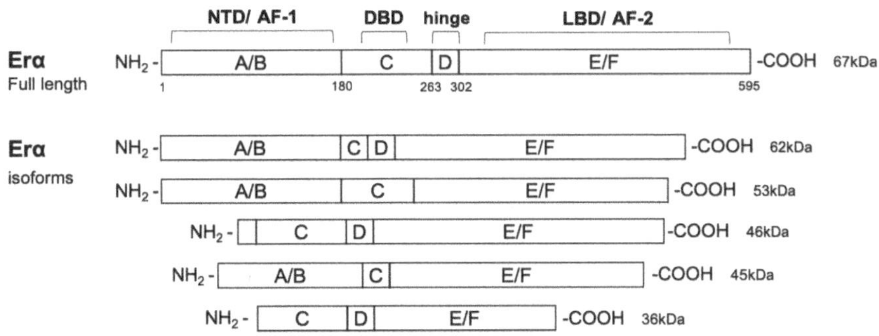

Fig. 5 Estrogen receptor alpha (ERα) isoforms. The domain organization of the full-length 595 amino acid ERα (67 kDa), and truncated shorter isoforms (62 kDa, 53 kDa, 46 kDa, 45 kDa, and 36 kDa) resulting from alternative splicing and/or alternate translation start sites are illustrated. Protein domains are labeled as A to F with numbering denoting amino acid sequence number based on the full-length protein (595 aa). ERα domains: N-terminal (NTD, A/B domains, AF-1), DNA binding domain (DBD, C domain), hinge (D) domain, and C-terminal region containing the ligand binding domain (LBD, E/F domain, AF-2).

the full-length ERα isoform (66 kDa), several shorter isoforms (36 kDa, 46 kDa) have been identified as a result of the presence of alternate start codons, or as productos of alternative splicing (Fig. 5). Some of these shorter isoforms do not have the NTD and thus lack the AF-1 domain. Therefore, they cannot activate transcription. Instead, they are able to form heterodimers with the full-length ERα and inhibit its ability to control transcriptional. The shorter isoform, ERα-36, lacks both AF-1 and AF-2 transcriptional activation domains, and it has been shown to exert membrane-initiated signaling events upon binding to estradiol, estriol, and estretrol (Gu et al., 2014), as well as to medicate GPER1 responses (Arnal et al., 2017; Romano & Gorelick, 2018).

On the other hand, ERβ is encoded by the ESR2 gene located in chromosome 14 (14q23–24), and has five known isoforms (Enmark et al., 1997) (Fig. 6). The main difference between the full-length ERβ and the shorter ERβ isoforms is on the C-terminal LBD. Therefore, ERβ isoforms that have no transcriptional activity can also suppress ERα signaling by dimerizing with ERα (Vrtačnik, Ostanek, Mencej-Bedrač, & Marc, 2014).

Finally, the gene coding for the membrane receptor GPER1 is located in chromosome 7 (locus 7p22.3). In terms of structure, GPER1 does not share similarities with ERα or ERβ. As a typical G protein coupled receptor, its structure consists of 7 transmembrane α-helical regions, 4 extracellular segments, and 4 cytosolic segments (Barton et al., 2018). This receptor

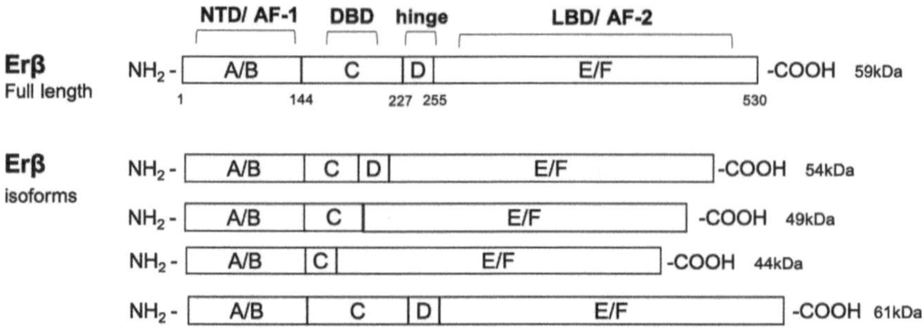

Fig. 6 Estrogen receptor beta (ERβ) isoforms. The domain organization of the full-length 530 amino acid ERβ (59 kDa), truncated shorter isoforms (54 kDa, 49 kDa, and 44 kDa), and elongated isoform (61 kDa), resulting from alternative splicing and/or alternate translation start sites are illustrated. Protein domains are labeled as A to F with numbering denoting amino acid sequence number based on the full-length protein (595 aa). ERβ domains: N-terminal (NTD, A/B domains, AF-1), DNA binding domain (DBD, C domain), hinge (D) domain, and C-terminal region containing the ligand binding domain (LBD, E/F domain, AF-2).

has low binding affinity (17B-estradiol) when compared to other estrogen receptors (Prossnitz & Barton, 2014). However, this may be important as GPER1 is accountable for rapid responses to estrogen, and activation of intracellular signaling cascades mediated by second messengers (Filardo & Thomas, 2012).

7. Mechanims of estrogen receptor signaling

As a steroid hormone, estrogen can enter the plasma membrane and interact with intracellular ERα and ERβ to exert direct effects by binding to DNA sequences. Alternatively, estrogen can activate intracellular signaling cascades via interaction with the GPER1 and/or ERα and ERβ. Due to differences in the cellular and molecular events leading to gene expression regulation in which estrogen-receptor complexes can either bind directly or indirectly to DNA, estrogen-mediated signaling events ca be divided into genomic and non-genomic. Genomic effects are those involving migration of the estrogen-receptor complexes to the cell nucleus, and direct interaction with chromatin at specific DNA sequences known as estrogen response elements (EREs). While EREs have been identified in several gene promoters and regulatory regions, it has been reported than more than one third of human genes regulated by estrogen receptors do not contain ERE sequence elements (O'Lone, Frith, Karlsson, & Hansen, 2004). On the other hand, non-genomic effects involve indirect regulation of gene expression through a variety of intracellular signaling events. The known mechanisms for genomic and non-genomic control of gene expression by estrogens are described below.

8. Nuclear estrogen receptors: Direct genomic signaling

Direct genomic signaling is known as the classical mechanism of estrogen signaling. In this process, the nuclear estrogen receptors ERα and ERβ act as ligand-activated transcription factors (Marino, Galluzzo, & Ascenzi, 2006; O'Malley, 2005). Upon binding of estradiol to ERα or ERβ in the cytoplasm, a conformational change occurs inducing receptor dimerization (Le Dily & Beato, 2018(Fig. 7). This complex is then translocated to the nucleus, where it binds to the chromatin at ERE sequences, enhancer regions within or close to promoters, and/or 3′-untranlated regions of target genes (Klinge, 2001).

Recent advances in computational biology have facilitated the identification of EREs in many gene promoters, and allowed prediction of genes regulated by estrogen and other hormones in the genomes of many species (Bajic et al., 2003; Bourdeau et al., 2004). A recent genome-wide screening study identified over 70,000 EREs in the human and mouse genomes (Bourdeau et al., 2004). Interestingly, 17,000 of these EREs were located near mRNA transcriptional start sites, and only 660 were conserved sites. The efficacy of this computational approach was further supported by functional validation of estrogen receptor interaction sites (Carroll & Brown, 2006). While these elements share a high degree of sequence similarity, it is important to recognize that the intrinsic sequence composition of the EREs can alter the affinity of the receptor to bind DNA. For example, ERα has a high binding affinity for the canonical ERE sequence located within the vitellogenin A2 gene, but with less affinity for the EREs located in the oxytocin gene (Sausville, Carney, & Battey, 1985). This moderately explains why differences in ERE sequences, such as those resulting from inter-individual gene variability or mutations, can affect the activation of gene expression (Loven, Wood, & Nardulli, 2001; Yi et al., 2002). In addition, specific ERE sequences can cause allosteric changes in the receptor's structure, and thus alter the ability of the complex to recruit coactivators and transcription factors that may contribute to ER biological activity (Hall, McDonnell, & Korach, 2002; Yaşar, Ayaz, User, Güpür, & Muyan, 2017).

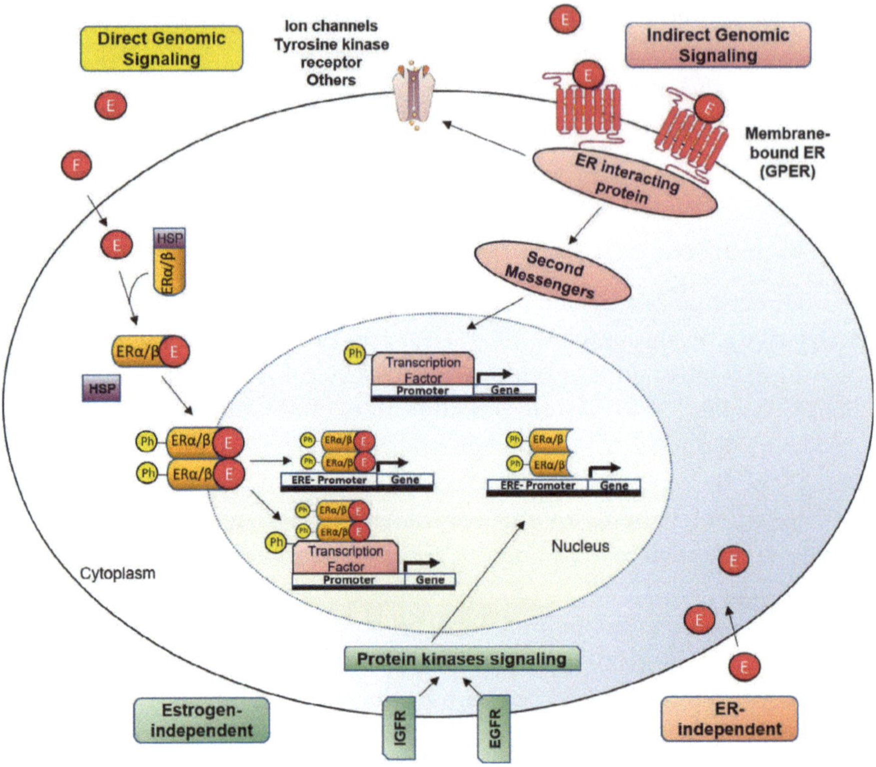

Fig. 7 Genomic and non-genomic estrogen signaling pathways. There are different estrogen-mediated signaling mechanisms. (1) Direct genomic signaling: estrogen binds to ERs. The complex dimerizes and translocate to the nucleus inducing transcriptional changes in estrogen-responsive genes with or without EREs. (2) Indirect genomic signaling: the membrane bound receptor induces cytoplasmic events such as modulation of membrane-based ion channels, second-messenger cascades and transcription factors. (3) ER-independent: estrogen exerts antioxidant effects in an ER-independent manner. (4) Estrogen independent: ligand-independent genomic events.

9. Nuclear estrogen receptors: Indirect genomic signaling

As mentioned earlier, the transcription of several genes that do not contain EREs in their promoter regions can also be regulated by estradiol, without direct binding of the estrogen receptors to the DNA. According to the most recent reports, an estimated 35% of genes targeted by estrogen lack ERE-like sequences (Marino et al., 2006; Vrtačnik et al., 2014). In these, the mechanisms by which estrogen affects gene expression are collectively known as "indirect genomic signaling" or "transcriptional cross-talk", and are based on activation of gene expression by estrogen receptors not binding DNA directly. Rather, the estrogen receptor complexes act through protein-protein interactions with other transcription factors and response elements (Aranda & Pascual, 2001; Göttlicher, Heck, & Herrlich, 1998). In this way, estrogens indirect signaling influences activation or suppression of target gene expression.

An important mediator of indirect genomic signaling is the stimulating protein-1 (Sp-1). Binding of this transcription factor to promoter regions at GC-rich sites is enhanced by the presence of estrogen receptors (Bajic et al., 2003; O'Lone et al., 2004). Examples of genes induced by estrogen via the Sp-1 mechanism are: low-density lipoprotein (LDL) receptor (Li, Briggs, Ahlborn, Kraemer, & Liu, 2001), progesterone receptor B (O'Lone et al., 2004), endothelial nitric oxide synthase (eNOS) (Chambliss & Shaul, 2002), GATA binding protein 1 (GATA1), signal transducer and activator of transcription 5 (STAT5) (Björnström & Sjöberg, 2005), and the retinoic acid receptor-1α genes (Sun, Porter, & Safe, 1998). A few studies have shown that ERα can also interact with the c-rel subunit of the nuclear factor-κB (NF-κB) complex, preventing NF-κB from binding to cytokine genes promoters (Galien & Garcia, 1997; Kalaitzidis & Gilmore, 2005). Moreover, ERα can also interact with other transcriptional modulators such as the activating transcription factor (ATF)-2, c-jun, the ATF-1/cAMP response element binding protein (ATF-1/CREB), and the nuclear transcription factor-Y (NF-Y) (O'Lone et al., 2004).

The nuclear estrogen receptors also induce the expression of genes containing the activator protein-1 (AP-1) sites though protein-protein interactions (Gaub, Bellard, Scheuer, Chambon, & Sassone-Corsi, 1990). AP-1 is a transcription factor that regulates key cellular processes such as cell differentiation, proliferation, and apoptosis. The structure of AP-1 consists of a heterodimer composed of proteins belonging to the c-Fos, c-Jun, ATF, and the Jun dimerization partners (JDP) families (Piu, Aronheim, Katz, & Karin, 2001). The ERα also interacts with c-Fos and c-Jun at these binding regions (O'Lone et al., 2004). Some examples of genes induced by ERα via the AP-1 mechanism are insulin-like growth factor-1 (IGF1), collagenase, IGF1-receptor, ovalbumin, and cyclin D1 (Fujimoto, Honda, & Kitamura, 2004; Marino, Acconcia, Bresciani, Weisz, & Trentalance, 2002). However, previous studies have shown that ERα and ERβ signal in different ways depending on the ligand and response elements present at the AP-1 sites. In fact, 17β-estradiol activates AP-1-dependent transcription via ERα, whereas ERβ inhibits this mechanism (Paech et al., 1997). Likewise, 17β-estradiol binding to ERα induces transcription when linked to Sp-1 in GC-rich regions, but not when 17β-estradiol is bound to ERβ. One example of this is the contrasting action of ERα and ERβ on the control of cyclin D1 gene expression (Liu et al., 2002), where estrogen-bound ERβ suppresses cyclin D1 expression (Marino et al., 2006) and blocks ERα-mediated production when both receptors are present (Acconcia et al., 2005; Matthews & Gustafsson, 2003). The diversity of mechanisms of transcriptional regulation in different cells by the two estrogen receptors and their interactions with local transcription factors may explain the differences observed in tissue specific biologic responses to estrogens.

10. Membrane receptor: Indirect non-genomic signaling

As mentioned above, not all estrogen responses fit the classical genomic model of steroid action. The observation of excessively fast estrogen-induced biological responses led to the development of the hypothesis that estrogen could be acting by mechanisms not involving direct target gene transcription and protein synthesis, and the subsequent discovery of the GPER1 (Prossnitz & Barton, 2011). Non-genomic actions of estrogen often involve activation of signal-transduction mechanisms with the subsequent production of intracellular second messengers, cAMP regulation and protein-kinase activation of signaling cascades that result in indirect changes in gene expression (Lösel & Wehling, 2003) (Fig. 7). The protein-kinase cascades can be classified into four major ones: (1) the phospholipase C (PLC)/protein kinase C (PKCs) pathway (Marino, Pallottini, & Trentalance, 1998), (2) the Ras/Raf/MAPK cascade (Dos Santos et al., 2002; Watters, Campbell, Cunningham, Krebs, & Dorsa, 1997), (3) the phosphatidyl inositol 3 kinase (PI3K)/Akt kinase cascade (Marino, Acconcia, & Trentalance, 2003), and (4) the cAMP/protein kinase A (PKA) signaling pathway (Gu & Moss, 1996; Picotto, Massheimer, & Boland, 1996). Additionally, GPER1 binding to estrogens promotes estrogen-dependent activation of adenylyl cyclase and epidermal growth factor receptor (EGFR). Subsequent phosphorylation of transcription factors by the protein kinases mentioned above can alter their function and ability to bind to genomic sequences to affect gene expression. Examples of transcription factors that are affected by these signaling mechanisms include: Elk-1, CREB, CCAAT-enhancer-binding protein beta (C/EBPβ), the NF-κB complex, and the signal transducer and activator of transcription (STAT) family (Cavalcanti, Lucas, Lazari, & Porto, 2015;

Fox, Andrade, & Shupnik, 2009; Furth, 2014; Kousteni et al., 2003; Laliotis et al., 2013; Ozes et al., 1999; Romashkova & Makarov, 1999). Thus, by activating these non-genomic to genomic mechanisms, the estrogen receptors ERα and ERβ indirectly regulate gene transcription at alternative DNA response elements, in addition to the abovementioned genomic effects involving direct binding to EREs (Fig. 7). Another interesting fact is that both ERα and ERβ are also targets for phosphorylation by protein kinases including MAPKs, indicating that non-genomic actions of estrogens may also involve self-regulation of receptor expression (de Leeuw, Neefjes, & Michalides, 2011; Kato et al., 1995).

Both the membrane bound estrogen receptor GPER1, and some variants of the ERα and ERβ have been associated to non-genomic estrogen signaling (Barton et al., 2018; Filardo & Thomas, 2012). It has been suggested that non-genomic actions of the ERα and ERβ could be mediated through a sub-population of receptors that located at the cell membrane and can activate intracellular signaling cascades (Razandi, Pedram, Merchenthaler, Greene, & Levin, 2004). At the cell membrane, the ERα and ERβ can interact with scaffold proteins such as caveolin-1 and MNAR/PELP-1 (modulator of non-genomic activity of estrogen receptor) (Chambliss et al., 2000; Cheskis et al., 2008; Shaul & Anderson, 1998). By proximity, the ERα and ERβ also interact with G proteins, various membrane receptors (e.g., tyrosine kinase, insulin growth factor 1, and epidermal growth factor receptors), and signaling molecules including ras, Src and PI3 kinases, ErbB2 (HER-2/neu) and Shc that are located at or near the membrane (Boonyaratanakornkit, 2011; Li et al., 2007; Migliaccio et al., 1996; Song et al., 2010; Song, Zhang, Chen, Bao, & Santen, 2007; Song, Zhang, & Santen, 2005). Interactions with these molecules promotes intracellular activation of mitogen activated protein kinases (MAPK) and protein kinase B (Akt) signaling pathways that can affect transcriptional regulation (Li et al., 2010). While there is no clear consensus among the experts in the field about binding of ERα and ERβ to the plasma membrane, it appears that the mechanisms described above are cell-type specific and activated under certain physiological events, and by specific receptor variants (Li, Haynes, & Bender, 2003).

11. Genomic and non-genomic signaling crosstalk

As exemplified in the previous sections, it is evident that the mechanisms of action of estrogen in the various cell targets represent a combination of complex multifactorial processes. Besides the independent genomic and non-genomic pathways described above, many authors have proposed the existence of additional convergent pathways involving both genomic and non-genomic factors that result in regulation of gene transcription (Björnström & Sjöberg, 2005; Silva, Kabil, & Kortenkamp, 2010; Vrtačnik et al., 2014). Two mechanisms of "cross-talk" have been described, and involve protein-protein interactions of components of both pathways. In one mechanism, estrogen-bound nuclear estrogen receptor complexes are dimerized and translocated to the nucleus, where they bind to phosphorylated transcription factors resulting from GPER1-mediated signaling. The complexes then bind to either ERE sequences via the nuclear estrogen receptors, or to AP-1, STATs, ATF-2/c-Jun, Sp1, and/or NF-κB cognate DNA binding sites (Björnström & Sjöberg, 2005). In the second mechanism, interaction of GPER1 and ERα and ERβ located at the plasma membrane activate protein kinase cascades that result in phosphorylation of AP-1, STATs, Elk-1, CREB, and NF-κB, and other transcription factors, as well as estrogen receptors themselves, that can then interact with DNA sequences to regulate transcription (Björnström & Sjöberg, 2005). Thus, convergence of the two classical estrogen receptor regulation pathways can result in enhanced transcriptional activity in specific tissues and physiological processes.

12. Estrogen receptor ligand independent signaling

An interesting phenomenon observed in many cells is that estrogen receptors can actually be activated in the absence of estrogens or other receptor agonists (Bennesch & Picard, 2015; Maggi, 2011; Vrtačnik et al., 2014). This ligand-independent estrogen receptor activation is mainly triggered by phosphorylation on specific residues (e.g., serine and tyrosine) in the receptors themselves, or their association with coregulators (described below). This independent mechanism requires the action of regulatory molecules necessary for phosphorylation, such as protein kinase A (PKA), protein kinase C (PKC), MAPK phosphorylation cascade components, as well as inflammatory cytokines (e.g., interleukin-2), cell adhesion molecules (e.g., heregulin), cell cycle regulators (e.g., RAS p21 protein activator cyclins A and D1), and peptide growth factors including EGF, insulin, IGF1, and transforming growth factor beta (TGFβ) (Nilsson et al., 2001).

13. Estrogen receptor coregulators and transcriptional control

In addition to the regulatory pathways described above, the cell also expresses a battery of coregulators that can either enhance or decrease transcriptional activity of steroid hormone receptors. These are called estrogen receptor coactivators and corepressors, respectively. Coregulators are involved in many steps of the gene expression process, including chromatin modification and remodeling, transcription initiation, elongation of RNA chains, mRNA splicing, mRNA translation, miRNA processing, and degradation of the activated NR-coregulator complexes (Lonard & O'malley, 2007). Currently, there are hundreds of coregulators of nuclear receptors described that play a key role in promoting gene expression and transcriptional activity. Coregulators are a dynamic group of proteins able to act as integrators of signals from steroid hormones, and have been linked to many diseases affected by sex hormones, such as cancer (Lonard & O'Malley, 2006). One of the first coregulators of ERα, known as steroid receptor coactivator (SRC-1), was identified in 1995 (Oñate, Tsai, Tsai, & O'Malley, 1995). Since then, many additional coregulators have been discovered for ERα, although very few are known for ERβ (Lonard & O'Malley, 2006). Coregulators for ERα comprise members of the steroid receptor coactivator (SRC)/p160 group, the histone acetyltransferase cAMP responsive element binding protein (CREB)-binding protein (CBP)/p300, ATP-dependent chromatin remodeling complexes like SWI/SNF, E3 ubiquitin-protein ligases, and steroid RNA activator (SRA) (Lonard & O'Malley, 2006; Manavathi, Samanthapudi, & Gajulapalli, 2014). Therefore, as indicated above, even though both nuclear estrogen receptors are able to use estradiol as their physiological ligand, they exert multiple effects and functions in different cells and tissues that are mediated by several intermediaries and differential utilization of coregulators (Manavathi et al., 2014).

The mechanisms by which coregulators control the actions of estrogen receptors are still a topic of ongoing research. From studies in cancer cells, we have learned that a large group of coregulators have specific structural motifs that than affect their contact with ER ligand-binding domains (Heery, Kalkhoven, Hoare, & Parker, 1997). The specific motifs are called NR boxes or LXXLL motifs (X, any amino acid; L, leucine). On the other hand, we know that corepressors block ER-mediated gene transcription via

(1) direct interaction with unbound estrogen receptors; (2) using their corepressor nuclear receptor box; (3) competing with coactivators (Hu & Lazar, 1999). It has also been reported that the concentration of several coregulators depends on estrogen induced-transcriptional regulation via the estrogen receptors (Mishra, Balasenthil, Nguyen, & Vadlamudi, 2004). Additionally, several post-translational modifications such as phosphorylation, methylation, ubiquitination, SUMOylation, and acetylation can impact the action of coregulators targeting gene expression (Han, Lonard, & O'Malley, 2009; Lonard & O'malley, 2007; O'Malley & McKenna, 2008).

14. Endogenous and exogenous estrogen receptors ligands

Apart from the estrogens that are naturally produced by gonadal and other tissues in the body, there is a diverse variety of organic and inorganic molecules that are able to recognize the estrogen receptors ligand-binding domains in a precise manner (Table 1). Most of these ligands display higher selectivity toward ERα, however, several selective compounds for ERβ have recently been described (Farooq, 2015). There are five main classes of ER ligands: endoestrogens, phytoestrogens, xenoestrogens, selective estrogen receptor modulators (SERMs) and metalloestrogens.

Endoestrogens are physiological estrogens that are endogenously produced by the body. Most endoestrogens (i.e., estradiol, estriol, estretrol, and estrone) were previously discussed in the chapter. Briefly, endoestrogens are steroidal compounds produced from cholesterol in the male and female gonads and other organs (Farooq, 2015). In contrast, phytoestrogens are non-steroidal compounds produced by plants. There are three known groups of phytoestrogens: isoflavones, coumestans, and lignans (Basu & Maier, 2018). Because phytoestrogens are chemically and structurally similar to estradiol, they can participate in both estrogenic and antiestrogenic effects through activation or blocking of the estrogen receptor ligand-binding domains (Turner, Agatonovic-Kustrin, & Glass, 2007). Interestingly, the phytoestrogens genistein, coumestrol, and liquiritigenin have been reported to display more affinity toward ERβ than to ERα, but the implications of these differences remain unknown (Kuiper et al., 1998; Manas, Xu, Unwalla, & Somers, 2004; Mersereau et al., 2008; Nilsson, Kuiper, & Gustafsson, 1998).

Xenoestrogens are another group of ligands that comprise an extensive variety of non-natural synthetic chemical compounds with estrogenic effects.

Table 1 Types of estrogen receptor ligands.

Endoestrogens	Phytoestrogens	Xenoestrogens	SERMs	Metalloestrogens
Estrone 17β-estradiol estriol estretrol	Isoflavones: genistein, daidzein, formononetin, glycitein Coumestans: coumestrol, repensol, trifoliol Lignans: lariciresinol, matairesinol, pinoresinol, secoisolariciresinol, podophyllotoxin, steganacin	Medicinal drugs: diethylstilbestrol, ethinyl estradiol Food additives: butylated hydroxyanisole, erythrosine Body cosmetics: 4-methylbenzylidene camphor, methylparaben, ethylparaben, propylparaben Environmental pesticides: atrazine, dichlorodiphenyldichloroethyle ne, dichlorodiphenyltrichloroethan e, methoxychlor, dieldrin, endosulfan, heptachlor, lindane Industrial chemicals: bisphenol A, nonylphenol, monochlorobiphenyl and dichlorobiphenyl, di-2-ethylhexyl phthalate, diisodecyl phthalate, diisononyl phthalate	Tamoxifen Clomifene Toremifene Raloxifene Ormeloxifene	Cations: aluminum (Al^{3+}), antimony (Sb^{3+}), barium (Ba^{2+}), cadmium (Cd^{2+}), chromium (Cr^{2+}), cobalt (Co^{2+}), copper (Cu^{2+}), lead (Pb^{2+}), mercury (Hg^{2+}), nickel (Ni^{2+}) Anions: arsenite ($AsO_3{}^{3-}$), selenite ($SeO_3{}^{2-}$), vanadate ($VO_4{}^{3-}$)

The family of xenoestrogens can be divided into five major types: medicinal drugs, food additives, body cosmetics, environmental pesticides, and industrial chemicals (Farooq, 2015). Drugs such as diethylstilbestrol (DES) and ethinyl estradiol were specifically synthesized to mimic the action of endoestrogens, and have been extensively to treat many conditions in women (Gennari, Merlotti, Valleggi, Martini, & Nuti, 2007; Maximov, Lee, & Jordan, 2013). However, it has been found that these compounds can affect cellular and molecular processes leading to severe effects on health, and their use in medical therapeutics remains controversial (Aravindakshan, Gregory, Marcogliese, Fournier, & Cyr, 2004; Aravindakshan et al., 2004; Arukwe, Celius, Walther, & Goksøyr, 2000; Christin et al., 2004; Golden et al., 1998; Iorga et al., 2017; Vajda et al., 2008; Williams, Lech, & Buhler, 1998). In the past few years, a wealth of evidence has been accumulated demonstrating that estrogens regulate many facets of the inflammatory response and the immune system via complex molecular mechanisms that are also sex dependent (Khan & Ansar Ahmed, 2015). It is now plausible that any immune cell that expresses estrogen receptors can potentially respond to ligand binding in a context-dependent manner, which will affect the outcome of the overall immune response. Thus, given the known spatial and temporal expression of the estrogen receptors, it is important to consider this aspect when designing potential therapeutic therapies targeting the estrogen receptor signaling pathways (Arnal et al., 2017). Additionally, precise timing of treatment initiation and duration may be required to determine the true efficacy of estrogen treatment (Burns & Korach, 2012; Hamilton, Hewitt, Arao, & Korach, 2017).

The selective estrogen receptor modulators (SERMs) are another type of estrogen receptor ligands. The main difference between SERMs and xenoestrogens relies on the fact that SERMs present functional duality and are able to act both as agonists and antagonists of the estrogen receptors in different tissues (Martinkovich, Shah, Planey, & Arnott, 2014; Shang & Brown, 2002; Smith & O'Malley, 2004). At the molecular level, SERMs employ their antagonistic actions by competing with estradiol for binding to an inner hydrophobic pocket within the ligand-binding domain of ERα (Bourguet, Germain, & Gronemeyer, 2000; Shiau et al., 1998; Wärnmark et al., 2002). Binding of this estradiol agonist induces a conformational change in the LBD that results in sealing the ligand binding pocket. Some of the most important SERMs include tamoxifen, raloxifene, clomifene, ormeloxifene, and toremifene (Farooq, 2015). One of the most used SERMs in the treatment of breast cancer, tamoxifen, acts as an antagonist in breast tissue, but as an agonist in the uterus. Therefore, while tamoxifen is often the selected treatment for ER-positive breast cancer, it can also stimulate endometrial cell growth leading to uterine cancer (Hu, Hilakivi-Clarke, & Clarke, 2015). While most SERMs are mainly selective for ERα, there are a few synthetic steroidal analogs that can regulate the actions of ERβ, or both receptors (Blizzard et al., 2007; Blizzard et al., 2006; Blizzard et al., 2007; Papapetropoulos, 2007).

Finally, in addition to the organic ligands mentioned above, there are also inorganic compounds in the form of heavy metal ions that present estrogenic activity. These are collectively known as metalloestrogens. Examples of these include: aluminum (Al^{3+}), antimony (Sb^{3+}), barium (Ba^{2+}), cadmium (Cd^{2+}), chromium (Cr^{2+}), cobalt (Co^{2+}), copper (Cu^{2+}), lead (Pb^{2+}), mercury (Hg^{2+}), nickel (Ni^{2+}), arsenite ($AsO_3{}^{3-}$), selenite ($SeO_3{}^{2-}$) and vanadate ($VO_4{}^{3-}$) (Farooq, 2015). Studies have shown that these metalloestrogens are able to coordinate to specific amino acid residues within the ligand-binding domain of the nuclear estrogen receptors, thus blocking binding of estradiol in a non-competitive manner (Stoica, Katzenellenbogen, & Martin, 2000; Stoica, Pentecost, & Martin, 2000a, 2000b).

15. Discussion

Estrogen receptors regulate a multitude of biological and physiological processes. These are tightly controlled by complex mechanisms involving either genomic nuclear direct binding to specific DNA sequences, or activation of intracellular cascades resulting in non-genomic control of transcription. Over the past 60 years since the discovery of the first nuclear estrogen receptors, and the almost 20 years since the discovery of the membrane receptor, multiple mechanisms of action have been discovered and characterized. These involve a multitude of intracellular kinases, transcription and growth factors, membrane receptors, coregulators, and natural and synthetic ligands. The information obtained in these studies has helped in the design of therapeutic strategies for diseases involving the estrogen receptors such as many cancers, as well as in the treatment of endocrine conditions affecting fertility and resulting from menopause. While there are still many diseases for which estrogens have been implicated but the role of their receptors has not been elucidated, the knowledge gained in the past six decades together with new advances in precision medicine and molecular diagnostic techniques will allow for the development of more personalized strategies to prevent and treat conditions that are affected by estrogens and other steroid hormones.

Acknowledgments

This work was supported by funding from NIH K01HL133520 (PS), NIH R03HL141618 (PS), Center for Research on Women and Newborn Health (PS), and the American Physiological Society Porter Physiology Development Program (NF).

References

Acconcia, F., Totta, P., Ogawa, S., Cardillo, I., Inoue, S., Leone, S., et al. (2005). Survival versus apoptotic 17beta-estradiol effect: Role of ER alpha and ER beta activated non-genomic signaling. *Journal of Cellular Physiology*, *203*(1), 193–201. Retrieved from https://www.ncbi.nlm.nih.gov/pubmed/15389627. https://doi.org/10.1002/jcp.20219.

Allen, E., & Doisy, E. A. (1983). Landmark article Sept 8, 1923. An ovarian hormone. Preliminary report on its localization, extraction and partial purification, and action in test animals. By Edgar Allen and Edward A. Doisy. *JAMA*, *250*(19), 2681–2683. Retrieved from https://www.ncbi.nlm.nih.gov/pubmed/6355545.

Aranda, A., & Pascual, A. (2001). Nuclear hormone receptors and gene expression. *Physiological Reviews*, *81*(3), 1269–1304. Retrieved from https://www.ncbi.nlm.nih.gov/pubmed/11427696. https://doi.org/10.1152/physrev.2001.81.3.1269.

Aravindakshan, J., Gregory, M., Marcogliese, D. J., Fournier, M., & Cyr, D. G. (2004). Consumption of xenoestrogen-contaminated fish during lactation alters adult male reproductive function. *Toxicological Sciences: An Official Journal of the Society of Toxicology*, *81*(1), 179–189. Retrieved from https://www.ncbi.nlm.nih.gov/pubmed/15159524. https://doi.org/10.1093/toxsci/kfh174.

Aravindakshan, J., Paquet, V., Gregory, M., Dufresne, J., Fournier, M., Marcogliese, D. J., et al. (2004). Consequences of xenoestrogen exposure on male reproductive function in spottail shiners (Notropis hudsonius). *Toxicological Sciences: An Official Journal of the Society of Toxicology*, *78*(1), 156–165. Retrieved from https://www.ncbi.nlm.nih.gov/pubmed/14657511. https://doi.org/10.1093/toxsci/kfh042.

Arnal, J. F., Lenfant, F., Metivier, R., Flouriot, G., Henrion, D., Adlanmerini, M., et al. (2017). Membrane and nuclear estrogen receptor alpha actions: From tissue specificity to medical implications. *Physiological Reviews*, *97*(3), 1045–1087. Retrieved from https://www.ncbi.nlm.nih.gov/pubmed/28539435. https://doi.org/10.1152/physrev.00024.2016.

Arukwe, A., Celius, T., Walther, B. T., & Goksøyr, A. (2000). Effects of xenoestrogen treatment on zona radiata protein and vitellogenin expression in Atlantic salmon (Salmo salar). *Aquatic Toxicology*, *49*(3), 159–170. Retrieved from https://www.ncbi.nlm.nih.gov/pubmed/10856602.

Bajic, V. B., Tan, S. L., Chong, A., Tang, S., Ström, A., Gustafsson, J. A., et al. (2003). Dragon ERE Finder version 2: A tool for accurate detection and analysis of estrogen response elements in vertebrate genomes. *Nucleic Acids Research*, *31*(13), 3605–3607. Retrieved from https://www.ncbi.nlm.nih.gov/pubmed/12824376.

Barakat, R., Oakley, O., Kim, H., Jin, J., & Ko, C. J. (2016). Extra-gonadal sites of estrogen biosynthesis and function. *BMB Reports*, *49*(9), 488–496. Retrieved from https://www.ncbi.nlm.nih.gov/pubmed/27530684.

Barton, M., Filardo, E. J., Lolait, S. J., Thomas, P., Maggiolini, M., & Prossnitz, E. R. (2018). Twenty years of the G protein-coupled estrogen receptor GPER: Historical and personal perspectives. *The Journal of Steroid Biochemistry and Molecular Biology*, *176*, 4–15. Retrieved from https://www.ncbi.nlm.nih.gov/pubmed/28347854. https://doi.org/10.1016/j.jsbmb.2017.03.021.

Basu, P., & Maier, C. (2018). Phytoestrogens and breast cancer: In vitro anticancer activities of isoflavones, lignans, coumestans, stilbenes and their analogs and derivatives. *Biomedicine & Pharmacotherapy*, *107*, 1648–1666. Retrieved from https://www.ncbi.nlm.nih.gov/pubmed/30257383. https://doi.org/10.1016/j.biopha.2018.08.100.

Belfiore, C. J., Hawkins, D. E., Wiltbank, M. C., & Niswender, G. D. (1994). Regulation of cytochrome P450scc synthesis and activity in the ovine corpus luteum. *The Journal of Steroid Biochemistry and Molecular Biology*, *51*(5–6), 283–290. Retrieved from https://www.ncbi.nlm.nih.gov/pubmed/7826890.

Bennesch, M. A., & Picard, D. (2015). Minireview: Tipping the balance: Ligand-independent activation of steroid receptors. *Molecular Endocrinology*, *29*(3), 349–363. Retrieved from https://www.ncbi.nlm.nih.gov/pubmed/25625619. https://doi.org/10.1210/me.2014-1315.

Björnström, L., & Sjöberg, M. (2005). Mechanisms of estrogen receptor signaling: Convergence of genomic and nongenomic actions on target genes. *Molecular Endocrinology*, *19*(4), 833–842. Retrieved from https://www.ncbi.nlm.nih.gov/pubmed/15695368. https://doi.org/10.1210/me.2004-0486.

Blizzard, T. A., Gude, C., Chan, W., Birzin, E. T., Mojena, M., Tudela, C., et al. (2007). Bridged androstenediol analogs as ER-beta selective SERMs. *Bioorganic & Medicinal Chemistry Letters*, *17*(10), 2944–2948. Retrieved from https://www.ncbi.nlm.nih.gov/pubmed/17448656. https://doi.org/10.1016/j.bmcl.2006.12.053.

Blizzard, T. A., Gude, C., Morgan, J. D., Chan, W., Birzin, E. T., Mojena, M., et al. (2006). Androstenediol analogs as ER-beta-selective SERMs. *Bioorganic & Medicinal Chemistry Letters*, *16*(4), 834–838. Retrieved from https://www.ncbi.nlm.nih.gov/pubmed/16309907. https://doi.org/10.1016/j.bmcl.2005.11.014.

Blizzard, T. A., Gude, C., Morgan, J. D., Chan, W., Birzin, E. T., Mojena, M., et al. (2007). Androstene-3,5-dienes as ER-beta selective SERMs. *Bioorganic & Medicinal Chemistry Letters*, *17*(22), 6295–6298. Retrieved from https://www.ncbi.nlm.nih.gov/pubmed/17890084. https://doi.org/10.1016/j.bmcl.2007.09.001.

Boonyaratanakornkit, V. (2011). Scaffolding proteins mediating membrane-initiated extra-nuclear actions of estrogen receptor. *Steroids*, *76*(9), 877–884. Retrieved from https://www.ncbi.nlm.nih.gov/pubmed/21354435. https://doi.org/10.1016/j.steroids.2011.02.017.

Boonyaratanakornkit, V., & Edwards, D. P. (2007). Receptor mechanisms mediating non-genomic actions of sex steroids. *Seminars in Reproductive Medicine*, *25*(3), 139–153. Retrieved from https://www.ncbi.nlm.nih.gov/pubmed/17447204. https://doi.org/10.1055/s-2007-973427.

Bourdeau, V., Deschênes, J., Métivier, R., Nagai, Y., Nguyen, D., Bretschneider, N., et al. (2004). Genome-wide identification of high-affinity estrogen response elements in human and mouse. *Molecular Endocrinology*, *18*(6), 1411–1427. Retrieved from https://www.ncbi.nlm.nih.gov/pubmed/15001666. https://doi.org/10.1210/me.2003-0441.

Bourguet, W., Germain, P., & Gronemeyer, H. (2000). Nuclear receptor ligand-binding domains: Three-dimensional structures, molecular interactions and pharmacological implications. *Trends in Pharmacological Sciences*, *21*(10), 381–388. Retrieved from https://www.ncbi.nlm.nih.gov/pubmed/11050318.

Bulun, S. E., Zeitoun, K., Sasano, H., & Simpson, E. R. (1999). Aromatase in aging women. *Seminars in Reproductive Endocrinology*, *17*(4), 349–358. Retrieved from https://www.ncbi.nlm.nih.gov/pubmed/10851574. https://doi.org/10.1055/s-2007-1016244.

Burns, K. A., & Korach, K. S. (2012). Estrogen receptors and human disease: An update. *Archives of Toxicology*, *86*(10), 1491–1504. Retrieved from https://www.ncbi.nlm.nih.gov/pubmed/22648069. https://doi.org/10.1007/s00204-012-0868-5.

Carmeci, C., Thompson, D. A., Ring, H. Z., Francke, U., & Weigel, R. J. (1997). Identification of a gene (GPR30) with homology to the G-protein-coupled receptor superfamily associated with estrogen receptor expression in breast cancer. *Genomics*, *45*(3), 607–617. Retrieved from https://www.ncbi.nlm.nih.gov/pubmed/9367686. https://doi.org/10.1006/geno.1997.4972.

Carr, B. R., MacDonald, P. C., & Simpson, E. R. (1982). The role of lipoproteins in the regulation of progesterone secretion by the human corpus luteum. *Fertility and Sterility*, *38*(3), 303–311. Retrieved from https://www.ncbi.nlm.nih.gov/pubmed/7117556.

Carroll, J. S., & Brown, M. (2006). Estrogen receptor target gene: An evolving concept. *Molecular Endocrinology*, *20*(8), 1707–1714. Retrieved from https://www.ncbi.nlm.nih.gov/pubmed/16396959. https://doi.org/10.1210/me.2005-0334.

Cavalcanti, F. N., Lucas, T. F., Lazari, M. F., & Porto, C. S. (2015). Estrogen receptor ESR1 mediates activation of ERK1/2, CREB, and ELK1 in the corpus of the epididymis. *Journal of Molecular Endocrinology*, *54*(3), 339–349. Retrieved from https://www.ncbi.nlm.nih.gov/pubmed/26069273. https://doi.org/10.1530/JME-15-0086.

Chambliss, K. L., & Shaul, P. W. (2002). Rapid activation of endothelial NO synthase by estrogen: Evidence for a steroid receptor fast-action complex (SRFC) in caveolae. *Steroids*, *67*(6), 413–419. Retrieved from https://www.ncbi.nlm.nih.gov/pubmed/11960616.

Chambliss, K. L., Yuhanna, I. S., Mineo, C., Liu, P., German, Z., Sherman, T. S., et al. (2000). Estrogen receptor alpha and endothelial nitric oxide synthase are organized into a functional signaling module in caveolae. *Circulation Research*, *87*(11), E44–E52. Retrieved from https://www.ncbi.nlm.nih.gov/pubmed/11090554.

Cheng, Z., Rios, G. R., King, C. D., Coffman, B. L., Green, M. D., Mojarrabi, B., et al. (1998). Glucuronidation of catechol estrogens by expressed human UDP-glucuronosyltransferases (UGTs) 1A1, 1A3, and 2B7. *Toxicological Sciences: An Official Journal of the Society of Toxicology*, *45*(1), 52–57. Retrieved from https://www.ncbi.nlm.nih.gov/pubmed/9848110. https://doi.org/10.1006/toxs.1998.2494.

Cheskis, B. J., Greger, J., Cooch, N., McNally, C., Mclarney, S., Lam, H. S., et al. (2008). MNAR plays an important role in ERa activation of Src/MAPK and PI3K/Akt signaling pathways. *Steroids*, *73*(9–10), 901–905. Retrieved from https://www.ncbi.nlm.nih.gov/pubmed/18261753. https://doi.org/10.1016/j.steroids.2007.12.028.

Christin, M. S., Ménard, L., Gendron, A. D., Ruby, S., Cyr, D., Marcogliese, D. J., et al. (2004). Effects of agricultural pesticides on the immune system of Xenopus laevis and Rana pipiens. *Aquatic Toxicology*, *67*(1), 33–43. Retrieved from https://www.ncbi.nlm.nih.gov/pubmed/15019249. https://doi.org/10.1016/j.aquatox.2003.11.007.

Coelingh Bennink, F., Holinka, C. F., Visser, M., & Coelingh Bennink, H. J. (2008). Maternal and fetal estetrol levels during pregnancy. *Climacteric: The Journal of the International Menopause Society*, *11*(Suppl. 1), 69–72. Retrieved from https://www.ncbi.nlm.nih.gov/pubmed/18464026. https://doi.org/10.1080/13697130802056321.

de Leeuw, R., Neefjes, J., & Michalides, R. (2011). A role for estrogen receptor phosphorylation in the resistance to tamoxifen. *International Journal of Breast Cancer*, *2011*, 232435. Retrieved from https://www.ncbi.nlm.nih.gov/pubmed/22295213. https://doi.org/10.4061/2011/232435.

Dos Santos, E. G., Dieudonne, M. N., Pecquery, R., Le Moal, V., Giudicelli, Y., & Lacasa, D. (2002). Rapid nongenomic E2 effects on p42/p44 MAPK, activator protein-1, and cAMP response element binding protein in rat white adipocytes. *Endocrinology*, *143*(3), 930–940. Retrieved from https://www.ncbi.nlm.nih.gov/pubmed/11861515. https://doi.org/10.1210/endo.143.3.8678.

Enmark, E., Pelto-Huikko, M., Grandien, K., Lagercrantz, S., Lagercrantz, J., Fried, G., et al. (1997). Human estrogen receptor beta-gene structure, chromosomal localization, and expression pattern. *The Journal of Clinical Endocrinology and Metabolism*, *82*(12), 4258–4265. Retrieved from https://www.ncbi.nlm.nih.gov/pubmed/9398750. https://doi.org/10.1210/jcem.82.12.4470.

Farooq, A. (2015). Structural and functional diversity of estrogen receptor ligands. *Current Topics in Medicinal Chemistry*, *15*(14), 1372–1384. Retrieved from https://www.ncbi.nlm.nih.gov/pubmed/25866274.

Feng, Y., & Gregor, P. (1997). Cloning of a novel member of the G protein-coupled receptor family related to peptide receptors. *Biochemical and Biophysical Research Communications*, *231*(3), 651–654. Retrieved from https://www.ncbi.nlm.nih.gov/pubmed/9070864. https://doi.org/10.1006/bbrc.1997.6161.

Filardo, E. J., Quinn, J. A., Bland, K. I., & Frackelton, A. R. (2000). Estrogen-induced activation of Erk-1 and Erk-2 requires the G protein-coupled receptor homolog, GPR30, and occurs via trans-activation of the epidermal growth factor receptor through release of HB-EGF. *Molecular Endocrinology*, *14*(10), 1649–1660. Retrieved from https://www.ncbi.nlm.nih.gov/pubmed/11043579. https://doi.org/10.1210/mend.14.10.0532.

Filardo, E., Quinn, J., Pang, Y., Graeber, C., Shaw, S., Dong, J., et al. (2007). Activation of the novel estrogen receptor G protein-coupled receptor 30 (GPR30) at the plasma membrane. *Endocrinology*, *148*(7), 3236–3245. Retrieved from https://www.ncbi.nlm.nih.gov/pubmed/17379646. https://doi.org/10.1210/en.2006-1605.

Filardo, E. J., & Thomas, P. (2012). Minireview: G protein-coupled estrogen receptor-1, GPER-1: Its mechanism of action and role in female reproductive cancer, renal and vascular physiology. *Endocrinology*, *153*(7), 2953–2962. Retrieved from https://www.ncbi.nlm.nih.gov/pubmed/22495674. https://doi.org/10.1210/en.2012-1061.

Fox, E. M., Andrade, J., & Shupnik, M. A. (2009). Novel actions of estrogen to promote proliferation: Integration of cytoplasmic and nuclear pathways. *Steroids*, *74*(7), 622–627. Retrieved from https://www.ncbi.nlm.nih.gov/pubmed/18996136. https://doi.org/10.1016/j.steroids.2008.10.014.

Fujimoto, N., Honda, H., & Kitamura, S. (2004). Effects of environmental estrogenic chemicals on AP1 mediated transcription with estrogen receptors alpha and beta. *The Journal of Steroid Biochemistry and Molecular Biology*, *88*(1), 53–59. Retrieved from https://www.ncbi.nlm.nih.gov/pubmed/15026083. https://doi.org/10.1016/j.jsbmb.2003.10.006.

Furth, P. A. (2014). STAT signaling in different breast cancer sub-types. *Molecular and Cellular Endocrinology*, *382*(1), 612–615. Retrieved from https://www.ncbi.nlm.nih.gov/pubmed/23562422. https://doi.org/10.1016/j.mce.2013.03.023.

Galien, R., & Garcia, T. (1997). Estrogen receptor impairs interleukin-6 expression by preventing protein binding on the NF-kappaB site. *Nucleic Acids Research*, *25*(12), 2424–2429. Retrieved from https://www.ncbi.nlm.nih.gov/pubmed/9171095.

Garbacz, W. G., Jiang, M., & Xie, W. (2017). Sex-dependent role of estrogen sulfotransferase and steroid sulfatase in metabolic homeostasis. *Advances in Experimental Medicine and Biology*, *1043*, 455–469. Retrieved from https://www.ncbi.nlm.nih.gov/pubmed/29224107. https://doi.org/10.1007/978-3-319-70178-3_21.

Gaub, M. P., Bellard, M., Scheuer, I., Chambon, P., & Sassone-Corsi, P. (1990). Activation of the ovalbumin gene by the estrogen receptor involves the fos-jun complex. *Cell*, *63*(6), 1267–1276. Retrieved from https://www.ncbi.nlm.nih.gov/pubmed/2124518.

Gennari, L., Merlotti, D., Valleggi, F., Martini, G., & Nuti, R. (2007). Selective estrogen receptor modulators for postmenopausal osteoporosis: Current state of development. *Drugs & Aging*, *24*(5), 361–379. Retrieved from https://www.ncbi.nlm.nih.gov/pubmed/17503894. https://doi.org/10.2165/00002512-200724050-00002.

Golden, R. J., Noller, K. L., Titus-Ernstoff, L., Kaufman, R. H., Mittendorf, R., Stillman, R., et al. (1998). Environmental endocrine modulators and human health: An assessment of the biological evidence. *Critical Reviews in Toxicology*, *28*(2), 109–227. Retrieved from https://www.ncbi.nlm.nih.gov/pubmed/9557209. https://doi.org/10.1080/10408449891344191.

Gosden, J. R., Middleton, P. G., & Rout, D. (1986). Localization of the human oestrogen receptor gene to chromosome 6q24——q27 by in situ hybridization. *Cytogenetics and Cell Genetics*, *43*(3–4), 218–220. Retrieved from https://www.ncbi.nlm.nih.gov/pubmed/3802924. https://doi.org/10.1159/000132325.

Göttlicher, M., Heck, S., & Herrlich, P. (1998). Transcriptional cross-talk, the second mode of steroid hormone receptor action. *Journal of Molecular Medicine (Berlin)*, *76*(7), 480–489. Retrieved from https://www.ncbi.nlm.nih.gov/pubmed/9660166.

Green, S., Walter, P., Kumar, V., Krust, A., Bornert, J. M., Argos, P., et al. (1986). Human oestrogen receptor cDNA: Sequence, expression and homology to v-erb-A. *Nature*, *320*(6058), 134–139. Retrieved from https://www.ncbi.nlm.nih.gov/pubmed/3754034. https://doi.org/10.1038/320134a0.

Greene, G. L., Gilna, P., Waterfield, M., Baker, A., Hort, Y., & Shine, J. (1986). Sequence and expression of human estrogen receptor complementary DNA. *Science (New York, N.Y.)*, *231*(4742), 1150–1154. Retrieved from https://www.ncbi.nlm.nih.gov/pubmed/3753802.

Gruber, C. J., Tschugguel, W., Schneeberger, C., & Huber, J. C. (2002). Production and actions of estrogens. *The New England Journal of Medicine*, *346*(5), 340–352. Retrieved from https://www.ncbi.nlm.nih.gov/pubmed/11821512. https://doi.org/10.1056/NEJMra000471.

Gu, Y., Chen, T., López, E., Wu, W., Wang, X., Cao, J., et al. (2014). The therapeutic target of estrogen receptor-alpha36 in estrogen-dependent tumors. *Journal of Translational Medicine*, *12*, 16. Retrieved from https://www.ncbi.nlm.nih.gov/pubmed/24447535. https://doi.org/10.1186/1479-5876-12-16.

Gu, Q., & Moss, R. L. (1996). 17 beta-Estradiol potentiates kainate-induced currents via activation of the cAMP cascade. *The Journal of Neuroscience*, *16*(11), 3620–3629. Retrieved from https://www.ncbi.nlm.nih.gov/pubmed/8642406.

Hall, J. M., McDonnell, D. P., & Korach, K. S. (2002). Allosteric regulation of estrogen receptor structure, function, and coactivator recruitment by different estrogen response elements. *Molecular Endocrinology*, *16*(3), 469–486. Retrieved from https://www.ncbi.nlm.nih.gov/pubmed/11875105. https://doi.org/10.1210/mend.16.3.0814.

Hamilton, K. J., Hewitt, S. C., Arao, Y., & Korach, K. S. (2017). Estrogen hormone biology. *Current Topics in Developmental Biology*, *125*, 109–146. Retrieved from https://www.ncbi.nlm.nih.gov/pubmed/28527569. https://doi.org/10.1016/bs.ctdb.2016.12.005.

Han, S. J., Lonard, D. M., & O'Malley, B. W. (2009). Multi-modulation of nuclear receptor coactivators through posttranslational modifications. *Trends in Endocrinology and Metabolism*, *20*(1), 8–15. Retrieved from https://www.ncbi.nlm.nih.gov/pubmed/19019695. https://doi.org/10.1016/j.tem.2008.10.001.

Heery, D. M., Kalkhoven, E., Hoare, S., & Parker, M. G. (1997). A signature motif in transcriptional co-activators mediates binding to nuclear receptors. *Nature*, *387*(6634), 733–736. Retrieved from https://www.ncbi.nlm.nih.gov/pubmed/9192902. https://doi.org/10.1038/42750.

Holinka, C. F., Diczfalusy, E., & Coelingh Bennink, H. J. (2008). Estetrol: A unique steroid in human pregnancy. *The Journal of Steroid Biochemistry and Molecular Biology*, *110*(1–2), 138–143. Retrieved from https://www.ncbi.nlm.nih.gov/pubmed/18462934. https://doi.org/10.1016/j.jsbmb.2008.03.027.

Hu, R., Hilakivi-Clarke, L., & Clarke, R. (2015). Molecular mechanisms of tamoxifen-associated endometrial cancer (Review). *Oncology Letters*, *9*(4), 1495–1501. Retrieved from https://www.ncbi.nlm.nih.gov/pubmed/25788989. https://doi.org/10.3892/ol.2015.2962.

Hu, X., & Lazar, M. A. (1999). The CoRNR motif controls the recruitment of corepressors by nuclear hormone receptors. *Nature*, *402*(6757), 93–96. Retrieved from https://www.ncbi.nlm.nih.gov/pubmed/10573424. https://doi.org/10.1038/47069.

Iorga, A., Cunningham, C. M., Moazeni, S., Ruffenach, G., Umar, S., & Eghbali, M. (2017). The protective role of estrogen and estrogen receptors in cardiovascular disease and the controversial use of estrogen therapy. *Biology of Sex Differences*, *8*(1), 33. Retrieved from https://www.ncbi.nlm.nih.gov/pubmed/29065927. https://doi.org/10.1186/s13293-017-0152-8.

ITTRICH, G., & NEUMANN, H. (1963). PARTICIPATION OF THE FETUS IN ESTROGEN METABOLISM OF THE MOTHER. *Ceskoslovenská Gynekologie*, *28*, 454–456. Retrieved from https://www.ncbi.nlm.nih.gov/pubmed/14091393.

Jensen, E. V., Desombre, E. R., Kawashima, T., Suzuki, T., Kyser, K., & Jungblut, P. W. (1967). Estrogen-binding substances of target tissues. *Science (New York, N.Y.)*, *158*(3800), 529–530. Retrieved from https://www.ncbi.nlm.nih.gov/pubmed/17749092. https://doi.org/10.1126/science.158.3800.529-c.

Jensen, E. V., Suzuki, T., Kawashima, T., Stumpf, W. E., Jungblut, P. W., & DeSombre, E. R. (1968). A two-step mechanism for the interaction of estradiol with rat uterus. *Proceedings of the National Academy of Sciences of the United States of America*, *59*(2), 632–638. Retrieved from https://www.ncbi.nlm.nih.gov/pubmed/5238991.

Kalaitzidis, D., & Gilmore, T. D. (2005). Transcription factor cross-talk: The estrogen receptor and NF-kappaB. *Trends in Endocrinology and Metabolism*, *16*(2), 46–52. Retrieved from https://www.ncbi.nlm.nih.gov/pubmed/15734144. https://doi.org/10.1016/j.tem.2005.01.004.

Kato, S., Endoh, H., Masuhiro, Y., Kitamoto, T., Uchiyama, S., Sasaki, H., et al. (1995). Activation of the estrogen receptor through phosphorylation by mitogen-activated protein kinase. *Science (New York, N.Y.)*, *270*(5241), 1491–1494. Retrieved from https://www.ncbi.nlm.nih.gov/pubmed/7491495.

Khan, D., & Ansar Ahmed, S. (2015). The immune system is a natural target For estrogen action: Opposing effects of estrogen in two prototypical autoimmune diseases. *Frontiers in Immunology*, *6*, 635. Retrieved from https://www.ncbi.nlm.nih.gov/pubmed/26779182. https://doi.org/10.3389/fimmu.2015.00635.

Klinge, C. M. (2001). Estrogen receptor interaction with estrogen response elements. *Nucleic Acids Research*, *29*(14), 2905–2919. Retrieved from https://www.ncbi.nlm.nih.gov/pubmed/11452016.

Koos, R. D. (2011). Minireview: Putting physiology back into estrogens' mechanism of action. *Endocrinology*, *152*(12), 4481–4488. Retrieved from https://www.ncbi.nlm.nih.gov/pubmed/21952239. https://doi.org/10.1210/en.2011-1449.

Kousteni, S., Han, L., Chen, J. R., Almeida, M., Plotkin, L. I., Bellido, T., et al. (2003). Kinase-mediated regulation of common transcription factors accounts for the bone-protective effects of sex steroids. *The Journal of Clinical Investigation*, *111*(11), 1651–1664. Retrieved from https://www.ncbi.nlm.nih.gov/pubmed/12782668. https://doi.org/10.1172/JCI17261.

Kuiper, G. G., Enmark, E., Pelto-Huikko, M., Nilsson, S., & Gustafsson, J. A. (1996). Cloning of a novel receptor expressed in rat prostate and ovary. *Proceedings of the National Academy of Sciences of the United States of America*, *93*(12), 5925–5930. Retrieved from https://www.ncbi.nlm.nih.gov/pubmed/8650195.

Kuiper, G. G., Lemmen, J. G., Carlsson, B., Corton, J. C., Safe, S. H., van der Saag, P. T., et al. (1998). Interaction of estrogenic chemicals and phytoestrogens with estrogen receptor beta. *Endocrinology*, *139*(10), 4252–4263. Retrieved from https://www.ncbi.nlm.nih.gov/pubmed/9751507. https://doi.org/10.1210/endo.139.10.6216.

Kumar, R., Zakharov, M. N., Khan, S. H., Miki, R., Jang, H., Toraldo, G., et al. (2011). The dynamic structure of the estrogen receptor. *Journal of Amino Acids*, *2011*, 812540. Retrieved from https://www.ncbi.nlm.nih.gov/pubmed/22312471. https://doi.org/10.4061/2011/812540.

Kvingedal, A. M., & Smeland, E. B. (1997). A novel putative G-protein-coupled receptor expressed in lung, heart and lymphoid tissue. *FEBS Letters*, *407*(1), 59–62. Retrieved from https://www.ncbi.nlm.nih.gov/pubmed/9141481.

Labrie, F., Bélanger, A., Luu-The, V., Labrie, C., Simard, J., Cusan, L., et al. (1998). DHEA and the intracrine formation of androgens and estrogens in peripheral target tissues: Its role during aging. *Steroids*, *63*(5–6), 322–328. Retrieved from https://www.ncbi.nlm.nih.gov/pubmed/9618795.

Lakhani, N. J., Venitz, J., Figg, W. D., & Sparreboom, A. (2003). Pharmacogenetics of estrogen metabolism and transport in relation to cancer. *Current Drug Metabolism*, *4*(6), 505–513. Retrieved from https://www.ncbi.nlm.nih.gov/pubmed/14683478.

Laliotis, A., Vrekoussis, T., Kafousi, M., Sanidas, E., Askoxilakis, J., Melissas, J., et al. (2013). Immunohistochemical study of pElk-1 expression in human breast cancer: Association with breast cancer biologic profile and clinicopathologic features. *Breast*, *22*(1), 89–95. Retrieved from https://www.ncbi.nlm.nih.gov/pubmed/23127278. https://doi.org/10.1016/j.breast.2012.09.013.

Le Dily, F., & Beato, M. (2018). Signaling by steroid hormones in the 3D nuclear space. *International Journal of Molecular Sciences*, *19*(2), 306. https://doi.org/10.3390/ijms19020306Retrieved from https://www.ncbi.nlm.nih.gov/pubmed/29360755.

Li, C., Briggs, M. R., Ahlborn, T. E., Kraemer, F. B., & Liu, J. (2001). Requirement of Sp1 and estrogen receptor alpha interaction in 17beta-estradiol-mediated transcriptional activation of the low density lipoprotein receptor gene expression. *Endocrinology*, *142*(4), 1546–1553. Retrieved from https://www.ncbi.nlm.nih.gov/pubmed/11250935. https://doi.org/10.1210/endo.142.4.8096.

Li, L., Haynes, M. P., & Bender, J. R. (2003). Plasma membrane localization and function of the estrogen receptor alpha variant (ER46) in human endothelial cells. *Proceedings of the National Academy of Sciences of the United States of America*, *100*(8), 4807–4812. Retrieved from https://www.ncbi.nlm.nih.gov/pubmed/12682286. https://doi.org/10.1073/pnas.0831079100.

Li, L., Hisamoto, K., Kim, K. H., Haynes, M. P., Bauer, P. M., Sanjay, A., et al. (2007). Variant estrogen receptor-c-Src molecular interdependence and c-Src structural requirements for endothelial NO synthase activation. *Proceedings of the National Academy of Sciences of the United States of America*, *104*(42), 16468–16473. Retrieved from https://www.ncbi.nlm.nih.gov/pubmed/17921256. https://doi.org/10.1073/pnas.0704315104.

Li, Y., Wang, J. P., Santen, R. J., Kim, T. H., Park, H., Fan, P., et al. (2010). Estrogen stimulation of cell migration involves multiple signaling pathway interactions. *Endocrinology*, *151*(11), 5146–5156. Retrieved from https://www.ncbi.nlm.nih.gov/pubmed/20861240. https://doi.org/10.1210/en.2009-1506.

Liang, J., & Shang, Y. (2013). Estrogen and cancer. *Annual Review of Physiology*, *75*, 225–240. Retrieved from https://www.ncbi.nlm.nih.gov/pubmed/23043248. https://doi.org/10.1146/annurev-physiol-030212-183708.

Liu, M. M., Albanese, C., Anderson, C. M., Hilty, K., Webb, P., Uht, R. M., et al. (2002). Opposing action of estrogen receptors alpha and beta on cyclin D1 gene expression. *The Journal of Biological Chemistry*, *277*(27), 24353–24360. Retrieved from https://www.ncbi.nlm.nih.gov/pubmed/11986316. https://doi.org/10.1074/jbc.M201829200.

Lonard, D. M., & O'Malley, B. W. (2006). The expanding cosmos of nuclear receptor coactivators. *Cell*, *125*(3), 411–414. Retrieved from https://www.ncbi.nlm.nih.gov/pubmed/16678083. https://doi.org/10.1016/j.cell.2006.04.021.

Lonard, D. M., & O'malley, B. W. (2007). Nuclear receptor coregulators: Judges, juries, and executioners of cellular regulation. *Molecular Cell*, *27*(5), 691–700. Retrieved from https://www.ncbi.nlm.nih.gov/pubmed/17803935. https://doi.org/10.1016/j.molcel.2007.08.012.

Lösel, R., & Wehling, M. (2003). Nongenomic actions of steroid hormones. *Nature Reviews. Molecular Cell Biology*, *4*(1), 46–56.
Loven, M. A., Wood, J. R., & Nardulli, A. M. (2001). Interaction of estrogen receptors alpha and beta with estrogen response elements. *Molecular and Cellular Endocrinology*, *181*(1–2), 151–163. Retrieved from https://www.ncbi.nlm.nih.gov/pubmed/11476949.
Maggi, A. (2011). Liganded and unliganded activation of estrogen receptor and hormone replacement therapies. *Biochimica et Biophysica Acta*, *1812*(8), 1054–1060. Retrieved from https://www.ncbi.nlm.nih.gov/pubmed/21605666. https://doi.org/10.1016/j.bbadis.2011.05.001.
Manas, E. S., Xu, Z. B., Unwalla, R. J., & Somers, W. S. (2004). Understanding the selectivity of genistein for human estrogen receptor-beta using X-ray crystallography and computational methods. *Structure*, *12*(12), 2197–2207. Retrieved from https://www.ncbi.nlm.nih.gov/pubmed/15576033. https://doi.org/10.1016/j.str.2004.09.015.
Manavathi, B., Samanthapudi, V. S., & Gajulapalli, V. N. (2014). Estrogen receptor coregulators and pioneer factors: The orchestrators of mammary gland cell fate and development. *Frontiers in Cell and Development Biology*, *2*, 34. Retrieved from https://www.ncbi.nlm.nih.gov/pubmed/25364741. https://doi.org/10.3389/fcell.2014.00034.
Marino, M., Acconcia, F., Bresciani, F., Weisz, A., & Trentalance, A. (2002). Distinct nongenomic signal transduction pathways controlled by 17beta-estradiol regulate DNA synthesis and cyclin D(1) gene transcription in HepG2 cells. *Molecular Biology of the Cell*, *13*(10), 3720–3729. Retrieved from https://www.ncbi.nlm.nih.gov/pubmed/12388769. https://doi.org/10.1091/mbc.e02-03-0153.
Marino, M., Acconcia, F., & Trentalance, A. (2003). Biphasic estradiol-induced AKT phosphorylation is modulated by PTEN via MAP kinase in HepG2 cells. *Molecular Biology of the Cell*, *14*(6), 2583–2591. Retrieved from https://www.ncbi.nlm.nih.gov/pubmed/12808053. https://doi.org/10.1091/mbc.e02-09-0621.
Marino, M., Galluzzo, P., & Ascenzi, P. (2006). Estrogen signaling multiple pathways to impact gene transcription. *Current Genomics*, *7*(8), 497–508. Retrieved from https://www.ncbi.nlm.nih.gov/pubmed/18369406.
Marino, M., Pallottini, V., & Trentalance, A. (1998). Estrogens cause rapid activation of IP3-PKC-alpha signal transduction pathway in HEPG2 cells. *Biochemical and Biophysical Research Communications*, *245*(1), 254–258. Retrieved from https://www.ncbi.nlm.nih.gov/pubmed/9535818. https://doi.org/10.1006/bbrc.1998.8413.
Martinkovich, S., Shah, D., Planey, S. L., & Arnott, J. A. (2014). Selective estrogen receptor modulators: Tissue specificity and clinical utility. *Clinical Interventions in Aging*, *9*, 1437–1452. Retrieved from https://www.ncbi.nlm.nih.gov/pubmed/25210448. https://doi.org/10.2147/CIA.S66690.
Matthews, J., & Gustafsson, J. A. (2003). Estrogen signaling: A subtle balance between ER alpha and ER beta. *Molecular Interventions*, *3*(5), 281–292. Retrieved from https://www.ncbi.nlm.nih.gov/pubmed/14993442. https://doi.org/10.1124/mi.3.5.281.
Maximov, P. Y., Lee, T. M., & Jordan, V. C. (2013). The discovery and development of selective estrogen receptor modulators (SERMs) for clinical practice. *Current Clinical Pharmacology*, *8*(2), 135–155. Retrieved from https://www.ncbi.nlm.nih.gov/pubmed/23062036.
Mersereau, J. E., Levy, N., Staub, R. E., Baggett, S., Zogovic, T., Zogric, T., et al. (2008). Liquiritigenin is a plant-derived highly selective estrogen receptor beta agonist. *Molecular and Cellular Endocrinology*, *283*(1–2), 49–57. Retrieved from https://www.ncbi.nlm.nih.gov/pubmed/18177995. https://doi.org/10.1016/j.mce.2007.11.020.
Migliaccio, A., Di Domenico, M., Castoria, G., de Falco, A., Bontempo, P., Nola, E., et al. (1996). Tyrosine kinase/p21ras/MAP-kinase pathway activation by estradiol-receptor complex in MCF-7 cells. *The EMBO Journal*, *15*(6), 1292–1300. Retrieved from https://www.ncbi.nlm.nih.gov/pubmed/8635462.

Miller, W. L. (2017). Steroidogenesis: Unanswered Questions. *Trends in Endocrinology and Metabolism*, *28*(11), 771–793. Retrieved from https://www.ncbi.nlm.nih.gov/pubmed/29031608. https://doi.org/10.1016/j.tem.2017.09.002.
Miller, W. L., & Strauss, J. F. (1999). Molecular pathology and mechanism of action of the steroidogenic acute regulatory protein, StAR. *The Journal of Steroid Biochemistry and Molecular Biology*, *69*(1–6), 131–141. Retrieved from https://www.ncbi.nlm.nih.gov/pubmed/10418987.
Mishra, S. K., Balasenthil, S., Nguyen, D., & Vadlamudi, R. K. (2004). Cloning and functional characterization of PELP1/MNAR promoter. *Gene*, *330*, 115–122. Retrieved from https://www.ncbi.nlm.nih.gov/pubmed/15087130. https://doi.org/10.1016/j.gene.2004.01.011.
Molina, L., Figueroa, C. D., Bhoola, K. D., & Ehrenfeld, P. (2017). GPER-1/GPR30 a novel estrogen receptor sited in the cell membrane: Therapeutic coupling to breast cancer. *Expert Opinion on Therapeutic Targets*, *21*(8), 755–766. Retrieved from https://www.ncbi.nlm.nih.gov/pubmed/28671018. https://doi.org/10.1080/14728222.2017.1350264.
Nelson, L. R., & Bulun, S. E. (2001). Estrogen production and action. *Journal of the American Academy of Dermatology*, *45*(3 Suppl), S116–S124. Retrieved from https://www.ncbi.nlm.nih.gov/pubmed/11511861.
Nilsson, S., Kuiper, G., & Gustafsson, J. A. (1998). ERbeta: A novel estrogen receptor offers the potential for new drug development. *Trends in Endocrinology and Metabolism*, *9*(10), 387–395. Retrieved from https://www.ncbi.nlm.nih.gov/pubmed/18406312.
Nilsson, S., Mäkelä, S., Treuter, E., Tujague, M., Thomsen, J., Andersson, G., et al. (2001). Mechanisms of estrogen action. *Physiological Reviews*, *81*(4), 1535–1565. Retrieved from https://www.ncbi.nlm.nih.gov/pubmed/11581496. https://doi.org/10.1152/physrev.2001.81.4.1535.
O'Dowd, B. F., Nguyen, T., Marchese, A., Cheng, R., Lynch, K. R., Heng, H. H., et al. (1998). Discovery of three novel G-protein-coupled receptor genes. *Genomics*, *47*(2), 310–313. Retrieved from https://www.ncbi.nlm.nih.gov/pubmed/9479505. https://doi.org/10.1006/geno.1998.5095.
O'Lone, R., Frith, M. C., Karlsson, E. K., & Hansen, U. (2004). Genomic targets of nuclear estrogen receptors. *Molecular Endocrinology*, *18*(8), 1859–1875. Retrieved from https://www.ncbi.nlm.nih.gov/pubmed/15031323. https://doi.org/10.1210/me.2003-0044.
O'Malley, B. W. (2005). A life-long search for the molecular pathways of steroid hormone action. *Molecular Endocrinology*, *19*(6), 1402–1411. Retrieved from https://www.ncbi.nlm.nih.gov/pubmed/15914709. https://doi.org/10.1210/me.2004-0480.
O'Malley, B. W., & McKenna, N. J. (2008). Coactivators and corepressors: What's in a name? *Molecular Endocrinology*, *22*(10), 2213–2214. Retrieved from https://www.ncbi.nlm.nih.gov/pubmed/18701638. https://doi.org/10.1210/me.2008-0201.
Oñate, S. A., Tsai, S. Y., Tsai, M. J., & O'Malley, B. W. (1995). Sequence and characterization of a coactivator for the steroid hormone receptor superfamily. *Science (New York, N.Y.)*, *270*(5240), 1354–1357. Retrieved from https://www.ncbi.nlm.nih.gov/pubmed/7481822.
Owman, C., Blay, P., Nilsson, C., & Lolait, S. J. (1996). Cloning of human cDNA encoding a novel heptahelix receptor expressed in Burkitt's lymphoma and widely distributed in brain and peripheral tissues. *Biochemical and Biophysical Research Communications*, *228*(2), 285–292. Retrieved from https://www.ncbi.nlm.nih.gov/pubmed/8920907. https://doi.org/10.1006/bbrc.1996.1654.
Ozes, O. N., Mayo, L. D., Gustin, J. A., Pfeffer, S. R., Pfeffer, L. M., & Donner, D. B. (1999). NF-kappaB activation by tumour necrosis factor requires the Akt serine-threonine kinase. *Nature*, *401*(6748), 82–85. Retrieved from https://www.ncbi.nlm.nih.gov/pubmed/10485710. https://doi.org/10.1038/43466.

Paech, K., Webb, P., Kuiper, G. G., Nilsson, S., Gustafsson, J., Kushner, P. J., et al. (1997). Differential ligand activation of estrogen receptors ERalpha and ERbeta at AP1 sites. *Science (New York, N.Y.)*, *277*(5331), 1508–1510. Retrieved from https://www.ncbi.nlm.nih.gov/pubmed/9278514.

Papapetropoulos, A. (2007). A ginseng-derived oestrogen receptor beta (ERbeta) agonist, Rb1 ginsenoside, attenuates capillary morphogenesis. *British Journal of Pharmacology*, *152*(2), 172–174. Retrieved from https://www.ncbi.nlm.nih.gov/pubmed/17603551. https://doi.org/10.1038/sj.bjp.0707360.

Picotto, G., Massheimer, V., & Boland, R. (1996). Acute stimulation of intestinal cell calcium influx induced by 17 beta-estradiol via the cAMP messenger system. *Molecular and Cellular Endocrinology*, *119*(2), 129–134. Retrieved from https://www.ncbi.nlm.nih.gov/pubmed/8807632.

Piu, F., Aronheim, A., Katz, S., & Karin, M. (2001). AP-1 repressor protein JDP-2: Inhibition of UV-mediated apoptosis through p53 down-regulation. *Molecular and Cellular Biology*, *21*(9), 3012–3024. Retrieved from https://www.ncbi.nlm.nih.gov/pubmed/11287607. https://doi.org/10.1128/MCB.21.9.3012-3024.2001.

Prossnitz, E. R., & Barton, M. (2011). The G-protein-coupled estrogen receptor GPER in health and disease. *Nature Reviews. Endocrinology*, 7(12), 715–726. Retrieved from https://www.ncbi.nlm.nih.gov/pubmed/21844907. https://doi.org/10.1038/nrendo.2011.122.

Prossnitz, E. R., & Barton, M. (2014). Estrogen biology: New insights into GPER function and clinical opportunities. *Molecular and Cellular Endocrinology*, *389*(1–2), 71–83. Retrieved from https://www.ncbi.nlm.nih.gov/pubmed/24530924. https://doi.org/10.1016/j.mce.2014.02.002.

Purohit, A., & Reed, M. J. (2002). Regulation of estrogen synthesis in postmenopausal women. *Steroids*, *67*(12), 979–983. Retrieved from https://www.ncbi.nlm.nih.gov/pubmed/12398994.

Purohit, A., Tutill, H. J., Day, J. M., Chander, S. K., Lawrence, H. R., Allan, G. M., et al. (2006). The regulation and inhibition of 17beta-hydroxysteroid dehydrogenase in breast cancer. *Molecular and Cellular Endocrinology*, *248*(1–2), 199–203. Retrieved from https://www.ncbi.nlm.nih.gov/pubmed/16414180. https://doi.org/10.1016/j.mce.2005.12.003.

Razandi, M., Pedram, A., Merchenthaler, I., Greene, G. L., & Levin, E. R. (2004). Plasma membrane estrogen receptors exist and functions as dimers. *Molecular Endocrinology*, *18*(12), 2854–2865. Retrieved from https://www.ncbi.nlm.nih.gov/pubmed/15231873. https://doi.org/10.1210/me.2004-0115.

Revankar, C. M., Cimino, D. F., Sklar, L. A., Arterburn, J. B., & Prossnitz, E. R. (2005). A transmembrane intracellular estrogen receptor mediates rapid cell signaling. *Science (New York, N.Y.)*, *307*(5715), 1625–1630. Retrieved from https://www.ncbi.nlm.nih.gov/pubmed/15705806. https://doi.org/10.1126/science.1106943.

Romano, S. N., & Gorelick, D. A. (2018). Crosstalk between nuclear and G protein-coupled estrogen receptors. *General and Comparative Endocrinology*, *261*, 190–197. Retrieved from https://www.ncbi.nlm.nih.gov/pubmed/28450143. https://doi.org/10.1016/j.ygcen.2017.04.013.

Romashkova, J. A., & Makarov, S. S. (1999). NF-kappaB is a target of AKT in anti-apoptotic PDGF signalling. *Nature*, *401*(6748), 86–90. Retrieved from https://www.ncbi.nlm.nih.gov/pubmed/10485711. https://doi.org/10.1038/43474.

RYAN, K. J. (1959). Biological aromatization of steroids. *The Journal of Biological Chemistry*, *234*(2), 268–272. Retrieved from https://www.ncbi.nlm.nih.gov/pubmed/13630892.

Samavat, H., & Kurzer, M. S. (2015). Estrogen metabolism and breast cancer. *Cancer Letters*, *356*(2 Pt. A), 231–243. Retrieved from https://www.ncbi.nlm.nih.gov/pubmed/24784887. https://doi.org/10.1016/j.canlet.2014.04.018.

Sausville, E., Carney, D., & Battey, J. (1985). The human vasopressin gene is linked to the oxytocin gene and is selectively expressed in a cultured lung cancer cell line. *The Journal of Biological Chemistry*, *260*(18), 10236–10241. Retrieved from https://www.ncbi.nlm.nih.gov/pubmed/2991279.

Scheidereit, C., Krauter, P., von der Ahe, D., Janich, S., Rabenau, O., Cato, A. C., et al. (1986). Mechanism of gene regulation by steroid hormones. *Journal of Steroid Biochemistry*, *24*(1), 19–24. Retrieved from https://www.ncbi.nlm.nih.gov/pubmed/3009974.

Schulster, M., Bernie, A. M., & Ramasamy, R. (2016). The role of estradiol in male reproductive function. *Asian Journal of Andrology*, *18*(3), 435–440. Retrieved from https://www.ncbi.nlm.nih.gov/pubmed/26908066. https://doi.org/10.4103/1008-682X.173932.

Schwabe, J. W., & Teichmann, S. A. (2004). Nuclear receptors: The evolution of diversity. *Science's STKE: Signal Transduction Knowledge Environment*, *2004*(217), pe4. Retrieved from https://www.ncbi.nlm.nih.gov/pubmed/14747695. https://doi.org/10.1126/stke.2172004pe4.

Shang, Y., & Brown, M. (2002). Molecular determinants for the tissue specificity of SERMs. *Science (New York, N.Y.)*, *295*(5564), 2465–2468. Retrieved from https://www.ncbi.nlm.nih.gov/pubmed/11923541. https://doi.org/10.1126/science.1068537.

Sharma, G., & Prossnitz, E. R. (2016). GPER/GPR30 Knockout Mice: Effects of GPER on Metabolism. *Methods in Molecular Biology*, *1366*, 489–502. Retrieved from https://www.ncbi.nlm.nih.gov/pubmed/26585159. https://doi.org/10.1007/978-1-4939-3127-9_38.

Shaul, P. W., & Anderson, R. G. (1998). Role of plasmalemmal caveolae in signal transduction. *The American Journal of Physiology*, *275*(5 Pt. 1), L843–L851. Retrieved from https://www.ncbi.nlm.nih.gov/pubmed/9815100.

Shiau, A. K., Barstad, D., Loria, P. M., Cheng, L., Kushner, P. J., Agard, D. A., et al. (1998). The structural basis of estrogen receptor/coactivator recognition and the antagonism of this interaction by tamoxifen. *Cell*, *95*(7), 927–937. Retrieved from https://www.ncbi.nlm.nih.gov/pubmed/9875847.

Silva, E., Kabil, A., & Kortenkamp, A. (2010). Cross-talk between non-genomic and genomic signalling pathways—Distinct effect profiles of environmental estrogens. *Toxicology and Applied Pharmacology*, *245*(2), 160–170. Retrieved from https://www.ncbi.nlm.nih.gov/pubmed/20206645. https://doi.org/10.1016/j.taap.2010.02.015.

Simpson, E. R. (2003). Sources of estrogen and their importance. *The Journal of Steroid Biochemistry and Molecular Biology*, *86*(3–5), 225–230. Retrieved from https://www.ncbi.nlm.nih.gov/pubmed/14623515.

Simpson, E. R., Clyne, C., Rubin, G., Boon, W. C., Robertson, K., Britt, K., et al. (2002). Aromatase—A brief overview. *Annual Review of Physiology*, *64*, 93–127. Retrieved from https://www.ncbi.nlm.nih.gov/pubmed/11826265. https://doi.org/10.1146/annurev.physiol.64.081601.142703.

Simpson, E. R., Misso, M., Hewitt, K. N., Hill, R. A., Boon, W. C., Jones, M. E., et al. (2005). Estrogen—The good, the bad, and the unexpected. *Endocrine Reviews*, *26*(3), 322–330. Retrieved from https://www.ncbi.nlm.nih.gov/pubmed/15817672. https://doi.org/10.1210/er.2004-0020.

Smith, C. L., & O'Malley, B. W. (2004). Coregulator function: A key to understanding tissue specificity of selective receptor modulators. *Endocrine Reviews*, *25*(1), 45–71. Retrieved from https://www.ncbi.nlm.nih.gov/pubmed/14769827. https://doi.org/10.1210/er.2003-0023.

Song, R. X., Chen, Y., Zhang, Z., Bao, Y., Yue, W., Wang, J. P., et al. (2010). Estrogen utilization of IGF-1-R and EGF-R to signal in breast cancer cells. *The Journal of Steroid Biochemistry and Molecular Biology*, *118*(4–5), 219–230. Retrieved from https://www.ncbi.nlm.nih.gov/pubmed/19815064. https://doi.org/10.1016/j.jsbmb.2009.09.018.

Song, R. X., Zhang, Z., Chen, Y., Bao, Y., & Santen, R. J. (2007). Estrogen signaling via a linear pathway involving insulin-like growth factor I receptor, matrix metalloproteinases, and epidermal growth factor receptor to activate mitogen-activated protein kinase in MCF-7 breast cancer cells. *Endocrinology*, *148*(8), 4091–4101. Retrieved from https://www.ncbi.nlm.nih.gov/pubmed/17525128. https://doi.org/10.1210/en.2007-0240.

Song, R. X., Zhang, Z., & Santen, R. J. (2005). Estrogen rapid action via protein complex formation involving ERalpha and Src. *Trends in Endocrinology and Metabolism*, *16*(8), 347–353. Retrieved from https://www.ncbi.nlm.nih.gov/pubmed/16126407. https://doi.org/10.1016/j.tem.2005.06.010.

Stoica, A., Katzenellenbogen, B. S., & Martin, M. B. (2000). Activation of estrogen receptor-alpha by the heavy metal cadmium. *Molecular Endocrinology*, *14*(4), 545–553. Retrieved from https://www.ncbi.nlm.nih.gov/pubmed/10770491. https://doi.org/10.1210/mend.14.4.0441.

Stoica, A., Pentecost, E., & Martin, M. B. (2000a). Effects of arsenite on estrogen receptor-alpha expression and activity in MCF-7 breast cancer cells. *Endocrinology*, *141*(10), 3595–3602. Retrieved from https://www.ncbi.nlm.nih.gov/pubmed/11014213. https://doi.org/10.1210/endo.141.10.7704.

Stoica, A., Pentecost, E., & Martin, M. B. (2000b). Effects of selenite on estrogen receptor-alpha expression and activity in MCF-7 breast cancer cells. *Journal of Cellular Biochemistry*, *79*(2), 282–292. Retrieved from https://www.ncbi.nlm.nih.gov/pubmed/10967555.

Sun, G., Porter, W., & Safe, S. (1998). Estrogen-induced retinoic acid receptor alpha 1 gene expression: Role of estrogen receptor-Sp1 complex. *Molecular Endocrinology*, *12*(6), 882–890. Retrieved from https://www.ncbi.nlm.nih.gov/pubmed/9626663. https://doi.org/10.1210/mend.12.6.0125.

Takada, Y., Kato, C., Kondo, S., Korenaga, R., & Ando, J. (1997). Cloning of cDNAs encoding G protein-coupled receptor expressed in human endothelial cells exposed to fluid shear stress. *Biochemical and Biophysical Research Communications*, *240*(3), 737–741. Retrieved from https://www.ncbi.nlm.nih.gov/pubmed/9398636. https://doi.org/10.1006/bbrc.1997.7734.

Thomas, P., Pang, Y., Filardo, E. J., & Dong, J. (2005). Identity of an estrogen membrane receptor coupled to a G protein in human breast cancer cells. *Endocrinology*, *146*(2), 624–632. Retrieved from https://www.ncbi.nlm.nih.gov/pubmed/15539556. https://doi.org/10.1210/en.2004-1064.

Tora, L., White, J., Brou, C., Tasset, D., Webster, N., Scheer, E., et al. (1989). The human estrogen receptor has two independent nonacidic transcriptional activation functions. *Cell*, *59*(3), 477–487. Retrieved from https://www.ncbi.nlm.nih.gov/pubmed/2805068.

Truss, M., & Beato, M. (1993). Steroid hormone receptors: Interaction with deoxyribonucleic acid and transcription factors. *Endocrine Reviews*, *14*(4), 459–479. Retrieved from https://www.ncbi.nlm.nih.gov/pubmed/8223341. https://doi.org/10.1210/edrv-14-4-459.

Tsuchiya, Y., Nakajima, M., & Yokoi, T. (2005). Cytochrome P450-mediated metabolism of estrogens and its regulation in human. *Cancer Letters*, *227*(2), 115–124. Retrieved from https://www.ncbi.nlm.nih.gov/pubmed/16112414. https://doi.org/10.1016/j.canlet.2004.10.007.

Turner, J. V., Agatonovic-Kustrin, S., & Glass, B. D. (2007). Molecular aspects of phytoestrogen selective binding at estrogen receptors. *Journal of Pharmaceutical Sciences*, *96*(8), 1879–1885. Retrieved from https://www.ncbi.nlm.nih.gov/pubmed/17518366. https://doi.org/10.1002/jps.20987.

Vajda, A. M., Barber, L. B., Gray, J. L., Lopez, E. M., Woodling, J. D., & Norris, D. O. (2008). Reproductive disruption in fish downstream from an estrogenic wastewater effluent. *Environmental Science & Technology*, *42*(9), 3407–3414. Retrieved from https://www.ncbi.nlm.nih.gov/pubmed/18522126.

Voogt, J. L. (1978). Control of hormone release during lactation. *Clinics in Obstetrics and Gynaecology*, *5*(2), 435–455. Retrieved from https://www.ncbi.nlm.nih.gov/pubmed/361330.

Vrtačnik, P., Ostanek, B., Mencej-Bedrač, S., & Marc, J. (2014). The many faces of estrogen signaling. *Biochemica Medica*, *24*(3), 329–342. Retrieved from https://www.ncbi.nlm.nih.gov/pubmed/25351351. 10.11613/BM.2014.035.

Wärnmark, A., Treuter, E., Gustafsson, J. A., Hubbard, R. E., Brzozowski, A. M., & Pike, A. C. (2002). Interaction of transcriptional intermediary factor 2 nuclear receptor box peptides with the coactivator binding site of estrogen receptor alpha. *The Journal of Biological Chemistry*, *277*(24), 21862–21868. Retrieved from https://www.ncbi.nlm.nih.gov/pubmed/11937504. https://doi.org/10.1074/jbc.M200764200.

Watson, C. S., Jeng, Y. J., & Kochukov, M. Y. (2008). Nongenomic actions of estradiol compared with estrone and estriol in pituitary tumor cell signaling and proliferation. *The FASEB Journal*, *22*(9), 3328–3336. Retrieved from https://www.ncbi.nlm.nih.gov/pubmed/18541692. https://doi.org/10.1096/fj.08-107672.

Watters, J., Campbell, J., Cunningham, M., Krebs, E., & Dorsa, D. (1997). Rapid membrane effects of steroids in neuroblastoma cells: Effects of estrogen on mitogen activated protein kinase signalling cascade and c-fos immediate early gene transcription. *Endocrinology*, *138*(9), 4030–4033.

Williams, D. E., Lech, J. J., & Buhler, D. R. (1998). Xenobiotics and xenoestrogens in fish: Modulation of cytochrome P450 and carcinogenesis. *Mutation Research*, *399*(2), 179–192. Retrieved from https://www.ncbi.nlm.nih.gov/pubmed/9672659.

Wilson, R., Eriksson, G., & Diczfalusy, E. (1964). Oestriol metabolism in pregnant women. *Acta Endocrinologica*, *46*, 525–543. Retrieved from https://www.ncbi.nlm.nih.gov/pubmed/14202469.

Yaşar, P., Ayaz, G., User, S. D., Güpür, G., & Muyan, M. (2017). Molecular mechanism of estrogen-estrogen receptor signaling. *Reproductive Medicine and Biology*, *16*(1), 4–20. Retrieved from https://www.ncbi.nlm.nih.gov/pubmed/29259445. https://doi.org/10.1002/rmb2.12006.

Yi, P., Driscoll, M. D., Huang, J., Bhagat, S., Hilf, R., Bambara, R. A., et al. (2002). The effects of estrogen-responsive element- and ligand-induced structural changes on the recruitment of cofactors and transcriptional responses by ER alpha and ER beta. *Molecular Endocrinology*, *16*(4), 674–693. Retrieved from https://www.ncbi.nlm.nih.gov/pubmed/11923465. https://doi.org/10.1210/mend.16.4.0810.

CHAPTER FOUR

Intracellular signaling of the AMP-activated protein kinase

Miribane Dërmaku-Sopjani[a], Mentor Sopjani[b,*]
[a]Department of Chemistry, University of Prishtina, Prishtina, Kosova
[b]Faculty of Medicine, University of Prishtina, Prishtina, Kosova
*Corresponding author: e-mail addresses: msopjani@yahoo.com; mentor.sopjani@uni-pr.edu

Contents

Abstract

AMP-activated protein kinase (AMPK) is an essential cellular energy sensor that senses the cellular energy status and maintains cellular energy balance. The AMPK coordinates cellular and whole-body energy homeostasis through stimulating catabolic ATP-producing and suppressing anabolic ATP-consuming intracellular signaling pathways. AMPK induces autophagy and inhibits cell growth in response to starvation, a process that involves regulating certain intracellular signaling molecules. Recent advances demonstrated the AMPK to exert tumor suppressor activity realized through various signaling molecules by stimulating different cellular processes such as apoptosis, autophagy and cell growth and proliferation. AMPK can also be used to protect against metabolic syndrome. AMPK has previously been reported to be either directly or indirectly involved in the regulation of many different cellular transport proteins of high importance for cellular physiology and pathophysiology. Thus, AMPK provides a necessary link between cellular energy metabolism and cellular transport activities. A better understanding of the AMPK role in intracellular signaling under physiological and pathological conditions may represent a potential strategy for developing therapies for treating many different human diseases and disorders, in which AMPK plays a key role.

Advances in Protein Chemistry and Structural Biology, Volume 116
ISSN 1876-1623
https://doi.org/10.1016/bs.apcsb.2018.12.001

1. The protein enzyme AMPK

Adenosine monophosphate (AMP)-activated protein kinase (AMPK) is a phylogenetically conserved serine/threonine kinase. This ubiquitously expressed kinase is a protein complex that senses the energy state of the cell (Hardie & Alessi, 2013) and has many essential roles in many metabolic processes (Herzig & Shaw, 2018). AMPK is a molecule that works as a cellular fuel gauge, being activated during energy depletion. AMPK maintains energy homeostasis by activating catabolic pathways involved in ATP generation, such as glucose uptake, glycolysis, mitochondrial biogenesis, and fatty acid oxidation. At the same time, AMPK inhibits anabolic pathways involved in ATP and NADPH consumption (Jeon, Chandel, & Hay, 2012), such as glycogen synthesis, protein synthesis (Hardie & Alessi, 2013; McBride, Ghilagaber, Nikolaev, & Hardie, 2009), gluconeogenesis (Kim et al., 2008), and fatty acid synthesis (Hardie & Alessi, 2013; McFadden & Corl, 2009).

Pharmacological activation of AMPK by metformin and other agents presents a unique therapeutic potential to reverse metabolic abnormalities such as diabetes mellitus type 2 (DMT2) (Fryer et al., 2002; Langelueddecke et al., 2012) and nonalcoholic fatty liver disease (NAFDL) (Ix & Sharma, 2010) that is strongly associated with insulin resistance and metabolic syndrome (obesity). Nowadays, altered energy metabolism to support the rapid growth of tumor cells is widely recognized and AMPK may be an emerging hallmark and potential target (Imamura, Ogura, Kishimoto, Kaminishi, & Esumi, 2001; Jones et al., 2005) to be used in cancer-preventive and/or cancer-treatment strategies.

1.1 The structure of the AMPK

The mammalian AMPK is a heterotrimeric protein complex composed of a catalytic subunit (AMPK-α (alpha)) and two regulatory subunits (AMPK-β (beta) and AMPK-γ (gamma)).

In humans and other mammals, there are two α-subunits, α1 and α2 (Stapleton et al., 1996), two β-subunits, β1 and β2 (Thornton, Snowden, & Carling, 1998), and three γ-subunits, γ1, γ2 and γ3 (Cheung, Salt, Davies, Hardie, & Carling, 2000) (Fig. 1). In humans α1 and α2 are encoded by the genes PRKAA1 and PRKAA2 (Stapleton et al., 1996), β1 and β2 are encoded by PRKAB1 and PRKAB2 (Thornton et al., 1998), whereas γ1, γ2, and γ3 are encoded by PRKAG1, PRKAG2 and PRKAG3 (Cheung et al., 2000).

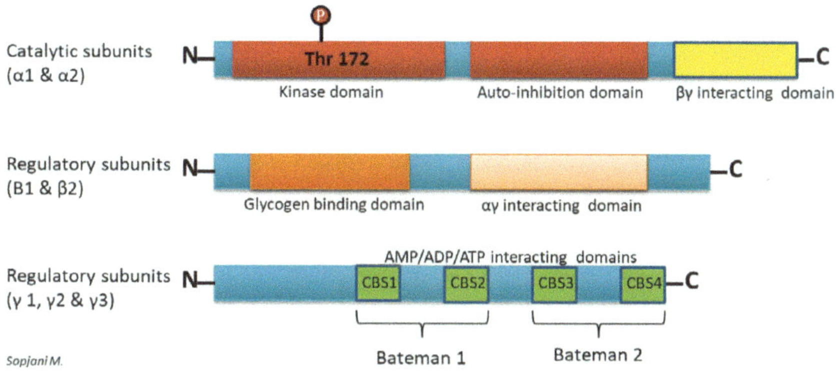

Fig. 1 Schematic diagram representing the domain structure of the AMPK. The AMP-activated protein kinase is a heterotrimeric complex composed of a catalytic subunit (AMPK-α) and two regulatory subunits (AMPK-β and AMPK-γ). The subunit contains the kinase domain in the NH_2-terminal that can be phosphorylated by the upstream protein kinases, AMPK kinases (AMPKK), which phosphorylates it on the conserved threonine residue Thr-172 within the catalytic subunit. This subunit also has an auto-inhibition domain and a domain for interacting with the co-functional regulatory subunits β and γ. The AMPK β subunit has a glycogen-binding domain (GBD) at the N-terminal and an interacting domain for subunits α and γ at the C-terminal. The γ subunit of AMPK has four tandem repeats of a cystathionine-β-synthase (CBS) motifs organized in a pair of structures (CBS1/2 and CBS3/4) called Bateman domains, responsible for interacting with adenosine-containing ligands such as AMP, ADP, and ATP.

All combinations are possible in each heterotrimeric AMPK complex, thus potentially creating 12 distinct AMPK complexes (Ross, Mackintosh, & Hardie, 2016). However, it is not known yet whether these different AMPK combinations exhibit functionally distinct heterotrimeric complexes with specific substrate recognition profiles and/or subcellular localization.

1.2 AMPK subunit composition and their main roles

The α-subunit contains the serine-threonine kinase domain and an essential residue, Thr^{172} whose phosphorylation by upstream kinases leads to the AMPK activation.

The β subunit at its N-terminal has a glycogen-binding domain (GBD) that allows the enzyme to associate with glycogen. GBD is responsible for targeting the kinase to the membrane (Warden et al., 2001).

The γ subunits of AMPK enables the kinase to respond to changes in the cellular ATP:AMP ratio as it has four tandem repeats sequences called CBS (cystathionine-β-synthase) motifs, named after similar sequences in cystathionine β-synthase initially recognized by Alex Bateman

(Bateman, 1997) (Fig. 1). The γ subunit CBS domains can bind adenine nucleotides. Binding of AMP, and to a lesser extent ADP, to the AMPK γ-subunit stimulates its activity. CBS's act in pairs. Pairs of CBS sequences within an enzyme fold into so named "Bateman domains" responsible for binding the adenosine-containing ligands (Xiao et al., 2007). Mutations in genes encoding γ subunits are associated with GSD (glycogen storage diseases), an inherited disorder in animals (Arad et al., 2002; Park et al., 2003).

Last decades, the AMPK function has emerged as a key and powerful regulator involved in linking energy metabolism to numerous physiological processes. Regulation of AMPK may be a key factor in multiple health benefits related to diet and exercise.

1.3 Mechanisms of the AMPK activation

It seems that all eukaryotes have homologs of AMPK proteins, but the number of subunits nay varies between the organisms. In the mammalian genome seven genes encoding AMPK exist; i.e., two α (α1 and α2)-, two β (β1 and β2)-, and three γ (γ1, γ2, and γ3)-subunit isoforms, and all resulting in ≤12 αβγ heterotrimeric possible combinations present in bacteria or in mammalian cells (Hardie, 2011), thus increasing the diversification of its essential roles related to the regulating energy balance. Even though the consequences of distinct heterotrimeric complex activation are not elucidated, some reports indicate that specific heterotrimeric subunit compositions may allow specific AMPK complexes to respond to specific types of stress stimuli in different cell types (Jensen et al., 2015).

The α subunits have a typical serine/threonine kinase domain at the N terminus that is active only when phosphorylated by any of several upstream kinases at the activation loop (Thr^{172} in human α1) (Hawley et al., 1996). This activation site is widely used experimentally as a biomarker for inducing AMPK activation (Fig. 2).

The AMPK is regulated by certain regulators. The liver kinase B1 (LKB1), a tumor suppressor protein molecule, is the major upstream kinase that phosphorylates Thr^{172} (AMPK α) with subsequent activation of AMPK (Woods, Johnstone, et al., 2003; Woods, Vertommen, et al., 2003). This, in turn, inhibits cell growth and proliferation, effects possibly mediated by AMPK (Woods, Johnstone, et al., 2003). The β subunit has a GBD mediating the binding with glycogen particles (Bendayan et al., 2009) and thus through sensing the structural state of glycogen (McBride et al., 2009) the kinase senses cellular energy status (Xiao et al., 2007). The regulatory

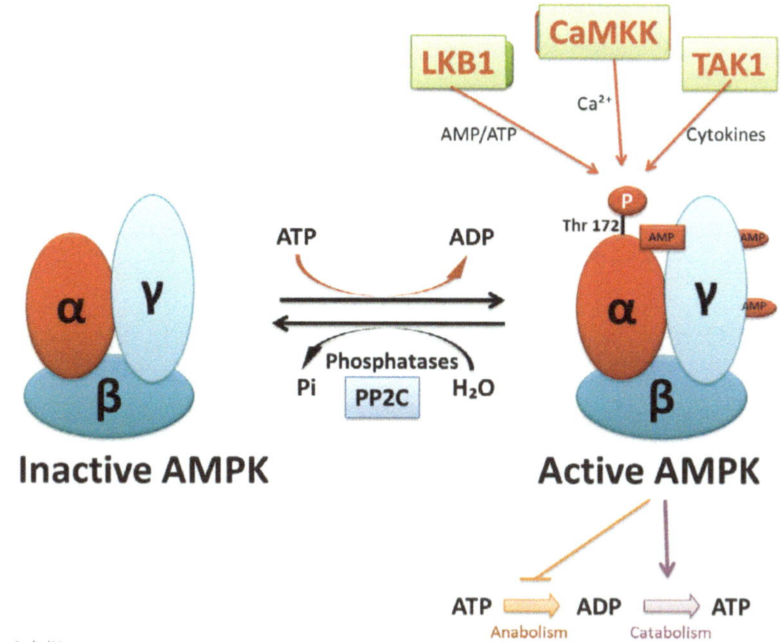

Fig. 2 Schematic representation of AMPK activation and its role in maintaining energy homeostasis. The AMPK is activated by phosphorylation of threonine 172 (Thr^{172}) within the catalytic subunit catalyzed by any of the AMPK kinases (AMPKK), such as by LKB1 (for activity is required binding of MO25 and STE-20 related adaptor protein (STRAD) forming a heterotrimeric complex) in response to an increase in the AMP/ATP ratio, or by CaMKKβ in Ca^{2+}-dependent manner, or by TAK1 in response to cytokines. Upon activation, AMPK affects many other downstream targets, generally by switching-off anabolism signaling pathways, while, at the same time, switching-on catabolism signaling pathways and therefore restoring energy homeostasis. AMPK is inactivated by Thr^{172} dephosphorylation catalyzed by protein phosphatase 2C (PP2C). See text for further details.

γ subunit has two nucleotide binding sites for the AMP and ATP in an antagonistic way (Xiao et al., 2007). The binding of AMP to the γ-subunit increases AMPK activity through three specific mechanisms. First, AMP binding to the γ-subunit promotes phosphorylation of Thr^{172} *via* directly stimulating the activity of the upstream kinases (Hawley et al., 1996) or *via* a direct allosteric mechanism that would return AMPK a better substrate for the upstream kinases (Hardie, Scott, Pan, & Hudson, 2003). However, these models have been challenged by other reports based on *in vitro* studies (Hardie & Alessi, 2013; Oakhill et al., 2010). Second, AMP binding to the γ-subunit protects the AMPK from dephosphorylation of Thr^{172} in the

α-subunit by phosphatases (Davies, Helps, Cohen, & Hardie, 1995). Third and last, AMP causes significant allosteric activation of kinase already phosphorylated on Thr^{172} residue (Gowans, Hawley, Ross, & Hardie, 2013). During energy depletion ATP consumption is always followed by an increase in cellular AMP and ADP. Even though both ADP and AMP have the affinity to bind to the CBS domains with subsequent stimulation of the AMPK activity, however, AMP appear to be the major contributor in AMPK activity *in vivo* (Gowans et al., 2013). It should be mentioned that ATP may also bind to the γ-subunit CBS domains (Hardie, Carling, & Gamblin, 2011), which allows the kinase to respond to ATP-to-AMP ratio changes and thus to act as a sensor of energy status that maintains cellular energy homeostasis. Of note, LKB1 continuously phosphorylates Thr^{172} within the activation loop of the AMPK α subunit, but it is also immediately dephosphorylated in the absence of AMP, a signal of low cellular energy status. However, the binding of AMP results in a kinase conformational change that inhibits the AMPK dephosphorylation. In addition, binding of AMP to the γ- non-catalytic AMPK subunit causes a direct allosteric activation of the kinase (Sanders, Grondin, Hegarty, Snowden, & Carling, 2007).

Another AMPK regulating upstream kinase is the calmodulin-dependent kinase kinase β (CaMKKβ), which phosphorylates AMPK at threonine 172 residue triggered by an increase of cytosolic Ca^{2+} concentration (Hurley et al., 2005; Sopjani, Kunert, Czarkowski, et al., 2010) (Fig. 2). AMPK is also activated by the transforming growth factor β-associated kinase 1 (TAK1) (Hardie, 2011) as well as by glucosamine (Kim et al., 2010).

Stimulation of AMPK is not only reported following increase in the AMP/ATP ratio and cytosolic Ca^{2+} activity, but as well following a decrease of oxygen (O_2) levels (Evans et al., 2005), or after exposure of skeletal muscle cells to nitric oxide (NO) (Lira et al., 2007). Thus, alterations of cellular AMP/ATP ratio, cytosolic Ca^{2+} levels, O_2 concentration, and NO formation may influence AMPK-mediated cellular signaling including autophagy, tumor suppression processes and regulation of cellular transport.

2. AMPK-mediated downstream targets

Upon activation under energy starvation, AMPK stimulates ATP-generating catabolic pathways and inhibits ATP-consuming anabolic pathways and other ATP-consuming processes, thus restoring the energy homeostasis (Sopjani, Bhavsar, Fraser, et al., 2010). Accordingly, AMPK effects predominantly serve to replenish cellular energy sources, ATP levels respectively. ATP is the leading energy "currency" of the cell.

AMPK participates in the regulation of numerous cellular processes, such as stimulation of autophagy and phagocytosis, regulation of apoptosis, inhibition of cell growth and proliferation, and counteracts hypertrophy (Dermaku-Sopjani, Kolgeci, Abazi, & Sopjani, 2013; Foller, Sopjani, Koka, et al., 2009; Hardie, 2011; Lang, Eylenstein, & Shumilina, 2012). Similar to apoptosis of nucleated cells, AMPK participates in the regulation of eryptosis (Foller, Sopjani, Koka, et al., 2009), a suicidal erythrocyte death, which is considered to be a window to systemic disease (Foller, Sopjani, Mahmud, & Lang, 2008; Foller, Sopjani, Schlemmer, Claussen, & Lang, 2009; Sopjani, Foller, Dreischer, & Lang, 2008; Sopjani, Foller, Gulbins, & Lang, 2008; Sopjani, Foller, Haendeler, Gotz, & Lang, 2009; Sopjani, Foller, & Lang, 2008). In addition to its role in controlling numerous metabolic pathways, many studies reported AMPK is also involved in the regulation of various different cell membrane transport systems (Dermaku-Sopjani, Abazi, Faggio, Kolgeci, & Sopjani, 2014), and therefore potentially coupling cellular transport activity to cellular stress and energy status.

2.1 AMPK-mediated autophagy and cell growth processes

Autophagy (self-eating) is an important evolutionarily conserved catabolic process that utilizes lysosomal activity to degrade cytoplasmic materials, including cytosol proteins and cellular organelles.

Mammalian cells can undergo three morphologically and mechanistically distinct types of autophagy: macroautophagy, microautophagy, and chaperone-mediated autophagy. Unless otherwise specified, the term "autophagy" usually mean macroautophagy, which is thought to be the predominant and the most characterized type of autophagy (Mizushima & Komatsu, 2011; Tsukamoto et al., 2014). Macroautophagy delivers cargo to the membrane-bound intracellular organelles, lysosomes, for degradation through the formation and transport of specific intracellular double-membrane vesicles named autophagosomes (Mizushima & Komatsu, 2011; Watanabe, Kuma, & Mizushima, 2011). It has been demonstrated that AMPK has a key role in the autophagy process. Activation of AMPK promotes the process of autophagy by phosphorylating tuberous sclerosis 2 (TSC2) and regulatory-associated protein of mTOR (RPTOR) that leads to inactivation of the mammalian target of rapamycin (mTOR)

(Hong-Brown et al., 2010), an inhibitor of ULK1 (Unc51-like kinase, hATG1) complex, a protein enzyme that in humans is encoded by the *ULK1* gene. Through inhibition the mTOR the AMPK indirectly activates the ULK1 complex, which is a serine/threonine-protein kinase that plays an important function in inducing autophagy in response to starvation (Jung, Seo, Otto, & Kim, 2011). On the other side, mTOR is involved and have a key role in cell growth and cell proliferation by stimulating many anabolic processes (Kempe et al., 2010; Tyan et al., 2010; Wu et al., 2014) (Fig. 3). The mTOR integrates mitogen and nutrient signals to regulate cell growth and cell division. By contrast, AMPK inhibits mTOR by phosphorylating Raptor and TSC2, resulting in decreased anabolic processes.

Some other studies reported that AMPK interact directly to ULK1 resulting in ULK1 phosphorylation in certain sites and in a nutrient-dependent fashion (Egan et al., 2011; Kim, Kundu, Viollet, & Guan, 2011; Shang et al., 2011). ULK1 phosphorylation sites are attributed to either mTOR or AMPK. Noteworthy, mTOR is a key cell-growth regulator that integrates growth factor and nutrient signals. mTOR is also involved in many other cellular processes, such as metabolism and apoptosis as well as in many diseases, including diabetes and cancer diseases (Kempe et al., 2010; Tyan et al., 2010). AMPK phosphorylation site Ser^{555} is supposed to recruit phospho-binding protein 14-3-3 to the ULK1 complex (Foller, Sopjani, Koka, et al., 2009; Gehring et al., 2009), whereas mTOR site Ser^{757} is required for AMPK binding, as its mutation is confirmed to disrupt ULK1-AMPK interaction (Kim et al., 2011; Shang et al., 2011). However, it should be noted that the role of AMPK-mediated phosphorylation of ULK1 complex on autophagy process is not well clarified since there are several conflicting reports (Kim et al., 2011; Shang et al., 2011) on whether it results in the stimulation or inhibition of autophagy. For instance, a study reported that under glucose depletion, AMPK stimulates autophagy by directly activating ULK1 *via* phosphorylation of its Ser^{317} and Ser^{777} resulting in ULK1 activation (Kim et al., 2011). Thus, the kinase activity of ULK1 (measured by the levels of ULK1 autophosphorylation) increases upon glucose depletion and is associated with AMP-activated protein kinase activation. It has been also reported that under nutrient sufficiency, high mTOR activity prevents the activation of ULK1 by phosphorylating ULK1 Ser^{757} and therefore disrupting the interaction between ULK1 and AMPK. This coordinated phosphorylation of ULK1 through mTOR and AMPK is important for the ULK1 role and hence to autophagy

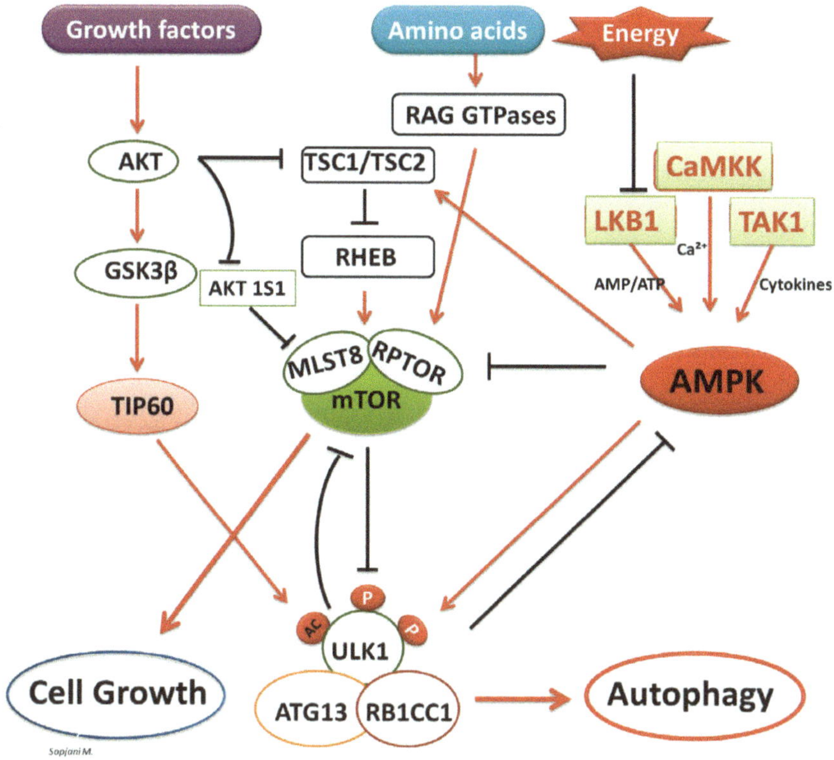

Fig. 3 Intracellular signaling showing the role of the AMPK in the regulation of autophagy and cell growth. Several different nutrient-sensing signaling pathways including growth factors, amino acids, and energy converge on ULK1 (Unc-51 Like Autophagy Activating Kinase 1) *via* unique specific molecule modifications, which at the end orchestrate the function of autophagy-stimulation complex. On one hand, the mTOR inhibits the autophagy stimulation through regulation of the ULK1 complex, a key initiator for mammalian autophagy, which undergoes general dephosphorylation under starvation conditions. The acetyltransferase TIP60, which is regulated by the growth factor-sensitive kinase GSK3beta, catalyzes acetylation of ULK1 to bolster its kinase activity. On the other hand, the AMP-activated protein kinase, which interacts with ULK1 in a nutrient-dependent fashion, may stimulate autophagy *via* directly phosphorylating many different protein substrates involved in its initiation phases, such as TSC2, Raptor, and ULK1. Generally, the AMPK-mediated autophagy process can be either through the inhibition of mTOR or through regulation of the ULK1 complex. The red arrows indicate stimulatory effects, while a black line inhibitory effects. Read the text for further details.

induction in response to nutrient signaling (Fig. 3). However, another study (Shang et al., 2011) identified AMPK sites that are dephosphorylated under amino acid starvation leading to the conclusion that AMPK phosphorylation sites on ULK1 may have an inhibitory effect to autophagy process induction

(Shang et al., 2011). These differences in their observations may be a reflection of distinct autophagy functions of AMPK when sensing by various triggers such as glucose starvation, amino acid starvation respectively. That could also be due to different phosphorylation sites monitored in these two studies.

Another report identified the biochemical connection between AMPK and the critical regulator of autophagy, ULK1 (Mack, Zheng, Asara, & Thomas, 2012). Results suggested that the AMPK-mediated phosphorylation of ULK1 is involved in regulation of the autophagy machinery. Specifically, the AMPK together with ULK1 can regulate the localization of a critical component of the phagophore, ATG9 (Mack et al., 2012). The AMPK, as an ULK1 kinase, *via* AMPK-mediated phosphorylation of ULK1 is identified to increase binding of the adaptor protein YWHAZ/14-3-3ζ. This binding alters ULK1 activation through its phosphorylation. Both AMPK and ULK1 regulate the intracellular trafficking of the putative autophagosomal membrane carrier protein molecule, ATG9, and thereby providing some clues to explain new mechanistic insights into the regulation of autophagy. In yet another study the AMPK regulation of ULK1 was associated with mitophagy and AMPK phosphorylation on ULK1 is needed for ULK1 function in autophagy (Egan et al, 2011). The loss of AMPK or ULK1 leads to aberrant accumulation of the autophagy adaptor p62 and subsequently defective mitophagy as it has been demonstrated in experiments with mammalian. Reconstitution of deficient cells for ULK1 with a mutant ULK1 that cannot be phosphorylated by AMPK revealed that such phosphorylation is needed for mitochondrial balance and cell survival during starvation. Thus, suggesting the coupling nutrient status with autophagy and cell survival. Taken together, many reports by different research groups have identified AMPK as a ULK1 kinase that can directly interact and phosphorylates of ULK1, therefore its activator role on the ULK1 complex. The fact that AMPKα1/α2 double knockout (AMPK DKO) mouse embryo fibroblasts (MEFs) can still undergo autophagy, rules out a decisive essential autophagic function for AMPK. This at least under conditions of amino acid starvation (Kim et al., 2011; Mack et al., 2012). Accordingly, autophagy process-induced following AMPK regulation of ULK1 activity may also, at least partly, be regulated by other regulators.

Intriguingly there have also been reports on the potential reverse action of ULK1 on mTOR and AMPK activity regulation. *In vitro* and *in vivo* experiments revealed that ULK1 can bind to and induce RPTOR (regulatory-associated protein of mTOR, or raptor) phosphorylation to downregulate mTOR activity. The ULK1 contributes to the mammalian/mechanistic target of rapamycin complex 1 (mTORC1) inhibition, whereas knockdown of ULK1 leads to enhanced mTORC1 signaling, even in the absence of TSC2, as evaluated by the phosphorylation of mTORC1 substrates, thus, suggesting to exist an interplay between mTORC1 and ULK1 (Dunlop, Hunt, Acosta-Jaquez, Fingar, & Tee, 2011; Jung et al., 2011). This negative feedback circuit of mTORC1 inhibition by ULK1 could be important to coordinately regulate cell growth and autophagy process with optimized use of cellular energy (Jung et al., 2011) when nutrient supplies are limiting (Dunlop et al., 2011). Consistent with this, through its interaction with TSC1 the RB1CC1 is also reported to regulate mTOR (Gan, Melkoumian, Wu, Guan, & Guan, 2005; Gan et al., 2006), a tumor suppressor protein that interacts with various intracellular signaling pathways to coordinately regulate many cell functions including cell growth and proliferation, apoptosis, autophagy, and cell migration. Regarding the AMPK, ULK1 has been reported to be able to phosphorylate all three AMPK subunits (Loffler et al., 2011). The phosphorylation of AMPK by ULK1 is a negative feedback loop and the ULK1 does not only take part in the stimulation of autophagy process, but it also terminates the AMPK-mediated autophagy process.

Additionally, ULK1 can act directly by the acetyltransferase TIP60 (Lin et al., 2012). The GSK3β (glycogen synthase kinase-3 beta) activates TIP60 through phosphorylating TIP60-Ser^{86}, which directly acetylates and stimulates the ULK1. On another side, protein kinase B (Akt or PKB), which is activated under growth factor deprivation, stimulates GSK3β (Sopjani et al., 2018) and inhibits TSC1/TSC2 (Foller et al., 2011). The PKB-GSK3β-TIP60-ULK1 intracellular signaling pathway (Lin et al., 2012) integrates protein acetylation and phosphorylation to link growth factor deprivation to autophagy (Fig. 3).

2.2 AMPK-mediated tumor suppressor activities

The metabolic stress sensor, AMPK, is regulated by the tumor suppressor LKB1 (Hardie, 2011). The tumor suppressor activities of LKB1, at least in part, are interfered by AMPK. The tumor suppressor activities of AMPK are demonstrated in many tumor types. It has been reported that genetic ablation of the only catalytic subunit of AMPK expressed in B cells, AMPK-α1, leads to accelerating the development of lymphomas, a group of cancers of the lymphatic system, in transgenic mice overexpressing c-Myc in the B cells (Faubert et al., 2013). Specifically, AMPK inhibits aerobic glycolysis (the Warburg effect) in cancer cells and suppresses tumor growth *in vivo* (Faubert et al., 2013), and therefore prevents tumor development.

The AMPK have an important role in tumor suppressor, downregulation of tumorigenesis and tumor metabolism. These functions are accomplished by different mechanisms. First, the AMPK activation results in cell cycle arrest associated with stabilization of p53 through phosphorylation on serine^{15} (Imamura et al., 2001), and the cyclin-dependent kinase (CDK) inhibitors p21WAF1 and p27CIP1 (Faubert et al., 2013; Jones et al., 2005; Liang et al., 2007). The mammalian cell-cycle checkpoint responds to glucose availability. A glucose-sensitive checkpoint occurs at the G (1)/S boundary and is regulated by the AMPK (Jones et al., 2005). AMPK-sensitive cell-cycle arrest mediated by p53 activation enables cell survival in response to energy depletion; however, during energy restoration cell can rapidly reenter the cell cycle (Jones et al., 2005). Second, AMPK activation under stress conditions inhibits the synthesis of most macromolecules, including fatty acids, glycogen, triglycerides, cholesterol, ribosomal RNA and proteins (Hardie, 2011), which all are needed for cell growth, proliferation and survival of mammalian cells (Liang et al., 2007), especially for rapidly growing tumor cells. In addition, a particularly important target of AMPK is the mTORC1, which is inhibited by phosphorylating of both, its upstream regulator TSC2 (Inoki, Zhu, & Guan, 2003) and its regulatory subunit, i.e., mTOR binding partner Raptor (Gwinn et al., 2008), and subsequently suppressing cell growth and biosynthetic processes, including translation of many proteins necessary for fast cell growth of tumor cells. The inhibition of mTORC1 by AMPK results in cell cycle arrest caused by energy stress and thus the AMP-activated protein kinase mediates its role as a metabolic checkpoint coordinating energy status with cell growth.

There are reported many different interactions between AMPK and cancer. Among the proteins that their translation inhibition is mediated by AMPK is the hypoxia-inducible factor-1α (HIF-1α) (Faubert et al., 2013), which effect is attributed to the third potential tumor suppressor effect of the kinase, "anti-Warburg" action respectively. AMPK inhibits "the Warburg effect" in cancer cells with subsequent inhibition of tumor growth *in vivo* (Faubert et al., 2013). The Warburg effect is a metabolic shift to aerobic glycolysis, elevated allocation of glucose carbon into lipids, and biomass accumulation that is a characteristic for most tumor cells. These metabolic effects require HIF-1α, suggesting that AMPK activity opposes tumor development (Faubert et al., 2013). This latter function contributes to the third potential tumor suppressor role of the kinase -the "anti-Warburg" effect. More specifically, the Warburg effect is the metabolic switch from oxidative metabolism as in most normal cells to rapid glucose uptake, anaerobic glycolysis and lactate output that produces energy by a high rate and is a particular feature of most tumor cells (Vander Heiden, Cantley, & Thompson, 2009).

Nonetheless, AMPK activation can acutely facilitate the uptake of glucose into cells through inducing glucose transporters expression (Barnes et al., 2002; Lira et al., 2007; Sopjani, Bhavsar, Fraser, et al., 2010; Walker et al., 2005) and glycolysis (Ganapathy, Thangaraju, & Prasad, 2009; Malaisse, Conget, Sener, & Rorsman, 1994; Marsin, Bouzin, Bertrand, & Hue, 2002), in the longer term it promotes the more efficient energy generation *via* oxidative metabolism. This is realized by stimulating mitochondrial biogenesis (Canto et al., 2009) and increased expression of oxidative enzymes (Evans et al., 2005; Vander Heiden et al., 2009) while inhibiting the glycolytic pathway by inhibiting mTORC1 (Dunlop et al., 2011; Hong-Brown et al., 2010). The more detail explanation of AMPK-dependent glucose transporters regulation can be found in the last part of this chapter.

Numerous papers reported that AMPK activation effectively inhibits cell proliferation in non-malignant cells as well as in tumor cells. AMPK participates in the regulations of many tumors, including its role as a tumor suppressor as well as tumorigenesis functions (Fig. 4). AMPK activity suppresses tumor development and the kinase loss induces tumor progression in part by regulating cellular metabolic pathways that support cell growth and cell proliferation such as inhibiting aerobic glycolysis (the Warburg effect) in cancer cells. Genetic ablation of the AMPKα1 accelerates c-Myc overexpression in B cells as demonstrated *in vivo*, as well as in lung and colon cancer cells in culture *in vitro* (Faubert et al., 2013). These metabolic functions require HIF-1α as its silencing reverses those functions.

Growth factors and starvation induce phosphorylation with subsequent activation of AMPK through the DNA damage sensor ataxia-telangiectasia mutated (ATM). The ionizing radiation (IR) is also demonstrated to activate AMPK in human cancer cells in an LKB1-independent fashion, resulting in the induction of p21(waf/cip) and regulation of the cell cycle and cell survival (Fig. 4). The AMPK-mediated activity influencing the increased mRNA levels of the apoptotic genes p53 and p21 were also observed in MCF7 human breast cancer cells when treated with pomolic acid (PA) (Youn et al., 2012), an inducer of AMPK and inhibitor of the cancer cells growth. The stimulation of tumor suppressor genes (p53, p21, and p27)

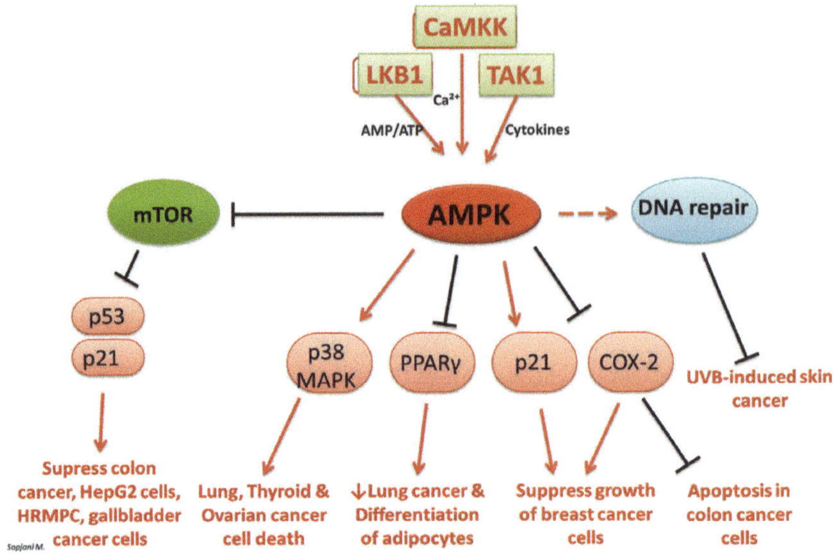

Fig. 4 Intracellular signaling indicating the AMPK role in tumor suppression. Upon activation by upstream kinases, the AMPK activation also leads to the inhibition of mTOR and through it participates in the regulation of tumor cell activity, an effect mediated by mTOR. The AMPK also increases cell death of lung, thyroid and ovarian cancer cells, decreases the differentiation of adipocytes, suppresses the breast cancer cells grown as well as regulates the apoptosis of colon cancer cells. All these regulations in a p38/MAPK-, PRAPγ-, p21-, COX-2-mediated manner. Further, AMPK indirectly participates in DNA repair process and through it, the kinase protects against UVB-induced skin cancer.

mediated by AMPK are reported in esophageal cancer induced by both oleic acid (OA) and alpha-linolenic acid (ALA) (Moon, Batriel, & Mantzoros, 2014), while these functions are abolished by AMPK siRNA administration. Further, AMPK is an upstream signal of prostaglandin-endoperoxide synthase 2 (COX-2, or cyclooxygenase 2). The kinase is essential for the anti-cancer effect of quercetin and that the AMPK mediated decrease of COX-2 expression is reported to participate in quercetin-mediated cancer control (Lee, Park, Kim, Lee, & Park, 2009). AMPK-COX-2 signaling pathway seems to be important in curcumin controlling adipocytes and MCF-7 breast tumor cells (Lee, Lee, et al., 2009). Noteworthy, AMPK inhibits the peroxisome proliferator-activated receptor-α and γ (PPAR-α and -γ), a key regulators of short- and long-term fatty acid oxidation. This inhibition can allow for short-term processes to increase energy generation before the cells devote resources to enhance their capacity for fatty acid oxidation (Sozio, Lu, Zeng, Liangpunsakul, & Crabb, 2011).

This signaling pathway inhibits lung cancer cell growth (Hann et al., 2014) and adipocyte differentiation *via* AMPK-mediated PPARγ inhibition. This signaling also prevents metabolic disorders such as obesity (Hwang et al., 2007) (Fig. 4), and a potential contributor to improving human health.

In cancer treatment, the induction of apoptosis in cancer cells represents one of the most important therapeutic strategies. The quercetin-induced apoptosis is realized by the activation of the AMPK/p38/MAPK (Mitogen-activated protein kinases) signaling pathway in the HT-29 colon cancer cells treated with quercetin (Kim, Lee, Kim, & Kim, 2014), and this process is also reported to be dependent on sestrin 2 as well as accompanied with intracellular ROS production. As presented in the Fig. 4, the AMPK and p38/MAPK signaling may be involved in the inhibition of cell proliferation and survival of thyroid cancer cells (Kim, Choi, Kim, Shong, & Kim, 2014), therefore suggesting that these signaling pathways may be useful as therapeutic targets to treat thyroid and colon cancers.

Skin cancers are one of the most dangerous forms of cancers that are mainly caused due to the DNA-damage by ultraviolet B (UVB) radiation. The AMPK has a pivotal role in reducing skin cancer incidence (Wu et al., 2013). This tumor suppressor activity of AMPK goes through stimulating the gene expression of the DNA repair protein *xeroderma pigmentosum* C (XPC), which results in preventing the UVB-induced skin tumorigenesis (Wu et al., 2013) and inducing DNA repair and controlling the cell proliferation in AMPK wild-type (WT) cells but not in AMPK-deficient cells (Fig. 4). This clearly indicates that AMPK-dependent cellular mechanism for skin tumors suppression suggests the AMPK to be a potential tool for cancer prevention and cancer therapy.

Berberine is an isoquinoline alkaloid that has several roles including. It is considered to be a safe anti-cancer agent. Berberine is reported to induce apoptosis in colorectal cancer cell lines in mice. Berberine also activates AMPK and inhibits another AMPK-downstream target mTOR with subsequent p53 phosphorylation. Berberine may also inhibit the activity of nuclear factor kappa-B (NF-κB). The tumor suppressor properties of berberine in suppressing colon epithelial proliferation and tumorigenesis are realized through AMPK-mediated inhibition of mTOR activity and AMPK-independent inhibition of NF-κB (Li et al., 2014). Activation of AMPK with subsequent inhibition of mTOR is also reported as a key mediator of plumbagin's anti-tumor activity in cultured human colon cancer cells (Chen et al., 2013). This pathway is involved in other cancer types as well, including gallbladder cancer cells (Wu et al., 2014), hormone-refractory metastatic prostate cancer (HRMPC) (Hsu et al., 2014) and hepatocellular carcinoma (HepG2 cells) (Yu et al., 2014).

2.3 AMPK role in the regulation of cellular transport

2.3.1 AMPK-mediated stimulatory effects on cellular transport

The molecule of glucose is an essential energy source that has a role in cellular energy balance. A cellular uptake of glucose depends on the presence of glucose transporters on the plasma membrane (Dermaku-Sopjani et al., 2014). Glucose transporters have been reported to be regulated by the AMPK (Sopjani, Bhavsar, Fraser, et al., 2010). In this regard, several studies reported AMPK as a powerful regulator of glucose uptake mediated by several facilitative glucose transporters such as Glut1 (Fryer et al., 2002; Pehmoller et al., 2009), Glut2 (Walker et al., 2005), Glut3 (Weisova, Concannon, Devocelle, Prehn, & Ward, 2009), and Glut4 (Fryer et al., 2002; Liu et al., 2013) as well as by Na^+-coupled glucose transporter SGLT1 (Sopjani, Bhavsar, Fraser, et al., 2010) and thus the kinase does not only upregulate passive but also secondary-active cellular entry of glucose.

The AMPK enhances the cell surface abundance of SGLT1 (Sopjani, Bhavsar, Fraser, et al., 2010) and Glut transporters (Fryer et al., 2002; Iwabuchi & Kawahara, 2013; Walker et al., 2005; Weisova et al., 2009), therefore it stimulates cellular glucose uptake. This mechanism, as in case of Glut4 transporter is reported (Pehmoller et al., 2009), involves the phosphorylation with subsequent inhibition of TBC1D1, a Rab GTPase activating protein (GAP) that regulates Glut4 translocation to the cell surface (Pehmoller et al., 2009) and/or PIKfyve (phosphatidylinositol 3-phosphate 5-kinase) phosphorylation by the AMPK. PIKfyve in turn phosphorylates phosphatidylinositol 3-phosphate (PtdIns3P) to phosphatidylinositol 3,5-bisphosphate (PtdIns(3,5)P2). Increased PtdIns(3,5)P2 levels help Glut4 translocation at the plasma membrane. Accordingly, AMPK participates in the insulin-induced glucose uptake (Langelueddecke et al., 2012; Liu et al., 2013). Noteworthy, cellular transporters are very important for various cellular processes (Almilaji et al., 2014; Dermaku-Sopjani et al., 2014; Fedorenko et al., 2009; Gehring et al., 2009; Kempe et al., 2010; Rexhepaj et al., 2010; Sopjani, Konjufca, Rinnerthaler, Rexhepaj, & Dermaku-Sopjani, 2016; Sopjani, Rinnerthaler, Almilaji, Ahmeti, & Dermaku-Sopjani, 2014).

Consistent with its complex function in cellular energy balance, AMPK-dependent signal pathways mediates acute exercise-induced increase of gene expression of the monocarboxylate transporters, MCT1 and MCT4, required for lactate output (Kitaoka et al., 2014; Takimoto, Takeyama, & Hamada, 2013), indicating a crucial role of the AMPK in lactic acid efflux and metabolism. However, the underlying mechanism is poorly understood.

Observations using human embryonic kidney (HEK) cells demonstrated that AMPK regulates the cytosolic pH, maintains neural pH balance respectively. The kinase stimulates the ubiquitously expresses Na^+/H^+ exchanger 1 (NHE1) (Rotte et al., 2010), NH3 (Pieri, Christian, Wilkins, Boyd, & Meredith, 2010) and NH5 (Jinadasa, Szabo, Numata, & Orlowski, 2014), thus it increases the highly pH-sensitive glycolytic flux (Rotte et al., 2010), a prerequisite for AMPK-sensitive induction of energy production by glycolysis. AMPK directly target and phosphorylate NHE5, while the mechanism of AMPK action on NHE1 and NHE3 remains to be determined.

In contrast to the AMPK general inhibitory activity found in numerous other ion channels as are described below, the evidence indicates the AMPK has a stimulation function in the cardiac voltage-gated sodium channel Nav1.5 (Light, Wallace, & Dyck, 2003). The authors revealed that Nav1.5 is a target of the AMPK.

In the erythroid cells are expressed several members of the canonical transient receptor potential canonical (TRPC) family of ion channels including the TRPC3, which is important for erythropoietin (Epo)-mediated calcium influx, response to Epo respectively (Sopjani, Foller, & Lang, 2008). This action implies the AMPK binding and phosphorylation site on TRPC3 C2 terminus (Hirschler-Laszkiewicz et al., 2011).

The ATP-sensitive potassium (K_{ATP}) channels play important roles in several biological processes, including insulin secretion and thus glucose homeostasis. One of the signaling molecules that targets K_{ATP} channels is the AMPK (Kudo et al., 1996). K_{ATP} channels are hetero-octameric complexes expressed in most of the tissues, including the heart, brain, kidney, skeletal and smooth muscle, and pancreatic β-cells (Budas, Jovanovic, Crawford, & Jovanovic, 2004; Lim et al., 2009; Yoshida et al., 2012). These channels are activated both by decreasing cellular ATP levels and increasing ADP levels during energy starvation. In the rat cardiomyocytes is revealed that AMPK causes channel activation through physically direct associating with K_{ATP} channel leading to an increase of K_{ATP} current in cardiomyocytes (Yoshida et al., 2012). However, in studies using pancreatic β cells, AMPK was reported both to stimulate (Lim et al., 2009) and to inhibit (Chang et al., 2009; Wang et al., 2005) K_{ATP} channel activity. The reason for this variance is still not known and needs further efforts to be solved.

The voltage-gated potassium ion channels Kv2.1 are important in the brain for the regulation of action potential frequency in central neurons (Misonou, Mohapatra, & Trimmer, 2005). The activation of AMPK in neurons under metabolic stress induces activation of ion channel Kv2.1 with consequent inhibition of neuronal excitability (Ikematsu et al., 2011). Reduced neuronal excitability conserves energy during metabolic stress (Sopjani, Rinnerthaler, Kruja, & Dermaku-Sopjani, 2015). Specifically, AMPK directly phosphorylates this channel at two specific sites (Ser^{440} and Ser^{537}) (Ikematsu et al., 2011) indicating the kinase protective role in conserving energy during metabolic stress. A summary of the AMPK-mediated transporter regulation is given in Table 1.

2.3.2 AMPK-mediated inhibitory effects on cellular transport

AMPK participates in the downregulation of numerous cell transporting molecules. For these inhibitory effects of AMPK, several mechanisms have been proposed. There are predominantly two main mechanisms, either by phosphorylation with subsequent inhibition or by Nedd4-2 promoting plasma membrane protein ubiquitination and endocytosis with subsequent degradation (Fig. 5). Of note, many cell transporters reported to being inhibited by the AMPK the underlying mechanisms are still elusive. This may lead to the new intracellular mechanisms in which the AMPK may be involved.

2.3.2.1 Downregulation of transport proteins by direct phosphorylation

Cystic fibrosis transmembrane conductance regulator (CFTR) acts as a cAMP-dependent chloride channel, expressed in various cell types and localized in the apical membrane of epithelial cells. The gene of CFTR is mutated in patients with cystic fibrosis (*CF*), a lung disease. It has been demonstrated that AMPK binds to and directly phosphorylates CFTR, primary at the Ser^{768} site of the CFTR R domain residue, considered to be a regulatory PKA site (Hallows, Raghuram, Kemp, Witters, & Foskett, 2000; King et al., 2009) (Fig. 5).

Another AMPK regulated target is the vacuolar H^+-ATPase (V-ATPase) that mediates ATP-driven H^+ cell transport across membranes. This ATPase is the major contributor for luminal acidification in the kidney collecting duct and non-volatile acid excretion (Breton, Smith, Lui, & Brown, 1996; Sopjani et al., 2011, 2015). The AMPK can directly bind to and phosphorylate the vacuolar H^+-ATPase A subunit at Ser^{384} that is a new regulatory mechanism of the H^+-ATPase by AMPK (Alzamora et al., 2013).

Table 1 Summary of cellular transporters regulated by AMPK.

Cell transporter	AMPK function	Effective through	Cell type	Ref.
Stimulatory effects of AMPK				
Nav1.5	Slowing of channel inactivation	Unknown	Rat ventricular myocytes	Light et al. (2003)
KATP ion channel	Direct channel activation and enhancing the ion channel plasma membrane expression	Unknown	Rat cardiomyocytes	Sukhodub et al. (2007), Yoshida et al. (2012)
BK ion channels	Upregulating the ion channel plasma membrane abundance	Unknown	*Xenopus* oocytes	Foller et al. (2012)
Kv2.1	Facilitating activation of the ion channel	Direct phosphorylation at defined sites of Ser-440 and Ser-537	HEK293 cells	Ikematsu et al. (2011)
Glut1	Upregulating of the transporter protein expression	Unclear	Muscle cells	Fryer et al. (2002)
Glut4	Upregulating of the transporter protein expression	Phosphorylation of TBC1D1 and/or PIKfyve	Muscle cells	Fryer et al. (2002), Liu et al. (2013)
Glut2	Upregulating of the transporter protein expression	Unclear	Jejunum tissue	Walker et al. (2005)
Glut3	Upregulating of the transporter protein expression	Activation of ATP-dependent P2Y receptor signaling, resulting in an increase in c-Fos and c-Jun proteins and by NF-κB mediated signaling	Astrocytes and neurons	Iwabuchi and Kawahara (2013), Weisova et al. (2009)
SGLT1	Upregulating of the transporter protein expression	Increased plasma membrane protein insertion	*Xenopus* oocytes	Sopjani, Bhavsar, Fraser, et al. (2010)

Continued

Table 1 Summary of cellular transporters regulated by AMPK.—cont'd

Cell transporter	AMPK function	Effective through	Cell type	Ref.
MCT1	Upregulating the mRNA transporter expression	Unclear	Muscle cells	Takimoto et al. (2013)
MCT4	Upregulating the mRNA transporter expression	Unclear	Muscle cells	Takimoto et al. (2013)
Na+/Ca^{2+} exchanger	Upregulating of the exchanger activity	Unclear	Dendritic cells	Nurbaeva et al. (2012)
Na+/H+ exchanger 1	Upregulating of the exchanger activity	Unclear	HEK cells	Rotte et al. (2010)
Na+/H+ exchanger 5	Upregulating of the exchanger protein plasma membrane abundance	Direct phosphorylation	Hippocampal neurons	Jinadasa et al. (2014)
TRPC6	Upregulating of ion channel activity	Unclear	hPASMC	Tang, To, Meng, Wang, and Gu (2010)
Na^+/K^+-ATPase	Upregulating of pump protein plasma membrane abundance and activity	Phosphatase PP2Ac induced-dephosphorylation of the pump α_1-subunit at Ser-18	Rat L6 myotubes and primary mouse muscle cells	Benziane et al. (2012)
Inhibitory effects of AMPK				
ENaC	Reducing of the ion channel plasma membrane abundance	Activation of Nedd4-2-mediated endocytosis	*Xenopus* oocytes	Almaca et al. (2009), Bhalla et al. (2006), Carattino et al. (2005)
KCNQ1/KCNE1	Reducing of the ion channel plasma membrane abundance	Activation of Nedd4-2-mediated endocytosis	*Xenopus* oocytes	Alesutan, Foller, et al. (2011)
Kv1.5	Reducing of the ion channel plasma membrane abundance	Activation of Nedd4-2-mediated endocytosis	*Xenopus* oocytes	Mia et al. (2012)

Kir2.1 (KCNJ2)	Reducing of the ion channel plasma membrane abundance	Activation of Nedd4-2-mediated endocytosis	*Xenopus* oocytes	Alesutan, Munoz, et al. (2011)
Kv7.1	Reducing of the ion channel plasma membrane abundance	Activation of Nedd4-2-mediated endocytosis	*Xenopus* oocytes	Andersen, Krzystanek, Jespersen, Olesen, and Rasmussen (2012)
hERG	Reducing of the ion channel plasma membrane abundance	Activation of Nedd4-2-mediated endocytosis	*Xenopus* oocytes and RD cells	Almilaji et al. (2013)
Orai1	Reducing of the ion channel plasma membrane abundance	Activation of Nedd4-2-mediated endocytosis	HEK293 cells	Eylenstein et al. (2011)
ROMK	Reducing of the ion channel plasma membrane abundance	Unclear	*Xenopus* oocytes	Siraskar et al. (2013)
Na^+/K^+ ATPase	Reducing of the protein plasma membrane abundance	Unclear	Alveolar epithelial cells	Gusarova et al. (2011)
NaPi-IIa	Reducing of the carrier plasma membrane abundance	Unclear	*Xenopus* oocytes	Dermaku-Sopjani, Almilaji, et al. (2013)
SMIT	Reducing of the transporter activity	Unclear	*Xenopus* oocytes	Munoz et al. (2012)
BGT1	Reducing of the transporter activity	Unclear	*Xenopus* oocytes	Munoz et al. (2012)
PepT1	Reducing of the carrier activity	Unclear	Caco-2 cells	Pieri et al. (2010)
EAAT3	Reducing of the carrier plasma membrane abundance without modifying carrier affinity	Activation of Nedd4-2-mediated endocytosis	*Xenopus* oocytes	Sopjani, Alesutan, Dermaku-Sopjani, et al. (2010)
EAAT4	Reducing of the carrier plasma membrane abundance without modifying carrier affinity	Activation of Nedd4-2-mediated endocytosis	*Xenopus* oocytes	Sopjani, Alesutan, Dermaku-Sopjani, et al. (2010)

Continued

Table 1 Summary of cellular transporters regulated by AMPK.—cont'd

Cell transporter	AMPK function	Effective through	Cell type	Ref.
CRT	Reducing of the $V_{(max)}$	Indirect effect (potentially) through mTOR pathway	*Xenopus* oocytes	Li et al. (2010)
TREK-1	Downregulating of the transporter activity	Channel phosphorylation at Ser^{300} and Ser^{333}	HEK293 cells	Kreneisz, Benoit, Bayliss, and Mulkey (2009)
TREK-2	Downregulating of the transporter activity	Channel phosphorylation at Ser-326 and Ser-359	HEK293 cells	Kreneisz et al. (2009)
Kir6.2	Downregulating of the ion channel activity	Channel phosphorylation at Ser-385	Rat islet beta cells	Chang et al. (2009)
BKCa channel	Downregulating of the ion channel activity	Direct alpha subunit channel phosphorylation	Rat carotid body type I cell	Wyatt et al. (2007)
$K_{Ca}3.1$	Downregulating of the transporter activity	Direct phosphorylation at the C-terminus region of the channel	Human airway epithelial cells	Klein et al. (2009)
Connexin 26	Inhibiting of the connexin plasma membrane abundance	Unknown	*Xenopus* oocytes	Alesutan, Sopjani, et al. (2011)
H^+-ATPase	Inhibiting of the protein plasma membrane abundance	Directly binds to and phosphorylate the H^+-ATPase A subunit at Ser-384	Rabbit kidney cells	Alzamora et al. (2013)

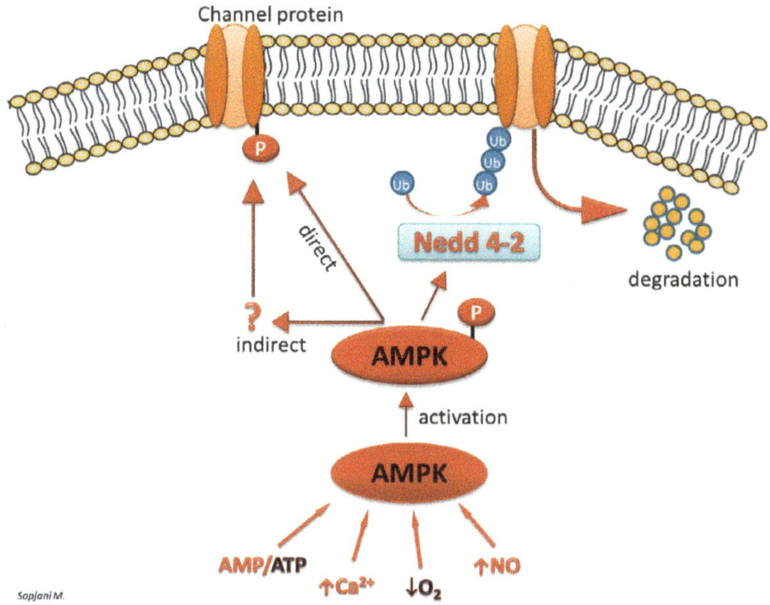

Fig. 5 A tentative model illustrating the downregulation of ion channels by the AMPK. AMPK activation can be stimulated by the variety of stimulators such as an increased AMP/ATP ratio, enhanced cytosolic Ca^{2+} activity, decreased of oxygen (O_2) levels, or increased nitric oxide (NO) production. Upon activation, the AMPK may further stimulate the ubiquitin ligase Nedd4-2, which, in turn, ubiquitinates a wide variety of membrane transports proteins. Alternatively, activated AMPK may also, at least partially, be effective through the phosphorylation of various plasma membrane transport proteins, either directly or indirectly.

Similarly, the large conductance calcium-activated potassium (BK_{Ca}) channels are also reported to be downregulated by the AMPK through direct phosphorylation (Wyatt et al., 2007). However, another study performed in *Xenopus* oocytes as well as $ampk^{-/-}$ and $ampk^{+/+}$ mice have shown AMPK as an upregulator of BK channel by enhancing membrane channel protein abundance (Foller et al., 2012). Nonetheless, the particular role of AMPK in BK channels in different cell types still needs further efforts for clarification. Further, AMPK takes part in the regulation of several O_2-sensitive K^+ channels. Specifically, it downregulates O_2 sensitive K^+ channels TREK and TASK (Dallas et al., 2009). The AMPK-mediated downregulation of TREK goes through serine residues phosphorylation (Ser^{300} and Ser^{333} for TREK-1 and Ser^{326} and Ser^{359} for TREK-2) in the C-terminus of the channel that are also sensitive for PKA and PKC channel modulation (Kreneisz et al., 2009). Moreover, the AMPK can phosphorylate and induce inhibition of the inwardly rectifying K^+ channel subunit (Kir6.2) in serine residue at position 385 in rat islet beta cells which may play a role in insulin secretion (Chang et al., 2009). Interestingly, AMPK has been demonstrated to phosphorylate and inhibit other K^+ channels, including the Ca^{2+}-activated K^+ channels $K_{Ca}3.1$. Experiments in human airway epithelial cells revealed a protein/protein interaction between the C-terminus region of the channel and the γ1-AMPK and for the formation of this complex the amino acids extending from Asp (380) to Ala (400) in C-terminus are reported to be essential (Klein et al., 2009).

2.3.2.2 Downregulation of transport proteins by Nedd4-2 activation

The amiloride-sensitive epithelial sodium channel (ENaC) is an important ion channel for sodium balance, water absorption in the lung, and blood pressure (Menniti et al., 2010; Sopjani & Dermaku-Sopjani, 2016), and is regulated by several regulators including AMPK (Bhalla et al., 2006). The ENaC, which mediates salt and water reabsorption in lung epithelia, is indirectly inhibited by AMPK by decreasing its cell membrane expression through induced ubiquitin ligase Nedd4-2 (neuronal precursor cells expressed developmentally downregulated)-dependent ENaC endocytosis and subsequent degradation as demonstrated in experiments using *Xenopus* oocytes and HEK293 cells (Bhalla et al., 2006) (Fig. 5). The AMPK-mediated downregulation of ENaC requires an interaction, with subsequent phosphorylation, between cytoplasmic tail binding motifs of the ENaC β-subunit for the Nedd4-2. However, the exact AMPK phosphorylation sites in Nedd4-2 are still unknown.

Except its effect in ENaC, there are also several other channels similarly, at least partially, regulated by AMPK in a Nedd4-2-mediated regulation mode, such as KCNQ1/KCNE1 (Alesutan, Foller, et al., 2011), Kir2.1 (Alesutan, Munoz, et al., 2011), Kv 7.1 (Andersen et al., 2012), Kv 1.5 (Mia et al., 2012), hERG (Almilaji et al., 2013) and Orai1/STIM1 (Eylenstein et al., 2011). All these channels are involved in a diverse range of cellular functions (Dermaku-Sopjani et al., 2014; Sopjani & Dermaku-Sopjani, 2016; Sopjani et al., 2016, 2014, 2015).

Importantly, several other ion channels (SCN5A, Kv1.3, ClC-2, Kv4.3) and carriers (SN1, NaPiIIb, EAAT1, EAAT2) have previously been shown to be regulated by Nedd4-2 (Bongiorno, Schuetz, Poronnik, & Adams, 2011; Lang et al., 2006, 2012; Palmada et al., 2004), may similarly be a possible target of AMPK regulation *via* Nedd4-2. Thereby, AMPK-dependent Nedd4-2 cellular transport regulation may represent a crucial mechanism for cellular transport regulation under cellular stress. The AMPK role may offer a potential for improved therapies in various human diseases and disorders through modulating the transport proteins involved in these abnormalities.

2.3.2.3 Downregulation of transport proteins by unclear mechanisms

Many other membrane transport proteins are reported to be regulated by the AMPK but through unclear mechanisms. One of this AMPK targets is the Na^+/K^+ ATPase. AMPK inhibits Na^+/K^+ ATPase activity and its cell surface abundance during hypoxia (Benziane et al., 2009; Gusarova et al., 2011). Since Na^+/K^+ ATPase is the most important intracellular ATP-consuming transport protein located in the plasma membrane, its inhibition upon AMPK activation has a profound effect on the cellular energy balance. AMPK-mediated inhibition of Na^+/K^+ ATPase decreases the driving force for Na^+-coupled transport.

The Na^+-coupled phosphate transporter, NaPi-IIa, is the most important cellular carrier for phosphate transport across the apical brush border membrane of proximal renal tubules (Bhandaru et al., 2011; Dermaku-Sopjani, Kolgeci, et al., 2013; Dermaku-Sopjani et al., 2011; Foller et al., 2011) that is responsible for renal phosphate reabsorption. NaPi-IIa uses the electrochemical gradient for Na+, which is maintained by the energy-consuming Na^+/K^+ ATPase in the basolateral cell membrane (Foller et al., 2011). The AMPK is a powerful downregulator of NaPi-IIa and thus of renal tubular phosphate transport (Dermaku-Sopjani, Almilaji, et al., 2013). Accordingly, the AMPK-sensitive downregulation of NaPi-IIa appears to be part of the protective effect of AMPK during energy depletion. Other Na^+-coupled carriers regulated by the AMPK (Sopjani, Alesutan, Dermaku-Sopjani, et al., 2010) are the excitatory amino acid transporters (EAAT), EAAT3 and EAAT4. The AMPK inhibition of the transport rate of both EAAT3 and EAAT4 is, at least partially, due to a decrease in their plasma membrane protein abundance without modifying carrier affinity.

Through undefined mechanisms, AMPK has also been reported to downregulate the cell membrane creatine transporter CreaT (Li et al., 2010), H^+-coupled di- and tripeptide transporter, PepT1 (Pieri et al., 2010), an important carrier for dietary protein absorption (Rexhepaj et al., 2009). The mechanism by which the AMPK regulates PepT1 is unclear, but certainly not through transporter phosphorylation since this transporter does not have AMPK phosphorylation sites (Hardie et al., 2003).

The cell volume is an important feature of the cell (Almilaji et al., 2014; Foller, Sopjani, Koka, et al., 2009). In the cell volume regulatory mechanisms are involved several ion channels (Alesutan, Munoz, et al., 2011; Sopjani, Kunert, Czarkowski, et al., 2010), exchangers (Nurbaeva et al., 2012), Na^+/K^+ ATPase (Sopjani, Alesutan, Wilmes, et al., 2010), and many carriers including osmolyte transporters, such as SMIT1 and BGT1 (Munoz et al., 2012). The AMPK downregulates the activity of both SMIT, a Na^+-coupled transporter of myo-inositol, and BGT1, betaine/GABA transporter, transporters (Munoz et al., 2012), and, thus, serves as a defense mechanism against cell swelling during energy depletion.

AMPK further downregulates the renal outer medullary K^+ channel ROMK (Siraskar et al., 2013), a potassium secretory channel located at the apical membrane of renal epithelial cells and involved in renal K^+ secretion and recycling (Garcia et al., 2014).

The AMPK inhibits the connexins (Cx) activity, including Cx26 (Alesutan, Sopjani, et al., 2011) and Cx43 (Alesutan et al., 2015). Connexins provide intercellular connections for passage of ions and small organic molecules and enabling intercellular communication. Timely closure of connexins in an ATP-depleted cell may be crucial for the survival of the neighboring cells, which may jeopardize their survival, thus, potentiating a significant role of AMPK in cells survival. Nonetheless, AMPK plays essential roles in the regulation of Ca^{2+} homeostasis by inhibiting of SOCE and Na^+/Ca^{2+} exchangers in bone marrow-derived dendritic cells (DCs), thereof their migration inhibition (Nurbaeva et al., 2012). Accordingly, AMPK may function as a modulator regulating Ca^{2+} levels in the cell in response to energy depletion. Many cellular functions are mediated by Ca^{2+} signaling.

3. Conclusions and perspectives

AMPK-mediated intracellular signaling is fairly complex and has a multitude of roles (Egan et al., 2011; Foller, Sopjani, Koka, et al., 2009; Hardie & Alessi, 2013; Jung et al., 2011; Kim et al., 2011; Lang et al., 2012; Winder et al., 1997; Woods, Johnstone, et al., 2003; Wu et al., 2014). This indicates to how important the effects of this kinase are indeed. It seems that AMPK has a crucial role in cell growth and autophagy processes. AMPK activation induces autophagy by phosphorylating TSC2 and RPTOR leading to the mTOR inactivation (Hong-Brown et al., 2010), thus, indirectly activating the ULK1 complex that induces autophagy in response to starvation (Jung et al., 2011). AMPK inhibits cell growth in response to energy depletion (Hardie et al., 2003; Wu et al., 2014). Further, AMPK activation enhances cancer cell apoptosis, autophagy and inhibits cell growth and proliferation and therefore exert its tumor suppressor activities (Imamura et al., 2001; Wu et al., 2013; Yu et al., 2014). AMPK is reported to protect against metabolic syndromes including diabetes, obesity, cardiovascular diseases, and inflammatory diseases (Eylenstein et al., 2012; Lang et al., 2012; Salminen, Hyttinen, & Kaarniranta, 2011). One of the major functions of the AMPK is in the regulation of cellular transport in the cell membrane (Sopjani, Alesutan, Dermaku-Sopjani, et al., 2010; Takimoto et al., 2013; Yoshida et al., 2012). AMPK-dependent cellular transport regulation imply a wide range of cell membrane transporter proteins, such as ion channels (Bongiorno et al., 2011; Dallas et al., 2009), cellular carriers (Munoz et al., 2012; Sopjani, Bhavsar, Fraser, et al., 2010) and pumps (Alzamora et al., 2013; Benziane et al., 2012), while the mode of action may be either stimulatory or inhibitory and realized by direct (Yoshida et al., 2012) and/or indirect (Li et al., 2010) kinase actions, influencing cellular transporter activity, genetic controls and protein expression (Andersen et al., 2012; Eylenstein et al., 2012). All these in accordance with the function of AMPK as a metabolic sensor that helps maintain energy balance. There is no doubt that future studies will reveal even more transport proteins and other target molecules to be regulated by AMPK. This along with elucidating the underlying molecular mechanisms involved in AMPK-dependent effects may point out AMPK as an interesting therapeutic target with great potential.

References

Alesutan, I., Foller, M., Sopjani, M., Dermaku-Sopjani, M., Zelenak, C., Frohlich, H., et al. (2011). Inhibition of the heterotetrameric K+ channel KCNQ1/KCNE1 by the AMP-activated protein kinase. *Molecular Membrane Biology*, *28*, 79–89.

Alesutan, I., Munoz, C., Sopjani, M., Dermaku-Sopjani, M., Michael, D., Fraser, S., et al. (2011). Inhibition of Kir2.1 (KCNJ2) by the AMP-activated protein kinase. *Biochemical and Biophysical Research Communications*, *408*, 505–510.

Alesutan, I., Sopjani, M., Munoz, C., Fraser, S., Kemp, B. E., Foller, M., et al. (2011). Inhibition of connexin 26 by the AMP-activated protein kinase. *The Journal of Membrane Biology*, *240*, 151–158.

Alesutan, I., Voelkl, J., Stockigt, F., Mia, S., Feger, M., Primessnig, U., et al. (2015). AMP-activated protein kinase alpha1 regulates cardiac gap junction protein connexin 43 and electrical remodeling following pressure overload. *Cellular Physiology and Biochemistry*, *35*, 406–418.

Almaca, J., Kongsuphol, P., Hieke, B., Ousingsawat, J., Viollet, B., Schreiber, R., et al. (2009). AMPK controls epithelial Na(+) channels through Nedd4-2 and causes an epithelial phenotype when mutated. *Pflügers Archiv*, *458*, 713–721.

Almilaji, A., Munoz, C., Elvira, B., Fajol, A., Pakladok, T., Honisch, S., et al. (2013). AMP-activated protein kinase regulates hERG potassium channel. *Pflügers Archiv*, *465*, 1573–1582.

Almilaji, A., Sopjani, M., Elvira, B., Borras, J., Dermaku-Sopjani, M., Munoz, C., et al. (2014). Upregulation of the creatine transporter Slc6A8 by Klotho. *Kidney & Blood Pressure Research, 39*, 516–525.

Alzamora, R., Al Bataineh, M. M., Liu, W., Gong, F., Li, H., Thali, R. F., et al. (2013). AMP-activated protein kinase regulates the vacuolar H+-ATPase via direct phosphorylation of the A subunit (ATP6V1A) in the kidney. *American Journal of Physiology. Renal Physiology, 305*, F943–F956.

Andersen, M. N., Krzystanek, K., Jespersen, T., Olesen, S. P., & Rasmussen, H. B. (2012). AMP-activated protein kinase downregulates Kv7.1 cell surface expression. *Traffic, 13*, 143–156.

Arad, M., Benson, D. W., Perez-Atayde, A. R., McKenna, W. J., Sparks, E. A., Kanter, R. J., et al. (2002). Constitutively active AMP kinase mutations cause glycogen storage disease mimicking hypertrophic cardiomyopathy. *The Journal of Clinical Investigation, 109*, 357–362.

Barnes, K., Ingram, J. C., Porras, O. H., Barros, L. F., Hudson, E. R., Fryer, L. G., et al. (2002). Activation of GLUT1 by metabolic and osmotic stress: Potential involvement of AMP-activated protein kinase (AMPK). *Journal of Cell Science, 115*, 2433–2442.

Bateman, A. (1997). The structure of a domain common to archaebacteria and the homocystinuria disease protein. *Trends in Biochemical Sciences, 22*, 12–13.

Bendayan, M., Londono, I., Kemp, B. E., Hardie, G. D., Ruderman, N., & Prentki, M. (2009). Association of AMP-activated protein kinase subunits with glycogen particles as revealed in situ by immunoelectron microscopy. *The Journal of Histochemistry and Cytochemistry, 57*, 963–971.

Benziane, B., Bjornholm, M., Lantier, L., Viollet, B., Zierath, J. R., & Chibalin, A. V. (2009). AMP-activated protein kinase activator A-769662 is an inhibitor of the Na(+)-K(+)-ATPase. *American Journal of Physiology. Cell Physiology, 297*, C1554–C1566.

Benziane, B., Bjornholm, M., Pirkmajer, S., Austin, R. L., Kotova, O., Viollet, B., et al. (2012). Activation of AMP-activated protein kinase stimulates Na+, K+-ATPase activity in skeletal muscle cells. *The Journal of Biological Chemistry, 287*, 23451–23463.

Bhalla, V., Oyster, N. M., Fitch, A. C., Wijngaarden, M. A., Neumann, D., Schlattner, U., et al. (2006). AMP-activated kinase inhibits the epithelial Na+ channel through functional regulation of the ubiquitin ligase Nedd4-2. *The Journal of Biological Chemistry, 281*, 26159–26169.

Bhandaru, M., Kempe, D. S., Rotte, A., Capuano, P., Pathare, G., Sopjani, M., et al. (2011). Decreased bone density and increased phosphaturia in gene-targeted mice lacking functional serum- and glucocorticoid-inducible kinase 3. *Kidney International, 80*, 61–67.

Bongiorno, D., Schuetz, F., Poronnik, P., & Adams, D. J. (2011). Regulation of voltage-gated ion channels in excitable cells by the ubiquitin ligases Nedd4 and Nedd4-2. *Channels (Austin, Tex.), 5*, 79–88.

Breton, S., Smith, P. J., Lui, B., & Brown, D. (1996). Acidification of the male reproductive tract by a proton pumping (H+)-ATPase. *Nature Medicine, 2*, 470–472.

Budas, G. R., Jovanovic, S., Crawford, R. M., & Jovanovic, A. (2004). Hypoxia-induced preconditioning in adult stimulated cardiomyocytes is mediated by the opening and trafficking of sarcolemmal KATP channels. *The FASEB Journal, 18*, 1046–1048.

Canto, C., Gerhart-Hines, Z., Feige, J. N., Lagouge, M., Noriega, L., Milne, J. C., et al. (2009). AMPK regulates energy expenditure by modulating NAD+ metabolism and SIRT1 activity. *Nature, 458*, 1056–1060.

Carattino, M. D., Edinger, R. S., Grieser, H. J., Wise, R., Neumann, D., Schlattner, U., et al. (2005). Epithelial sodium channel inhibition by AMP-activated protein kinase in oocytes and polarized renal epithelial cells. *The Journal of Biological Chemistry, 280*, 17608–17616.

Chang, T. J., Chen, W. P., Yang, C., Lu, P. H., Liang, Y. C., Su, M. J., et al. (2009). Serine-385 phosphorylation of inwardly rectifying K+ channel subunit (Kir6.2) by AMP-dependent protein kinase plays a key role in rosiglitazone-induced closure of the K(ATP) channel and insulin secretion in rats. *Diabetologia*, *52*, 1112–1121.
Chen, M. B., Zhang, Y., Wei, M. X., Shen, W., Wu, X. Y., Yao, C., et al. (2013). Activation of AMP-activated protein kinase (AMPK) mediates plumbagin-induced apoptosis and growth inhibition in cultured human colon cancer cells. *Cellular Signalling*, *25*, 1993–2002.
Cheung, P. C., Salt, I. P., Davies, S. P., Hardie, D. G., & Carling, D. (2000). Characterization of AMP-activated protein kinase gamma-subunit isoforms and their role in AMP binding. *The Biochemical Journal*, *346*(Pt. 3), 659–669.
Dallas, M. L., Scragg, J. L., Wyatt, C. N., Ross, F., Hardie, D. G., Evans, A. M., et al. (2009). Modulation of O(2) sensitive K (+) channels by AMP-activated protein kinase. *Advances in Experimental Medicine and Biology*, *648*, 57–63.
Davies, S. P., Helps, N. R., Cohen, P. T., & Hardie, D. G. (1995). 5'-AMP inhibits dephosphorylation, as well as promoting phosphorylation, of the AMP-activated protein kinase. Studies using bacterially expressed human protein phosphatase-2C alpha and native bovine protein phosphatase-2AC. *FEBS Letters*, *377*, 421–425.
Dermaku-Sopjani, M., Abazi, S., Faggio, C., Kolgeci, J., & Sopjani, M. (2014). AMPK-sensitive cellular transport. *Journal of Biochemistry*, *155*, 147–158.
Dermaku-Sopjani, M., Almilaji, A., Pakladok, T., Munoz, C., Hosseinzadeh, Z., Blecua, M., et al. (2013). Down-regulation of the Na-coupled phosphate transporter NaPi-IIa by AMP-activated protein kinase. *Kidney & Blood Pressure Research*, *37*, 547–556.
Dermaku-Sopjani, M., Kolgeci, S., Abazi, S., & Sopjani, M. (2013). Significance of the anti-aging protein Klotho. *Molecular Membrane Biology*, *30*, 369–385.
Dermaku-Sopjani, M., Sopjani, M., Saxena, A., Shojaiefard, M., Bogatikov, E., Alesutan, I., et al. (2011). Downregulation of NaPi-IIa and NaPi-IIb Na-coupled phosphate transporters by coexpression of Klotho. *Cellular Physiology and Biochemistry*, *28*, 251–258.
Dunlop, E. A., Hunt, D. K., Acosta-Jaquez, H. A., Fingar, D. C., & Tee, A. R. (2011). ULK1 inhibits mTORC1 signaling, promotes multisite Raptor phosphorylation and hinders substrate binding. *Autophagy*, *7*, 737–747.
Egan, D. F., Shackelford, D. B., Mihaylova, M. M., Gelino, S., Kohnz, R. A., Mair, W., et al. (2011). Phosphorylation of ULK1 (hATG1) by AMP-activated protein kinase connects energy sensing to mitophagy. *Science*, *331*, 456–461.
Evans, A. M., Mustard, K. J., Wyatt, C. N., Peers, C., Dipp, M., Kumar, P., et al. (2005). Does AMP-activated protein kinase couple inhibition of mitochondrial oxidative phosphorylation by hypoxia to calcium signaling in O_2-sensing cells? *The Journal of Biological Chemistry*, *280*, 41504–41511.
Eylenstein, A., Gehring, E. M., Heise, N., Shumilina, E., Schmidt, S., Szteyn, K., et al. (2011). Stimulation of Ca2+-channel Orai1/STIM1 by serum- and glucocorticoid-inducible kinase 1 (SGK1). *The FASEB Journal*, *25*, 2012–2021.
Eylenstein, A., Schmidt, S., Gu, S., Yang, W., Schmid, E., Schmidt, E. M., et al. (2012). Transcription factor NF-kappaB regulates expression of pore-forming Ca2+ channel unit, Orai1, and its activator, STIM1, to control Ca2+ entry and affect cellular functions. *The Journal of Biological Chemistry*, *287*, 2719–2730.
Faubert, B., Boily, G., Izreig, S., Griss, T., Samborska, B., Dong, Z., et al. (2013). AMPK is a negative regulator of the Warburg effect and suppresses tumor growth in vivo. *Cell Metabolism*, *17*, 113–124.
Fedorenko, O., Tang, C., Sopjani, M., Foller, M., Gehring, E. M., Strutz-Seebohm, N., et al. (2009). PIP5K2A-dependent regulation of excitatory amino acid transporter EAAT3. *Psychopharmacology*, *206*, 429–435.

Foller, M., Jaumann, M., Dettling, J., Saxena, A., Pakladok, T., Munoz, C., et al. (2012). AMP-activated protein kinase in BK-channel regulation and protection against hearing loss following acoustic overstimulation. *The FASEB Journal, 26*, 4243–4253.

Foller, M., Kempe, D. S., Boini, K. M., Pathare, G., Siraskar, B., Capuano, P., et al. (2011). PKB/SGK-resistant GSK3 enhances phosphaturia and calciuria. *Journal of the American Society of Nephrology, 22*, 873–880.

Foller, M., Sopjani, M., Koka, S., Gu, S., Mahmud, H., Wang, K., et al. (2009). Regulation of erythrocyte survival by AMP-activated protein kinase. *The FASEB Journal, 23*, 1072–1080.

Foller, M., Sopjani, M., Mahmud, H., & Lang, F. (2008). Vanadate-induced suicidal erythrocyte death. *Kidney & Blood Pressure Research, 31*, 87–93.

Foller, M., Sopjani, M., Schlemmer, H. P., Claussen, C. D., & Lang, F. (2009). Triggering of suicidal erythrocyte death by radiocontrast agents. *European Journal of Clinical Investigation, 39*, 576–583.

Fryer, L. G., Foufelle, F., Barnes, K., Baldwin, S. A., Woods, A., & Carling, D. (2002). Characterization of the role of the AMP-activated protein kinase in the stimulation of glucose transport in skeletal muscle cells. *The Biochemical Journal, 363*, 167–174.

Gan, B., Melkoumian, Z. K., Wu, X., Guan, K. L., & Guan, J. L. (2005). Identification of FIP200 interaction with the TSC1-TSC2 complex and its role in regulation of cell size control. *The Journal of Cell Biology, 170*, 379–389.

Gan, B., Peng, X., Nagy, T., Alcaraz, A., Gu, H., & Guan, J. L. (2006). Role of FIP200 in cardiac and liver development and its regulation of TNFalpha and TSC-mTOR signaling pathways. *The Journal of Cell Biology, 175*, 121–133.

Ganapathy, V., Thangaraju, M., & Prasad, P. D. (2009). Nutrient transporters in cancer: Relevance to Warburg hypothesis and beyond. *Pharmacology & Therapeutics, 121*, 29–40.

Garcia, M. L., Priest, B. T., Alonso-Galicia, M., Zhou, X., Felix, J. P., Brochu, R. M., et al. (2014). Pharmacologic inhibition of the renal outer medullary potassium channel causes diuresis and natriuresis in the absence of kaliuresis. *The Journal of Pharmacology and Experimental Therapeutics, 348*, 153–164.

Gehring, E. M., Zurn, A., Klaus, F., Laufer, J., Sopjani, M., Lindner, R., et al. (2009). Regulation of the glutamate transporter EAAT2 by PIKfyve. *Cellular Physiology and Biochemistry, 24*, 361–368.

Gowans, G. J., Hawley, S. A., Ross, F. A., & Hardie, D. G. (2013). AMP is a true physiological regulator of AMP-activated protein kinase by both allosteric activation and enhancing net phosphorylation. *Cell Metabolism, 18*, 556–566.

Gusarova, G. A., Trejo, H. E., Dada, L. A., Briva, A., Welch, L. C., Hamanaka, R. B., et al. (2011). Hypoxia leads to Na,K-ATPase downregulation via Ca(2+) release-activated Ca(2+) channels and AMPK activation. *Molecular and Cellular Biology, 31*, 3546–3556.

Gwinn, D. M., Shackelford, D. B., Egan, D. F., Mihaylova, M. M., Mery, A., Vasquez, D. S., et al. (2008). AMPK phosphorylation of raptor mediates a metabolic checkpoint. *Molecular Cell, 30*, 214–226.

Hallows, K. R., Raghuram, V., Kemp, B. E., Witters, L. A., & Foskett, J. K. (2000). Inhibition of cystic fibrosis transmembrane conductance regulator by novel interaction with the metabolic sensor AMP-activated protein kinase. *The Journal of Clinical Investigation, 105*, 1711–1721.

Hann, S. S., Tang, Q., Zheng, F., Zhao, S., Chen, J., & Wang, Z. (2014). Repression of phosphoinositide-dependent protein kinase 1 expression by ciglitazone via Egr-1 represents a new approach for inhibition of lung cancer cell growth. *Molecular Cancer, 13*, 149.

Hardie, D. G. (2011). AMP-activated protein kinase: An energy sensor that regulates all aspects of cell function. *Genes & Development, 25*, 1895–1908.

Hardie, D. G., & Alessi, D. R. (2013). LKB1 and AMPK and the cancer-metabolism link—Ten years after. *BMC Biology, 11*, 36.

Hardie, D. G., Carling, D., & Gamblin, S. J. (2011). AMP-activated protein kinase: Also regulated by ADP? *Trends in Biochemical Sciences*, *36*, 470–477.
Hardie, D. G., Scott, J. W., Pan, D. A., & Hudson, E. R. (2003). Management of cellular energy by the AMP-activated protein kinase system. *FEBS Letters*, *546*, 113–120.
Hawley, S. A., Davison, M., Woods, A., Davies, S. P., Beri, R. K., Carling, D., et al. (1996). Characterization of the AMP-activated protein kinase kinase from rat liver and identification of threonine 172 as the major site at which it phosphorylates AMP-activated protein kinase. *The Journal of Biological Chemistry*, *271*, 27879–27887.
Herzig, S., & Shaw, R. J. (2018). AMPK: Guardian of metabolism and mitochondrial homeostasis. *Nature Reviews. Molecular Cell Biology*, *19*, 121–135.
Hirschler-Laszkiewicz, I., Tong, Q., Waybill, K., Conrad, K., Keefer, K., Zhang, W., et al. (2011). The transient receptor potential (TRP) channel TRPC3 TRP domain and AMP-activated protein kinase binding site are required for TRPC3 activation by erythropoietin. *The Journal of Biological Chemistry*, *286*, 30636–30646.
Hong-Brown, L. Q., Brown, C. R., Kazi, A. A., Huber, D. S., Pruznak, A. M., & Lang, C. H. (2010). Alcohol and PRAS40 knockdown decrease mTOR activity and protein synthesis via AMPK signaling and changes in mTORC1 interaction. *Journal of Cellular Biochemistry*, *109*, 1172–1184.
Hsu, J. L., Liu, S. P., Lee, C. C., Hsu, L. C., Ho, Y. F., Huang, H. S., et al. (2014). A unique amidoanthraquinone derivative displays antiproliferative activity against human hormone-refractory metastatic prostate cancers through activation of LKB1-AMPK-mTOR signaling pathway. *Naunyn-Schmiedeberg's Archives of Pharmacology*, *387*, 979–990.
Hurley, R. L., Anderson, K. A., Franzone, J. M., Kemp, B. E., Means, A. R., & Witters, L. A. (2005). The Ca2+/calmodulin-dependent protein kinase kinases are AMP-activated protein kinase kinases. *The Journal of Biological Chemistry*, *280*, 29060–29066.
Hwang, J. T., Kim, S. H., Lee, M. S., Kim, S. H., Yang, H. J., Kim, M. J., et al. (2007). Anti-obesity effects of ginsenoside Rh2 are associated with the activation of AMPK signaling pathway in 3T3-L1 adipocyte. *Biochemical and Biophysical Research Communications*, *364*, 1002–1008.
Ikematsu, N., Dallas, M. L., Ross, F. A., Lewis, R. W., Rafferty, J. N., David, J. A., et al. (2011). Phosphorylation of the voltage-gated potassium channel Kv2.1 by AMP-activated protein kinase regulates membrane excitability. *Proceedings of the National Academy of Sciences of the United States of America*, *108*, 18132–18137.
Imamura, K., Ogura, T., Kishimoto, A., Kaminishi, M., & Esumi, H. (2001). Cell cycle regulation via p53 phosphorylation by a 5'-AMP activated protein kinase activator, 5-aminoimidazole- 4-carboxamide-1-beta-D-ribofuranoside, in a human hepatocellular carcinoma cell line. *Biochemical and Biophysical Research Communications*, *287*, 562–567.
Inoki, K., Zhu, T., & Guan, K. L. (2003). TSC2 mediates cellular energy response to control cell growth and survival. *Cell*, *115*, 577–590.
Iwabuchi, S., & Kawahara, K. (2013). Extracellular ATP-prinoceptor signaling and AMP-activated protein kinase regulate astrocytic glucose transporter 3 in an in vitro ischemia. *Neurochemistry International*, *63*, 259–268.
Ix, J. H., & Sharma, K. (2010). Mechanisms linking obesity, chronic kidney disease, and fatty liver disease: The roles of fetuin-A, adiponectin, and AMPK. *Journal of the American Society of Nephrology*, *21*, 406–412.
Jensen, T. E., Ross, F. A., Kleinert, M., Sylow, L., Knudsen, J. R., Gowans, G. J., et al. (2015). PT-1 selectively activates AMPK-gamma1 complexes in mouse skeletal muscle, but activates all three gamma subunit complexes in cultured human cells by inhibiting the respiratory chain. *The Biochemical Journal*, *467*, 461–472.
Jeon, S. M., Chandel, N. S., & Hay, N. (2012). AMPK regulates NADPH homeostasis to promote tumour cell survival during energy stress. *Nature*, *485*, 661–665.

Jinadasa, T., Szabo, E. Z., Numata, M., & Orlowski, J. (2014). Activation of AMP-activated protein kinase regulates hippocampal neuronal pH by recruiting Na+/H+ exchanger NHE5 to the cell surface. *The Journal of Biological Chemistry*, *289*, 20879–20897.

Jones, R. G., Plas, D. R., Kubek, S., Buzzai, M., Mu, J., Xu, Y., et al. (2005). AMP-activated protein kinase induces a p53-dependent metabolic checkpoint. *Molecular Cell*, *18*, 283–293.

Jung, C. H., Seo, M., Otto, N. M., & Kim, D. H. (2011). ULK1 inhibits the kinase activity of mTORC1 and cell proliferation. *Autophagy*, *7*, 1212–1221.

Kempe, D. S., Dermaku-Sopjani, M., Frohlich, H., Sopjani, M., Umbach, A., Puchchakayala, G., et al. (2010). Rapamycin-induced phosphaturia. *Nephrology, Dialysis, Transplantation*, *25*, 2938–2944.

Kim, W. G., Choi, H. J., Kim, T. Y., Shong, Y. K., & Kim, W. B. (2014). The effect of 5-aminoimidazole-4-carboxamide-ribonucleoside was mediated by p38 mitogen activated protein kinase signaling pathway in FRO thyroid cancer cells. *The Korean Journal of Internal Medicine*, *29*, 474–481.

Kim, J., Kundu, M., Viollet, B., & Guan, K. L. (2011). AMPK and mTOR regulate autophagy through direct phosphorylation of Ulk1. *Nature Cell Biology*, *13*, 132–141.

Kim, G. T., Lee, S. H., Kim, J. I., & Kim, Y. M. (2014). Quercetin regulates the sestrin 2-AMPK-p38 MAPK signaling pathway and induces apoptosis by increasing the generation of intracellular ROS in a p53-independent manner. *International Journal of Molecular Medicine*, *33*, 863–869.

Kim, Y. D., Park, K. G., Lee, Y. S., Park, Y. Y., Kim, D. K., Nedumaran, B., et al. (2008). Metformin inhibits hepatic gluconeogenesis through AMP-activated protein kinase-dependent regulation of the orphan nuclear receptor SHP. *Diabetes*, *57*, 306–314.

Kim, Y. K., Park, J. H., Park, S. H., Lim, B., Baek, W. K., Suh, S. I., et al. (2010). Protective role of glucagon-like peptide-1 against glucosamine-induced cytotoxicity in pancreatic beta cells. *Cellular Physiology and Biochemistry*, *25*, 211–220.

King, J. D., Jr., Fitch, A. C., Lee, J. K., McCane, J. E., Mak, D. O., Foskett, J. K., et al. (2009). AMP-activated protein kinase phosphorylation of the R domain inhibits PKA stimulation of CFTR. *American Journal of Physiology. Cell Physiology*, *297*, C94–101.

Kitaoka, Y., Takahashi, Y., Machida, M., Takeda, K., Takemasa, T., & Hatta, H. (2014). Effect of AMPK activation on monocarboxylate transporter (MCT)1 and MCT4 in denervated muscle. *The Journal of Physiological Sciences*, *64*, 59–64.

Klein, H., Garneau, L., Trinh, N. T., Prive, A., Dionne, F., Goupil, E., et al. (2009). Inhibition of the KCa3.1 channels by AMP-activated protein kinase in human airway epithelial cells. *American Journal of Physiology. Cell Physiology*, *296*, C285–C295.

Kreneisz, O., Benoit, J. P., Bayliss, D. A., & Mulkey, D. K. (2009). AMP-activated protein kinase inhibits TREK channels. *The Journal of Physiology*, *587*, 5819–5830.

Kudo, N., Gillespie, J. G., Kung, L., Witters, L. A., Schulz, R., Clanachan, A. S., et al. (1996). Characterization of 5'AMP-activated protein kinase activity in the heart and its role in inhibiting acetyl-CoA carboxylase during reperfusion following ischemia. *Biochimica et Biophysica Acta*, *1301*, 67–75.

Lang, F., Bohmer, C., Palmada, M., Seebohm, G., Strutz-Seebohm, N., & Vallon, V. (2006). (Patho)physiological significance of the serum- and glucocorticoid-inducible kinase isoforms. *Physiological Reviews*, *86*, 1151–1178.

Lang, F., Eylenstein, A., & Shumilina, E. (2012). Regulation of Orai1/STIM1 by the kinases SGK1 and AMPK. *Cell Calcium*, *52*, 347–354.

Langelueddecke, C., Jakab, M., Ketterl, N., Lehner, L., Hufnagl, C., Schmidt, S., et al. (2012). Effect of the AMP-kinase modulators AICAR, metformin and compound C on insulin secretion of INS-1E rat insulinoma cells under standard cell culture conditions. *Cellular Physiology and Biochemistry*, *29*, 75–86.

Lee, Y. K., Lee, W. S., Hwang, J. T., Kwon, D. Y., Surh, Y. J., & Park, O. J. (2009). Curcumin exerts antidifferentiation effect through AMPKalpha-PPAR-gamma in 3T3-L1 adipocytes and antiproliferatory effect through AMPKalpha-COX-2 in cancer cells. *Journal of Agricultural and Food Chemistry*, *57*, 305–310.

Lee, Y. K., Park, S. Y., Kim, Y. M., Lee, W. S., & Park, O. J. (2009). AMP kinase/cyclooxygenase-2 pathway regulates proliferation and apoptosis of cancer cells treated with quercetin. *Experimental & Molecular Medicine*, *41*, 201–207.

Li, W., Hua, B., Saud, S. M., Lin, H., Hou, W., Matter, M. S., et al. (2014). Berberine regulates AMP-activated protein kinase signaling pathways and inhibits colon tumorigenesis in mice. *Molecular Carcinogenesis*, *54*, 1096–1109.

Li, H., Thali, R. F., Smolak, C., Gong, F., Alzamora, R., Wallimann, T., et al. (2010). Regulation of the creatine transporter by AMP-activated protein kinase in kidney epithelial cells. *American Journal of Physiology. Renal Physiology*, *299*, F167–F177.

Liang, J., Shao, S. H., Xu, Z. X., Hennessy, B., Ding, Z., Larrea, M., et al. (2007). The energy sensing LKB1-AMPK pathway regulates p27(kip1) phosphorylation mediating the decision to enter autophagy or apoptosis. *Nature Cell Biology*, *9*, 218–224.

Light, P. E., Wallace, C. H., & Dyck, J. R. (2003). Constitutively active adenosine monophosphate-activated protein kinase regulates voltage-gated sodium channels in ventricular myocytes. *Circulation*, *107*, 1962–1965.

Lim, A., Park, S. H., Sohn, J. W., Jeon, J. H., Park, J. H., Song, D. K., et al. (2009). Glucose deprivation regulates KATP channel trafficking via AMP-activated protein kinase in pancreatic beta-cells. *Diabetes*, *58*, 2813–2819.

Lin, S. Y., Li, T. Y., Liu, Q., Zhang, C., Li, X., Chen, Y., et al. (2012). GSK3-TIP60-ULK1 signaling pathway links growth factor deprivation to autophagy. *Science*, *336*, 477–481.

Lira, V. A., Soltow, Q. A., Long, J. H., Betters, J. L., Sellman, J. E., & Criswell, D. S. (2007). Nitric oxide increases GLUT4 expression and regulates AMPK signaling in skeletal muscle. *American Journal of Physiology. Endocrinology and Metabolism*, *293*, E1062–E1068.

Liu, Y., Lai, Y. C., Hill, E. V., Tyteca, D., Carpentier, S., Ingvaldsen, A., et al. (2013). Phosphatidylinositol 3-phosphate 5-kinase (PIKfyve) is an AMPK target participating in contraction-stimulated glucose uptake in skeletal muscle. *The Biochemical Journal*, *455*, 195–206.

Loffler, A. S., Alers, S., Dieterle, A. M., Keppeler, H., Franz-Wachtel, M., Kundu, M., et al. (2011). Ulk1-mediated phosphorylation of AMPK constitutes a negative regulatory feedback loop. *Autophagy*, *7*, 696–706.

Mack, H. I., Zheng, B., Asara, J. M., & Thomas, S. M. (2012). AMPK-dependent phosphorylation of ULK1 regulates ATG9 localization. *Autophagy*, *8*, 1197–1214.

Malaisse, W. J., Conget, I., Sener, A., & Rorsman, P. (1994). Insulinotropic action of AICA riboside. II. Secretory, metabolic and cationic aspects. *Diabetes Research*, *25*, 25–37.

Marsin, A. S., Bouzin, C., Bertrand, L., & Hue, L. (2002). The stimulation of glycolysis by hypoxia in activated monocytes is mediated by AMP-activated protein kinase and inducible 6-phosphofructo-2-kinase. *The Journal of Biological Chemistry*, *277*, 30778–30783.

McBride, A., Ghilagaber, S., Nikolaev, A., & Hardie, D. G. (2009). The glycogen-binding domain on the AMPK beta subunit allows the kinase to act as a glycogen sensor. *Cell Metabolism*, *9*, 23–34.

McFadden, J. W., & Corl, B. A. (2009). Activation of AMP-activated protein kinase (AMPK) inhibits fatty acid synthesis in bovine mammary epithelial cells. *Biochemical and Biophysical Research Communications*, *390*, 388–393.

Menniti, M., Iuliano, R., Foller, M., Sopjani, M., Alesutan, I., Mariggio, S., et al. (2010). 60 kDa lysophospholipase, a new Sgk1 molecular partner involved in the regulation of ENaC. *Cellular Physiology and Biochemistry*, *26*, 587–596.

Mia, S., Munoz, C., Pakladok, T., Siraskar, G., Voelkl, J., Alesutan, I., et al. (2012). Down-regulation of Kv1.5 K channels by the AMP-activated protein kinase. *Cellular Physiology and Biochemistry*, *30*, 1039–1050.

Misonou, H., Mohapatra, D. P., & Trimmer, J. S. (2005). Kv2.1: A voltage-gated k+ channel critical to dynamic control of neuronal excitability. *Neurotoxicology*, *26*, 743–752.

Mizushima, N., & Komatsu, M. (2011). Autophagy: Renovation of cells and tissues. *Cell*, *147*, 728–741.

Moon, H. S., Batriel, S., & Mantzoros, C. S. (2014). Alpha linolenic acid and oleic acid additively down-regulate malignant potential and positively cross-regulate AMPK/S6 axis in OE19 and OE33 esophageal cancer cells. *Metabolism: Clinical and Experimental*, *63*, 1447–1454.

Munoz, C., Sopjani, M., Dermaku-Sopjani, M., Almilaji, A., Foller, M., & Lang, F. (2012). Downregulation of the osmolyte transporters SMIT and BGT1 by AMP-activated protein kinase. *Biochemical and Biophysical Research Communications*, *422*, 358–362.

Nurbaeva, M. K., Schmid, E., Szteyn, K., Yang, W., Viollet, B., Shumilina, E., et al. (2012). Enhanced Ca(2)(+) entry and Na+/Ca(2)(+) exchanger activity in dendritic cells from AMP-activated protein kinase-deficient mice. *The FASEB Journal*, *26*, 3049–3058.

Oakhill, J. S., Chen, Z. P., Scott, J. W., Steel, R., Castelli, L. A., Ling, N., et al. (2010). Beta-subunit myristoylation is the gatekeeper for initiating metabolic stress sensing by AMP-activated protein kinase (AMPK). *Proceedings of the National Academy of Sciences of the United States of America*, *107*, 19237–19241.

Palmada, M., Dieter, M., Speil, A., Bohmer, C., Mack, A. F., Wagner, H. J., et al. (2004). Regulation of intestinal phosphate cotransporter NaPi IIb by ubiquitin ligase Nedd4-2 and by serum- and glucocorticoid-dependent kinase 1. *American Journal of Physiology Gastrointestinal and Liver Physiology*, *287*, G143–G150.

Park, H. B., Marklund, S., Jeon, J. T., Mickelson, J. R., Valberg, S. J., Sandberg, K., et al. (2003). Molecular characterization and mutational screening of the PRKAG3 gene in the horse. *Cytogenetic and Genome Research*, *102*, 211–216.

Pehmoller, C., Treebak, J. T., Birk, J. B., Chen, S., Mackintosh, C., Hardie, D. G., et al. (2009). Genetic disruption of AMPK signaling abolishes both contraction- and insulin-stimulated TBC1D1 phosphorylation and 14-3-3 binding in mouse skeletal muscle. *American Journal of Physiology. Endocrinology and Metabolism*, *297*, E665–E675.

Pieri, M., Christian, H. C., Wilkins, R. J., Boyd, C. A., & Meredith, D. (2010). The apical (hPepT1) and basolateral peptide transport systems of Caco-2 cells are regulated by AMP-activated protein kinase. *American Journal of Physiology. Gastrointestinal and Liver Physiology*, *299*, G136–G143.

Rexhepaj, R., Dermaku-Sopjani, M., Gehring, E. M., Sopjani, M., Kempe, D. S., Foller, M., et al. (2010). Stimulation of electrogenic glucose transport by glycogen synthase kinase 3. *Cellular Physiology and Biochemistry*, *26*, 641–646.

Rexhepaj, R., Rotte, A., Kempe, D. S., Sopjani, M., Foller, M., Gehring, E. M., et al. (2009). Stimulation of electrogenic intestinal dipeptide transport by the glucocorticoid dexamethasone. *Pflügers Archiv*, *459*, 191–202.

Ross, F. A., Mackintosh, C., & Hardie, D. G. (2016). AMP-activated protein kinase: A cellular energy sensor that comes in 12 flavours. *The FEBS Journal*, *283*, 2987–3001.

Rotte, A., Pasham, V., Eichenmuller, M., Bhandaru, M., Foller, M., & Lang, F. (2010). Upregulation of Na+/H+ exchanger by the AMP-activated protein kinase. *Biochemical and Biophysical Research Communications*, *398*, 677–682.

Salminen, A., Hyttinen, J. M., & Kaarniranta, K. (2011). AMP-activated protein kinase inhibits NF-kappaB signaling and inflammation: Impact on healthspan and lifespan. *Journal of Molecular Medicine (Berlin, Germany)*, *89*, 667–676.

Sanders, M. J., Grondin, P. O., Hegarty, B. D., Snowden, M. A., & Carling, D. (2007). Investigating the mechanism for AMP activation of the AMP-activated protein kinase cascade. *The Biochemical Journal, 403*, 139–148.

Shang, L., Chen, S., Du, F., Li, S., Zhao, L., & Wang, X. (2011). Nutrient starvation elicits an acute autophagic response mediated by Ulk1 dephosphorylation and its subsequent dissociation from AMPK. *Proceedings of the National Academy of Sciences of the United States of America, 108*, 4788–4793.

Siraskar, B., Huang, D. Y., Pakladok, T., Siraskar, G., Sopjani, M., Alesutan, I., et al. (2013). Downregulation of the renal outer medullary K(+) channel ROMK by the AMP-activated protein kinase. *Pflügers Archiv, 465*, 233–245.

Sopjani, M., Alesutan, I., Dermaku-Sopjani, M., Fraser, S., Kemp, B. E., Foller, M., et al. (2010). Down-regulation of Na+-coupled glutamate transporter EAAT3 and EAAT4 by AMP-activated protein kinase. *Journal of Neurochemistry, 113*, 1426–1435.

Sopjani, M., Alesutan, I., Dermaku-Sopjani, M., Gu, S., Zelenak, C., Munoz, C., et al. (2011). Regulation of the Na^+/K^+ ATPase by Klotho. *FEBS Letters, 585*, 1759–1764.

Sopjani, M., Alesutan, I., Wilmes, J., Dermaku-Sopjani, M., Lam, R. S., Koutsouki, E., et al. (2010). Stimulation of Na^+/K^+ ATPase activity and Na+ coupled glucose transport by beta-catenin. *Biochemical and Biophysical Research Communications, 402*, 467–470.

Sopjani, M., Bhavsar, S. K., Fraser, S., Kemp, B. E., Foller, M., & Lang, F. (2010). Regulation of Na+-coupled glucose carrier SGLT1 by AMP-activated protein kinase. *Molecular Membrane Biology, 27*, 137–144.

Sopjani, M., & Dermaku-Sopjani, M. (2016). Klotho-dependent cellular transport regulation. *Vitamins and Hormones, 101*, 59–84.

Sopjani, M., Foller, M., Dreischer, P., & Lang, F. (2008). Stimulation of eryptosis by cadmium ions. *Cellular Physiology and Biochemistry, 22*, 245–252.

Sopjani, M., Foller, M., Gulbins, E., & Lang, F. (2008). Suicidal death of erythrocytes due to selenium-compounds. *Cellular Physiology and Biochemistry, 22*, 387–394.

Sopjani, M., Foller, M., Haendeler, J., Gotz, F., & Lang, F. (2009). Silver ion-induced suicidal erythrocyte death. *Journal of Applied Toxicology, 29*, 531–536.

Sopjani, M., Foller, M., & Lang, F. (2008). Gold stimulates Ca2+ entry into and subsequent suicidal death of erythrocytes. *Toxicology, 244*, 271–279.

Sopjani, M., Konjufca, V., Rinnerthaler, M., Rexhepaj, R., & Dermaku-Sopjani, M. (2016). The relevance of JAK2 in the regulation of cellular transport. *Current Medicinal Chemistry, 23*, 578–588.

Sopjani, M., Kunert, A., Czarkowski, K., Klaus, F., Laufer, J., Foller, M., et al. (2010). Regulation of the Ca(2+) channel TRPV6 by the kinases SGK1, PKB/Akt, and PIKfyve. *The Journal of Membrane Biology, 233*, 35–41.

Sopjani, M., Millaku, L., Nebija, D., Emini, M., Dermaku-Sopjani, M., & Rifati-Nixha, A. (2018). The glycogen synthase kinase-3 in the regulation of ion channels and cellular carriers. *Current Medicinal Chemistry.*

Sopjani, M., Rinnerthaler, M., Almilaji, A., Ahmeti, S., & Dermaku-Sopjani, M. (2014). Regulation of cellular transport by klotho protein. *Current Protein & Peptide Science, 15*, 828–835.

Sopjani, M., Rinnerthaler, M., Kruja, J., & Dermaku-Sopjani, M. (2015). Intracellular signaling of the aging suppressor protein Klotho. *Current Molecular Medicine, 15*, 27–37.

Sozio, M. S., Lu, C., Zeng, Y., Liangpunsakul, S., & Crabb, D. W. (2011). Activated AMPK inhibits PPAR-{alpha} and PPAR-{gamma} transcriptional activity in hepatoma cells. *American Journal of Physiology. Gastrointestinal and Liver Physiology, 301*, G739–G747.

Stapleton, D., Mitchelhill, K. I., Gao, G., Widmer, J., Michell, B. J., Teh, T., et al. (1996). Mammalian AMP-activated protein kinase subfamily. *The Journal of Biological Chemistry, 271*, 611–614.

Sukhodub, A., Jovanovic, S., Du, Q., Budas, G., Clelland, A. K., Shen, M., et al. (2007). AMP-activated protein kinase mediates preconditioning in cardiomyocytes by regulating activity and trafficking of sarcolemmal ATP-sensitive K(+) channels. *Journal of Cellular Physiology*, *210*, 224–236.

Takimoto, M., Takeyama, M., & Hamada, T. (2013). Possible involvement of AMPK in acute exercise-induced expression of monocarboxylate transporters MCT1 and MCT4 mRNA in fast-twitch skeletal muscle. *Metabolism*, *62*, 1633–1640.

Tang, C., To, W. K., Meng, F., Wang, Y., & Gu, Y. (2010). A role for receptor-operated Ca2+ entry in human pulmonary artery smooth muscle cells in response to hypoxia. *Physiological Research*, *59*, 909–918.

Thornton, C., Snowden, M. A., & Carling, D. (1998). Identification of a novel AMP-activated protein kinase beta subunit isoform that is highly expressed in skeletal muscle. *The Journal of Biological Chemistry*, *273*, 12443–12450.

Tsukamoto, S., Hara, T., Yamamoto, A., Kito, S., Minami, N., Kubota, T., et al. (2014). Fluorescence-based visualization of autophagic activity predicts mouse embryo viability. *Scientific Reports*, *4*, 4533.

Tyan, L., Sopjani, M., Dermaku-Sopjani, M., Schmid, E., Yang, W., Xuan, N. T., et al. (2010). Inhibition of voltage-gated K+ channels in dendritic cells by rapamycin. *American Journal of Physiology. Cell Physiology*, *299*, C1379–C1385.

Vander Heiden, M. G., Cantley, L. C., & Thompson, C. B. (2009). Understanding the Warburg effect: The metabolic requirements of cell proliferation. *Science*, *324*, 1029–1033.

Walker, J., Jijon, H. B., Diaz, H., Salehi, P., Churchill, T., & Madsen, K. L. (2005). 5-aminoimidazole-4-carboxamide riboside (AICAR) enhances GLUT2-dependent jejunal glucose transport: A possible role for AMPK. *The Biochemical Journal*, *385*, 485–491.

Wang, C. Z., Wang, Y., Di, A., Magnuson, M. A., Ye, H., Roe, M. W., et al. (2005). 5-amino-imidazole carboxamide riboside acutely potentiates glucose-stimulated insulin secretion from mouse pancreatic islets by KATP channel-dependent and -independent pathways. *Biochemical and Biophysical Research Communications*, *330*, 1073–1079.

Warden, S. M., Richardson, C., O'Donnell, J., Jr., Stapleton, D., Kemp, B. E., & Witters, L. A. (2001). Post-translational modifications of the beta-1 subunit of AMP-activated protein kinase affect enzyme activity and cellular localization. *The Biochemical Journal*, *354*, 275–283.

Watanabe, T., Kuma, A., & Mizushima, N. (2011). Physiological role of autophagy in metabolism and its regulation mechanism. *Nihon Rinsho*, *69*(Suppl. 1), 775–781.

Weisova, P., Concannon, C. G., Devocelle, M., Prehn, J. H., & Ward, M. W. (2009). Regulation of glucose transporter 3 surface expression by the AMP-activated protein kinase mediates tolerance to glutamate excitation in neurons. *The Journal of Neuroscience*, *29*, 2997–3008.

Winder, W. W., Wilson, H. A., Hardie, D. G., Rasmussen, B. B., Hutber, C. A., Call, G. B., et al. (1997). Phosphorylation of rat muscle acetyl-CoA carboxylase by AMP-activated protein kinase and protein kinase A. *Journal of Applied Physiology*, *82*, 219–225.

Woods, A., Johnstone, S. R., Dickerson, K., Leiper, F. C., Fryer, L. G., Neumann, D., et al. (2003). LKB1 is the upstream kinase in the AMP-activated protein kinase cascade. *Current Biology*, *13*, 2004–2008.

Woods, A., Vertommen, D., Neumann, D., Turk, R., Bayliss, J., Schlattner, U., et al. (2003). Identification of phosphorylation sites in AMP-activated protein kinase (AMPK) for upstream AMPK kinases and study of their roles by site-directed mutagenesis. *The Journal of Biological Chemistry*, *278*, 28434–28442.

Wu, W. D., Hu, Z. M., Shang, M. J., Zhao, D. J., Zhang, C. W., Hong, D. F., et al. (2014). Cordycepin down-regulates multiple drug resistant (MDR)/HIF-1alpha through regulating AMPK/mTORC1 signaling in GBC-SD gallbladder cancer cells. *International Journal of Molecular Sciences*, *15*, 12778–12790.

Wu, C. L., Qiang, L., Han, W., Ming, M., Viollet, B., & He, Y. Y. (2013). Role of AMPK in UVB-induced DNA damage repair and growth control. *Oncogene*, *32*, 2682–2689.
Wyatt, C. N., Mustard, K. J., Pearson, S. A., Dallas, M. L., Atkinson, L., Kumar, P., et al. (2007). AMP-activated protein kinase mediates carotid body excitation by hypoxia. *The Journal of Biological Chemistry*, *282*, 8092–8098.
Xiao, B., Heath, R., Saiu, P., Leiper, F. C., Leone, P., Jing, C., et al. (2007). Structural basis for AMP binding to mammalian AMP-activated protein kinase. *Nature*, *449*, 496–500.
Yoshida, H., Bao, L., Kefaloyianni, E., Taskin, E., Okorie, U., Hong, M., et al. (2012). AMP-activated protein kinase connects cellular energy metabolism to KATP channel function. *Journal of Molecular and Cellular Cardiology*, *52*, 410–418.
Youn, S. H., Lee, J. S., Lee, M. S., Cha, E. Y., Thuong, P. T., Kim, J. R., et al. (2012). Anticancer properties of pomolic acid-induced AMP-activated protein kinase activation in MCF7 human breast cancer cells. *Biological & Pharmaceutical Bulletin*, *35*, 105–110.
Yu, R., Zhang, Z. Q., Wang, B., Jiang, H. X., Cheng, L., & Shen, L. M. (2014). Berberine-induced apoptotic and autophagic death of HepG2 cells requires AMPK activation. *Cancer Cell International*, *14*, 49.

CHAPTER FIVE

Relationship between mitofusin 2 and cancer

Alessandro Allegra*, Vanessa Innao, Andrea Gaetano Allegra, Caterina Musolino
Division of Hematology, Department of Human Pathology in Adulthood and Childhood "Gaetano Barresi", University of Messina, Messina, Italy
*Corresponding author: e-mail address: aallegra@unime.it

Contents

Abstract

Mitochondria are dynamic organelles whose actions are fundamental for cell viability. Within the cell, the mitochondrial system is incessantly modified via the balance between fusion and fission processes. Among other proteins, mitofusin 2 is a central protagonist in all these mitochondrial events (fusion, trafficking, contacts with other organelles), the balance of which causes the correct mitochondrial action, shape, and distribution within the cell. Here we examine the structural and functional characteristics of mitofusin 2, underlining its essential role in numerous intracellular pathways, as well as in the pathogenesis of cancer.

1. Mitochondria as signaling organelles

The mitochondrion is a double membrane-bound organelle detected in spermatocytes more than a century ago (Benda, 1898). Mitochondria are endosymbiont-derived structures (Poole & Gribaldo, 2014) occupying various actions in cell survival and eukaryotic energy metabolism (Blackstone, 2015; Wang, Yang, & Shen, 2015) (Fig. 1).

Advances in Protein Chemistry and Structural Biology, Volume 116
ISSN 1876-1623
https://doi.org/10.1016/bs.apcsb.2018.11.009

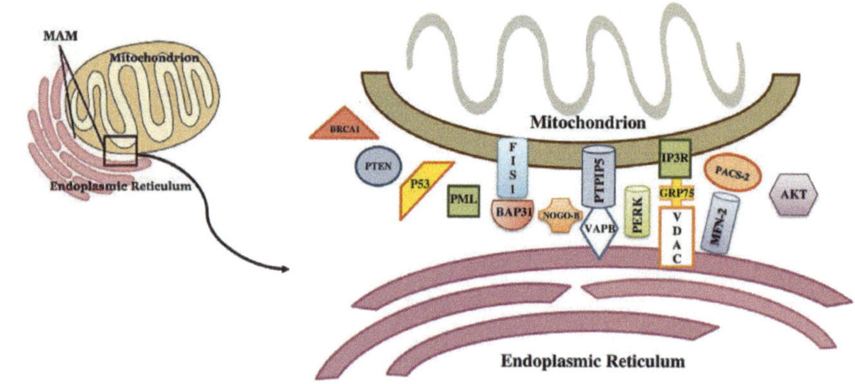

Fig. 1 Mitochondrion structure.

The inclusion of mitochondria into the primordial eukaryotic cell was a central moment in evolution. Without the amount of energy delivered by mitochondria, the development of higher organisms would not have been achievable. Each cell contains a diverse number of mitochondria. The number present is dependent upon how much energy the cell requires. The more energy a cell needs the more mitochondria will be present. Cells have the capability of producing a larger or a smaller quantity of mitochondria than required. They can also combine mitochondria to make larger ones. Moreover, these structures not only supply energy but are also implicated in essential metabolic pathways, such as steroidogenesis, urea cycle, fatty acid oxidation, gluconeogenesis, and synthesis of heme (Table 1). Furthermore, mitochondria regulate apoptosis. Equilibrium between anti- and proapoptotic elements of the Bcl-2 family regulates the delivery of Smac/DIABLO, endonuclease G cytochrome *c*, and apoptosis induction factor from mitochondria. When maintained in the mitochondrion, these elements contribute to cellular metabolism (e.g., cytochrome *c* shuttles electrons in the respiratory chain). When discharged into the cytoplasm, these elements operate as proapoptotic elements, either by an enzymatic activity (endonuclease G) or by stimulating apoptosome formation (cytochrome *c*). Lastly, mitochondria collaborate with the endoplasmic reticulum (ER) not only to regulate calcium homeostasis but also to produce diverse intracellular signals (Naon & Scorrano, 2014; Rahman & Rahman, 2018).

The mitochondrion comprises mitochondrial DNA (mtDNA), a small DNA of almost 16,569 bp (Attardi, 1985), which codes for 22 tRNAs, 2 rRNAs (12S and 16S), and 13 proteins subunits of four of the structures

Table 1 Vital functions performed by mitochondrion.

• Production of ATP molecules
• Control of calcium ions within the compartments of the cell
• Production of hormones like testosterone and estrogen
• Generation of free radicals
• The liver cells mitochondria have enzymes that detoxify ammonia
• Control of apoptosis or programmed cell death
• Heme synthesis reactions
• Heat production
• Cellular metabolism
• Signaling mechanisms

of the respiratory chain. Thus, the majority of mitochondrial proteins (about 1500) are produced by nuclear DNA, transcribed and translated in the cytoplasm and then transferred in the mitochondrion. The non-coding (D-loop) region of mtDNA is approximately 1.1 kbp long and it is the most variable portion of the genome; it controls regulatory signals for transcription and replication (Chen & Butow, 2005).

An emergent notion is the action of mitochondria as signaling structures in continual contact with the nucleus and sub-cellular sections (Chandel, 2014).

In fact, the mitochondrial system is extremely dynamic and is transformed under stress or change of cell metabolism to adjust the energetic requirements of the cell (Hyde, Twig, & Shirihai, 2010). For instance, the mitochondrial system morphology changes depending on growth on respiratory or glycolytic carbon sources (Mishra, Carelli, Manfredi, & Chan, 2014; Rossignol et al., 2004). Moreover, mitochondrial system interconnections are significantly augmented under starvation. The increase of connectivity of the mitochondrial system has been intended to block mitochondrial degradation by autophagy (Gomes, Di Benedetto, & Scorrano, 2011; Rambold, Kostelecky, Elia, & Lippincott-Schwartz, 2011). In contrast, mitochondrial system fragmentation under apoptotic situations enhances mitochondrial destruction (Frank et al., 2001; Karbowski, Norris, Cleland, Jeong, & Youle, 2006; Pyakurel, Savoia, Hess, & Scorrano, 2015) followed by production of proapoptotic elements from mitochondria mediated by BCL2-antagonist/killer (BAK) and BCL2-associated X protein (BAX) (Cassidy-Stone et al., 2008).

In conclusion, the mitochondrial system is endowed with extreme plasticity.

2. Mitochondrial fission and fusion

In eukaryotic cells, mitochondria are organized in an ample multiplicity of forms, varying from individual spheres to long interconnected tubules (Bereiter-Hahn & Veoth, 1994; Chen et al., 2003; Koning, Lum, Williams, & Wright, 1993; Liesa, Palacin, & Zorzano, 2009).

In fact, the mitochondrial morphology is the consequence of balanced fusion and fission regulated by components of the dynamin-related protein (DRP) family. Mitochondrial fission implicates the fragmentation or division of one healthy mitochondrion into two, while mitochondrial fusion implicates the integration of two mitochondria (malfunctioning or healthy) into an elongated one. Via the two antagonistic events, mitochondria can modify its dimension, form, and position in a few seconds (Bereiter-Hahn & Veoth, 1994). Fusion can reduce stress by mixing the contents of partly injured mitochondria as a system of complementation. Fission is not only to produce novel mitochondria but also participates in quality control by eliminating the injured mitochondria (Youle & Van Der Bliek, 2012). Fission and fusion also support mitochondria to modify the cytosolic localization, like gathering where high quantities of ATP needed, or where Ca^{2+} signaling requires an adjustment (Campello & Scorrano, 2010).

New biochemical, structural, and microscopic studies have been conducted in order to describe the fundamental system of mitochondrial fusion and fission (Held & Houtkooper, 2015; Lin & Sheng, 2015). These processes have the mitochondrial membranes and the ER system as protagonists (Fig. 2).

The inner mitochondrial membrane (IMM) defines the matrix and intermembrane space (IMS), whereas the outer mitochondrial membrane (OMM) divides the IMS from the cytosol. The cristae invaginations of the inner mitochondrial membrane contain the oxidative phosphorylation system (OXPHOS) (Appelhans et al., 2012; Busch, Deckers-Hebestreit, Hanke, & Mulkidjanian, 2013; Gilkerson, Selker, & Capaldi, 2003). This system is constituted of two functional components, the respiratory chain (RC) and the phosphorylation system, which contains the ATP synthase and membrane carriers such as the ATP/ADP carrier and the phosphate carrier. The RC is

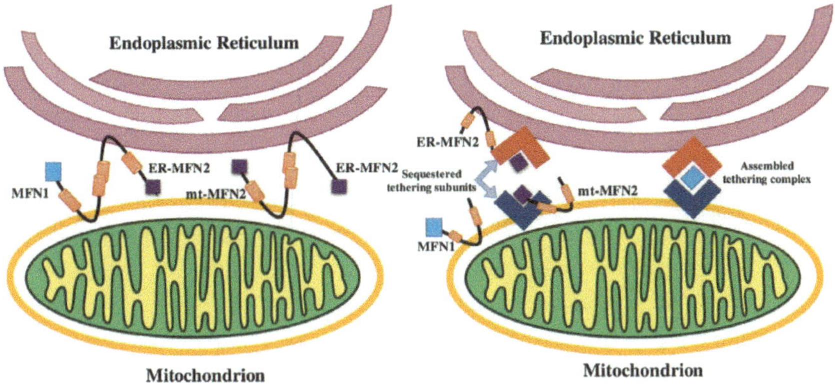

Fig. 2 Mitochondrial membranes and ER system.

described as comprising the mobile electron carriers coenzyme Q and cytochrome *c* and four respiratory complexes, designated complexes I–IV, which execute substrate oxidation to drive proton extrusion from the mitochondrial matrix to the intermembrane space (Mitchell, 1961).

The presence of particular areas where ER is in intimate connection with mitochondria is recognized since the electron microscopy (EM) investigations in the 50s (Bernhard & Rouiller, 1956). In these areas, OMM and ER membranes are tethered by proteinaceous filaments (Csordas et al., 2006), whose molecular characteristics are still disputed. The two membranes come in close contiguity but never join each other and are separated by a hiatus of thickness from ~10 nm up to 30–40 nm (Giacomello & Pellegrini, 2016). Greater juxtapositions (up to 80–100 nm) have been also described (Filadi et al., 2015).

Normal mitochondrial fission is controlled by highly conserved two GTPase genes: Fis1 and Drp1. Drp1 is a cytosolic protein and is translocated to the outer membrane of the mitochondria and interacts with multiple proteins, including Fis1, mitochondrial fission factor (Mff), and MiD49 and Mid51 (Reddy, 2014; Wakabayashi et al., 2009).

Ishihara et al. also demonstrated that DRP-1 knockout mice have a brain hypoplasia (Ishihara et al., 2009). This event is due to an imbalance between mitochondrial fusion and fission (Ishihara et al., 2009). DRP-1 possesses an N-terminal GTPase domain and a C-terminal GTPase effector domain, which is implicated in intramolecular and intermolecular interactions and influences the GTPase activity. Cytosolic DRP-1 is engaged on the mitochondrial surface in the fission foci by adaptor proteins, including FIS-1. FIS-1 is an integral protein of the OMM and includes a cytosolic hydrophobic tetratricopeptide repeat (TPR). The TPR is believed to act with DRP-1 (Carlucci, Lignitto, & Feliciello, 2008; Hoppins, 2014). At fission sites, DRP-1 creates spiral chains around membrane constriction sites. GTP hydrolysis provides energy to produce the mechanical force needed for fission (Carlucci et al., 2008). However, there are several adaptors that can engage DRP-1 to the mitochondria membrane, i.e., the mitochondrial fission factor (MFF) and the mitochondrial dynamic proteins 49 and 51 kDa (MiD49 and MiD51). These proteins can trigger mitochondrial fission through a FIS-1-independent process (Palmer et al., 2013).

The following paragraph will provide an outline of how several proteins are involved in the fusion processes and of the particular relevant action played by mitofusins (MTFs).

3. Mitofusin 1 and mitofusin 2

Two isoforms of mitofusins (MFNs), MFN1 and MFN2, can be found in mammals. Both MFNs are transmembrane GTPases, situated in the OMM. Both MFNs are ubiquitously expressed (Eura, Ishihara, Yokota, & Mihara, 2003; Rojo, Legros, Chateau, & Lombes, 2002; Santel et al., 2003; Santel & Fuller, 2001).

MFNs have a central action in the tethering and fusion of the OMMs. Nevertheless, the GTPase and tethering actions of MFN1 are bigger than that those of MFN2. As a consequence, MFN1 is believed to be the principal GTP-dependent membrane tethering protein for mitochondrial fusion.

MFN2 is a 757 aminoacids protein. The MFN2 comprises a large cytosolic GTPase domain at the -terminal, followed by a coiled-coil heptad-repeat (HR1) domain, a proline-rich (PR) region, two sequential transmembrane (TM) domains crossing the OMM and a second cytosolic heptad-repeat (HR2) domain at the C-terminal (Fig. 3).

Remarkably, MFN2 shares ~80% sequence-identity with its homolog MFN1. The PR domain, on the other hand, is only present in MFN2 and is supposed to be accountable for particular protein–protein interactions (Filadi, Greotti, & Pizzo, 2018).

In fact, although MFN1 and MFN2 genes are essentially equal to each other, the alteration of either MFN1 or MFN2 causes different inhibition of the fusion reaction (Chen et al., 2003; Eura et al., 2003). Embryonic fibroblasts deficient in MFN1 or MFN2 show different forms of fragmented mitochondria indicating that MFN1 and MFN2 are not functionally redundant. In cells missing MFN1, mitochondria fail to bind, implying that MFN1

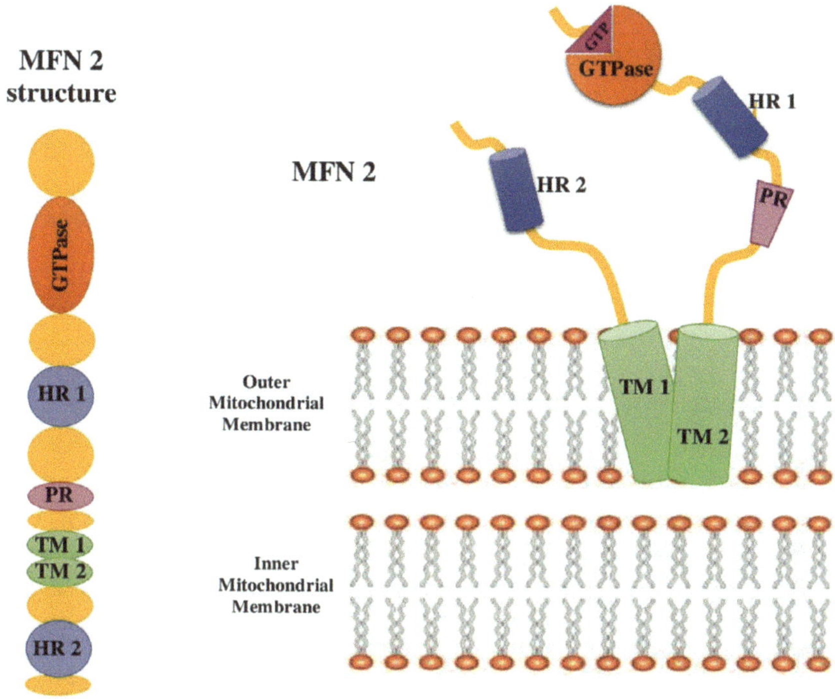

Fig. 3 MFN2 structure.

works in mitochondrial tethering while MFN2 operates in a later process of the fusion reaction, coherent with a minor amount of mitochondrial fragmentation detected in Mfn2$^{-/-}$ cells as compared to the Mfn1$^{-/-}$ ones. As a result of its superior GTP hydrolysis aptitude, MFN1 fully complements mitochondrial morphology in Mfn1$^{-/-}$ and Mfn2$^{-/-}$ cells. In contrast, MFN2 completely complements MFN2$^{-/-}$ but only 25% of MFN1$^{-/-}$ fusion activity (Cipolat, de Martins, Dal, & Scorrano, 2004). Probably, the equilibrium between the two MFNs controls the contact between mitochondria and the rough and smooth ER (rER/sER). More specifically, MFN1 seems to block the formation of sER–mitochondria contacts, while MFN2 seems to disturb the formation of rER–mitochondria contacts in cells with augmented MFN1 concentrations (Li et al., 2015; Wang, Garcin, et al., 2015).

Mice lacking MFN1 are seemingly fit, whereas the loss of MFN2 provokes mouse lethality in the early period and causes cerebellar atrophy with severe alterations in movement (Chen, McCaffery, & Chan, 2007). Furthermore, partial deficiency of mitochondrial fusion caused by the lack of MFN2 in diverse mammalian tissues harshly alters the physiological function and the metabolism. In addition to its action in mitochondrial fusion, there is diverse evidence supporting supplementary actions of MFN2.

The performing differences between MFN1 and MFN2 are further substantiated by in vivo studies in mice, displaying that MFN2 is needed for glucose homeostasis (Sebastián et al., 2012), steroidogenesis (Duarte et al., 2012), cerebellar development, and function (Chen et al., 2007), for maintaining axonal projections of dopaminergic neurons (Lee et al., 2012; Pham, Meng, Chu, & Chan, 2012), to sustain autophagy (Zhao et al., 2012), and coenzyme Q homeostasis (Mourier et al., 2015).

The precise molecular machinery and the exact succession of the actions accountable for MFNs mediated fusion of the OMM, have still not been completely identified. In the standard model, MFNs amass in the areas of close juxtaposition between contiguous mitochondria (Eura et al., 2003; Rojo et al., 2002; Santel et al., 2003), crossing from the opposite OMM and interacting in trans to develop antiparallel hetero- or homodimers of their HR2 domains. The HR2-HR2 contact, consisting in a 95 Å coiled-coil system, is sufficient to tether the OMM of juxtaposed mitochondria, but it is not enough to achieve their fusion that necessitates the MFNs' GTPase domains (Koshiba et al., 2004). MFN1 homodimers effectively fuse opposing membranes, MFN1/2 heterodimers show an intermediate efficiency, whereas MFN2 homodimers are weakly efficient (Ishihara, Eura, & Mihara, 2004). Furthermore, MFN1 complexes of ~250 KDa in *cis* and of ~450 KDa in *trans* have been observed (Ishihara et al., 2004). Pondering the described MFNs molecular weight of ~80 KDa, this remark suggest that the stoichiometry of the interaction is not 1:1, and/or that supplementary proteins assemble within the MFNs complex (Carlucci et al., 2008; Zorzano, Liesa, Sebastian, Segales, & Palacin, 2010).

An alternative model for MFN2-mediated ER–mitochondria tethering has been proposed. According to this view, MFN2 on both the ER and the OMM interacts with and sequesters still unknown tethering subunits, hindering their assembling into a functional tethering complex (Filadi, Pendin, & Pizzo, 2018).

In a recently published paper, the presence of two diverse conformations of MFN2 has been proposed. In inactive conditions, a tethering non-permissive state (characterized by an intramolecular, antiparallel interaction between the HR1 and HR2 domains, as well as by the strict adherence of the GTPase domain to the OMM) hinders the interaction between MFNs and partner proteins on the opposing OMM. The alteration of the intramolecular HR1-HR2 interaction is accountable for the shift toward the tethering-permissive state, in which the HR2 domain is free to recruit in trans interactions the HR2 of the opposing MFNs. The bending of the HR2 domains and a GTPase-dependent mechanism enables OMM fusion (Daumke & Roux, 2017; Qi et al., 2016).

The MFN2 complexes at the ER–mitochondria interface comprise numerous other proteins. A mitochondrial putative tumor suppressor gene linked to intermediate filaments (IFs) was identified in a proteomic screening of MFN2 interacting with proteins trichoplein/mitostatin (TpMs) (Cerqua et al., 2010). TpMs negatively control ER–mitochondria tethering via its interaction with MFN2. By detaching the two organelles, TpMs defend against Ca^{2+}-dependent apoptotic stimuli. MFN2 interacts also with miro. This interaction has been implicated in the MFN2-deficient mitochondria motility alteration (Friedman, Webster, Mastronarde, Verhey, & Voeltz, 2010).

A diverse controller of MFN2-dependent ER–mitochondria tethering has been recognized in MITOL, a component of the membrane-associated RING-CH E3 ubiquitin ligase family located in OMM and MAMs. MITOL ubiquitinates K192 of MFN2 (absent in MFN1) and the decreased ubiquitination of MFN2 described in MITOL-deficient cells correlates with decreased MFN2-dependent ER–mitochondria tethering and ER–mitochondria Ca^{2+} transfer (Sugiura et al., 2013).

Finally, several works associate MFN2 in tethering of mitochondria with lipid droplets (Wasilewski et al., 2012), melanosomes (Daniele et al., 2014), and with the miro/kinesin system (Misko, Jiang, Wegorzewska, Milbrandt, & Baloh, 2010; Misko, Sasaki, Tuck, Milbrandt, & Baloh, 2012).

4. Mitofusin 2 and cancer

Mitochondrial alteration can cause a large multiplicity of diseases (Chakraborty et al., 2018).

These are usually due to a deficiency of energy creation in the affected cells. Therefore, these alterations usually present themselves as skeletal muscle or neuromuscular disorders. Some common mitochondrial diseases comprise LHON, MELAS, and MIRAS (Khan et al., 2017).

In addition to hypertension, MFN2 disorders have also been identified in multiple disorders such as Charcot-Marie-Tooth disease (Zuchner et al., 2004), obesity (Bach et al., 2003), diabetes (Hernandez-Allvarez et al., 2010), or atherosclerosis (Chien & Hoshijima, 2004).

Moreover, mitochondrial disorder can also cause cancer (Singh & Modica-Napolitano, 2017) and management of mitochondrial function is acquiring relevance in the field of tumor therapeutics (Choudhary et al., 2011; Jiang, Zu, Wang, Zhang, & Fu, 2011; Karbowski & Youle, 2003; Yingkun, Lvsong, & Huimin, 2010).

Several mechanisms may justify the role played by MFN2 in the onset and progression of neoplastic disease, and in particular its effects on intracellular Ca^{++} signaling, the generation of growth-promoting factors, the actions on the metabolic systems, on cell cycle, mitophagy, cell death, and apoptosis. It is however to be kept in mind that the action of MFN2 is strictly specific for each tissue and tumor and that totally different effects can be found in different neoplastic diseases.

As mitochondria play a central function in apoptosis by influencing caspase activation, any modification in the mitochondrial fusion system changes the mitochondria's perception of apoptotic stimuli.

Furthermore, MFN2 has been described to interact with specific Bcl-2 proapoptotic family components such as Bax and Bak (Suen, Norris, & Youle, 2008). In fact, Karbowski demonstrated that MFN2 colocalized with the proapoptotic molecule, Bax, on mitochondria, suggesting a potential controlling action in apoptosis (Karbowski, 2002). Numerous studies demonstrate that MFN2-induced apoptosis is likely to happen via the PI3K-Akt pathway (Datta et al., 1997; Del Peso, Gonzalez Garcia, Page, Herrera, & Nunez, 1997; Kim, Kim, Park, Son, & Bae, 2007).

Sugioka demonstrated that inhibition of MFNs augmented sensitivity to apoptosis while overexpression of MFN's caused a suspension in the apoptotic response (Frezza et al., 2006; Sugioka, Shimizu, & Tsujimoto, 2004).

However, results from a recent work proved that MFN2 could block proliferation and provoke apoptosis in a dose- and time-dependent manner in HeLa cells. The membrane potential of the mitochondria reduced in a dose-dependent manner concentrations of cleaved caspase-3 and cleaved caspase-9 augmented in HeLa cells after Adv-MFN2 treatment. Wang et al. also described significant augments in the Bax level and a reduced Bcl-2 level with Adv-Mfn2 treatment. Moreover, they reported that MFN2 could significantly block the proliferation of the cervical tumor in the xenografted cervical carcinoma mouse model. After a 9-day therapy, the growth of the Adv-MFN2 group was blocked and it was induced into apoptosis. The data proved that the overexpression of MFN2 could not only augment the concentrations of Bax and Bid in tumor cells but also reduce the phosphorylation of Bad (Wang, Liu, Guo, & Quan, 2018).

It was also observed that MFN2 provoked a cell cycle arrest in the G0/G1 phase. The p21 (Ras) signature motif in the N-terminal region of MFN2 is relevant in the control of the cell cycle. A yeast two-hybrid test proved that MFN2 acts with a Ras analog (Chen et al., 2004).

A close relationship has been identified between p53, P21, and apoptosis (Polyak, Xia, Zweier, Kinzler, & Vogelstein, 1997). Western blot analysis demonstrated a p53 and p21 increase in SGC7901 cells transfected with MFN2.

A supplementary mechanism through which MFN2 could intervene in neoplastic disease is its action on autophagy. In fact, morphological changes in mitochondria, due to fusion, are connected with mitophagy (Chan, 2006; Okamoto & Shaw, 2005), a form of autophagy that causes the selective clearance of injured mitochondria and has been involved in the mechanisms of cell proliferation and death, as well as in cancer cells (Parsons & Green, 2010). While injured mitochondria stimulate apoptotic cell death through the production of cytochrome *c*, mitochondrial clearance causes cell survival (Mozdy & Shaw, 2003).

The central role of MFN2 in the regulation of calcium cannot be underestimated. In the recent past, several results have demonstrated the ability of tumor cells to influence intracellular Ca^{2+} signaling, promoting their growth and survival. Among various mechanisms, increased resistance to apoptosis of tumor cells involves regulation of the ER–mitochondria Ca^{2+} cross-talk (Marchi et al., 2014; Stewart, Yapa, & Monteith, 2015), while MAM-localized oncoproteins and tumor suppressors certainly interact with Ca^{2+} handling proteins and regulate their activity. Coherent with this supposition, the phosphatase tensin homolog (PTEN) (Bononi et al., 2013), the kinase Akt (Betz et al., 2013), breast/ovarian cancer susceptibility gene 1 (BRCA1) (Hedgepeth et al., 2015), and the promyelocytic leukemia (PML) protein (Giorgi et al., 2015, 2010; Missiroli et al., 2016) are all locate to mitochondria or to ER–mitochondria contacts, and here they exercise their cancer-regulating actions by interacting with Ca^{2+}-handling proteins and either increase ER–mitochondria Ca^{2+} flux (tumor suppressors) or decrease it (oncoproteins) (Missiroli et al., 2016).

Furthermore, the action of MFN2 on the flow of Ca^{++} could also be relevant in the process of metastatization. Ca^{2+} is among the main regulators of cell migration. In the progression of metastasis, the specific temporal and spatial organization of Ca^{2+} signaling play a central role in focal adhesion mechanisms and cytoskeletal reorganization (Prevarskaya, Skryma, & Shuba, 2011). A modified action of MFS 2 on mitochondrial fission and fusion-mediated cytosolic Ca^{2+} signaling could stimulate the reprogramming of focal adhesion dynamics and lamellipodia creation in cancer cells essentially by activating the CaMKII/ERK/FAK pathway (Sun et al., 2014).

Furthermore, the metabolic activity of the cancer cell appears able to modify MFN levels. The Warburg effect describes an enhanced tendency of cancer cells in generating lactic acid in the presence or absence of O_2 (Allegra, Innao, Gerace, Bianco, & Musolino, 2016). A largely accepted notion is that the Warburg effect is connected to energy. The Warburg effect serves a functional action in controlling acidic pericellular pH (pHe), which is due to a metabolic inversion between glycolytic rate substrate level phosphorylation and mitochondrial oxidative phosphorylation (OXPHOS) in order to control lactic acid production. In contrast, alkaline pHe initiated loss of MFN2, complexes II–IV (Sdhaf1, Uqcrq, Cox4i2, and Aldh1l2), aconitase, mitochondrial carrier triple repeat 1, and mt biosynthetic (Mazzio, Boukli, Rivera, & Soliman, 2012).

Lou et al. performed expression profiling in MFN2 knockdown A549 cells and found that cancer-related metabolic pathways were among the most susceptible pathways to MFN2 deficiency (Lou et al., 2016).

In conclusion, MFN2 could assume the role of an oncoprotein or tumor suppressor (Ma, Chang, Yu, He, & Liu, 2015; Wu et al., 2016; Zhou et al., 2016).

5. MFN 2 in solid tumors

Several researchers have demonstrated the possibility that a modification of MFN2 expression may play a relevant action in the onset of specific neoplastic diseases. However, results in the medical literature do not appear to be concordant, and MFN2 appears to exert differing actions in different tissues (Table 2). The extremely contrasting actions of MFN2 on diverse oncogenes might explain the paradoxical data regarding the roles of MFN2 in tumor development. Furthermore, it is conceivable that the overall effect of MFN2 might differ according to the environmental situations and the tumor types.

Mitochondrial fission is usually augmented in hepatocellular carcinoma (HCC) (Huang, Huang, et al., 2016; Huang, Zhan, et al., 2016). In a recent investigation, MFN2 immunostaining was very low in HCC tissue and was significantly correlated with cancer dimension. Moreover, HCC subjects with minor MFN2 expression had a poorer prognosis. Overexpression of MFN2 provoked HepG2 cells apoptosis, decreased the mitochondrial

Table 2 MFN2 in solid cancers.

Type of tumor	MFN2 action	Pro- or anti-carcinogenesis mechanisms	References
Hepatocellular carcinoma	Anticancer action	Cells apoptosis decrease of membrane potential decrease of endoplasmic reticulum calcium ion levels miR-761 action	Huang, Huang, et al. (2016), Huang, Zhan, et al. (2016), Wang et al. (2012), Wang, Garcin, et al. (2015), Wang, Xie, et al. (2015); Wang, Yang, et al. (2015), and Zhou et al. (2016)
Breast cancer	Anticancer action	Rapamycin complex 2 (mTORC2)/ Akt signaling pathway Crispr/Cas9 Cyclin A expression Transition of the cell cycle	Cheng et al. (2016), Xu et al. (2017), and Ma et al. (2015)
Bladder cancer	Anticancer action	Transition of the cell cycle Cell apoptosis Cleavage of poly (ADP-ribose) polymerase PARP	Cormio et al. (2017)
Lung cancer	Anticancer action	Cell apoptosis Action on oncogenes (ITGA2, RAP1A, and RALB)	Rehman et al. (2012) and Lou et al. (2015)
Gastric cancer	Uncertain action	? Downregulation of MMP-2 and MMP-9 block of cell cycle cell apoptosis P21 and PI3K/Akt signaling	Fang et al. (2017) and Zhang et al. (2013)

membrane potential and endoplasmic reticulum calcium ion levels, and increased intracellular reactive oxygen species (Wang, Xie, et al., 2015). Authors observed that MFN2 causes apoptosis in HCC cells by stimulating influx of Ca^{2+} into the mitochondria from the ER (Wang et al., 2012).

However, in HCC cells the regulatory action exerted by the MFN2 could at least partly take place via a diverse mechanism involving miRNAs. In fact, it was reported that MFN2 is a direct target of miR-761, which was seen to be upregulated in HCC tissues. Moreover, a miR-761 inhibitor altered mitochondrial function by upregulating MFN2 and efficiently reduced cancer growth and metastasis. These data support new insights into the mechanism underlying MFN2 control in HCC, making it a potential candidate for future use in HCC therapy (Zhou et al., 2016).

In patients with *breast cancer*, decreased expression of MFN2 related to poor prognosis. In a mouse xenograft model of human breast cancer, there was a correlation in the core region of cancers between MFN2 downregulation and the occurrence of extremely fragmented mitochondria (Cheng et al., 2016).

Recently, these results were confirmed by various other researchers. Xu et al. demonstrated that MFN2 knockout from MCF7 and A549 cells via Crispr/Cas9 highly stimulates cell viability, and invasion of tumor cells in vivo and in vitro. Signaling studies propose the mammalian target of rapamycin complex 2 (mTORC2)/Akt signaling pathway is extremely increased in MFN2 knockout cancer cells. The augmented mTORC2 stimulates tumor cell proliferation and metastasis via Akt^{S437} phosphorylation-mediated signaling pathway. Several investigations demonstrate that MFN2 reduces mTORC2 via a direct interaction by binding its domain HR1. Reduction of mTORC2 extremely reduces MFN2-deficient tumor proliferation (Xu et al., 2017).

In a different study authors tried to clarify the action of MFN2 in breast cancer. The data demonstrated that MFN2 reduces the growth of MCF-7 cells by controlling more cells at the G0/G1 phase and reducing proliferating of cell nuclear antigen and cyclin A expression. The data also confirmed that the PI3K/Akt signaling pathway is implicated in MFN2-regulated proliferation and apoptosis (Ma et al., 2015).

Cormio et al. investigated the potential action of mtDNA defects as disease markers in *Bladder Cancer* (BC). Cancer-specific mtDNA mutations and RNAs delivered encouraging results for early detection of BC in body fluids. Furthermore, MFN2, mitochondrial proteins Lon protease, and TFAM may have prognostic value and may be potential therapeutic targets (Cormio et al., 2017).

MFN2 overexpression appears to block cell growth by stopping the transition of the cell cycle from the G1 to the S phase and to provoke apoptosis by upregulating active caspase-3 and augmenting cleavage of poly (ADP-ribose) polymerase PARP. The MFN2 gene could be a potential BC tumor gene that promotes apoptosis and blocks the growth of BC cells. MFN2 could become a central therapeutic target for BC.

Human *lung cancer* cell lines display an alteration of Drp-1/MFN2 expression, which stimulates a state of mitochondrial fission. Lung tumor tissue samples from subjects exhibited an analogous augment in Drp-1 and a reduction in MFN2 when compared to a healthy lung. Complementary methods to reestablish mitochondrial network formation in lung cancer cells by overexpression of MFN2 caused an evident decrease of tumor cell growth and an augment in apoptosis. In a xenotransplantation model, MFN2 gene therapy or Drp-1 inhibition can regress cancer proliferation. The cancer size is reduced from 205.6 59 to 70.6 15 mm^3 with MFN2 overexpression and from 186.0 19 to 87.0 6 mm^3 with Drp-1 inhibition (Rehman et al., 2012).

On the contrary, Lou et al. studied MFN2 expression in 30 lung adenocarcinoma samples and discovered that the expression of MFN2 was significantly greater in lung adenocarcinoma tissues as compared to adjacent normal tissues (Lou et al., 2015). Moreover, they evaluated the effect of MFN2 knockdown on A549 human lung adenocarcinoma cells and demonstrated that cell growth and invasion comportment were all altered by MFN2 knockdown. An alteration of the cell cycle pathway was also verified. Moreover, diverse oncogenes such as ITGA2, RAP1A, and RALB were oppositely regulated by MFN2, which provided molecular evidences for the actions of MFN2 in cancer expansion. These data demonstrate that the action of MFN2 in tumor progression might be more complex than supposed.

Data regarding the role of MFN2 in the genesis of *gastric cancer* (GC) found in scientific literature are particularly discordant. In a recent study, the overexpression of MFN2 was positively correlated with stage and depth of invasion. Subjects with increased expression of MFN2 had a significantly poorer overall survival (OS) rate and disease-free survival rate compared to those with decreased expression. Investigations performed in GC cells demonstrated that the knockdown of MFN2 blocks cell growth and invasion. Overexpression of MFN2 could be utilized as a marker to predict the outcome of subjects with GC (Fang et al., 2017).

On the contrary, Zhang et al. reported that MFN2 expression was lower in tumor tissue than in normal gastric mucosal tissue and that it was negatively correlated with cancer dimension, suggesting an anticancer action for MFN2. In vitro experiments demonstrated that MFN2 overexpression reduced gastric cancer cell growth and colony formation, reduced the progression and migratory capability of tumor cells by downregulating MMP-2 and MMP-9, blocked the cell cycle and provoked apoptosis. Western blotting suggested the participation of P21 and PI3K/Akt signaling. Therefore, MFN2 could be a potential anti-tumor gene and a potential therapeutic target for treating gastric cancer (Zhang et al., 2013).

Methodological and sampling differences could be the basis of the discrepancy of the results, and further analysis is needed in order to confirm the real role played by MFN2 in these patients.

Interesting studies are underway to evaluate the possible role of MFN2 in the pathogenesis of *melanoma*, in *cancer of the cervix and pancreas* (Ahn, Li, Zhang, & Hyun, 2018; Lee et al., 2018; Pan, Zhou, Yin, Bai, & Liu, 2018).

5.1 Mitofusin 2 and cancer cachexia

Cancer cachexia (CC) is a destroying situation associated with numerous sorts of malignant tumors (Thoresen et al., 2013). It is characterized by a reduction of body mass, with or without decrease of adipose tissue, systemic inflammation, and negative protein and energy balance (Aoyagi, Terracina, Raza, Matsubara, & Takabe, 2015).

Mitochondrial function is central for the preservation of the skeletal muscle (Romanello et al., 2010), and it has been reported that an alteration of the mitochondrial function is enough to provoke muscle atrophy in animals (Huang, Galloway, & Yoon, 2011; Soriano et al., 2006).

Mitochondrial alteration is implicated in the reduction of muscle featuring CC. However, it is uncertain whether mitochondrial quality control (MQC) is modified in skeletal myocytes of subjects with CC. In a recent study, biopsy samples of the rectus abdominis muscle were obtained from subjects with GC (nine with CC and nine non-CC) and nine controls. MFN2 expression was decreased in GC subjects compared with controls, independent of CC. Fission protein 1 (Fis1) was instead increased in CC subjects with respect to the other groups. As a result, the "fusion index" (MFN2/Fis1 ratio) was minor in GC subjects. These data suggest that CC is associated with alterations of the muscular MQC axis at numerous checkpoints: mitophagy signaling, mitochondrial dynamics, and mitochondrial tagging for disposal (Marzetti et al., 2016).

Nevertheless, in a different study, Xi et al. evaluated tumor subjects with or without CC, and displayed that MFN2 was downregulated in patients with CC. In vitro cell models revealed that a loss of MFN2 was related to atrophy of the C2C12 mouse myoblast cell line. In vivo animal models revealed that overexpression of MFN2 was able to partially reduce CC-induced gastrocnemius muscle loss (Xi et al., 2016).

Remarkably, although the molecular mechanisms accountable for the onset of CC have been studied, it is not quite clear if tumor treatments also have a causative action in the progress of CC. Results demonstrate that chemotherapy can increase CC enhancement regardless of its action on cancer development (Damrauer et al., 2008; Garcia, Cata, Dougherty, & Smith, 2008).

For instance, Folfox and Folfiri, which are chemotherapeutics used for the therapy of tumors, may influence the development of CC by disturbing the muscle mitochondrial pool (Barreto et al., 2016). A significant modification was found in 386 proteins in the muscle of Folfiri-treated mice. Pathway analyses revealed mitochondrial alterations, with decreased expression of mediators of mitochondrial fission (DRP-1), fusion (MFN2, OPA-1), and biogenesis (cytochrome *c*, PGC-1α).

All these data suggested that MFN2 is implicated in CC-induced muscle loss and could be a potential target for CC treatment.

5.2 Mitofusin and cardioprotection

An abnormal mitochondrial dynamic process has been involved in heart failure. Recent studies demonstrated that the ratio of MFN 2 and Drp-1 was reduced during heart failure, indicating augmented mitophagy (Givvimani, Pushpakumar, Veeranki, & Tyagi, 2014). Therapy with mitochondrial division inhibitor 1 (Mdivi-1), a specific inhibitor of DRP-1, has been proved to better cardiac function.

Anthracycline drugs, such as doxorubicin (DOX), are useful anticancer drugs frequently prescribed to cure malignancies, but their usage is limited by a cumulative, dose-dependent cardiotoxicity. This is characterized by relevant modifications in cardiomyocyte biology and in the extracellular matrix. In these patients we have an alteration of the process of cardiac remodeling, which causes chamber dilatation, contractile alterations, and chronic heart failure (Bloom et al., 2016).

It is sure that the involvement of MFN2 in control of cardiomyocyte death is determined by its level of expression (Schrepfer & Scorrano, 2016). However, numerous results have revealed that whereas DOX through NADPH oxidase signaling can have a disadvantageous action on cardiomyocyte survival, a specific Nox2-stimulated pathway comprising MFN2 may be necessary to conserve mitochondrial biogenesis and cardiomyocyte survival (McLaughlin et al., 2017).

A recent study attempted to clarify the action of the MFN2 in DOX-induced cardiomyocyte apoptosis. Cultured neonatal rat cardiomyocytes were employed. Concentrations of MFN2 in cardiomyocytes were reduced after the cells were tested with DOX which was connected with augmented mitochondrial fission and mitochondrial ROS generation. An augment in cardiomyocyte concentrations of MFN2 reduced the DOX-induced cardiomyocyte mitochondrial ROS generation. A pretreatment of cardiomyocytes with an anti-oxidant, also reduced the DOX-induced cardiomyocyte apoptosis. These data suggest that DOX causes a reduced cardiomyocyte MFN2 expression which provokes mitochondrial fission and ROS production further conducts to cardiomyocyte apoptosis (Tang et al., 2017).

A pharmacological treatment including MFN2 that could theoretically block degeneration of cardiac cells appears to be an appealing approach. Such approach may be facilitated by the detection of a small molecule, 15-oxospiramilactone, which, via the reduction of a mitochondria-localized deubiquitinase, augments MFN2 action (Yue et al., 2014), although the efficacy has yet to be verified in appropriate models.

6. Future perspective

Recent studies have proved that regulatory proteins inside and outside mitochondria could cause or reduce tumorigenesis. A site to perform such a function could be the MAM. This intracellular signaling nucleus controls Ca^{2+} exchange between the ER and mitochondria that is necessitated for mitochondrial dehydrogenases and, thus, oxidative phosphorylation. A modified Ca^{2+} signaling at the MAM is identified as a feature of tumor cells that changes their metabolism to glycolysis and augments their resistance to cell death.

Consequently, one method by which cancer cells could silence mitochondria metabolism and apoptotic signaling would be by changing the ER lipid domain formation that may disturb normal MAM rafts.

Coherent with such an eventuality, several methods are presently in progress to change ER stress-related apoptosis in tumor cells (Beloribi-Djefaflia, Vasseur, & Guillaumond, 2016). In this sense, the possibility of intervening on the mechanisms of fusion and mitochondrial fission, by acting on the MFN2 concentrations appears to be an interesting possibility in the context of anti-neoplastic strategies.

Numerous therapeutic substances of natural origin have shown antineoplastic activities (Allegra et al., 2017, 2018).

Phyllanthus urinaria, a commonly utilized herbal medicine, has been described to have several biological characteristics comprising anti-virus, anti-bacteria, and anti-tumor. The water extract of *P. urinaria* not only dramatically decreases the cell viability of several tumor cell lines but also reduces cancer development in C57BL/6J mice after implantation of Lewis lung carcinoma cells (Huang et al., 2006). The anti-tumor action of the *P. urinaria* extract is essentially due to the induced apoptosis of tumor cells as proved by DNA fragmentation and augmented caspase-3 activity through both intrinsic and extrinsic pathways (Huang, Pang, Yang, & Wu, 2010).

In a different study, the use of *P. urinaria* caused a relevant reduction in the mitochondrial fusion proteins, including MFN1 and MFN2 and Opa1, as well as an augment in the fission protein Drp1 (Huang, Huang, et al., 2016; Huang, Zhan, et al., 2016).

Benzyl isothiocyanate (BITC) is a different and extremely interesting phytochemical present in cruciferous vegetables with preclinical proofs of in vivo effectiveness against breast tumor in xenograft mouse models. Mammary cancer chemoprevention by BITC is connected with apoptotic cell death. A recent work demonstrated that modified mitochondrial dynamics are key events in BITC-induced apoptosis in breast cancer cells (Sehrawat et al., 2016). Exposure of MCF-7 and MDA-MB-231 cells to doses of BITC caused a fast collapse of the mitochondrial filamentous network. BITC therapy also reduced polyethyleneglycol-induced mitochondrial fusion. A normal human mammary epithelial cell line (MCF-10A), on the other and was resistant to BITC-mediated modifications in mitochondrial system as well as apoptosis. After BITC treatment, a reduction of the protein concentrations involved in the control of mitochondrial fission and fusion was distinctly discernible.

In a different approach, Wu et al. demonstrated that an adenovirus vector encoding MFN2 (human HSG) (Ad5-hHSG) had an anticancer action in a large variety of tumor cell lines. The researchers used the lung tumor cell line A549 and the colon tumor cell line HT-29 and reported that Ad5-hHSG caused apoptosis both in vitro and in vivo. Confocal laser scanning microscopy and electron microscopy demonstrated that cells infected with Ad5-hHSG provoked dose-dependent perinuclear clusters of fused mitochondria. Adenovirus-mediated hHSG overexpression caused apoptosis, cell cycle arrest, and mitochondrial membrane potential reduction in vitro. Overexpression of hHSG also reduced the growth of subcutaneous tumors in nude mice both ex vivo and in vivo (Wu et al., 2008).

In conclusion, we must now acknowledge that the research on the action of MFN2 in the determinism of neoplastic disease is incomplete and sometimes contradictory. The tumor microenvironment, the presence of proteins with oncogenic and oncosuppressive activity can probably change the role of MFN2 in the different types of tumors. However, in the future, a detailed analysis of MFN2 actions may offer insights into the role of MFN2 in cancer development and thus expand the knowledge about tumor treatment.

References

Ahn, S. Y., Li, C., Zhang, X., & Hyun, Y. M. (2018). Mitofusin-2 expression is implicated in cervical cancer pathogenesis. *Anticancer Research*, *38*(6), 3419–3426.

Allegra, A., Innao, V., Gerace, D., Bianco, O., & Musolino, C. (2016). The metabolomic signature of hematologic malignancies. *Leukemia Research*, *9*, 22–35.

Allegra, A., Innao, V., Russo, S., Gerace, D., Alonci, A., & Musolino, C. (2017). Anticancer activity of curcumin and its analogues: Preclinical and clinical studies. *Cancer Investigation*, *35*(1), 1–22.

Allegra, A., Speciale, A., Molonia, M. S., Guglielmo, L., Musolino, C., Ferlazzo, G., et al. (2018). Curcumin ameliorates the in vitro efficacy of carfilzomib in human multiple myeloma U266 cells targeting p53 and NF-κB pathways. *Toxicology In Vitro*, *47*, 186–194.

Aoyagi, T., Terracina, K. P., Raza, A., Matsubara, H., & Takabe, K. (2015). Cancer cachexia, mechanism and treatment. *World Journal of Gastrointestinal Oncology*, *7*, 17–29.

Appelhans, T., Richter, C. P., Wilkens, V., Hess, S. T., Piehler, J., & Busch, K. B. (2012). Nanoscale organization of mitochondrial microcompartments revealed by combining tracking and localization microscopy. *Nano Letters*, *12*, 610–616.

Attardi, G. (1985). Animal mitochondrial DNA: An extreme example of genetic economy. *International Review of Cytology*, *93*, 93–145.

Bach, D., Pich, S., Soriano, F. X., Vega, N., Baumgartner, B., Oriola, J., et al. (2003). Mitofusin-2 determines mitochondrial network architecture and mitochondrial metabolism. A novel regulatory mechanism altered in obesity. *The Journal of Biological Chemistry*, *278*, 17190–17197.

Barreto, R., Waning, D. L., Gao, H., Liu, Y., Zimmers, T. A., & Bonetto, A. (2016). Chemotherapy-related cachexia is associated with mitochondrial depletion and the activation of ERK1/2 and p38 MAPKs. *Oncotarget*, *7*, 43442–43460. https://doi.org/10.18632/oncotarget.9779.

Beloribi-Djefaflia, S., Vasseur, S., & Guillaumond, F. (2016). Lipid metabolic reprogramming in cancer cells. *Oncogene*, *5*, e189. https://doi.org/10.1038/oncsis.2015.49.

Benda, C. (1898). Ueber die Spermatogenese der Vertebraten und höherer Evertebraten, II. Theil: Die Histiogenese der Spermien. *Archives for Physiology*, *73*, 393–398.

Bereiter-Hahn, J., & Veoth, M. (1994). Dynamics of mitochondria in living cells: Shape changes, dislocations, fusion, and fission of mitochondria. *Microscopy Research and Technique*, *27*, 198–219.

Bernhard, W., & Rouiller, C. (1956). Close topographical relationship between mitochondria and ergastoplasm of liver cells in a definite phase of cellular activity. *The Journal of Biophysical and Biochemical Cytology*, *2*, 73–78.

Betz, C., Stracka, D., Prescianotto-Baschong, C., Frieden, M., Demaurex, N., & Hall, M. N. (2013). Feature article: mTOR complex 2-Akt signaling at mitochondria-associated endoplasmic reticulum membranes (MAM) regulates mitochondrial physiology. *Proceedings of the National Academy of Sciences of the United States of America*, *110*, 12526–12534.

Blackstone, N. W. (2015). The impact of mitochondrial endosymbiosis on the evolution of calcium signaling. *Cell Calcium*, *57*, 133–139.

Bloom, M. W., Hamo, C. E., Cardinale, D., Ky, B., Nohria, A., Baer, L., et al. (2016). Cancer therapy-related cardiac dysfunction and heart failure part 1: Definitions, pathophysiology, risk factors, and imaging. *Circulation. Heart Failure*, *9*, e002661.

Bononi, A., Bonora, M., Marchi, S., Missiroli, S., Poletti, F., Giorgi, C., et al. (2013). Identification of PTEN at the ER and MAMs and its regulation of Ca(2+) signaling and apoptosis in a protein phosphatase-dependent manner. *Cell Death and Differentiation*, *20*, 1631–1643.

Busch, K. B., Deckers-Hebestreit, G., Hanke, G. T., & Mulkidjanian, A. Y. (2013). Dynamics of bioenergetic microcompartments. *Biological Chemistry*, *394*, 163–188.

Campello, S., & Scorrano, L. (2010). Mitochondrial shape changes: Orchestrating cell pathophysiology. *EMBO Reports*, *11*, 678–684.

Carlucci, A., Lignitto, L., & Feliciello, A. (2008). Control of mitochondria dynamics and oxidative metabolism by cAMP, AKAPs and the proteasome. *Trends in Cell Biology*, *18*, 604–613.

Cassidy-Stone, A., Chipuk, J. E., Ingerman, E., Song, C., Yoo, C., Kuwana, T., et al. (2008). Chemical in-hibition of the mitochondrial division dynamin reveals its role in Bax/Bak-dependent mitochondrial outer membrane permeabilization. *Developmental Cell*, *14*, 193–204.

Cerqua, C., Anesti, V., Pyakurel, A., Liu, D., Naon, D., Wiche, G., et al. (2010). Trichoplein/mitostatin regulates endoplasmic reticulum-mitochondria juxtaposition. *EMBO Reports*, *11*, 854–860.

Chakraborty, S., Prasada, K. S., Singh, R. P., Thangaraj, K., Singh, K. K., & Satyamoorthy, K. (2018). Mitochondria in health and disease. *Mitochondrion*, *43*, 25–29. https://doi.org/10.1016/j.mito.2018.06.006.

Chan, D. C. (2006). Mitochondrial fusion and fission in mammals. *Annual Review of Cell and Developmental Biology*, *22*, 79–99.

Chandel, N. S. (2014). Mitochondria as signaling organelles. *BMC Biology*, *12*, 34.

Chen, X., & Butow, R. A. (2005). The organization and inheritance of the mitochondrial genome. *Nature Reviews. Genetics*, *6*, 815–825.

Chen, H., Detmer, S. A., Ewald, A. J., Griffin, E. E., Fraser, S. E., & Chan, D. C. (2003). Mitofusins Mfn1 and Mfn2 coordinately regulate mitochondrial fusion and are essential for embryonic development. *The Journal of Cell Biology*, *160*, 189–200.

Chen, K. H., Guo, X., Ma, D., Guo, Y., Li, Q., & Yang, D. (2004). Dysregulation of HSG triggers vascular proliferative disorders. *Nature Cell Biology*, *6*, 872–883.

Chen, H., McCaffery, J. M., & Chan, D. C. (2007). Mitochondrial fusion protects against neurodegeneration in the cerebellum. *Cell*, *130*, 548–562.

Cheng, C. T., Kuo, C. Y., Ouyang, C., Li, C. F., Chung, Y., David, C., et al. (2016). Metabolic stress-induced phosphorylation of KAP1 Ser473 blocks mitochondrial fusion in breast cancer cells. *Cancer Research*, *76*, 5006–5018.

Chien, K. R., & Hoshijima, M. (2004). Unravelling Ras signals in cardiovascular disease. *Nature Cell Biology*, *6*, 807–808.

Choudhary, V., Kaddour-Djebbar, I., Lakshmikanthan, V., et al. (2011). Novel role of androgens in mitochondrial fission and apoptosis. *Molecular Cancer Research*, *9*, 1067–1077.

Cipolat, S., de Martins, B. O., Dal, Z. B., & Scorrano, L. (2004). OPA1 requires mitofusin 1 to promote mitochondrial fusion. *Proceedings of the National Academy of Sciences of the United States of America*, *101*, 15927–15932.

Cormio, A., Sanguedolce, F., Musicco, C., Pesce, V., Calò, G., Bufo, P., et al. (2017). Mitochondrial dysfunctions in bladder cancer: Exploring their role as disease markers and potential therapeutic targets. *Critical Reviews in Oncology/Hematology*, *117*, 67–72. https://doi.org/10.1016/j.critrevonc.2017.07.001.

Csordas, G., Renken, C., Varnai, P., Walter, L., Weaver, D., Buttle, K. F., et al. (2006). Structural and functional features and significance of the physical linkage between ER and mitochondria. *Journal of Cell Biology, 174*, 915–921.

Damrauer, J. S., Stadler, M. E., Acharyya, S., Baldwin, A. S., Couch, M. E., & Guttridge, D. C. (2008). Chemotherapy-induced muscle wasting: Association with NF-κB and cancer cachexia. *Basic and Applied Myology, 18*, 139–148.

Daniele, T., Hurbain, I., Vago, R., Casari, G., Raposo, G., Tacchetti, C., et al. (2014). Mitochondria and melanosomes establish physical contacts modulated by Mfn2 and involved in organelle biogenesis. *Current Biology, 24*, 393–403.

Datta, S. R., Dudek, H., Tao, X., Masters, S., Fu, H., Gotoh, Y., et al. (1997). Akt phosphorylation of BAD couples survival signals to the cell-intrinsic death machinery. *Cell, 91*, 231–241.

Daumke, O., & Roux, A. (2017). Mitochondrial homeostasis: How do dimers of mitofusins mediate mitochondrial fusion? *Current Biology, 27*, R353–R356.

Del Peso, L., Gonzalez Garcia, M., Page, C., Herrera, R., & Nunez, G. (1997). Interleukin-3-induced phosphorylation of BAD through the protein kinase Akt. *Science, 278*, 687–689.

Duarte, A., Poderoso, C., Cooke, M., Soria, G., Cornejo Maciel, F., Gottifredi, V., et al. (2012). Mito-chondrial fusion is essential for steroid biosynthesis. *PLoS One*, 7, e45829.

Eura, Y., Ishihara, N., Yokota, S., & Mihara, K. (2003). Two mitofusin proteins, mammalian homo-logues of FZO, with distinct functions are both required for mitochondrial fusion. *Journal of Biochemistry, 134*, 333–344.

Fang, C. L., Sun, D. P., Chen, H. K., Lin, C. C., Hung, S. T., Uen, Y. H., et al. (2017). Overexpression of mitochondrial GTPase MFN2 represents a negative prognostic marker in human gastric cancer and its inhibition exerts anti-cancer effects. *Journal of Cancer, 8*, 1153–1161.

Filadi, R., Greotti, E., & Pizzo, P. (2018). Highlighting the endoplasmic reticulum-mitochondria connection: Focus on mitofusin 2. *Pharmacological Research, 128*, 42–51.

Filadi, R., Greotti, E., Turacchio, G., Luini, A., Pozzan, T., & Pizzo, P. (2015). Mitofusin 2 ablation increases endoplasmic reticulum-mitochondria coupling. *Proceedings of the National Academy of Sciences of the United States of America, 112*, E2174–E2181.

Filadi, R., Pendin, D., & Pizzo, P. (2018). Mitofusin 2: From functions to disease. *Cell Death & Disease, 9*(3), 330.

Frank, S., Gaume, B., Bergmann-Leitner, E. S., Leitner, W. W., Robert, E. G., Catez, F., et al. (2001). The role of dynamin-related protein 1, a mediator of mitochondrial fission, in apoptosis. *Developmental Cell, 1*, 515–525.

Frezza, C., Cipolat, S., Martins de Brito, O., Micaroni, M., Beznoussenko, G. V., Rudka, T., et al. (2006). OPA1 controls apoptotic cristae remodeling independently from mito-chondrial fusion. *Cell, 126*, 177–189.

Friedman, J. R., Webster, B. M., Mastronarde, D. N., Verhey, K. J., & Voeltz, G. K. (2010). ER sliding dynamics and ER-mitochondrial contacts occur on acetylated microtubules. *The Journal of Cell Biology, 190*, 363–375.

Garcia, J. M., Cata, J. P., Dougherty, P. M., & Smith, R. G. (2008). Ghrelin prevents cisplatin-induced mechanical hyperalgesia and cachexia. *Endocrinology, 149*, 455–460.

Giacomello, M., & Pellegrini, L. (2016). The coming of age of the mitochondria-ER contact: A matter of thickness. *Cell Death and Differentiation, 23*, 1417–1427.

Gilkerson, R. W., Selker, J. M. L., & Capaldi, R. A. (2003). The cristal membrane of mitochondria is the principal site of oxidative phosphorylation. *FEBS Letters, 546*, 355–358.

Giorgi, C., Bonora, M., Sorrentino, G., Missiroli, S., Poletti, F., Suski, J. M., et al. (2015). p53 at the endoplasmic reticulum regulates apoptosis in a Ca2+-dependent manner. *Proceedings of the National Academy of Sciences of the United States of America, 112*, 1779–1784.

Giorgi, C., Ito, K., Lin, H. K., Santangelo, C., Wieckowski, M. R., Lebiedzinska, M., et al. (2010). PML regulates apoptosis at endoplasmic reticulum by modulating calcium release. *Science*, *330*, 1247–1251.

Givvimani, S., Pushpakumar, S., Veeranki, S., & Tyagi, S. C. (2014). Dysregulation of Mfn2 and Drp-1 proteins in heart failure. *Canadian Journal of Physiology and Pharmacology*, *92*, 583–591.

Gomes, L. C., Di Benedetto, G., & Scorrano, L. (2011). During autophagy mitochondria elongate, are spared from degradation and sustain cell viability. *Nature Cell Biology*, *13*, 589–598.

Hedgepeth, S. C., Garcia, M. I., Wagner, L. E., II, Rodriguez, A. M., Chintapalli, S. V., Snyder, R. R., et al. (2015). The BRCA1 tumor suppressor binds to inositol 1,4,5-trisphosphate receptors to stimulate apoptotic calcium release. *The Journal of Biological Chemistry*, *290*, 7304–7313.

Held, N. M., & Houtkooper, R. H. (2015). Mitochondrial quality control pathways as determinants of metabolic health. *BioEssays*, *37*, 867–876.

Hernandez-Alvarez, M. I., Thabit, H., Burns, N., Shah, S., Brema, I., Hatunic, M., et al. (2010). Subjects with early-onset type 2 diabetes show defective activation of the skeletal muscle PGC-1 alpha/mitofusin-2 regulatory pathway in response to physical activity. *Diabetes Care*, *33*, 645–651.

Hoppins, S. (2014). The regulation of mitochondrial dynamics. *Current Opinion in Cell Biology*, *29*, 46–52.

Huang, P., Galloway, C. A., & Yoon, Y. (2011). Control of mitochondrial morphology through differential interactions of mitochondrial fusion and fission proteins. *PLoS One*, *6*, e20655, 2011.

Huang, S. T., Huang, C. C., Sheen, J. M., Lin, T. K., Liao, P. P., Huang, W. L., et al. (2016). *Phyllanthus urinaria*'s inhibition of human osteosarcoma xenografts growth in mice is associated with modulation of mitochondrial fission/fusion machinery. *The American Journal of Chinese Medicine*, 7, 1–17.

Huang, S. T., Pang, J. H., Yang, R. C., & Wu, H. T. (2010). Anti-cancer effect of *Phyllanthus urinaria* and its relevant mechanism. *Chang Gung Medical Journal*, *33*, 477–487.

Huang, S. T., Yang, R. C., Lee, P. N., Yang, S. H., Liao, S. K., Chen, T. Y., et al. (2006). Anti-tumor and anti-angiogenic effects of *Phyllanthus urinaria* in mice bearing Lewis lung carcinoma. *International Immunopharmacology*, *6*, 870–879.

Huang, Q., Zhan, L., Cao, H., et al. (2016). Increased mitochondrial fission promotes autophagy and hepatocellular carcinoma cell survival through the ROS-modulated coordinated regulation of the NFKB and TP53 pathways. *Autophagy*, *12*, 999–1014.

Hyde, B. B., Twig, G., & Shirihai, O. S. (2010). Organellar vs cellular control of mitochondrial dynamics. *Seminars in Cell & Developmental Biology*, *21*, 575–581.

Ishihara, N., Eura, Y., & Mihara, K. (2004). Mitofusin 1 and 2 play distinct roles in mitochondrial fusion reactions via GTPase activity. *Journal of Cell Science*, *117*, 6535–6546.

Ishihara, N., Nomura, M., Jofuku, A., Kato, H., Suzuki, S. O., Masuda, K., et al. (2009). Mitochondrial fission factor Drp1 is essential for embryonic development and synapse formation in mice. *Nature Cell Biology*, *11*, 958–966.

Jiang, S., Zu, Y., Wang, Z., Zhang, Y., & Fu, Y. (2011). Involvement of mitochondrial permeability transition pore opening in 7-xylosyl-10-deacetylpaclitaxel-induced apoptosis. *Planta Medica*, *77*, 1005–1012.

Karbowski, M. (2002). Spatial and temporal association of Bax with mitochondrial fission sites, Drp1, and Mfn2 during apoptosis. *The Journal of Cell Biology*, *159*, 931–938.

Karbowski, M., Norris, K. L., Cleland, M. M., Jeong, S.-Y., & Youle, R. J. (2006). Role of Bax and Bak in mitochondrial morphogenesis. *Nature*, *443*, 658–662.

Karbowski, M., & Youle, R. J. (2003). Dynamics of mitochondrial morphology in healthy cells and during apoptosis. *Cell Death and Differentiation*, *10*, 870–880.

Khan, N. A., Govindaraj, P., Soumittra, N., Sharma, S., Srilekha, S., Ambika, S., et al. (2017). Leber's hereditary optic neuropathy-specific mutation m.11778G>A exists on diverse mitochondrial haplogroups in India. *Investigative Ophthalmology & Visual Science, 58*, 3923–3930.

Kim, N. H., Kim, K., Park, W. S., Son, H. S., & Bae, Y. (2007). PKB/Akt inhibits ceramide-induced apoptosis in neuroblastoma cells by blocking apoptosis-inducing factor (AIF) translocation. *Journal of Cellular Biochemistry, 102*, 1160–1170.

Koning, A. J., Lum, P. Y., Williams, J. M., & Wright, R. (1993). DiOC6 staining reveals organelle structure and dynamics in living yeast cells. *Cell Motility and the Cytoskeleton, 25*, 111–128.

Koshiba, T., Detmer, S. A., Kaiser, J. T., Chen, H., McCaffery, J. M., & Chan, D. C. (2004). Structural basis of mitochondrial tethering by mitofusin complexes. *Science, 305*, 858–862.

Lee, Y. S., Jung, Y. Y., Park, M. H., Yeo, I. J., Im, H. S., Nam, K. T., et al. (2018). Deficiency of parkin suppresses melanoma tumor development and metastasis through inhibition of MFN2 ubiquitination. *Cancer Letters, 433*, 156–164.

Lee, S., Sterky, F. H., Mourier, A., Terzioglu, M., Cullheim, S., Olson, L., et al. (2012). Mitofusin 2 is necessary for striatal axonal projections of midbrain dopamine neurons. *Human Molecular Genetics, 21*, 4827–4835.

Li, L., Gao, G., Shankar, J., Joshi, B., Foster, L. J., & Nabi, I. R. (2015). p38 MAP kinase-dependent phosphorylation of the Gp78 E3 ubiquitin ligase controls ER-mitochondria association and mitochondria motility. *Molecular Biology of the Cell, 26*, 3828–3840.

Liesa, M., Palacín, M., & Zorzano, A. (2009). Mitochondrial dynamics in mammalian health and disease. *Physiological Reviews, 89*, 799–845.

Lin, M. Y., & Sheng, Z. H. (2015). Regulation of mitochondrial transport in neurons. *Experimental Cell Research, 334*, 35–44.

Lou, Y., Li, R., Liu, J., Zhang, Y., Zhang, X., Jin, B., et al. (2015). Mitofusin-2 over-expresses and leads to dysregulation of cell cycle and cell invasion in lung adenocarcinoma. *Medical Oncology, 32*, 132.

Lou, Y., Zhang, Y., Li, R., Gu, P., Xiong, L., Zhong, H., et al. (2016). Transcriptional profiling revealed the anti-proliferative effect of MFN2 deficiency and identified risk factors in lung adenocarcinoma. *Tumor Biology, 37*, 8643–8655.

Ma, L., Chang, Y., Yu, L., He, W., & Liu, Y. (2015). Pro-apoptotic and anti-proliferative effects of mitofusin-2 via PI3K/Akt signaling in breast cancer cells. *Oncology Letters, 10*, 3816–3822.

Marchi, S., Giorgi, C., Oparka, M., Duszynski, J., Wieckowski, M. R., & Pinton, P. (2014). Oncogenic and oncosuppressive signal transduction at mitochondria-associated endoplasmic reticulum membranes. *Molecular and Cellular Oncology, 1*, e956469.

Marzetti, E., Lorenzi, M., Landi, F., Picca, A., Rosa, F., Tanganelli, F., et al. (2016). Altered mitochondrial quality control signaling in muscle of old gastric cancer patients with cachexia. *Experimental Gerontology, 87*, 92–99. https://doi.org/10.1016/j.exger.2016.10.003.

Mazzio, E. A., Boukli, N., Rivera, N., & Soliman, F. K. (2012). Pericellular pH homeostasis is a primary function of the Warburg effect: Inversion of metabolic systems to control lactate steady state in tumor cells. *Cancer Science, 103*, 422–432.

McLaughlin, D., Zhao, Y., O'Neill, K. M., Edgar, K. S., Dunne, P. D., Kearney, A. M., et al. (2017). Signalling mechanisms underlying doxorubicin and Nox2 NADPH oxidase-induced cardiomyopathy: Involvement of mitofusin-2. *British Journal of Pharmacology, 174*, 3677–3695.

Mishra, P., Carelli, V., Manfredi, G., & Chan, D. C. (2014). Proteolytic cleavage of Opa1 stimulates mitochondrial inner membrane fusion and couples fusion to oxidative phosphorylation. *Cell Metabolism, 19*, 630–641.

Misko, A., Jiang, S., Wegorzewska, I., Milbrandt, J., & Baloh, R. H. (2010). Mitofusin 2 is necessary for transport of axonal mitochondria and interacts with the Miro/Milton complex. *The Journal of Neuroscience*, *30*, 4232–4240.

Misko, A. L., Sasaki, Y., Tuck, E., Milbrandt, J., & Baloh, R. H. (2012). Mitofusin2 mutations disrupt axo-nal mitochondrial positioning and promote axon degeneration. *The Journal of Neuroscience*, *32*, 4145–4155.

Missiroli, S., Bonora, M., Patergnani, S., Poletti, F., Perrone, M., Gafa, R., et al. (2016). PML at mitochondria-associated membranes is critical for the repression of autophagy and cancer development. *Cell Reports*, *16*, 2415–2427.

Mitchell, P. (1961). Coupling of phosphorylation to electron and hydrogen transfer by a chemi-osmotic type of mechanism. *Nature*, *191*, 144–148.

Mourier, A., Motori, E., Brandt, T., Lagouge, M., Atanassov, I., Galinier, A., et al. (2015). Mitofusin 2 is required to maintain mitochondrial coenzyme Q levels. *The Journal of Cell Biology*, *208*, 429–442.

Mozdy, A. D., & Shaw, J. M. (2003). A fuzzy mitochondrial fusion apparatus comes into focus. *Nature Reviews. Molecular Cell Biology*, *4*, 468–478.

Naon, D., & Scorrano, L. (2014). At the right distance: ER-mitochondria juxtaposition in cell life and death. *Biochimica et Biophysica Acta*, *1843*, 2184–2194.

Okamoto, K., & Shaw, J. M. (2005). Mitochondrial morphology and dynamics in yeast and multicellular eukaryotes. *Annual Review of Genetics*, *39*, 503–536.

Palmer, C. S., Elgass, K. D., Parton, R. G., Osellame, L. D., Stojanovski, D., & Ryan, M. T. (2013). Adaptor proteins MiD49 and MiD51 can act independently of Mff and Fis1 in Drp1 recruitment and are specific for mitochondrial fission. *The Journal of Biological Chemistry*, *288*, 27584–27593.

Pan, L., Zhou, L., Yin, W., Bai, J., & Liu, R. (2018). miR-125a induces apoptosis, metabolism disorder and migrationimpairment in pancreatic cancer cells by targeting Mfn2-related mitochondrial fission. *International Journal of Oncology*, *53*(1), 124–136.

Parsons, M. J., & Green, D. R. (2010). Mitochondria in cell death. *Essays in Biochemistry*, *47*, 99–114.

Pham, A. H., Meng, S., Chu, Q. N., & Chan, D. C. (2012). Loss of Mfn2 results in progressive, retrograde degeneration of dopaminergic neurons in the nigrostriatal circuit. *Human Molecular Genetics*, *21*, 4817–4826.

Polyak, K., Xia, Y., Zweier, J. L., Kinzler, K. W., & Vogelstein, B. (1997). A model for p53-induced apoptosis. *Nature*, *389*, 300–305.

Poole, A. M., & Gribaldo, S. (2014). Eukaryotic origins: How and when was the mitochondrion acquired? *Cold Spring Harbor Perspectives in Biology*, *6*, a015990.

Prevarskaya, N., Skryma, R., & Shuba, Y. (2011). Calcium in tumour metastasis: New roles for known actors. *Nature Reviews Cancer*, *11*, 609–618.

Pyakurel, A., Savoia, C., Hess, D., & Scorrano, L. (2015). Extracellular regulated kinase phosphorylates mitofusin 1 to control mitochondrial morphology and apoptosis. *Molecular Cell*, *58*, 244–254.

Qi, Y., Yan, L., Yu, C., Guo, X., Zhou, X., Hu, X., et al. (2016). Structures of human mitofusin 1 provide insight into mitochondrial tethering. *The Journal of Cell Biology*, *215*, 621–629.

Rahman, J., & Rahman, S. (2018). Mitochondrial medicine in the omics era. *Lancet*, *391*, 2560–2574. 2018 Jun 11. pii: S0140–6736(18)30727-X. https://doi.org/10.1016/S0140-6736(18)30727-X.

Rambold, A. S., Kostelecky, B., Elia, N., & Lippincott-Schwartz, J. (2011). Tubular network formation protects mitochondria from autophagosomal degradation during nutrient starvation. *Proceedings of the National Academy of Sciences of the United States of America*, *108*, 10190–10195.

Reddy, P. H. (2014). Inhibitors of mitochondrial fission as a therapeutic strategy for diseases with oxidative stress and mitochondrial dysfunction. *Journal of Alzheimer's Disease*, *40*, 245–256.

Rehman, J., Zhang, H. J., Toth, P. T., Zhang, Y., Marsboom, G., Hong, Z., et al. (2012). Inhibition of mitochondrial fission prevents cell cycle progression in lung cancer. *The FASEB Journal*, *26*, 2175–2186.

Rojo, M., Legros, F., Chateau, D., & Lombes, A. (2002). Membrane topology and mitochondrial targeting of mitofusins, ubiquitous mammalian homologs of the transmembrane GTPase Fzo. *Journal of Cell Science*, *115*, 1663–1674.

Romanello, V., Guadagnin, E., Gomes, L., Roder, I., Sandri, C., Petersen, Y., et al. (2010). Mitochondrial fission and remodelling contributes to muscle atrophy. *The EMBO Journal*, *29*, 1774–1785.

Rossignol, R., Gilkerson, R., Aggeler, R., Yamagata, K., Remington, S. J., & Capaldi, R. A. (2004). Energy substrate modulates mitochondrial structure and oxidative capacity in cancer cells. *Cancer Research*, *64*, 985–993.

Santel, A., Frank, S., Gaume, B., Herrler, M., Youle, R. J., & Fuller, M. T. (2003). Mitofusin-1 protein is a generally expressed mediator of mitochondrial fusion in mammalian cells. *Journal of Cell Science*, *116*, 2763–2774.

Santel, A., & Fuller, M. T. (2001). Control of mitochondrial morphology by a human mitofusin. *Journal of Cell Science*, *114*, 867–874.

Schrepfer, E., & Scorrano, L. (2016). Mitofusins, from mitochondria to metabolism. *Molecular Cell*, *61*, 683–694.

Sebastián, D., Hernández-Alvarez, M. I., Segalés, J., Sorianello, E., Muñoz, J. P., Sala, D., et al. (2012). Mitofusin 2 (Mfn2) links mitochondrial and endoplasmic reticulum function with insulin signaling and is essential for normal glucose homeostasis. *Proceedings of the National Academy of Sciences of the United States of America*, *109*, 5523–5528.

Sehrawat, A., St. Croix, C., Baty, C. J., Watkins, S., Tailor, D., Singh, R. P., et al. (2016). Inhibition of mitochondrial fusion is an early and critical event in breast cancer cell apoptosis by dietary chemopreventative benzyl isothiocyanate. *Mitochondrion*, *30*, 67–77.

Singh, K. K., & Modica-Napolitano, J. S. (2017). Special issue: Mitochondria in cancer. *Seminars in Cancer Biology*, *47*, 4–6.

Soriano, F. X., Liesa, M., Bach, D., Chan, D. C., Palacín, M., & Zorzano, A. (2006). Evidence for a mitochondrial regulatory pathway defined by peroxisome proliferator-activated receptor-gamma coactivator -1alpha, estrogen-related receptor-alpha, and mitofusin 2. *Diabetes*, *55*, 1783–1791.

Stewart, T. A., Yapa, K. T., & Monteith, G. R. (2015). Altered calcium signaling in cancer cells. *Biochimica et Biophysica Acta*, *1848*, 2502–2511.

Suen, D. F., Norris, K. L., & Youle, R. J. (2008). Mitochondrial dynamics and apoptosis. *Genes & Development*, *22*, 1577–1590.

Sugioka, R., Shimizu, S., & Tsujimoto, Y. (2004). Fzo1, a protein involved in mitochondrial fusion, inhibits apoptosis. *The Journal of Biological Chemistry*, *279*, 52726–52734.

Sugiura, A., Nagashima, S., Tokuyama, T., Amo, T., Matsuki, Y., Ishido, S., et al. (2013). MITOL regulates endoplas-mic reticulum-mitochondria contacts via mitofusin2. *Molecular Cell*, *51*, 20–34.

Sun, J., Lu, F., He, H., et al. (2014). STIM1- and Orai1-mediated Ca(2+) oscillation orchestrates invadopodium formation and melanoma invasion. *The Journal of Cell Biology*, *207*, 535–548.

Tang, H., Tao, A., Song, J., Liu, Q., Wang, H., & Rui, T. (2017). Doxorubicin-induced cardiomyocyte apoptosis: Role of mitofusin 2. *International Journal of Biochemistry and Cell Biology*, *88*, 55–59. https://doi.org/10.1016/j.biocel.2017.05.006.

Thoresen, L., Frykholm, G., Lydersen, S., Ulveland, H., Baracos, V., Prado, C. M., et al. (2013). Nutritional status, cachexia and survival in patients with advanced colorectal carcinoma. Different assessment criteria for nutritional status provide unequal results. *Clinical Nutrition*, *32*, 65–72.

Wakabayashi, J., Zhang, Z., Wakabayashi, N., Tamura, Y., Fukaya, M., Kensler, T. W., et al. (2009). The dynamin-related GTPase Drp1 is required for embryonic and brain development in mice. *The Journal of Cell Biology*, *186*, 805–816.

Wang, P. T., Garcin, P. O., Fu, M., Masoudi, M., St-Pierre, P., Pante, N., et al. (2015). Distinct mechanisms controlling rough and smooth endoplasmic reticulum contacts with mitochondria. *Journal of Cell Science*, *128*, 2759–2765.

Wang, W., Liu, X., Guo, X., & Quan, H. (2018). Mitofusin-2 triggers cervical carcinoma cell hela apoptosis via mitochondrial pathway in mouse model. *Cellular Physiology and Biochemistry*, *46*, 69–81.

Wang, W., Lu, J., Zhu, F., et al. (2012). Pro-apoptotic and anti-proliferative effects of mitofusin-2 via Bax signaling in hepatocellular carcinoma cells. *Medical Oncology*, *29*, 70–76.

Wang, W., Xie, Q., Zhou, X., Yao, J., Zhu, X., Huang, P., et al. (2015). Mitofusin-2 triggers mitochondria Ca^{2+} influx from the endoplasmic reticulum to induce apoptosis in hepatocellular carcinoma cells. *Cancer Letters*, *358*, 47–58.

Wang, L., Yang, X., & Shen, Y. (2015). Molecular mechanism of mitochondrial calcium uptake. *Cellular and Molecular Life Sciences*, *72*, 1489–1498.

Wasilewski, M., Semenzato, M., Rafelski, S. M., Robbins, J., Bakardjiev, A. I., & Scorrano, L. (2012). Optic atrophy 1-dependent mitochondrial remodeling controls steroidogenesis in trophoblasts. *Current Biology*, *22*, 1228–1234.

Wu, L., Li, Z., Zhang, Y., Zhang, P., Zhu, X., Huang, J., et al. (2008). Adenovirus-expressed human hyperplasia suppressor gene induces apoptosis in cancer cells. *Molecular Cancer Therapeutics*, *7*, 222–232.

Wu, Y., Zhou, D., Xu, X., Zhao, X., Huang, P., Zhou, X., et al. (2016). Clinical significance of mitofusin-2 and its signaling pathways in hepatocellular carcinoma. *World Journal of Surgical Oncology*, *14*, 179.

Xi, Q. L., Zhang, B., Jiang, Y., Zhang, H. S., Meng, Q. Y., Chen, Y., et al. (2016). Mitofusin-2 prevents skeletal muscle wasting in cancer cachexia. *Oncology Letters*, *12*, 4013–4020.

Xu, K., Chen, G., Li, X., Wu, X., Chang, Z., Xu, J., et al. (2017). MFN2 suppresses cancer progression through inhibition of mTORC2/Akt signaling. *Scientific Reports*, *7*, 41718.

Yingkun, N., Lvsong, Z., & Huimin, Y. (2010). Shikonin inhibits the proliferation and induces the apoptosis of human HepG2 cells. *Canadian Journal of Physiology and Pharmacology*, *88*, 1138–1146.

Youle, R. J., & Van Der Bliek, A. M. (2012). Mitochondrial fission, fusion, and stress. *Science*, *337*, 1062–1065.

Yue, W., Chen, Z., Liu, H., Yan, C., Chen, M., Feng, D., et al. (2014). A small natural molecule promotes mitochondrial fusion through inhibition of the deubiquitinase USP30. *Cell Research*, *24*, 482–498.

Zhang, G. E., Jin, H. L., Lin, X. K., Chen, C., Liu, X. S., Zhang, Q., et al. (2013). Antitumor effects of Mfn2 in gastric cancer. *International Journal of Molecular Sciences*, *14*, 13005–13021.

Zhao, T., Huang, X., Han, L., Wang, X., Cheng, H., Zhao, Y., et al. (2012). Central role of mitofusin 2 in autophagosome-lysosome fusion in cardiomyocytes. *The Journal of Biological Chemistry*, *287*, 23615–23625.

Zhou, X., Zhang, L., Zheng, B., Yan, Y., Zhang, Y., Xie, H., et al. (2016). MicroRNA-761 is upregulated in hepatocellular carcinoma and regulates tumorigenesis by targeting mitofusin-2. *Cancer Science*, *107*, 424–432.

Zorzano, A., Liesa, M., Sebastian, D., Segales, J., & Palacin, M. (2010). Mitochondrial fusion proteins: Dual regulators of morphology and metabolism. *Seminars in Cell & Developmental Biology*, *21*, 566–574.

Zuchner, S., Mersiyanova, I. V., Muglia, M., Bissar-Tadmouri, N., Rochelle, J., Dadali, E. L., et al. (2004). Mutations in the mitochondrial GTPase mitofusin 2 cause Charcot-Marie-Tooth neuropathy type 2A. *Nature Genetics*, *36*, 449–451.

Further reading

Debattisti, V., Pendin, D., Ziviani, E., Daga, A., & Scorrano, L. (2004). Reduction of endoplasmic reticu-lum stress attenuates the defects caused by Drosophila mitofusin depletion. *The Journal of Cell Biology*, *204*, 303–312.

Egner, A., Jakobs, S., & Hell, S. W. (2002). Fast 100-nm resolution three-dimensional microscope reveals structural plasticity of mitochondria in live yeast. *Proceedings of the National Academy of Sciences of the United States of America*, *99*, 3370–3375.

Jimenez, L., Laporte, D., Duvezin-Caubet, S., Courtout, F., & Sagot, I. (2014). Mitochondrial ATP synthases cluster as discrete domains that reorganize with the cellular demand for oxidative phosphorylation. *Journal of Cell Science*, *127*, 719–726.

CHAPTER SIX

Molecular signaling in bone cells: Regulation of cell differentiation and survival

Lilian I. Plotkin[a,b,c,*], **Angela Bruzzaniti**[a,b,d]
[a]Department of Anatomy and Cell Biology, Indiana University School of Medicine, Indianapolis, IN, United States
[b]Indiana Center for Musculoskeletal Health, Indianapolis, IN, United States
[c]Roudebush Veterans Administration Medical Center, Indianapolis, IN, United States
[d]Department of Biomedical and Applied Sciences, Indiana University School of Dentistry, Indianapolis, IN, United States
*Corresponding author: e-mail address: lplotkin@iupui.edu

Contents

Abstract

The achievement of proper bone mass and architecture, and their maintenance throughout life requires the concerted actions of osteoblasts, the bone forming cells, and osteoclasts, the bone resorbing cells. The differentiation and activity of osteoblasts and osteoclasts are regulated by molecules produced by matrix-embedded osteocytes, as well as by cross talk between osteoblasts and osteoclasts through secreted factors. In

Advances in Protein Chemistry and Structural Biology, Volume 116
ISSN 1876-1623
https://doi.org/10.1016/bs.apcsb.2019.01.002

addition, it is likely that direct contact between osteoblast and osteoclast precursors, and the contact of these cells with osteocytes and cells in the bone marrow, also modulates bone cell differentiation and function. With the advancement of molecular and genetic tools, our comprehension of the intracellular signals activated in bone cells has evolved significantly, from early suggestions that osteoblasts and osteoclasts have common precursors and that osteocytes are inert cells in the bone matrix, to the very sophisticated understanding of a network of receptors, ligands, intracellular kinases/phosphatases, transcription factors, and cell-specific genes that are known today. These advances have allowed the design and FDA-approval of new therapies to preserve and increase bone mass and strength in a wide variety of pathological conditions, improving bone health from early childhood to the elderly. We have summarized here the current knowledge on selected intracellular signal pathways activated in osteoblasts, osteocytes, and osteoclasts.

1. Introduction

The maintenance of bone mass throughout life is attained by the regulated actions of osteoblasts, the bone forming cells, and osteoclasts, the bone resorbing cells (Fig. 1) (Bellido, Plotkin, & Bruzzaniti, 2014). Osteoblasts are cuboidal cells present on the bone surface and are derived from

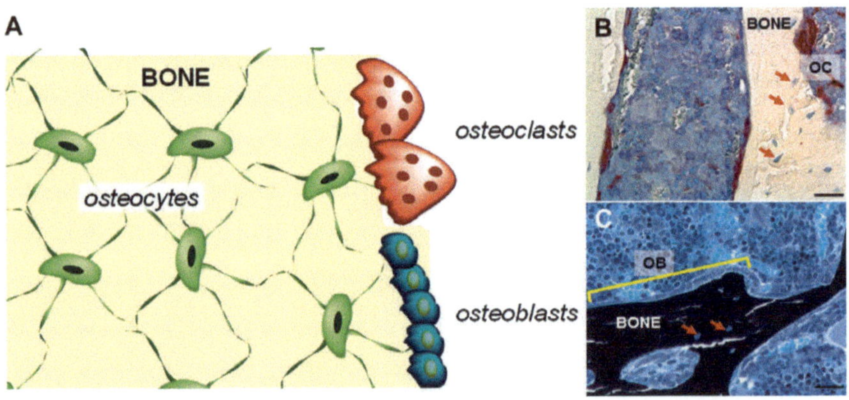

Fig. 1 (A). Schematic representation of bone containing osteocytes embedded in the matrix, and osteoblasts and osteoclasts on the surface. (B) and (C) Histologic sections of distal femur stained for (B) TRAP (osteoclasts, OC, red) and counterstained with toloduine blue and (C) von Kossa (mineralized bone, black) and counterstained with McNeal. A row of osteoblasts (OB) is indicated by the yellow line. Red arrows point at osteocytes within the bone matrix. Bone sections were prepared and stained at the ICMH Histology and Histomorphometry Core, Indianapolis, IN, USA. Scale bars indicate 25 μm.

mesenchymal stem cells (MSCs). Osteoblasts produce and secrete components of the bone matrix and regulate the deposit of hydroxyapatite, the mineral component of the mature bone. Osteoclasts are multinucleated cells derived from hematopoietic precursors also present on the bone surface. Osteoclasts produce enzymes that work in an acid environment and are able to dissolve bone mineral and proteins such as collagen molecules, removing bone and allowing for new bone to be formed. At the same time, osteoclastic bone resorption results in the release into the circulation of factors such as TGFβ, which are stored in bone and can now act on target cells and tissues.

Osteoblasts and osteoclast differentiation and activity are under the control of osteocytes, former osteoblasts embedded in the mineralized bone matrix (Bonewald, 2011). Over the past decade, our understanding of osteocyte biology has expanded and it is now clear that these cells are producers and targets of extracellular molecules that modulate cell activity and survival. In addition, all bone cells, as well as cells present in the bone marrow and in other tissues, produce factors that trigger intracellular signaling and modulate bone formation and resorption. Thus, a well-controlled network of cytokines and growth factors participate in the regulation of bone homeostasis. In addition to this paracrine/autocrine regulation, the generation, proliferation and survival of bone cells is regulated by systemic hormones, including parathyroid hormone, sex steroids (both androgens and estrogens) and glucocorticoids, reviewed elsewhere (Hachemi et al., 2018; Wein & Kronenberg, 2018; Zuo & Wan, 2017). In this chapter, we will describe the signaling pathways triggered in osteoblasts, osteocytes, and osteoclasts by selected regulatory molecules, as well as by direct cell-to-cell contact. The intracellular signaling mechanisms activated by additional bone-acting stimuli, as well as the biological consequences of these signaling pathways have been recently described in Plotkin and Bivi (2014), Verlinden, Vanderschueren, and Verstuyf (2016), Sims (2016), Strazzulla and Cronstein (2016), Grafe et al. (2018), Lisowska, Kosson, and Domaracka (2018), and Thompson, Rubin, and Rubin (2012).

2. Osteoblasts and osteocytes

Osteoblasts, which originate from MSC precursors, are the cells responsible for the synthesis of bone matrix proteins to form osteoid, and for matrix mineralization, to form mineralized, mature bone (Fig. 1) (Bellido et al., 2014). Under the influence of locally produced molecules,

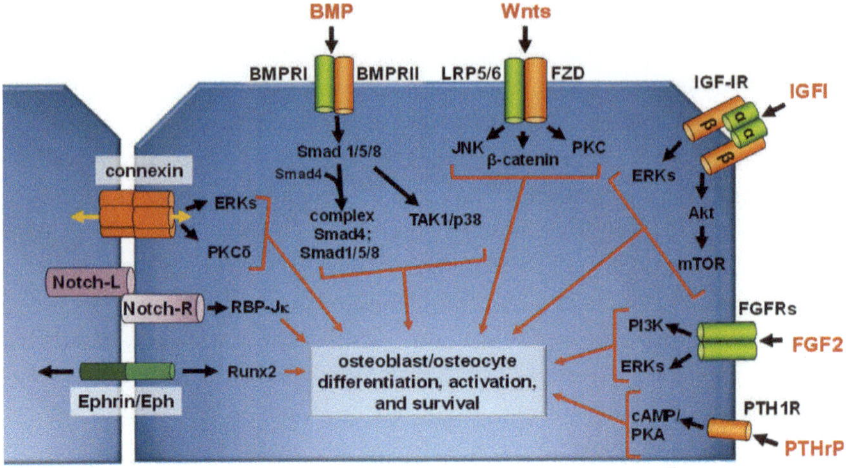

Fig. 2 Schematic representation of two neighboring osteoblastic cells, showing extracellular ligands and their corresponding receptors, and cell-to-cell signaling molecules. Selected downstream signaling molecules and the biological outcomes are also included. Notch-L, Notch ligand; Notch-R, Notch receptor.

the expression of lineage-specific transcription factors is stimulated and MSC undergo differentiation (Bradley, Westendorf, van Wijnen, & Dudakovic, 2018). These transcription factors include osterix, and Runt-related transcription factor 2 (Runx2) (Bellido et al., 2014) and deletion of these genes in mice results in defective osteoblast differentiation and lack of mineralized bone (Nakashima et al., 2002; Otto et al., 1997).

As osteoblast become surrounded by the bone matrix, they undergo changes in their morphology and the pattern of gene expression, transitioning into osteocytes, which are the most abundant and long-lived cells in bone. In spite the difference in morphology and gene expression pattern, most of the extracellular signaling molecules affect both osteoblasts and osteocytes, albeit with different biological consequences. We will describe the signaling pathways activated in both cell types, indicating when these pathways lead to different biological outcomes in osteoblast and osteocytes (Fig. 2).

2.1 Bone morphometric proteins

Bone morphogenetic protein (BMPs) are growth factors that belong to the transforming growth factor (TGF) β superfamily of proteins originally identified as inducers of ectopic bone formation (Katagiri & Watabe, 2016) and are now known to exert a potent osteogenic effect by promoting osteoblast differentiation and function. BMPs are synthesized as precursors, which are later cleaved and become active. Structurally, most of the BMPs exist as covalently-bound homodimers, although heterodimers have also been found.

BMPs bind to the BMP receptor (BMPR) type I (ALK-1-7) alone or in combination with a type II receptor (BMPR-II, ACTR-IIA, or ACTR-IIB). Type I and II receptors differ in cellular expression and ability to bind particular ligands, but all share kinase activity. Downstream of the receptors, BMPs phosphorylate Smad proteins. The so-called BMP-specific proteins include Smad1, Smad5 and Smad8, known as receptor-regulated Smads or R-Smads. R-Smad, in turn, bind to Smad4 and this complex translocates to the nucleus, where it can bind to transcription factors either stimulating or inhibiting gene transcription. In addition, Smad6 and 7 (inhibitory Smads or I-Smads) compete with R-Smad for Smad4 binding, therefore preventing Smad1/5/8 nuclear translocation and regulation of BMP-associated genes. The induction of gene transcription downstream of BMP2 requires connexin (Cx) 43 expression, and deletion of Cx43 from calvaria osteoblastic cells inhibits osteoblast differentiation induced by BMP-2 *in vitro* (Hashida et al., 2014). This effect of Cx43 requires direct cell-to-cell communication and cannot be restored by a Cx43 mutant that lacks the ability to form gap junction channels (see below for further description of connexins).

One of the transcription factors regulated by Smads downstream of BMP action is Runx2, which is required for osteoblast differentiation (Komori, 2006). Activation of the BMP/Smad signaling pathway results in the transcription of proteins involved in osteoblast differentiation and functions, such as osteocalcin, collagen alpha-1(I) chain, and alkaline phosphatase. In addition to activating Smads (the canonical BMP signaling pathway), BMPs also activate the non-canonical MAPKKK 7/TGF-β-activated kinase 1 (TAK1)-p38 pathway, independent of Smads (Gomez-Puerto, Iyengar, Garcia de Vinuesa, Ten Dijke, & Sanchez-Duffhues, 2018), a pathway required for bone and tooth development (Xu et al., 2008).

In addition to the intracellular regulators, BMP activity is regulated by extracellular antagonists and potentiators (Katagiri & Watabe, 2016). Antagonists include chordin, follistatin, gremlin, and noggin, among others, which sequester BMPs in the extracellular space or block ligand-receptor binding. BMPs potentiators include heparin, heparan-sulfate and dextran-sulfate, which bind to BMPs and potentiate the pro-osteoblastic effect of BMP-2, -4 and -7 (Irie, Habuchi, Kimata, & Sanai, 2003; Ruppert, Hoffmann, & Sebald, 1996; Takada et al., 2003). Other BMP potentiators have been described, although their role in osteoblastic cells is not known (Katagiri & Watabe, 2016).

The importance of BMP signaling for skeletal development, bone strength, and fracture repair has been demonstrated by human mutations in the ligands, receptors, antagonists, and intracellular signaling molecules (Wu, Chen, & Li, 2016). Further, animal models with both loss- and gain-of-function mutations have been developed, allowing for a better understanding of the role of the pathway and the consequences of its manipulation for human diseases.

2.2 Wnt-induced signaling

Wnts are a family of secrete glycoproteins originally described in *Drosophila*, as the Wingless (*Wg*) gene that is involved in wing development and in mice, as the *Int-1* gene that is involved in breast cancer development (Nusse et al., 1991; Nusse & Varmus, 1992). Wnts induce intracellular signaling by binding to members of the frizzled (FZD) family of receptors, which comprise 10 members (Schulte, 2015). FZDs are seven-transmembrane proteins, with a structure similar to G protein-coupled receptors. In addition, all members of the FZD family contain a cysteine-rich domain in the amino terminus that constitute the ligand-binding site, and a cytoplasmic tail that interacts with intracellular molecules. Wnts also interact with members of the low density lipoprotein receptor protein (LRP) family, which includes LRP4, LRP5 and LRP6. While LRP5 and 6 have been shown to mediate the activation of the Wnt signaling pathway, LRP4 binds to and facilitates the action of the Wnt antagonist sclerostin, the product of the *Sost* gene (Williams, 2017).

Through its binding to FZD receptors, Wnt ligands can activate four different signaling pathways: the canonical Wnt/β-catenin pathway, the non-canonical planar cell polarity (PCP) pathway, the Wnt-Ca2+, and the protein kinase A pathways (Baron & Kneissel, 2013). It has been shown that members of the Wnt family of proteins can activate one particular pathway in a cell context-dependent manner. Nevertheless, activation of both canonical and non-canonical Wnt signaling result from Wnt proteins binding to FZD-LRP5/6 co-receptor complex, followed by phosphorylation and activation of Disheveled (Dsh).

Activation of canonical Wnt signaling leads to the stabilization of β-catenin, followed by its translocation to the nucleus, where β-catenin binds to the transcription factor LEF and increases the transcription of Wnt target genes. In osteoblastic cells, osteoprotegerin (OPG) is one of the most studied Wnt target gene. OPG is involved in osteoclast differentiation and counteracts the pro-osteoclastogenic effects of receptor activator of NFκB ligand (RANKL), resulting in decreased osteoclast differentiation (see below for more details on osteoclastogenesis). Consistent with this, decreased bone mass induced by targeted deletion of the gene encoding β-catenin in osteoblasts and osteocytes results from enhanced bone resorption due to low OPG levels.

The role of Wnts in bone has been elucidated through understanding the genetic basis of human diseases associated with the pathway, and confirmed through genetic studies in which the different components of the signaling pathways have been modified either in cells of the osteoblastic lineage or in the germline (Baron & Kneissel, 2013). For example, both gain- and loss-of-function mutations have been described for LRP5, leading to high bone mass phenotypes and osteoporosis pseudoglioma, respectively (Williams, 2017). In addition, high bone mass phenotypes have also been described in individuals with loss-of-function mutations of the LRP4 gene (Leupin et al., 2011). Further, the absence of sclerostin due to null mutations or reduced expression of the protein due to absence of a regulatory region of the gene result in sclerosteosis or van Buchem's disease, respectively, conditions with high bone mass (Williams, 2017).

In addition, mutations in Wnt10b result in split hand/foot malformation 6 (Kantaputra, Kapoor, Verma, Intachai, & Ketudat Cairns, 2018; Zmuda et al., 2009) and deletion of this Wnt in mice leads to decreased bone marrow mesenchymal progenitors, low bone mass, and reduced expression of osteoblastic differentiation markers (Stevens et al., 2010). Further, osteoblast-derived Wnt16 suppresses osteoclast formation by increasing OPG levels (Alam et al., 2016; Moverare-Skrtic et al., 2014), and SNPs associated with loss of Wnt16 function are linked to low BMD and osteoporotic fractures (Koller et al., 2013).

The association of certain Wnts with particular FZD receptors lead to the phosphorylation of Dsh, followed by the activation of the PCP pathway, downstream of the small GTPases Rho and Rac, and the MAPK/c-Jun N-terminal kinase (JNK). The PCP pathway is involved in cytoskeletal organization, cell migration, and tissue pattering, and ultimately, in tissue morphogenesis. In bone, PCP is involved in the control of limb shape and dimensions. Thus, activation of PCP downstream of Wnt5a results in the transcription of the gene *Vangl2*, and absence of *Vangl2* leads to deficient limb bud development in mice. In addition, individuals with loss-of-function mutations in the Wnt5a gene or its co-receptor ROR2 exhibit brachydactyly type B and Robinow syndrome, conditions associated with shortening of long bones, and abnormal bone shape, among other abnormalities.

Further studies showed that the non-canonical members of the Wnt family of proteins Wnt4, Wnt5a, Wnt7b, and Wnt11 increase MSCs differentiation into the osteoblastic lineage. For example, Wnt5a stimulates osteogenic differentiation and inhibits adipogenesis in human adipose tissue-derived MSCs through activation of the Rho-associated protein kinase ROCK. In addition, the expression of Wnt5a and its co-receptor ROR2 are increased by mechanical stimulation, leading to RhoA activation, which is required for upregulation of Runx2 and the commitment of precursor cells to the osteogenic lineage induced by mechanical stimulation. As discussed later in this review, Wnt5a also plays a direct role in osteoclast differentiation.

2.2.1 Inhibitors of Wnt signaling

Activation of the Wnt signaling pathway is controlled by the regulation of expression and secretion of Wnt proteins, as well as by a network of inhibitors that act at different levels of the signaling cascade (Lerner & Ohlsson, 2015; Plotkin & Bivi, 2014). In bone, human mutations and animal experiments have demonstrated the importance of the inhibitors in the regulation of bone cell differentiation and function and, ultimately, bone mass and strength. Further, pharmacologic approaches are being developed to take advantage of the anabolic effect of Wnt signaling in bone, by blocking the Wnt antagonist, leading to increased bone mass (Baron & Gori, 2018).

2.2.1.1 Dickkopf

The Dickkopf (Dkk) proteins are members of a family of cysteine-rich domain containing proteins (Baron & Kneissel, 2013). Dkk1 and Dkk2 inhibit Wnt signaling by binding to LRP5/6 and Kremen (another Wnt inhibitor, see below). The expression of Dkk1, the most studied member of the family, is hormonally regulated, and it is also controlled by growth factors and mechanical stimulation. Dkk1 expression is required for limb morphogenesis and in its absence the head does not develop in Dkk1 knockout mice (Huang, Liu, & Liu, 2018). On the other hand, deletion of one Dkk1 allele results to high bone mass due to increased osteoblast number and activity, whereas Dkk1 overexpression results in reduced osteoblast number and activity and low bone mass. Consistent with an inhibitory role of Dkk1 on osteoblast function, high bone mass-associated LRP5 mutations render LRP5 proteins resistant to Dkk1-mediated inhibition of Wnt signaling. Dkk2 and Dkk4 have also been shown to inhibit canonical Wnt signaling in the presence of Kremen.

2.2.1.2 Kremen

The two members of this family of proteins, Kremen 1 and 2 are co-receptors for Dkk1, enhancing Dkk1-induced Wnt signaling inhibition (Baron & Kneissel, 2013; Plotkin & Bivi, 2014). Kremen 1 is widely expressed whereas Kremen 2 is predominantly expressed in bone cells. Kremen 2-deficient mice exhibit an age-dependent high bone mass phenotype, which can be detected at 2 years of age, and Kremen 1 and 2 double knockout mice exhibit increased Wnt signaling, bone mass, and bone formation. Conversely, transgenic mice overexpressing Kremen 2 exhibit reduced osteoblastic Wnt signaling and differentiation and activity and low bone mass.

2.2.1.3 Sclerostin

The product of the *Sost* gene, sclerostin, is expressed in osteocytes in bone and was originally described as an inhibitor of BMP due to its homology with known BMP antagonists (Baron & Kneissel, 2013; Plotkin & Bivi, 2014). It is now known that sclerostin binds with higher affinity to LRP5/6 and that LRP4 is required for the inhibition of Wnt signaling by sclerostin. As a consequence of its Wnt inhibitory effects, the absence of the Sost gene, reduced sclerostin levels or prevention of sclerostin association with LRP5/6 results in high bone mass. More recently, it has been shown that sclerostin can affect osteoclast differentiation indirectly, by stimulating RANKL expression in osteocytic cells *in vitro* and RANKL/OPG levels *in vivo* (Tu et al., 2015; Wijenayaka et al., 2011). However, whether this action of sclerostin depends on its binding to LRPs or on the regulation of Wnt signaling pathway, and the intracellular signaling molecules involved in RANKL/OPG regulation in osteoblastic cells remains unknown.

2.2.1.4 Secreted frizzled-related proteins

Secreted FZD-related proteins (sFRPs) are soluble proteins that bind to Wnts and prevent their association with the FZD receptors on the cell surface, inhibiting intracellular Wnt signaling (Baron & Kneissel, 2013; Plotkin & Bivi, 2014). Among the five members of the family of proteins, sFRP-1 was shown to be expressed in bone and to inhibit osteoblast differentiation and induce osteoblast and osteocyte apoptosis. Consistent with this, deletion of *Sfrp1* results in high bone mass whereas its overexpression decreased bone mass in transgenic mice. sFRP4 also inhibits bone formation by preventing osteoblast proliferation, and its overexpression in bone or liver lead to elevated sFRP4 circulating levels and low bone mass. Consistent with a role of the protein in bone, *SFRP4* polymorphisms have been associated with changes in bone mass in the hip and spine in humans.

2.2.1.5 Wise

Wise is another LRP5/6 binding protein that prevents Wnt signaling (Baron & Kneissel, 2013, Plotkin & Bivi, 2014). It has also been shown to associate with LRP4, and mutations of both Wise and LRP4 genes lead to similar abnormal teeth, suggesting that the two molecules interact during tooth development. The absence of Wise results in the activation of Wnt signaling and the modulation of the fibroblast growth factor (FGF) and sonic hedgehog pathways.

2.3 Insulin-like growth factor

Insulin-like growth factor (IGF) I and II, named based on their structural similarity with insulin, and their insulin-like properties, are produced mainly in the liver and also in bone and adipose tissue (Yakar, Werner, & Rosen, 2018). The two forms of the ligand share high homology. IGFI is expressed both during embryonic development and after birth, whereas IGFII is expressed in the embryo but is downregulated after birth. IGFs bind to and activate two receptors, IGF-IR and IGF-IIR, and can also bind to the insulin receptor, albeit with low affinity. IGF-IR is a tetramers formed by two extracellular ligand binding α subunits and two transmembrane β subunits, which have tyrosine kinase activity. IGF-IR activation results in the phosphorylation of adaptor insulin receptor substrate (IRS) proteins or of Shc (Bikle, 2008), followed by activation of the growth factor receptor binding protein 2 (Grb2)/Sos/Ras/Raf/MAPK extracellular signal-regulated kinase (MEK), resulting in extracellular signal-regulated kinase ERK activation. In addition, activation of the IGF-IR also results in the activation of PI3K, followed by phosphorylation and activation of Akt, and of the mammalian target of rapamycin (mTOR), eukaryotic translation initiation factor 4E (eIF4E), and p70S6 kinase (S6K). In general, activation of this receptor leads to cell proliferation and survival and has been associated with cancer cell metastatic potential. The IGF-IIR, also known as cation-independent mannose-6-phosphate receptor, has no intrinsic kinase activity and binds to IGF-II with high affinity, inducing IGF-II lysosomal degradation. IGF-IR is expressed in bone during development and in adults, whereas mRNA expression for IGF-IIR has been shown in embryos but not adult rat bones (Bikle et al., 1994). The activity of IGFs is modulated by the IGF binding proteins (IGFBP) 1–6, which regulate IGF bioavailability and can either enhance or reduce IGF effects (Bunn & Fowlkes, 2003). IGFBPs also exert direct effects on cell function, independent of IGFs. IGFs and IGFBPs form a complex with the acid-labile subunit (ALS), which is required for IGF/IGFBP circulation and IGF-I function (Boisclair, Rhoads, Ueki, Wang, & Ooi, 2001).

Both IGF-I and II are produced locally by osteoblastic cells and stored in the bone matrix as an inactive form bound to IGFBPs (Kasukawa, Miyakoshi, & Mohan, 2004). During bone resorption by osteoclasts, IGFs are released from the bone matrix and can act as signals to recruit osteoblast precursors to the newly eroded bone surface. Osteoblasts also produce several IGBPs, with a pattern that depends on the stage of osteoblast differentiation. The role of IGFs on osteoblast activity has been extensively studied *in vitro* and *in vivo*, using genetically modified mice in which the levels of the ligands and/or the receptors were manipulated globally or in a cell-specific manner (Yakar et al., 2018). In particular, it has been shown that IGFs induce osteoblast proliferation and differentiation and inhibit apoptosis (Bikle, 2008). The importance of IGF on bone development has been demonstrated by the profound skeletal defects observed in mice lacking IGF-I or IGF-IR, with impaired growth and delayed ossification, and reduced bone mass and osteoblast number and surface. The growth deficiency and delayed ossification observed in IGF-IR-deficient mice are reproduced in mice lacking Akt1/Akt2, a kinase activated by IGFs, demonstrating the importance of this signal transduction pathway for bone growth (Peng et al., 2003). In osteoblastic cells, IGF-I has been shown to activate the PI3K/PKD/Akt and Ras/Raf/MAPK pathways, both leading to activation of p70S6, which mediate the proliferative and anti-apoptotic actions of IGF-I (Govoni, 2012). A pathway activated by IGF-I via $G_{i\beta\gamma}$ leading to osteoblastic cell proliferation and survival independently of p70SK activation has also been proposed (Grey et al., 2003).

2.4 Parathyroid hormone-related peptide

Parathyroid hormone-related peptide (PTHrP) is a locally produced parathyroid hormone (PTH) analog expressed by osteoblasts and osteocytes (Schluter, 1999). Both proteins share high homology at the first 34 aminoacids and bind to the PTH receptor 1 (PTH1R) expressed in osteoblasts and osteocytes. However, they differ in pattern of expression and exhibit different functions; and while PTHrP deletion is lethal due to defective rib bone formation, PTH null mice show a milder phenotype and survive after birth. PTHrP is required for the commitment of osteoblast precursors to the osteoblastic lineage and for osteoblast maturation, and through actions on osteoblastic cells, on osteoclast differentiation via increases in RANKL expression.

Upon binding to the PTH1R, PTHrP induces the phosphorylation of the receptor and the activation of GTP-binding proteins. In osteoblasts, PTHrP binding to PTH1R results in the activation of the Gsα subunit of the G protein leading to cAMP/PKA signaling. PKA, in turn, phosphorylates and activates transcription factors, including CREB, c-Fos, c-Jun, and the osteoblastic transcription factor Runx2.

PTHrP can also activate of the G_q subunit downstream of the PTH1R. This results in activation of the 1-phosphatidylinositol 4,5-bisphosphate phosphodiesterase [phospholipase C (PLC)] and PI3K/diacylglycerol pathways, which increase intracellular $Ca2^+$ and stimulate PKC activity. Further, PTHrP can either activate or inhibit the MAPK ERK1/2 signaling pathway, depending on whether cells are undergoing proliferation or differentiation, respectively.

The activity of PTHrP is regulated by metalloproteinases that cleave the protein generating the $PTHrP_{1-17}$ peptide, which can stimulate calcium signaling but cannot activate cAMP. In addition, a $PTHrP_{12-48}$ proteolytic product regulates the bone marrow niche and acts as a suppressor of osteoclastogenesis. A $PTHrP_{107-139}$ peptide is also found *in vivo* and displays anti-resorptive and pro-osteoblastic properties when administered to animal models that lack sex steroids, have excess glucocorticoids, or have diabetes (Esbrit, Herrera, Portal-Nunez, Nogues, & Diez-Perez, 2016; Lozano et al., 2011).

Another way of regulating PTHrP signaling is through internalization of the receptor leading to desensitization of the cells to PTHrP (as well as PTH) effects. The internalization occurs upon ligand binding, by the recruitment of the scaffolding proteins β-arrestin-1 and -2. Binding to β-arrestins reduce the affinity of PTH1R for Gsα and suppresses cAMP-mediated responses. Receptors can be later recycled to the cell surface or targeted for degradation.

PTHrP also exert nuclear actions in osteoblastic cells, independent of PTH1R. Thus, PTHrP translocate to the nucleus where it can regulate apoptosis and cell proliferation by altering the pattern of gene expression (Garcia-Martin et al., 2014).

2.5 Fibroblast growth factors

Fibroblast growth factors (FGFs) are a family of proteins that exert both paracrine and endocrine functions (Beenken & Mohammadi, 2009). FGFs bind to the extracellular matrix until activated by proteases, when they are released to activate FGF receptors (FGFRs). Four FGFRs (1–4) have been described, all members of the tyrosine kinase receptor family, and FGFR1–3 are expressed in osteoblastic cells (Ornitz & Marie, 2015). Binding of FGFs to the receptors results in the activation of the intracellular signaling molecules PLC-γ1, MAPK, and PI3K/Akt (Yun et al., 2010). In addition to direct actions of FGFs through their receptors, the growth factors affect bone cells indirectly, through the upregulation of other local factors, such as TGFβ, IGF-I and VEGF (Ornitz & Marie, 2015).

Genetic studies in mice showed differing effects of different FGFR in osteoblastic cells. Thus, FGFR2 deletion in osteochondroprogenitors results in shorter axial and appendicular skeleton and reduced bone formation, whereas FGFR1 deletion in osteoprogenitors and differentiated osteoblasts results in increased bone mass, suggesting that the receptor mediates inhibitory actions on osteoblastic cells (Jacob, Smith, Partanen, & Ornitz, 2006; Yu et al., 2003).

Regarding FGFR ligands, FGF2 is required for proper osteoblast differentiation and activity, and its deletion results in decreased bone mass in mice (Yu et al., 2003). The mechanism of action of FGF2 and the consequence of overexpression of the growth factor depend on the molecular weight isoform, with the low molecular weight inducing activation of the WNT/β-catenin signaling pathway and increased bone formation (Xiao et al., 2009), and the high molecular isoform inducing the activation of the FGF23/FGFR/MAPK signaling pathway and leading to reduced bone formation *in vitro* (Xiao, Esliger, & Hurley, 2012). *In vitro* studies demonstrated that the pro-osteoblastogenic effects of FGF2 are potentiated by expression of Cx43 (Hebert & Stains, 2013; Lima, Niger, Hebert, & Stains, 2009; Niger et al., 2012; Niger, Hebert, & Stains, 2010; Niger et al., 2013) (see below). In addition, FGF2 promotes osteoblast survival through the activation of the PI3K signaling pathway (Park, Kim, Choi, Rhee, & Lim, 2009).

Other members of the FGF family of proteins with action on bone cells are FGF18, which is expressed in osteoblastic cells and induces osteogenic differentiation through activation of the FGFR1/2 and the MAPK and PI3K signaling pathways *in vitro* (Hamidouche et al., 2010), and FGF21, which is produced in pancreas, liver, and adipose tissue and decreases osteoblast surface, and bone mass and strength (Charoenphandhu et al., 2017).

Bone, and in particular, mature osteoblasts and osteocytes, are the main source of FGF23, a growth factor that reduces vitamin D levels and regulates phosphate homeostasis through inhibition of renal phosphate reabsorption (Yoshiko et al., 2007). FGF23 synthesis is, in turn, regulated by phosphate levels, vitamin D and PTH (Clinkenbeard & White, 2016). Activation of the FGFR by FGF23 requires the expression of the co-receptor Klotho, expressed in kidney (in the distal convoluted tubule) and the parathyroid gland (Clinkenbeard & White, 2016; Komaba & Lanske, 2018). More recently, it was shown that Klotho is also expressed in bone and, in particular, in osteoblasts and osteocytes (Rhee et al., 2011). Further studies showed that absence of Klotho from osteocytes results in increased bone mass and osteoblast activity in mice, whereas Klotho overexpression inhibits osteoblastic cell differentiation *in vitro* (Komaba & Lanske, 2018). The mechanism for the actions of FGF23 on osteoblastic cells is not completely understood, but it is believed to be mediated by induction of the expression of the Wnt/β-catenin inhibitor Dkk1, resulting on the inhibition of the bone anabolic Wnt pathway. In addition, FGF23 regulates its own synthesis through Klotho in osteoblastic cells. The effects of FGF23 on osteoblastic cells via FGFR1/Klotho stimulation result in the activation of ERKs, leading to modulation of gene expression.

3. Osteoclast signaling mechanisms

Osteoclasts are the predominant bone-degrading cell in the body (Fig. 1). The highly motile nature of osteoclasts, combined with their resorptive capacity, can result in extensive degradation of cortical bone, cancellous bone, alveolar bone, and tooth dentin. The molecular analysis of human forms of osteopetrosis caused by defects in osteoclast resorption, and the prevalence of low bone mass osteoporosis especially in the elderly which is due to excessive osteoclast activity, highlight the importance of understanding osteoclast biology for human disease. The catabolic effects of osteoclasts must therefore be tightly controlled, which occurs in part through the regulation of osteoclast formation and cell death, as well as the expression and secretion of bone degrading chemical and biological factors. The birth, life, and death of osteoclasts are regulated by a complex series of signaling molecules and pathways, many of which are known, while others still remain to be determined.

3.1 Osteoclast morphology and function

3.1.1 Regulation of osteoclast resorption by proton pump and chloride channel

Upon attachment to bone, osteoclasts change their morphology to become polarized multinucleated cells, which enables them to form two specialized surface domains, known as the sealing zone and the ruffled border membrane. The sealing zone surrounds the ruffled border and provides the machinery to enable attachment to the extracellular matrix. The ruffled border, on the other hand, is a highly convoluted membrane domain through which the hydrolytic enzymes and factors necessary for bone resorption are secreted (Baron et al., 1988; Baron, Neff, Louvard, & Courtoy, 1985; Blair, Teitelbaum, Ghiselli, & Gluck, 1989; Bruzzaniti & Baron, 2006; de Vernejoul, Horowitz, Demignon, Neff, & Baron, 1988; Vaananen, Zhao, Mulari, & Halleen, 2000).

The ruffled border contains an extensive amount of enfolded membrane, which is formed from the directed fusion of intracellular vesicles with the apical domain of the osteoclasts. The ruffled border membrane provides a large surface area through which bone dissolution and endocytosis of bone degradation products can occur. The degradation products of collagen and other matrix components are endocytosed at the ruffled border, transported through the cell to be degraded or secreted, or are released into the extracellular space during cell migration.

During bone resorption, the osteoclast subcellular lacunar compartment is acidified through the combined actions of the vacuolar ATPase (proton pump) (Feng et al., 2009; Lee et al., 2006; Scimeca et al., 2000; Wu, Xu, & Li, 2009; Yang et al., 2012), in conjunction with chloride ions secreted by the chloride channel CIC-7 (CLCN7) (Kasper et al., 2005; Kornak et al., 2001). Numerous mutations of the vacuolar ATPase or CIC-7 are known, which cause dysfunction in osteoclast resorption, leading to osteopetrosis in mice (Kornak et al., 2001) and humans (Cleiren et al., 2001; Waguespack et al., 2003). The targeted delivery of protons and chloride ions to the osteoclast resorption lacuna lowers the pH and dissolves hydroxyapatite while a mixture of proteases, including matrix metalloproteases such as MMP9 (Kim et al., 2016; Sundaram et al., 2007) and MMP14 (Sato et al., 1997) as well as tartrate-resistant acid phosphatase (TRAP) (Allen, Nuttleman, Ketcham, & Roberts, 1989; Marks & Grolman, 1987) and cathepsin K (Gelb, Shi, Chapman, & Desnick, 1996; Gowen et al., 1999; Lazner, Gowen, & Kola, 1999; Troen, 2004), a cysteine protease expressed predominately in osteoclasts, digest the organic bone matrix.

The extrusion of protons by the vATPase and chloride ions through the CIC-7 alters intracellular membrane polarity. Therefore, proton generation in mature osteoclasts is controlled by carbonic anhydrase II, an enzyme that catalyzes the interconversion of carbon dioxide and water to form protons and carbonic acid. In addition, electroneutrality within the osteoclast is maintained by the actions of CIC-7. These channels function as Cl-/H+ antiporters such that for every two chloride ions that are released into the resorption lacuna, one hydrogen ion flows in. The actions of CIC-7 also require OSTM1 (osteopetrosis-associated transmembrane protein 1) which is a transmembrane protein also localized to lysosomes which is thought to function as a β subunit of CIC-7, controlling its intracellular localization and protein stability (Chalhoub et al., 2003; Lange, Wartosch, Jentsch, & Fuhrmann, 2006; Maranda et al., 2008; Pangrazio et al., 2006).

3.1.2 Small GTP-binding proteins

Several small GTP-binding proteins are known to be expressed in osteoclasts and play a role in targeting the acid transporting vesicles to the ruffled border membrane (Itzstein, Coxon, & Rogers, 2011; Mulari, Vaaraniemi, & Vaananen, 2003; Mulari, Zhao, Lakkakorpi, & Vaananen, 2003; Sun, Buki, Ettala, Vaaraniemi, & Vaananen, 2005). For example, Rab5c is known to be associated with early recycling endosomes, while Rab11b is associated with cytosolic (perinuclear) recycling compartments and is thought to regulate turnover of the ruffled border membrane, and osteoclast motility. In addition, the GTP-binding proteins, Rab7 and Rab9, which are known to be associated with the late endosomes in other cells, were found to localize at the ruffled border membrane, suggesting that this specialized plasma membrane domain may behave similar to a late endosomal compartment. Several lysosomal proteins are also expressed at the ruffled border membrane and in the resorption lacuna, suggesting a lysosomal origin for these vesicles (Ferron et al., 2013; Lacombe, Karsenty, & Ferron, 2013). Indeed, the resorption lacuna acts much like an extracellular lysosome, providing the acidic environment as well as the proteases required for the degradation of the organic matrix of bone, mostly composed of type I collagen.

3.1.3 Osteoclast attachment and migration

Osteoclast attachment is mediated though integrins, which bind to arginyl-glycyl-aspartic acid (RGD) peptides at eroded bone surfaces (Fisher et al., 1993; Helfrich, Nesbitt, Dorey, & Horton, 1992; Horton, Taylor, Arnett, & Helfrich, 1991; Shankar et al., 1995). The integrins, as well as host of other cytoskeletal proteins, and filamentous actin, are localized to the sealing zone of osteoclasts. Osteoclasts resorption is linked with their motility, which is in part facilitated through the rapid polymerization and depolymerization of actin, as well as the assembly/disassembly of signaling molecules at adhesion "hot spots" in the sealing zone known as podosomes (Jurdic, Saltel, Chabadel, & Destaing, 2006). Podosomes are not just found in osteoclasts, but also in highly migratory cells such as macrophages and v-src transformed cancer cells (Destaing, Saltel, Geminard, Jurdic, & Bard, 2003; Linder & Aepfelbacher, 2003; Spinardi et al., 2004). Structurally, podosomes contain a central actin core surrounded by actin regulatory proteins, integrins, and the integrin-associated proteins such as talin, cortactin, dynamin (McNiven, Baldassarre, & Buccione, 2004; Ochoa et al., 2000). In addition, they contain a large number of signaling molecules such as Src, Pyk2, talin, cortactin, and others. Unlike focal adhesions, which are relatively more stable structures, podosome clusters assemble and disassemble within minutes in a process involving the rapid polymerization and depolymerization of the central actin core (McNiven et al., 2004; Ochoa et al., 2000). Thus, podosome turnover and targeted integrin engagement/disengagement in the sealing zone enables osteoclasts to rapidly migrate and, in the process, digest significant amounts of bone during their short life spans. Indeed, in studies published as far back as 1984, it was reported that rabbit osteoclasts in culture display an average resorption rate of 165 μm^3/h (Ali, Boyde, & Jones, 1984; Boyde, Ali, & Jones, 1984).

3.2 Osteoclast differentiation

Osteoclasts are derived from the hematopoietic, myeloid lineage. Many of the extracellular signals that drive osteoclast proliferation, differentiation, and fusion are secreted by the mesenchymal-derived osteoblasts and osteocytes. The osteoclast differentiation process can be recapitulated *in vitro* by the addition of two primary cytokines, macrophage colony-stimulating factor (M-CSF) and RANKL (Fig. 3) (Kong et al., 1999; Wiktor-Jedrzejczak et al., 1990), which are secreted by osteoblasts and osteocytes, as discussed previously in this review.

3.2.1 M-CSF/c-FMS

The critical role of M-CSF is demonstrated from deletions/mutations studies in mice (op/op) and rats (tl/tl) which have a point mutation in the *csf1* gene and express non-functional M-CSF protein (Felix, Cecchini, & Fleisch, 1990; Marks et al., 1992). As a result, osteoclasts are absent in these mice, leading to severe osteopetrosis. The predominant role of M-CSF is in supporting the proliferation and survival of osteoclasts. Binding of M-CSF

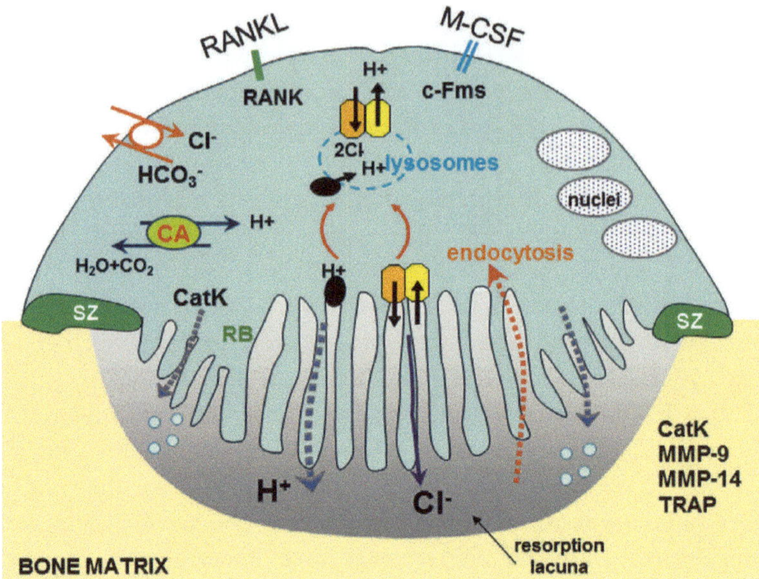

Fig. 3 Schematic representation of a resorbing osteoclast showing the sealing (SZ), ruffled border (RB) membrane, nuclei and resorption lacuna. Mature osteoclasts are polarized with the apical domain directed toward the bone surface. Protons generated by carbonic anhydrase (CA) are transported via the proton pump (black ellipse) to the ruffled border membrane. A coupled basolateral bicarbonate/chloride exchanger (solid white circle) maintains electroneutrality, avoiding changes in pH and/or membrane polarization of the cell. Secretion of protons via the proton pump (black circles) and chloride ions via ClC-7 (orange) in complex with Ostm1 (yellow) acidifies the resorption lacuna. Bone dissolution occurs through the actions of secreted cathepsin K (CatK), matrix metalloproteases and TRAP. Bone degradation products are released into the bone microenvironment during cell migration, or are internalized via recycling vesicles into the cell and are secreted, or degraded via the lysosomes (blue dashed ring). RANKL binding to RANK receptor and M-CSF binding to the c-Fms receptor are also indicated.

to its receptor c-Fms (or CSFR-1R) induces phosphorylation at key tyrosine sites in the cytoplasmic tail of c-Fms, which recruits the tyrosine kinase c-Src (Fig. 3). The c-Fms/c-Src protein complex then recruits the phosphatidylinositol 3-kinase (PI3K) and the E3 ubiquitin ligase, c-Cbl, in turn leading to the activation of the Akt pathway, and subsequently to the Grb2-mediated activation of ERK (Horne, Sanjay, Bruzzaniti, & Baron, 2005; Lee & States, 2000; Sanjay et al., 2001; Xiong et al., 2011; Yokouchi et al., 2001). Therefore, through a complex series of steps, M-CSF binding to c-Fms increases osteoclast precursor proliferation and survival, ultimately through the ERK and PI3K/Akt pathways.

3.2.2 Signaling by the RANK receptor

RANKL (also known as OPGL, ODF and TRANCE) binds to its cognate receptor, RANK (Fig. 3). As with c-Fms, genetic mouse models have demonstrated the critical role of the RANKL-RANK system in osteoclast formation, and mice deficient in either the ligand or the receptor develop osteopetrosis (Hanada, Hanada, & Penninger, 2010; Kong et al., 1999; Odgren et al., 2003; Walsh & Choi, 2003). The binding of RANKL to its receptor, RANK, on osteoclasts induces the association of RANK with the TRAF family members, including TRAF2 and TRAF6. Among these, TRAF6 has been most strongly associated with osteoclast formation and function and mice lacking TRAF6 develop osteoclast-deficient osteopetrosis (Lomaga et al., 1999; Naito et al., 1999). Downstream effectors of the RANK-TRAF6 complex in osteoclasts include NF-κB, Akt, and Nuclear Factor of Activated T-cells cytoplasmic 1 (NFATc1). Further, the activation of PI3K/Akt may be dependent on Src kinase activity, since in the absence of active c-Src, RANKL-mediated Akt activation is prevented and PI3K inhibitors block osteoclast formation. In addition to activating the PI3K/Akt pathway, TRAF6 binding to RANK, recruits TGF-β-activated kinase (TAK) 1 and the TAK-1-binding protein (TAB) 2, and this protein complex activates the MAPK pathways including ERK, JNK, and p38, promoting pre-osteoclast proliferation (Moon et al., 2012; Sugatani, Alvarez, & Hruska, 2003; Wong et al., 1999).

In contrast to the pro-osteoclastic effects of RANK-RANKL activation, an alternative receptor for RANKL in osteoclasts has recently been described which leads to the inhibition of osteoclast development. The leucine-rich repeat-containing G-protein coupled receptor 4 (LGR4, also known as GPR48) is a suppressor of osteoclast formation (Jin & Yang, 2016; Luo et al., 2016; Styrkarsdottir et al., 2013; Yi et al., 2013; Zhou et al., 2017; Zhu et al., 2016). RANKL binding to LGR4 activates Gαq, calcium signaling, and GSK3-β which lead to the inhibition of NFATc1 in the nucleus, blocking osteoclast differentiation. In support of the critical role of LRG4 in osteoclastogenesis, deletion of LRG4 in all tissues or specifically in monocytes leads to osteoporosis in mice, due to increases in osteoclast number and activity. Further, it was demonstrated that the soluble extracellular domain of LGR4 can bind RANKL, inhibiting osteoclast differentiation and increasing bone mass in several mouse models. Therefore, LGR4 acts as a second receptor for RANKL that negatively regulates osteoclast formation and activity.

3.2.3 Osteoclast transcription factors

Signaling through the c-Fms and RANK receptors leads to the activation of important osteoclasts transcription factors, including microphthalmia-induced transcription factor (MITF) and NFATc1 (Meadows et al., 2007). These transcription factors drive the expression of several genes necessary for osteoclast resorption. For example, in response to RANKL-RANK binding, MITF upregulates TRAP and cathepsin K, which are required for the ability of mature osteoclasts to degrade the bone matrix. MITF is also known to regulate the expression of ClC-7 and Ostm1.

RANKL also activates NF-κB through two distinct pathways, known as the classical pathway or the alternative pathways. NF-κB consists of five members, including p65 (RelA), RelB, c-Rel, p50/p105 (NF-κB1), and p52/p100 (NF-κB2). Activation of NF-κB can have both positive and negative effects on osteoclast activity (Boyce, Xiu, Li, Xing, & Yao, 2015). While TRAF6 activates only the classical NF-κB pathway, TRAF2 and TRAF5 activate both pathways (Hauer et al., 2005; Mizukami et al., 2002; Yi, Lin, Stunz, & Bishop, 2014). When activated by RANKL, NF-κB translocates to the nucleus and binds its target genes, leading to the upregulation of c-Fos. Subsequently, c-Fos binds to NFATc1. Thus, NF-kB positively regulates osteoclast differentiation and function by activating c-Fos and NFATc1, as well as by inhibiting specific NFATc1 repressor proteins. Conversely, NF-kB can inhibit osteoclast formation and function through TRAF3 and P110, and the suppressors of c-Fos/NFATc1 signaling, IRF8, and RBP-J. NFATc1 is also regulated by calcium/calmodulin and it is thought that RANKL-mediated phosphorylation of specialized motifs known as immunoreceptor tyrosine-based activation motifs (ITAMs). Phosphorylation of ITAMs results in the activation of Syk and PLCγ, which mobilizes intracellular calcium and, in turn activates calcineurin, a calmodulin-dependent phosphatase that directly dephosphorylates serine residues in NFATc1. Two specific tyrosine kinases, BTK and Tec, may provide the molecular link between ITAM and RANK receptor signaling leading to osteoclastogenesis (Pitcher & van Oers, 2003; Shinohara et al., 2008).

It is important to note that osteoclast differentiation can also be stimulated independently of RANKL, through tumor necrosis factor alpha (TNFα), and M-CSF is a common factor required for both RANKL and TNFα-induced osteoclast differentiation (Kobayashi et al., 2000). Osteoclast differentiation induced by TNFα occurs via the TNF receptor (TNFR) 1 and 2 which are expressed on osteoclast precursors. TNFR1 and TNFR2 also use TRAF2 as a common signal transducer, through which they promote NFATc1 activation and osteoclast differentiation (Kanazawa & Kudo, 2005). However, osteoclasts induced by TNFα formed resorption pits only in the presence of interleukin 1 alpha (IL-1α), and therefore play an important role in bone resorption associated with inflammatory bone diseases.

3.2.4 Osteoclast precursor fusion proteins

The formation of a mature multinucleated osteoclasts requires the fusion of mononuclear precursor cells. Two signaling proteins have been identified which regulate osteoclast fusion and multinucleation, namely DC-STAMP (dendritic-cell specific transmembrane protein) and OC-STAMP (osteoclast stimulatory transmembrane protein) (Kukita et al., 2004; Mensah, Ritchlin, & Schwarz, 2010; Miyamoto, 2006; Vignery, 2005; Yagi et al., 2005, 2007; Yang et al., 2008). DC-STAMP is expressed as a dimer on the cell surface of both human and murine osteoclast progenitors. It was discovered that mice lacking DC-STAMP form mononuclear osteoclast precursors that express TRAP, a key enzyme required for bone resorption, but these mononuclear precursors do not fuse and the DC-STAMP-deficient mice exhibit defects in osteoclast number and activity, leading to osteosclerosis. Conversely, DC-STAMP overexpression increases the formation of mature multinucleated osteoclasts. Although RANKL does not directly bind DC-STAMP it can modulate the expression of DC-STAMP in osteoclast precursors. In recent studies, it was reported that DC-STAMP contains an immunoreceptor tyrosine-based inhibitory motif (ITIM) in its cytoplasmic tail, and that DC-STAMP regulates osteoclast differentiation through the NFACTc1 signaling (Chiu et al., 2017).

The second osteoclast "fusogenic" protein, OC-STAMP, is also induced by exposure to RANKL (Kim, Park, Baek, Kim, & Kim, 2011; Yang et al., 2008). Unlike DC-STAMP, OC-STAMP appears to only be expressed in osteoclast lineage cells. In murine osteoclast progenitors, inhibition of OC-STAMP activity through the use of a neutralizing antibody or siRNA-mediated knockdown results in a marked, reduction in the number of mature multinucleated osteoclasts, whereas overexpression of OC-STAMP significantly increases the number of mature osteoclasts. Similar to mice lacking DC-STAMP, mice in which OC-STAMP is blocked contain TRAP-positive mononuclear cells, but few mature multinucleated osteoclasts (Kim et al., 2011, Yang et al., 2008).

3.3 Osteoclast survival and apoptosis

To prevent dysregulated bone resorption, osteoclast number is regulated at the level of cell differentiation and by limiting osteoclast life span via programed cell death, or apoptosis. Osteoclast survival is controlled by the same factors that are required for osteoclast activity, including the Ras-ERK pathway, and the PI3K/Akt signaling pathways. The mTOR protein is a target of PI3K activity and activation of this kinase is required for the anti-apoptotic actions of M-CSF, RANKL and TNFα in osteoclasts. Activation of the small GTPase Rac1, and the upstream effector, Vav3, is also involved in M-CSF induced cell survival.

The shortening of the osteoclast life span controls the extent and depth of osteoclast resorption activity. Osteoclast apoptosis is accelerated in the absence of supporting osteoblasts or bone marrow cells, and in conditions leading to low levels of RANKL, M-CSF or IL-1. Detachment of osteoclasts also induces apoptosis, presumably because integrins become disengaged from RGD peptides in the bone matrix.

The ability of osteoclasts to sense calcium in the bone microenvironment occurs through the calcium sensing receptor, CaSR. Activation of the CaSR in osteoclasts in response to calcium induces the PLC-mediated remodeling of their cytoskeleton, leading to decreased osteoclast bone resorbing activity (Bennett, Alvarez, & Hruska, 2001; Mentaverri et al., 2006). The PI3K/Akt pathway is also implicated in the response of osteoclasts to extracellular calcium (Bennett et al., 2001; Boudot et al., 2010). Thus, at high concentration of calcium, which mimics the microenvironment of demineralized bone, the activated CaSR drives osteoclast apoptosis (Lorget et al., 2000). Interestingly, at low concentrations of calcium in culture, osteoblasts produce more RANKL, thus driving osteoclastogenesis (Takeyama et al., 2000).

Currently, the majority of drug agents available to protect against bone loss are the anti-resorptive therapies (Stepan, Alenfeld, Boivin, Feyen, & Lakatos, 2003). These anti-osteoclastic therapies include estrogen and the selective estrogen receptor modulator (SERM) raloxifene, which both decrease osteoclast life span and promote apoptosis. Estrogens act via a complex signaling mechanism to inhibit the activation of T cells, decreasing their secretion of RANKL and TNFα. Estrogen also inhibits osteoclast differentiation and decreases life span, by regulating the expression of Fas ligand (FasL) in osteoblasts (Krum et al., 2008), which binds the death receptor Fas in osteoclasts (Nakamura, Imai, et al., 2007; Nakamura, Ly, et al., 2007), accelerating osteoclast cell death. Finally, the anti-resorptive amino-bisphosphonates also decrease bone remodeling by decreasing osteoclast activity and by inducing osteoclast apoptosis (Dunford, Rogers, Ebetino, Phipps, & Coxon, 2006; Tella & Gallagher, 2014; Tsubaki et al., 2014).

3.3.1 Bcl-2 proteins and osteoclast apoptosis

Apoptosis occurs through either the extrinsic (death receptor) pathway or the intrinsic (mitochondria) pathway (Hengartner, 2000). In the extrinsic pathway, activation of a TNFR induces apoptosis via the activation of caspase 8, an aspartate-specific cysteine protease. However, in the intrinsic pathway, cytochrome *c* release from the mitochondria is regulated by different Bcl-2 family members which are either anti-apoptotic or pro-apoptotic in action. For example, in mature osteoclasts, cytokine withdrawal leads to reduced expression of anti-apoptotic Bcl-2 and rapid apoptosis (Tanaka et al., 2010). Removal of RANKL signaling also leads to the activation of Bid-induced and caspase-3 induced osteoclast apoptosis (Vaira et al., 2008). M-CSF withdrawal also decreases Bcl-2 through inactivation of the transcription factor, MITF (Tanaka et al., 2010). Osteoclast apoptosis is also regulated by increased expression of the pro-apoptotic protein, Bim, a BH3-domain-only member of the Bcl-2 family proteins (Wakeyama et al., 2007). In the absence of Bim, mice have decreased osteoclast activity, despite increased cell survival (Tanaka et al., 2010). Bim expression, and therefore apoptosis, is kept in check through the c-Cbl-mediated ubiquitin/protease degradation pathway that is regulated by caspase-3, such that in the absence of caspase-3, the degradation of Bim is suppressed. The expression of Bim is also downregulated by IL-3 signaling through the Raf/ERK and/or PI3K/mTOR pathways. In contrast, upregulation of the anti-apoptotic protein Bcl-X_L, which decreases Src, inhibits the cleavage of procaspase-9, thereby preventing the initiation of osteoclast apoptosis.

3.3.2 FasL-induced osteoclast apoptosis

Osteoclast apoptosis is also believed to be regulated in part through the osteoblast-secreted factor FasL (or CD95L) (Kovacic, Grcevic, Katavic, Lukic, & Marusic, 2010). FasL is a membrane-bound protein belonging to the TNF family proteins. FasL expression in osteoblasts is regulated by estrogens, and MMP3-induced FASL cleavage in osteoblasts leads to the secretion of FasL, and subsequently to the osteoblast-mediated apoptosis of osteoclasts (Garcia et al., 2013; Krum et al., 2008). FasL binds the prototypical TNF family death receptor, Fas (FS-7-associated surface antigen; CD95; APO1; TNFR superfamily member 6, TNFRSF6) (Wu, Mckenna, Feng, Nagy, & Mcdonald, 2003). Fas is widely expressed in numerous cells, including macrophages and osteoclast progenitors. After Fas activation the adaptor protein FADD (Fas associated death domain) binds Fas, which leads to the recruitment of procaspase 8 and 10, forming a death-inducing signaling complex (DISC). The DISC complex then activates caspase 8, and consequently caspase 3 in the cytosol, promoting apoptosis. Although Fas is the main receptor of FasL, the decoy receptor 3 (DcR3), another member of the TNFR superfamily, also acts as a decoy receptor for FasL (Sheikh & Fornace, 2000; Yang et al., 2004). DcR3 is a secreted protein and competitively binds and inhibits several TNF family proteins, not just FasL. It was reported that DcR3 acts as a ligand to induce the differentiation of macrophages into osteoclasts. DcR3 also suppresses RANKL-induced osteoclast formation via the NF-kB and NFATc1 pathway (Cheng, Sheu, Sytwu, & Chang, 2013), thus functioning as a regulator of osteoclast apoptosis.

3.4 Osteoblast to osteoclast signaling

In addition to the well-known pro-osteoclastic factors, such as RANKL, M-CSF and IL-1, osteoblasts secrete factors that decrease osteoclast formation and activity. It has long been established that RANKL-mediated osteoclast formation is opposed by OPG, which is expressed by osteoblasts and osteocytes. OPG acts as a soluble decoy receptor for RANKL, effectively reducing the local level of RANKL, and consequently, decreasing osteoclast differentiation and cell number. Moreover, in the presence of OPG, increased rates of apoptosis of osteoclasts and osteoclast precursors have been observed which is due to activation of the classic FasL/Fas apoptosis pathway (Kovacic et al., 2007; Wang et al., 2015).

As discussed previously in this review, osteoblasts produce several Wnt family proteins. Wnt5a binds to two different receptor complexes in osteoclast precursors. Wnt5a binds to the FZD and LRP5/6 signaling complex and activates β-catenin in osteoclasts to promote osteoclastogenesis (Glass 2nd et al., 2005; Yamashita, Takahashi, & Udagawa, 2012). On the other hand, Wnt5a binding to FZD and Ror2 (receptor tyrosine kinase-like orphan receptor 2) upregulates RANK expression in osteoclasts, which increases RANKL-mediated osteoclast formation (Maeda et al., 2012).

Osteoblast to osteoclast signaling also occurs through the semaphorins family proteins. Semaphorin 3A (Sema3A) binds to a specific receptor complex on osteoclast precursors that is composed of neurophilin-1 (Nrp1) and plexin-A1, and activation of this complex inhibits osteoclast differentiation (Hayashi et al., 2012; Li et al., 2017). It has also been shown that another semaphorin, Sema6C/6D, binds to osteoclast precursor to stimulate ITAMs and DNAX-activating protein 12 (DAP12), which enhances RANK signaling and osteoclastogenesis (Kang et al., 2018; Koga et al., 2004; Takegahara et al., 2006). Of note, mice lacking the ITAM adaptor proteins, FcRγ (Fc receptor common gamma subunit) and DAP12 (DNAX-activating protein 12), exhibit defective osteoclast differentiation, leading to severe osteopetrosis. In osteoclast precursors, the FcRγ and DAP12 proteins can also activate calcium signaling through PLCγ which is also important for regulating osteoclast survival (Bennett et al., 2001).

4. Regulation of bone cell differentiation and function through cell-to-cell contact

4.1 Notch signaling

The Notch signaling pathway is activated by direct contact of a cell expressing the Notch ligands Delta-like proteins (Delta1, 3 and 4) and Jagged (1 and 2) and a cell expressing the receptors known as Notch proteins (Notch 1–4) (Artavanis-Tsakonas, Matsuno, & Fortini, 1995). Both ligands and receptors are transmembrane protein and their binding leads to activation of intracellular signaling in the cell expressing the receptor (unidirectional). Osteoblasts and osteocytes (as well as osteoclasts) express both components of the Notch signaling system (Canalis, 2018b). As with other molecules affecting osteoblastic cell differentiation and function, the relevance of Notch signaling for osteoblasts and osteocytes biology was put in evidence by the establishment of the link of human mutations with bone diseases (Canalis, 2018a; Zanotti & Canalis, 2012).

Upon ligand-receptor binding, the canonical Notch signaling pathway is activated, resulting in the proteolytic cleavage of the Notch receptor. This cleavage occurs in two steps, the first one as a result of the activity of the metalloproteinase ADAM17/TACE, which removes the extracellular domain of the receptor. The second step is carried out by the γ-secretase complex that is formed by presenilin-1 and -2, nicastrin, PEN-2, and

Aph-1, resulting in the release of the Notch intracellular domain or NICD. The NICD then translocates to the nucleus where it forms a complex with the transcription factors recombining binding protein suppressor of hairless (RBP-Jκ) and mastermind-like (MAML). This trimeric complex recruits co-activators and promotes the transcription of Notch target genes. As for BMP and Wnt signaling, a non-canonical Notch signaling pathway independent of RBP-Jκ has been described, although its role in bone cell is not known (Canalis, 2018b).

Activation of Notch signaling has profound effects on osteoblastogenesis, which depend on the stage of maturation of the cells of the mesenchymal/osteoblastic lineage. Thus, transgenic mice expressing the NICD (which acts as a constitutive active Notch receptor) in osteoblast precursors exhibit high bone mass and increased osteoblast number and proliferation, but impaired osteoblast maturation (Canalis, Parker, Feng, & Zanotti, 2013; Engin et al., 2008). As a result of this altered osteoblastogenesis, immature cells accumulate and produce disorganized, woven bone with accumulation of non-mineralized bone (osteoid). Further, NICD overexpression earlier in the osteoblast differentiation pathway results in reduced bone mass, associated with reduced osteoblast differentiation (Zanotti et al., 2008). On the other hand, NICD overexpression in osteocytes results in increased cancellous bone mass and cortical thickness associated with decreased osteoclast number and bone formation in cancellous bone and no change in osteoclasts but increased osteoblasts in cortical bone (Canalis, Adams, et al., 2013). The cellular changes are associated with a decrease in the RANKL/OPG ratio, which favors inhibition of osteoclast differentiation, and a decrease in the Wnt antagonist sclerostin and the consequent increase in Wnt/β-catenin signaling, which favors osteoblast differentiation. These pieces of evidence highlight the cross talk between different signaling pathways in bone cells. Further *in vitro* studies showed that the Notch/Wnt signaling cross talk is also involved in the osteoblast-osteocyte transition (Shao, Zhou, & Xiao, 2018). However, in this case it was shown that Notch activation inhibits Wnt/β-catenin signaling, which in turn results in the terminal differentiation of osteocytes. Conversely, activation of Wnt/β-catenin in osteocytes results in activation of Notch signaling (Tu et al., 2015).

Further studies show that double deletion of presenilin-1 and -2 (the enzymes that participate in the release of NICD) or of the Notch receptors-1 and -2 from MSCs in mice results in high bone mass (Hilton et al., 2008).

However, presenilin1/2-deficient mice undergo accelerated age-dependent bone loss by a combination of decreased bone formation and increased resorption. The increased resorption occurs indirectly, through increased RANKL/OPG ratio in cells of the osteoblastic lineage.

On the other hand, removal of presenilin-1 and -2 in mature osteoblasts results in decreased bone mass (Engin et al., 2008). However, the phenotype was not due to decreased osteoblast differentiation/activity, as bone formation parameters did not differ between knockout and control mice. Instead, osteoclast number/surface and eroded surface were increased in the mice lacking the presinilins indicating an indirect effect of Notch signaling in osteoblastic cells on osteoclastogenesis.

Osteocytic Notch1/2 deletion results in increased bone mass in both cancellous and cortical bone compartments, due to mechanisms that are both sex- and age-dependent (Canalis, Adams, et al., 2013). In addition, the same authors showed that conditional overexpression of Notch ICD in osteocytes results in inhibition of bone formation and resorption, leading to higher bone mass in a low remodeling state (Canalis, Adams, et al., 2013; Canalis, Parker, et al., 2013).

In addition to the indirect effect on osteoclasts through the control of the RANKL/OPG system, modulation of Notch signaling in osteoclast precursors directly regulates osteoclastogenesis (Canalis, 2018b). Studies have shown that deletion of the canonical Notch signaling transcription factor RBP-Jk in osteoclast precursors enhances osteoclast differentiation. However, the effect of activation of Notch in these precursors depends on the receptor and while Notch1 inhibits osteoclastogenesis, Notch2 activates it. The effect of Notch2 is associated with the interaction of NICD2 (the fragment of the Notch2 released upon receptor-ligand interaction) with NF-κB, resulting in NFATc1 transcription and osteoclast differentiation. In addition, a study showed that *Notch1* or *Notch3* deletion in bone marrow macrophages enhanced osteoclast differentiation and overexpression of the Notch ligand Jagged1 inhibited osteoclast formation *in vitro* (Bai et al., 2008). On the other hand, another study showed that silencing of Notch2, but not Notch1, inhibits osteoclastogenesis (Fukushima et al., 2008). The reason for the discrepancy between the two studies is not clear, although it is possible that the difference stems from the diverse methods used to remove Notch expression (silencing versus Cre-LoxP system/*in vivo* global gene deletion).

4.2 Ephrin-ephrin receptor system

The Ephrin-Ephrin receptor (Eph receptor) system mediates direct cell-to-cell communication through the interaction of the transmembrane proteins Ephrin with the Eph receptor (Cayuso, Xu, & Wilkinson, 2015). Unlike the unidirectional Notch signaling, signaling cascades are activated in cells expressing both Ephrin and Eph receptor, constituting a bidirectional signaling system. Two families of Ephrins have been described, ephrinA1–6 and ephrinB1–3 proteins, which bind to EphA1–10 and EphB1–6 receptors, respectively (Committee, 1997). Ephrin ligands are bound to the cell membranes through a glycosylphosphatidylinositol tail (ephrinA) or through a transmembrane domain (ephrinB). The receptors contain intracellular tyrosine kinase domains, which became activated upon ligand binding. The expression of Ephrins and Eph receptors has been demonstrated in osteoblasts and osteoclasts, with osteoblasts expressing ephrinA and B ligands and EphA and EphB receptors, and osteoclasts expressing only ephrinsA2, B1, and B2, but not EphB or EphA1, EphA2, or EphA4 receptors (Zhao et al., 2006). Osteocytes also express ephrinB2 and EphB4 (Sims & Vrahnas, 2014; Wang et al., 2014) and ephrinB1 and B2 (Sims & Martin, 2014). Activation of the Ephrin/Eph receptor system have been proposed to regulate the coupling of bone formation and resorption by triggering signaling in both cells expressing Ephrins and Eph receptors (Lindsey, Rundle, & Mohan, 2018). In particular, *in vitro* studies showed that osteoclast ephrinB2 binds to osteoblast EphB4 resulting in stimulation of osteoblast differentiation by activation of genes such as Runx2 (forward signaling) and inhibition of osteoclast differentiation through decreased expression of the osteoclastic transcription factors c-Fos and NFATc1 (reverse signaling) (Zhao et al., 2006). However, while the existence of this osteoblast-osteoclast interaction has been demonstrated *in vitro*, it is not clear whether it would exist *in vivo*, because osteoblasts and osteoclasts are not normally located in close proximity on the bone surface to allow for direct cell-to-cell contact (Sims & Martin, 2014). On the other hand, a more likely scenario has been proposed, in which ephrinB2/EphB4 is required for osteoblast differentiation and regulates osteoblast-osteoclast precursor interactions. Thus, blockade of ephrinB2/EphB4 interaction results in decreased osteoblast differentiation in the presence of PTH, and in increased osteoclast differentiation *in vitro* and *in vivo*. In addition to this interaction between two different cell types (heterotypic), the role of Eprhin/Eph interaction in osteoblastic cells (homotypic) was shown in mice lacking ephrinB1, which exhibit decreased osteoblast differentiation and activity (Tonna & Sims, 2014).

This phenotype results from the lack of interaction between ephrinB1 and EphB2 in osteoblastic cells, with reduced expression of the transcription factor osterix and of alkaline phosphatase. Conversely, stimulation of ephrinB1 reverse signaling by EphB2 results in increased osterix expression and alkaline phosphatase activity. Similarly, homotypic eprhinB2/EphB4 interaction is required for late stage osteoblast differentiation and blockade of this interaction results in the accumulation of osteoblasts and osteoid, but decreased mineralization. In addition, blockade of eprhinB2/EphB4 interaction leads to increased RANKL expression and osteoclast number in PTH-treated mice.

4.3 Connexins and intracellular signaling

Connexins are four transmembrane-spanning proteins that associate in hexamers known as connexons or hemichannels (Beyer, Paul, & Goodenough, 1990). Hemichannels present in neighboring cells align to form gap junction channels that allow the communication between the two cells. Hemichannels can also be found in unopposed cell membranes, where they mediate the interaction of the cell with the extracellular medium (Goodenough & Paul, 2003). In bone cells, hemichannels are involved in the transduction of mechanical, pharmacologic and hormonal stimuli into intracellular signaling (Plotkin, Davis, Cisterna, & Saez, 2017; Plotkin & Stains, 2015). In addition to their classic function as channel forming proteins, connexins are also mediators of intracellular signaling, mainly through interaction of the cytoplasmic connexin C-terminus region with structural proteins and enzymes, such as kinases and phosphatases (Leithe, Mesnil, & Aasen, 2018). In osteoblastic cells, the interactions between connexins and intracellular signaling molecules are required for prostaglandin release and anti-apoptosis induced by mechanical stimulation, differentiation induced by fibroblast growth factor 2, and cell survival induced by bisphosphonates and parathyroid hormone (Plotkin & Stains, 2015).

The expression and function of connexins in osteoblasts and osteocytes have been extensively studied (Plotkin & Stains, 2015). In particular, Cx43, Cx37, Cx45, and Cx46 are expressed in osteoblastic cells, being Cx43 the most highly expressed and best studied connexin in bone. *In vitro* and *in vivo* studies showed that Cx43 deletion reduces osteoblastogenesis, whereas its overexpression leads to increased osteoblast differentiation *in vitro*. Thus, overexpression of Cx43 in osteoblastic cells increases ERK signaling through a mechanism that requires direct cell-to-cell contact and intercellular communication (Niger et al., 2012; Stains & Civitelli, 2005), whereas expression of Cx45, which acts has partial dominant negative effects on Cx43 actions, reduces ERK and Akt activation in Cx43-expressing cells (Stains & Civitelli, 2005). Activation of ERK/PI3K pathways by Cx43 results in increased Sp1 phosphorylation and its recruitment to the promoter of osteoblastic genes such as osteocalcin and osterix, increasing their expression and osteoblast differentiation (Niger et al., 2011; Stains & Civitelli, 2005; Stains, Lecanda, Screen, Towler, & Civitelli, 2003).

Further studies demonstrated that increased Cx43 levels in osteoblastic cells *in vitro* enhances the activation of protein kinase C delta (PKCδ) induced by FGF2 (Lima et al., 2009; Niger et al., 2012), through direct binding of PKCδ with Cx43 C-terminal domain (Hebert & Stains, 2013; Niger et al., 2010). In addition, Cx43 increases the transcriptional activity of the master regulator of osteoblastogenesis Runx2 through activation of inositol polyphosphate/PKCδ signaling in osteoblastic cells (Lima et al., 2009; Niger et al., 2012, 2013).

In addition to its role on osteoblast differentiation and FGF2 actions, Cx43 also mediate the anti-apoptotic effects of bisphosphonates and PTH (Plotkin & Bellido, 2013). In the case of bisphosphonates, addition of the drug to cultured osteocytic cells results in opening of Cx43 hemichannel, and the activation of a signaling cascade that involves the formation of a complex comprising β-arrestin, Scr, MEK and ERKs (Plotkin, Aguirre, Kousteni, Manolagas, & Bellido, 2005; Plotkin, Manolagas, & Bellido, 2002). This leads to the phosphorylation of the cytoplasmic ERK target p90RSK, which in turn phosphorylates C/EBPβ and BAD, resulting in cell survival. This anti-apoptotic signaling pathway does not required nuclear actions of ERKs or ERK-mediated gene transcription and, instead, it is blocked by forced ERK nuclear accumulation.

Cx43 expression is also required for the anti-apoptotic effect of PTH on osteoblastic cells *in vitro* (Plotkin & Bellido, 2013). As with bisphosphonates, hemichannel but not gap junction channel activity is required for the effect of the hormone. Cx43 associates with β-arrestin through the phosphorylated serine 368 in the C-terminus domain of the connexin (Bivi, Lezcano, Romanello, Bellido, & Plotkin, 2011). Cx43-β-arrestin association releases the PTH1R from its binding to β-arrestin, allowing for cAMP accumulation, protein kinase A (PKA) activation and downstream gene transcription.

Cx43 also mediates the activation of intracellular signaling induced by mechanical stimulation in osteocytic cells (Plotkin & Bellido, 2013). In this case, *in vitro* studies showed that Cx43 C-terminus domain interacts with α5β1 integrin, leading to opening of Cx43 hemichannels in osteocytes subjected to fluid flow shear stress (Batra et al., 2012, 2014). Opened Cx43 hemichannels mediate the release of autocrine/paracrine factors such as ATP and prostaglandin E2 (PGE2) (Cherian et al., 2005; Genetos, Kephart, Zhang, Yellowley, & Donahue, 2007; Siller-Jackson et al., 2008). PGE2, in turn has been shown to mediate the survival effect of mechanical stimulation in osteocytes (Kitase, Barragan, Jiang, Johnson, & Bonewald, 2010). The protective effect of the prostaglandin is mediated by its interaction with EP2 and EP4 receptors, resulting in the activation of the cAMP/PKA and the Wnt/β-catenin pathways.

Reduced Cx43 levels in osteoblastic cells indirectly affect osteoclast differentiation (Bivi et al., 2012; Watkins et al., 2011; Zhang et al., 2011). Thus, deletion of Cx43 from osteoblast precursors, mature osteoblasts, or osteocytes in mice results in increased RANKL/OPG ratio and high osteoclast numbers *in vivo*. Cx43 is also expressed in osteoclastic cells from normal mice (Ilvesaro, Väänänen, & Tuukkanen, 2000) in osteoclasts from patients with Paget's disease of the bone and giant cell tumors of the bone, although the role of the connexin in these conditions has not been established (Schilling et al., 2008). On the other hand, *in vitro* studies showed that Cx43 blockade inhibits osteoclast fusion and function (Ilvesaro, Tavi, & Tuukkanen, 2001; Ilvesaro et al., 2000; Schilling et al., 2008), although whether Cx43 is also required for osteoclast differentiation and function *in vivo* remains to be determined. On the other hand, Cx37, another connexin member of the same family of gap junction proteins as Cx43, which is expressed in osteoclasts, is involved in osteoclast differentiation in mice *in vivo* (Pacheco-Costa et al., 2014). Thus, mice with global deletion of Cx37 exhibit increased bone mass due to reduced osteoclast differentiation. The impaired osteoclastogenesis in these mice is associated with accumulation of osteoclast precursors and activation of the Notch signaling pathway.

5. Concluding remarks

In this chapter, we described several of the signaling pathways triggered in osteoblasts, osteocytes, and osteoclasts by selected regulatory molecules, as well as by direct cell-to-cell contact. Through these interrelated signaling pathways, the generation, viability and function of osteoblasts, osteocytes and osteoclasts are controlled. Appropriate differentiation and activity of each of the bone cells is required for the maintenance of bone mass, architecture, and strength. Therefore, understanding the interrelated signaling stimuli that regulate these cells, and the intracellular signaling pathways activated, will allow for better design of strategies to improve bone mass in conditions associated with low bone mass and increased bone fragility.

Acknowledgments

This research was supported by the National Institutes of Health R01-AR053643 to L.I.P., and ICMH pilot grant funding to L.I.P. and A.B., and NIAMS R01-AR060332 to A.B. A.B. was also supported in part by the Indiana Clinical and Translational Sciences Institutes of Health, in part by grant number ULITR001108 from the National Center for Advancing Translational Sciences, Clinical and Translational Science Award.

References

Alam, I., Alkhouli, M., Gerard-O'riley, R. L., Wright, W. B., Acton, D., Gray, A. K., et al. (2016). Osteoblast-specific overexpression of human WNT16 increases both cortical and trabecular bone mass and structure in mice. *Endocrinology, 157*, 722–736.

Ali, N. N., Boyde, A., & Jones, S. J. (1984). Motility and resorption: Osteoclastic activity in vitro. *Anatomy and Embryology (Berlin), 170*, 51–56.

Allen, S. H., Nuttleman, P. R., Ketcham, C. M., & Roberts, R. M. (1989). Purification and characterization of human bone tartrate-resistant acid phosphatase. *Journal of Bone and Mineral Research, 4*, 47–55.

Artavanis-Tsakonas, S., Matsuno, K., & Fortini, M. E. (1995). Notch signaling. *Science, 268*, 225–232.

Bai, S., Kopan, R., Zou, W., Hilton, M. J., Ong, C. T., Long, F., et al. (2008). NOTCH1 regulates osteoclastogenesis directly in osteoclast precursors and indirectly via osteoblast lineage cells. *The Journal of Biological Chemistry, 283*, 6509–6518.

Baron, R., & Gori, F. (2018). Targeting WNT signaling in the treatment of osteoporosis. *Current Opinion in Pharmacology, 40*, 134–141.

Baron, R., & Kneissel, M. (2013). WNT signaling in bone homeostasis and disease: From human mutations to treatments. *Nature Medicine, 19*, 179–192.

Baron, R., Neff, L., Brown, W., Courtoy, P. J., Louvard, D., & Farquhar, M. G. (1988). Polarized secretion of lysosomal enzymes: Co-distribution of cation-independent mannose-6-phosphate receptors and lysosomal enzymes along the osteoclast exocytic pathway. *The Journal of Cell Biology, 106*, 1863–1872.

Baron, R., Neff, L., Louvard, D., & Courtoy, P. J. (1985). Cell-mediated extracellular acidification and bone resorption: Evidence for a low pH in resorbing lacunae and localization of a 100-kD lysosomal membrane protein at the osteoclast ruffled border. *The Journal of Cell Biology, 101*, 2210–2222.

Batra, N., Burra, S., Siller-Jackson, A. J., Gu, S., Xia, X., Weber, G. F., et al. (2012). Mechanical stress-activated integrin alpha5beta1 induces opening of connexin 43 hemichannels. *Proceedings of the National Academy of Sciences of the United States of America, 109*, 3359–3364.

Batra, N., Riquelme, M. A., Burra, S., Rekha, K., Gu, S., & Jiang, J. X. (2014). Direct regulation of osteocytic connexin 43 hemichannels through AKT kinase activated by mechanical stimulation. *The Journal of Biological Chemistry, 289*, 10582–10591.

Beenken, A., & Mohammadi, M. (2009). The FGF family: Biology, pathophysiology and therapy. *Nature Reviews. Drug Discovery*, *8*, 235–253.
Bellido, T., Plotkin, L. I., & Bruzzaniti, A. (2014). Bone cells. In B. David & A. Matthew (Eds.), *Basic and applied bone biology* (1st ed., pp. 27–45). Elsevier.
Bennett, B. D., Alvarez, U., & Hruska, K. A. (2001). Receptor-operated osteoclast calcium sensing. *Endocrinology*, *142*, 1968–1974.
Beyer, E. C., Paul, D. L., & Goodenough, D. A. (1990). Connexin family of gap junction proteins. *The Journal of Membrane Biology*, *116*, 187–194.
Bikle, D. D. (2008). Integrins, insulin like growth factors, and the skeletal response to load. *Osteoporosis International*, *19*, 1237–1246.
Bikle, D. D., Harris, J., Halloran, B. P., Roberts, C. T., Leroith, D., & Morey-Holton, E. (1994). Expression of the genes for insulin-like growth factors and their receptors in bone during skeletal growth. *The American Journal of Physiology*, *267*, E278–E286.
Bivi, N., Condon, K. W., Allen, M. R., Farlow, N., Passeri, G., Brun, L., et al. (2012). Cell autonomous requirement of connexin 43 for osteocyte survival: Consequences for endocortical resorption and periosteal bone formation. *Journal of Bone and Mineral Research*, *27*, 374–389.
Bivi, N., Lezcano, V., Romanello, M., Bellido, T., & Plotkin, L. I. (2011). Connexin43 interacts with barrestin: A pre-requisite for osteoblast survival induced by parathyroid hormone. *Journal of Cellular Biochemistry*, *112*, 2920–2930.
Blair, H. C., Teitelbaum, S. L., Ghiselli, R., & Gluck, S. (1989). Osteoclastic bone resorption by a polarized vacuolar proton pump. *Science*, *245*, 855–857.
Boisclair, Y. R., Rhoads, R. P., Ueki, I., Wang, J., & Ooi, G. T. (2001). The acid-labile subunit (ALS) of the 150 kDa IGF-binding protein complex: An important but forgotten component of the circulating IGF system. *The Journal of Endocrinology*, *170*, 63–70.
Bonewald, L. F. (2011). The amazing osteocyte. *Journal of Bone and Mineral Research*, *26*, 229–238.
Boudot, C., Saidak, Z., Boulanouar, A. K., Petit, L., Gouilleux, F., Massy, Z., et al. (2010). Implication of the calcium sensing receptor and the phosphoinositide 3-kinase/Akt pathway in the extracellular calcium-mediated migration of RAW 264.7 osteoclast precursor cells. *Bone*, *46*, 1416–1423.
Boyce, B. F., Xiu, Y., Li, J., Xing, L., & Yao, Z. (2015). NF-kappaB-mediated regulation of osteoclastogenesis. *Endocrinology and Metabolism (Seoul, Korea)*, *30*, 35–44.
Boyde, A., Ali, N. N., & Jones, S. J. (1984). Resorption of dentine by isolated osteoclasts in vitro. *British Dental Journal*, *156*, 216–220.
Bradley, E. W., Westendorf, J. J., van Wijnen, A. J., & Dudakovic, A. (2018). Osteoblasts. Function, development, and regulation. In J. Bilezikian, R. Bouillon, T. Clemens, & J. Compston (Eds.), *Primer on the metabolic bone diseases and disorders of mineral metabolism* (9th ed., pp. 31–37). Wiley.
Bruzzaniti, A., & Baron, R. (2006). Molecular regulation of osteoclast activity. *Reviews in Endocrine & Metabolic Disorders*, *7*, 123–139.
Bunn, R. C., & Fowlkes, J. L. (2003). Insulin-like growth factor binding protein proteolysis. *Trends in Endocrinology and Metabolism*, *14*, 176–181.
Canalis, E. (2018a). Clinical and experimental aspects of notch receptor signaling: Hajdu-Cheney syndrome and related disorders. *Metabolism*, *80*, 48–56.
Canalis, E. (2018b). Notch in skeletal physiology and disease. *Osteoporosis International*, *29*, 2611–2621.
Canalis, E., Adams, D. J., Boskey, A., Parker, K., Kranz, L., & Zanotti, S. (2013). Notch signaling in osteocytes differentially regulates cancellous and cortical bone remodeling. *The Journal of Biological Chemistry*, *288*, 25614–25625.
Canalis, E., Parker, K., Feng, J. Q., & Zanotti, S. (2013). Osteoblast lineage-specific effects of notch activation in the skeleton. *Endocrinology*, *154*, 623–634.

Cayuso, J., Xu, Q., & Wilkinson, D. G. (2015). Mechanisms of boundary formation by Eph receptor and ephrin signaling. *Developmental Biology*, *401*, 122–131.

Chalhoub, N., Benachenhou, N., Rajapurohitam, V., Pata, M., Ferron, M., Frattini, A., et al. (2003). Grey-lethal mutation induces severe malignant autosomal recessive osteopetrosis in mouse and human. *Nature Medicine*, *9*, 399–406.

Charoenphandhu, N., Suntornsaratoon, P., Krishnamra, N., Sa-Nguanmoo, P., Tanajak, P., Wang, X., et al. (2017). Fibroblast growth factor-21 restores insulin sensitivity but induces aberrant bone microstructure in obese insulin-resistant rats. *Journal of Bone and Mineral Metabolism*, *35*, 142–149.

Cheng, C. P., Sheu, M. J., Sytwu, H. K., & Chang, D. M. (2013). Decoy receptor 3 suppresses RANKL-induced osteoclastogenesis via down-regulating NFATc1 and enhancing cell apoptosis. *Rheumatology (Oxford)*, *52*, 609–622.

Cherian, P. P., Siller-Jackson, A. J., Gu, S., Wang, X., Bonewald, L. F., Sprague, E., et al. (2005). Mechanical strain opens connexin 43 hemichannels in osteocytes: A novel mechanism for the release of prostaglandin. *Molecular Biology of the Cell*, *16*, 3100–3106.

Chiu, Y. H., Schwarz, E., Li, D., Xu, Y., Sheu, T. R., Li, J., et al. (2017). Dendritic cell-specific transmembrane protein (DC-STAMP) regulates osteoclast differentiation via the Ca(2+)/NFATc1 axis. *Journal of Cellular Physiology*, *232*, 2538–2549.

Cleiren, E., Benichou, O., van Hul, E., Gram, J., Bollerslev, J., Singer, F. R., et al. (2001). Albers-Schonberg disease (autosomal dominant osteopetrosis, type II) results from mutations in the ClCN7 chloride channel gene. *Human Molecular Genetics*, *10*, 2861–2867.

Clinkenbeard, E. L., & White, K. E. (2016). Systemic control of bone homeostasis by FGF23 signaling. *Current Molecular Biology Reports*, *2*, 62–71.

Committee, E. N. (1997). Unified nomenclature for Eph family receptors and their ligands, the ephrins. Eph nomenclature committee. *Cell*, *90*, 403–404.

de Vernejoul, M. C., Horowitz, M., Demignon, J., Neff, L., & Baron, R. (1988). Bone resorption by isolated chick osteoclasts in culture is stimulated by murine spleen cell supernatant fluids (osteoclast-activating factor) and inhibited by calcitonin and prostaglandin E2. *Journal of Bone and Mineral Research*, *3*, 69–80.

Destaing, O., Saltel, F., Geminard, J. C., Jurdic, P., & Bard, F. (2003). Podosomes display actin turnover and dynamic self-organization in osteoclasts expressing actin-green fluorescent protein. *Molecular Biology of the Cell*, *14*, 407–416.

Dunford, J. E., Rogers, M. J., Ebetino, F. H., Phipps, R. J., & Coxon, F. P. (2006). Inhibition of protein prenylation by bisphosphonates causes sustained activation of Rac, Cdc42, and Rho GTPases. *Journal of Bone and Mineral Research*, *21*, 684–694.

Engin, F., Yao, Z., Yang, T., Zhou, G., Bertin, T., Jiang, M. M., et al. (2008). Dimorphic effects of notch signaling in bone homeostasis. *Nature Medicine*, *14*, 299–305.

Esbrit, P., Herrera, S., Portal-Nunez, S., Nogues, X., & Diez-Perez, A. (2016). Parathyroid hormone-related protein analogs as osteoporosis therapies. *Calcified Tissue International*, *98*, 359–369.

Felix, R., Cecchini, M. G., & Fleisch, H. (1990). Macrophage colony stimulating factor restores in vivo bone resorption in the op/op osteopetrotic mouse. *Endocrinology*, *127*, 2592–2594.

Feng, S., Deng, L., Chen, W., Shao, J., Xu, G., & Li, Y. P. (2009). Atp6v1c1 is an essential component of the osteoclast proton pump and in F-actin ring formation in osteoclasts. *The Biochemical Journal*, *417*, 195–203.

Ferron, M., Settembre, C., Shimazu, J., Lacombe, J., Kato, S., Rawlings, D. J., et al. (2013). A RANKL-PKCbeta-TFEB signaling cascade is necessary for lysosomal biogenesis in osteoclasts. *Genes & Development*, *27*, 955–969.

Fisher, J. E., Caulfield, M. P., Sato, M., Quartuccio, H. A., Gould, R. J., Garsky, V. M., et al. (1993). Inhibition of osteoclastic bone resorption in vivo by echistatin, an "arginyl-glycyl-aspartyl" (RGD)-containing protein. *Endocrinology*, *132*, 1411–1413.

Fukushima, H., Nakao, A., Okamoto, F., Shin, M., Kajiya, H., Sakano, S., et al. (2008). The association of Notch2 and NF-kappaB accelerates RANKL-induced osteoclastogenesis. *Molecular and Cellular Biology, 28*, 6402–6412.

Garcia, A. J., Tom, C., Guemes, M., Polanco, G., Mayorga, M. E., Wend, K., et al. (2013). ERalpha signaling regulates MMP3 expression to induce FasL cleavage and osteoclast apoptosis. *Journal of Bone and Mineral Research, 28*, 283–290.

Garcia-Martin, A., Ardura, J. A., Maycas, M., Lozano, D., Lopez-Herradon, A., Portal-Nunez, S., et al. (2014). Functional roles of the nuclear localization signal of parathyroid hormone-related protein (PTHrP) in osteoblastic cells. *Molecular Endocrinology, 28*, 925–934.

Gelb, B. D., Shi, G. P., Chapman, H. A., & Desnick, R. J. (1996). Pycnodysostosis, a lysosomal disease caused by cathepsin K deficiency. *Science, 273*, 1236–1238.

Genetos, D. C., Kephart, C. J., Zhang, Y., Yellowley, C. E., & Donahue, H. J. (2007). Oscillating fluid flow activation of gap junction hemichannels induces ATP release from MLO-Y4 osteocytes. *Journal of Cellular Physiology, 212*, 207–214.

Glass, D. A., 2nd, Bialek, P., Ahn, J. D., Starbuck, M., Patel, M. S., Clevers, H., et al. (2005). Canonical Wnt signaling in differentiated osteoblasts controls osteoclast differentiation. *Developmental Cell, 8*, 751–764.

Gomez-Puerto, M. C., Iyengar, P. V., Garcia de Vinuesa, A., Ten Dijke, P., & Sanchez-Duffhues, G. (2018). Bone morphogenetic protein receptor signal transduction in human diseases. *The Journal of Pathology, 247*, 9–20.

Goodenough, D. A., & Paul, D. L. (2003). Beyond the gap: Functions of unpaired connexon channels. *Nature Reviews. Molecular Cell Biology, 4*, 285–294.

Govoni, K. E. (2012). Insulin-like growth factor-I molecular pathways in osteoblasts: Potential targets for pharmacological manipulation. *Current Molecular Pharmacology, 5*, 143–152.

Gowen, M., Lazner, F., Dodds, R., Kapadia, R., Feild, J., Tavaria, M., et al. (1999). Cathepsin K knockout mice develop osteopetrosis due to a deficit in matrix degradation but not demineralization. *Journal of Bone and Mineral Research, 14*, 1654–1663.

Grafe, I., Alexander, S., Peterson, J. R., Snider, T. N., Levi, B., Lee, B., et al. (2018). TGF-beta family signaling in mesenchymal differentiation. *Cold Spring Harbor Perspectives in Biology, 10*, pii: a022202.

Grey, A., Chen, Q., Xu, X., Callon, K., & Cornish, J. (2003). Parallel phosphatidylinositol-3 kinase and p42/44 mitogen-activated protein kinase signaling pathways subserve the mitogenic and antiapoptotic actions of insulin-like growth factor I in osteoblastic cells. *Endocrinology, 144*, 4886–4893.

Hachemi, Y., Rapp, A. E., Picke, A. K., Weidinger, G., Ignatius, A., & Tuckermann, J. (2018). Molecular mechanisms of glucocorticoids on skeleton and bone regeneration after fracture. *Journal of Molecular Endocrinology, 61*, R75–R90.

Hamidouche, Z., Fromigue, O., Nuber, U., Vaudin, P., Pages, J. C., Ebert, R., et al. (2010). Autocrine fibroblast growth factor 18 mediates dexamethasone-induced osteogenic differentiation of murine mesenchymal stem cells. *Journal of Cellular Physiology, 224*, 509–515.

Hanada, R., Hanada, T., & Penninger, J. M. (2010). Physiology and pathophysiology of the RANKL/RANK system. *Biological Chemistry, 391*, 1365–1370.

Hashida, Y., Nakahama, K., Shimizu, K., Akiyama, M., Harada, K., & Morita, I. (2014). Communication-dependent mineralization of osteoblasts via gap junctions. *Bone, 61*, 19–26.

Hauer, J., Puschner, S., Ramakrishnan, P., Simon, U., Bongers, M., Federle, C., et al. (2005). TNF receptor (TNFR)-associated factor (TRAF) 3 serves as an inhibitor of TRAF2/5-mediated activation of the noncanonical NF-kappaB pathway by TRAF-binding TNFRs. *Proceedings of the National Academy of Sciences of the United States of America, 102*, 2874–2879.

Hayashi, M., Nakashima, T., Taniguchi, M., Kodama, T., Kumanogoh, A., & Takayanagi, H. (2012). Osteoprotection by semaphorin 3A. *Nature*, *485*, 69–74.

Hebert, C., & Stains, J. P. (2013). An intact connexin43 is required to enhance signaling and gene expression in osteoblast-like cells. *Journal of Cellular Biochemistry*, *114*, 2542–2550.

Helfrich, M. H., Nesbitt, S. A., Dorey, E. L., & Horton, M. A. (1992). Rat osteoclasts adhere to a wide range of RGD (Arg-Gly-Asp) peptide-containing proteins, including the bone sialoproteins and fibronectin, via a beta 3 integrin. *Journal of Bone and Mineral Research*, *7*, 335–343.

Hengartner, M. O. (2000). The biochemistry of apoptosis. *Nature*, *407*, 770–776.

Hilton, M. J., Tu, X., Wu, X., Bai, S., Zhao, H., Kobayashi, T., et al. (2008). Notch signaling maintains bone marrow mesenchymal progenitors by suppressing osteoblast differentiation. *Nature Medicine*, *14*, 306–314.

Horne, W. C., Sanjay, A., Bruzzaniti, A., & Baron, R. (2005). The role(s) of src kinase and Cbl proteins in the regulation of osteoclast differentiation and function. *Immunological Reviews*, *208*, 106–125.

Horton, M. A., Taylor, M. L., Arnett, T. R., & Helfrich, M. H. (1991). Arg-Gly-Asp (RGD) peptides and the anti-vitronectin receptor antibody 23C6 inhibit dentine resorption and cell spreading by osteoclasts. *Experimental Cell Research*, *195*, 368–375.

Huang, Y., Liu, L., & Liu, A. (2018). Dickkopf-1: Current knowledge and related diseases. *Life Sciences*, *209*, 249–254.

Ilvesaro, J., Tavi, P., & Tuukkanen, J. (2001). Connexin-mimetic peptide Gap 27 decreases osteoclastic activity. *BMC Musculoskeletal Disorders*, *2*, 10.

Ilvesaro, J., Väänänen, K., & Tuukkanen, J. (2000). Bone-resorbing osteoclasts contain gap-junctional connexin-43. *Journal of Bone and Mineral Research*, *15*, 919–926.

Irie, A., Habuchi, H., Kimata, K., & Sanai, Y. (2003). Heparan sulfate is required for bone morphogenetic protein-7 signaling. *Biochemical and Biophysical Research Communications*, *308*, 858–865.

Itzstein, C., Coxon, F. P., & Rogers, M. J. (2011). The regulation of osteoclast function and bone resorption by small GTPases. *Small GTPases*, *2*, 117–130.

Jacob, A. L., Smith, C., Partanen, J., & Ornitz, D. M. (2006). Fibroblast growth factor receptor 1 signaling in the osteo-chondrogenic cell lineage regulates sequential steps of osteoblast maturation. *Developmental Biology*, *296*, 315–328.

Jin, Y., & Yang, Y. (2016). LGR4: A new receptor for a stronger bone. *Science China. Life Sciences*, *59*, 735–736.

Jurdic, P., Saltel, F., Chabadel, A., & Destaing, O. (2006). Podosome and sealing zone: Specificity of the osteoclast model. *European Journal of Cell Biology*, *85*, 195–202.

Kanazawa, K., & Kudo, A. (2005). TRAF2 is essential for TNF-alpha-induced osteoclastogenesis. *Journal of Bone and Mineral Research*, *20*, 840–847.

Kang, S., Nakanishi, Y., Kioi, Y., Okuzaki, D., Kimura, T., Takamatsu, H., et al. (2018). Semaphorin 6D reverse signaling controls macrophage lipid metabolism and anti-inflammatory polarization. *Nature Immunology*, *19*, 561–570.

Kantaputra, P. N., Kapoor, S., Verma, P., Intachai, W., & Ketudat Cairns, J. R. (2018). Split hand-foot malformation and a novel WNT10B mutation. *European Journal of Medical Genetics*, *61*, 372–375.

Kasper, D., Planells-Cases, R., Fuhrmann, J. C., Scheel, O., Zeitz, O., Ruether, K., et al. (2005). Loss of the chloride channel ClC-7 leads to lysosomal storage disease and neurodegeneration. *The EMBO Journal*, *24*, 1079–1091.

Kasukawa, Y., Miyakoshi, N., & Mohan, S. (2004). The anabolic effects of GH/IGF system on bone. *Current Pharmaceutical Design*, *10*, 2577–2592.

Katagiri, T., & Watabe, T. (2016). Bone morphogenetic proteins. *Cold Spring Harbor Perspectives in Biology*, *8*, pii: a021899.

Kim, M. H., Park, M., Baek, S. H., Kim, H. J., & Kim, S. H. (2011). Molecules and signaling pathways involved in the expression of OC-STAMP during osteoclastogenesis. *Amino Acids*, *40*, 1447–1459.

Kim, K., Punj, V., Kim, J. M., Lee, S., Ulmer, T. S., Lu, W., et al. (2016). MMP-9 facilitates selective proteolysis of the histone H3 tail at genes necessary for proficient osteoclastogenesis. *Genes & Development*, *30*, 208–219.

Kitase, Y., Barragan, L., Jiang, J. X., Johnson, M. L., & Bonewald, L. F. (2010). Mechanical induction of PGE(2) in osteocytes blocks glucocorticoid induced apoptosis through both the beta-catenin and PKA pathways. *Journal of Bone and Mineral Research*, *25*, 2657–2668.

Kobayashi, K., Takahashi, N., Jimi, E., Udagawa, N., Takami, M., Kotake, S., et al. (2000). Tumor necrosis factor alpha stimulates osteoclast differentiation by a mechanism independent of the ODF/RANKL-RANK interaction. *The Journal of Experimental Medicine*, *191*, 275–286.

Koga, T., Inui, M., Inoue, K., Kim, S., Suematsu, A., Kobayashi, E., et al. (2004). Costimulatory signals mediated by the ITAM motif cooperate with RANKL for bone homeostasis. *Nature*, *428*, 758–763.

Koller, D. L., Zheng, H. F., Karasik, D., Yerges-Armstrong, L., Liu, C. T., Mcguigan, F., et al. (2013). Meta-analysis of genome-wide studies identifies WNT16 and ESR1 SNPs associated with bone mineral density in premenopausal women. *Journal of Bone and Mineral Research*, *28*, 547–558.

Komaba, H., & Lanske, B. (2018). Role of Klotho in bone and implication for CKD. *Current Opinion in Nephrology and Hypertension*, *27*, 298–304.

Komori, T. (2006). Regulation of osteoblast differentiation by transcription factors. *Journal of Cellular Biochemistry*, *99*, 1233–1239.

Kong, Y. Y., Yoshida, H., Sarosi, I., Tan, H. L., Timms, E., Capparelli, C., et al. (1999). OPGL is a key regulator of osteoclastogenesis, lymphocyte development and lymph-node organogenesis. *Nature*, *397*, 315–323.

Kornak, U., Kasper, D., Bosl, M. R., Kaiser, E., Schweizer, M., Schulz, A., et al. (2001). Loss of the ClC-7 chloride channel leads to osteopetrosis in mice and man. *Cell*, *104*, 205–215.

Kovacic, N., Grcevic, D., Katavic, V., Lukic, I. K., & Marusic, A. (2010). Targeting Fas in osteoresorptive disorders. *Expert Opinion on Therapeutic Targets*, *14*, 1121–1134.

Kovacic, N., Lukic, I. K., Grcevic, D., Katavic, V., Croucher, P., & Marusic, A. (2007). The Fas/Fas ligand system inhibits differentiation of murine osteoblasts but has a limited role in osteoblast and osteoclast apoptosis. *Journal of Immunology*, *178*, 3379–3389.

Krum, S. A., Miranda-Carboni, G. A., Hauschka, P. V., Carroll, J. S., Lane, T. F., Freedman, L. P., et al. (2008). Estrogen protects bone by inducing Fas ligand in osteoblasts to regulate osteoclast survival. *The EMBO Journal*, *27*, 535–545.

Kukita, T., Wada, N., Kukita, A., Kakimoto, T., Sandra, F., Toh, K., et al. (2004). RANKL-induced DC-STAMP is essential for osteoclastogenesis. *The Journal of Experimental Medicine*, *200*, 941–946.

Lacombe, J., Karsenty, G., & Ferron, M. (2013). Regulation of lysosome biogenesis and functions in osteoclasts. *Cell Cycle*, *12*, 2744–2752.

Lange, P. F., Wartosch, L., Jentsch, T. J., & Fuhrmann, J. C. (2006). ClC-7 requires Ostm1 as a beta-subunit to support bone resorption and lysosomal function. *Nature*, *440*, 220–223.

Lazner, F., Gowen, M., & Kola, I. (1999). An animal model for pycnodysostosis: The role of cathepsin K in bone remodelling. *Molecular Medicine Today*, *5*, 413–414.

Lee, S. H., Rho, J., Jeong, D., Sul, J. Y., Kim, T., Kim, N., et al. (2006). v-ATPase V0 subunit d2-deficient mice exhibit impaired osteoclast fusion and increased bone formation. *Nature Medicine*, *12*, 1403–1409.

Lee, A. W., & States, D. J. (2000). Both src-dependent and -independent mechanisms mediate phosphatidylinositol 3-kinase regulation of colony-stimulating factor 1-activated mitogen-activated protein kinases in myeloid progenitors. *Molecular and Cellular Biology*, *20*, 6779–6798.

Leithe, E., Mesnil, M., & Aasen, T. (2018). The connexin 43 C-terminus: A tail of many tales. *Biochimica et Biophysica Acta. Biomembranes*, *1860*, 48–64.

Lerner, U. H., & Ohlsson, C. (2015). The WNT system: Background and its role in bone. *Journal of Internal Medicine*, *277*, 630–649.

Leupin, O., Piters, E., Halleux, C., Hu, S., Kramer, I., Morvan, F., et al. (2011). Bone overgrowth-associated mutations in the LRP4 gene impair sclerostin facilitator function. *The Journal of Biological Chemistry*, *286*, 19489–19500.

Li, Z., Hao, J., Duan, X., Wu, N., Zhou, Z., Yang, F., et al. (2017). The role of semaphorin 3A in bone remodeling. *Frontiers in Cellular Neuroscience*, *11*, 40.

Lima, F., Niger, C., Hebert, C., & Stains, J. P. (2009). Connexin43 potentiates osteoblast responsiveness to fibroblast growth factor 2 via a protein kinase C-Delta/Runx2-dependent mechanism. *Molecular Biology of the Cell*, *20*, 2697–2708.

Linder, S., & Aepfelbacher, M. (2003). Podosomes: Adhesion hot-spots of invasive cells. *Trends in Cell Biology*, *13*, 376–385.

Lindsey, R. C., Rundle, C. H., & Mohan, S. (2018). Role of IGF1 and EFN-EPH signaling in skeletal metabolism. *Journal of Molecular Endocrinology*, *61*, T87–T102.

Lisowska, B., Kosson, D., & Domaracka, K. (2018). Lights and shadows of NSAIDs in bone healing: The role of prostaglandins in bone metabolism. *Drug Design, Development and Therapy*, *12*, 1753–1758.

Lomaga, M. A., Yeh, W. C., Sarosi, I., Duncan, G. S., Furlonger, C., Ho, A., et al. (1999). TRAF6 deficiency results in osteopetrosis and defective interleukin-1, CD40, and LPS signaling. *Genes & Development*, *13*, 1015–1024.

Lorget, F., Kamel, S., Mentaverri, R., Wattel, A., Naassila, M., Maamer, M., et al. (2000). High extracellular calcium concentrations directly stimulate osteoclast apoptosis. *Biochemical and Biophysical Research Communications*, *268*, 899–903.

Lozano, D., Fernandez-de-Castro, L., Portal-Nunez, S., Lopez-Herradon, A., Dapia, S., Gomez-Barrena, E., et al. (2011). The C-terminal fragment of parathyroid hormone-related peptide promotes bone formation in diabetic mice with low-turnover osteopaenia. *British Journal of Pharmacology*, *162*, 1424–1438.

Luo, J., Yang, Z., Ma, Y., Yue, Z., Lin, H., Qu, G., et al. (2016). LGR4 is a receptor for RANKL and negatively regulates osteoclast differentiation and bone resorption. *Nature Medicine*, *22*, 539–546.

Maeda, K., Kobayashi, Y., Udagawa, N., Uehara, S., Ishihara, A., Mizoguchi, T., et al. (2012). Wnt5a-Ror2 signaling between osteoblast-lineage cells and osteoclast precursors enhances osteoclastogenesis. *Nature Medicine*, *18*, 405–412.

Maranda, B., Chabot, G., Decarie, J. C., Pata, M., Azeddine, B., Moreau, A., et al. (2008). Clinical and cellular manifestations of OSTM1-related infantile osteopetrosis. *Journal of Bone and Mineral Research*, *23*, 296–300.

Marks, S. C., Jr., & Grolman, M. L. (1987). Tartrate-resistant acid phosphatase in mononuclear and multinuclear cells during the bone resorption of tooth eruption. *The Journal of Histochemistry and Cytochemistry*, *35*, 1227–1230.

Marks, S. C., Jr., Wojtowicz, A., Szperl, M., Urbanowska, E., Mackay, C. A., Wiktor-Jedrzejczak, W., et al. (1992). Administration of colony stimulating factor-1 corrects some macrophage, dental, and skeletal defects in an osteopetrotic mutation (toothless, tl) in the rat. *Bone*, *13*, 89–93.

McNiven, M. A., Baldassarre, M., & Buccione, R. (2004). The role of dynamin in the assembly and function of podosomes and invadopodia. *Frontiers in Bioscience*, *9*, 1944–1953.

Meadows, N. A., Sharma, S. M., Faulkner, G. J., Ostrowski, M. C., Hume, D. A., & Cassady, A. I. (2007). The expression of Clcn7 and Ostm1 in osteoclasts is coregulated by microphthalmia transcription factor. *The Journal of Biological Chemistry, 282*, 1891–1904.
Mensah, K. A., Ritchlin, C. T., & Schwarz, E. M. (2010). RANKL induces heterogeneous DC-STAMP(lo) and DC-STAMP(hi) osteoclast precursors of which the DC-STAMP (lo) precursors are the master fusogens. *Journal of Cellular Physiology, 223*, 76–83.
Mentaverri, R., Yano, S., Chattopadhyay, N., Petit, L., Kifor, O., Kamel, S., et al. (2006). The calcium sensing receptor is directly involved in both osteoclast differentiation and apoptosis. *The FASEB Journal, 20*, 2562–2564.
Miyamoto, T. (2006). The dendritic cell-specific transmembrane protein DC-STAMP is essential for osteoclast fusion and osteoclast bone-resorbing activity. *Modern Rheumatology, 16*, 341–342.
Mizukami, J., Takaesu, G., Akatsuka, H., Sakurai, H., Ninomiya-Tsuji, J., Matsumoto, K., et al. (2002). Receptor activator of NF-kappaB ligand (RANKL) activates TAK1 mitogen-activated protein kinase kinase kinase through a signaling complex containing RANK, TAB2, and TRAF6. *Molecular and Cellular Biology, 22*, 992–1000.
Moon, J. B., Kim, J. H., Kim, K., Youn, B. U., Ko, A., Lee, S. Y., et al. (2012). Akt induces osteoclast differentiation through regulating the GSK3beta/NFATc1 signaling cascade. *Journal of Immunology, 188*, 163–169.
Moverare-Skrtic, S., Henning, P., Liu, X., Nagano, K., Saito, H., Borjesson, A. E., et al. (2014). Osteoblast-derived WNT16 represses osteoclastogenesis and prevents cortical bone fragility fractures. *Nature Medicine, 20*, 1279–1288.
Mulari, M., Vaaraniemi, J., & Vaananen, H. K. (2003). Intracellular membrane trafficking in bone resorbing osteoclasts. *Microscopy Research and Technique, 61*, 496–503.
Mulari, M. T., Zhao, H., Lakkakorpi, P. T., & Vaananen, H. K. (2003). Osteoclast ruffled border has distinct subdomains for secretion and degraded matrix uptake. *Traffic, 4*, 113–125.
Naito, A., Azuma, S., Tanaka, S., Miyazaki, T., Takaki, S., Takatsu, K., et al. (1999). Severe osteopetrosis, defective interleukin-1 signalling and lymph node organogenesis in TRAF6-deficient mice. *Genes to Cells, 4*, 353–362.
Nakamura, T., Imai, Y., Matsumoto, T., Sato, S., Takeuchi, K., Igarashi, K., et al. (2007). Estrogen prevents bone loss via estrogen receptor alpha and induction of Fas ligand in osteoclasts. *Cell, 130*, 811–823.
Nakamura, A., Ly, C., Cipetic, M., Sims, N. A., Vieusseux, J., Kartsogiannis, V., et al. (2007). Osteoclast inhibitory lectin (OCIL) inhibits osteoblast differentiation and function in vitro. *Bone, 40*, 305–315.
Nakashima, K., Zhou, X., Kunkel, G., Zhang, Z., Deng, J. M., Behringer, R. R., et al. (2002). The novel zinc finger-containing transcription factor osterix is required for osteoblast differentiation and bone formation. *Cell, 108*, 17–29.
Niger, C., Buo, A. M., Hebert, C., Duggan, B. T., Williams, M. S., & Stains, J. P. (2012). ERK acts in parallel to PKC delta to mediate the connexin43-dependent potentiation of runx2 activity by FGF2 in MC3T3 osteoblasts. *American Journal of Physiology. Cell Physiology, 302*, C1035–C1044.
Niger, C., Hebert, C., & Stains, J. P. (2010). Interaction of connexin43 and protein kinase C-delta during FGF2 signaling. *BMC Biochemistry, 11*, 14.
Niger, C., Lima, F., Yoo, D. J., Gupta, R. R., Buo, A. M., Hebert, C., et al. (2011). The transcriptional activity of osterix requires the recruitment of Sp1 to the osteocalcin proximal promoter. *Bone, 49*, 683–692.
Niger, C., Luciotti, M. A., Buo, A. M., Hebert, C., Ma, V., & Stains, J. P. (2013). The regulation of runt-related transcription factor 2 by fibroblast growth factor-2 and connexin43 requires the inositol polyphosphate/protein kinase Cdelta cascade. *Journal of Bone and Mineral Research, 28*, 1468–1477.

Nusse, R., Brown, A., Papkoff, J., Scambler, P., Shackleford, G., Mcmahon, A., et al. (1991). A new nomenclature for int-1 and related genes: The Wnt gene family. *Cell*, *64*, 231.

Nusse, R., & Varmus, H. E. (1992). Wnt genes. *Cell*, *69*, 1073–1087.

Ochoa, G. C., Slepnev, V. I., Neff, L., Ringstad, N., Takei, K., Daniell, L., et al. (2000). A functional link between dynamin and the actin cytoskeleton at podosomes. *The Journal of Cell Biology*, *150*, 377–389.

Odgren, P. R., Kim, N., Mackay, C. A., Mason-Savas, A., Choi, Y., & Marks, S. C., Jr. (2003). The role of RANKL (TRANCE/TNFSF11), a tumor necrosis factor family member, in skeletal development: Effects of gene knockout and transgenic rescue. *Connective Tissue Research*, *44*(Suppl. 1), 264–271.

Ornitz, D. M., & Marie, P. J. (2015). Fibroblast growth factor signaling in skeletal development and disease. *Genes & Development*, *29*, 1463–1486.

Otto, F., Thornell, A. P., Crompton, T., Denzel, A., Gilmour, K. C., Rosewell, I. R., et al. (1997). Cbfa1, a candidate gene for cleidocranial dysplasia syndrome, is essential for osteoblast differentiation and bone development. see comments *Cell*, *89*, 765–771.

Pacheco-Costa, R., Hassan, I., Reginato, R. D., Davis, H. M., Bruzzaniti, A., Allen, M. R., et al. (2014). High bone mass in mice lacking Cx37 due to defective osteoclast differentiation. *The Journal of Biological Chemistry*, *289*, 8508–8520.

Pangrazio, A., Poliani, P. L., Megarbane, A., Lefranc, G., Lanino, E., di Rocco, M., et al. (2006). Mutations in OSTM1 (grey lethal) define a particularly severe form of autosomal recessive osteopetrosis with neural involvement. *Journal of Bone and Mineral Research*, *21*, 1098–1105.

Park, S. J., Kim, S. H., Choi, H. S., Rhee, Y., & Lim, S. K. (2009). Fibroblast growth factor 2-induced cytoplasmic asparaginyl-tRNA synthetase promotes survival of osteoblasts by regulating anti-apoptotic PI3K/Akt signaling. *Bone*, *45*, 994–1003.

Peng, X. D., Xu, P. Z., Chen, M. L., Hahn-Windgassen, A., Skeen, J., Jacobs, J., et al. (2003). Dwarfism, impaired skin development, skeletal muscle atrophy, delayed bone development, and impeded adipogenesis in mice lacking Akt1 and Akt2. *Genes and Developmeent*, *17*, 1352–1365.

Pitcher, L. A., & van Oers, N. S. (2003). T-cell receptor signal transmission: Who gives an ITAM? *Trends in Immunology*, *24*, 554–560.

Plotkin, L. I., Aguirre, J. I., Kousteni, S., Manolagas, S. C., & Bellido, T. (2005). Bisphosphonates and estrogens inhibit osteocyte apoptosis via distinct molecular mechanisms downstream of ERK activation. *The Journal of Biological Chemistry*, *280*, 7317–7325.

Plotkin, L. I., & Bellido, T. (2013). Beyond gap junctions: Connexin43 and bone cell signaling. *Bone*, *52*, 157–166.

Plotkin, L. I., & Bivi, N. (2014). Local regulation of bone cell function. In D. Burr & M. Allen (Eds.), *Basic and applied bone biology* (1st ed., pp. 47–74). Elsevier.

Plotkin, L. I., Davis, H. M., Cisterna, B. A., & Saez, J. C. (2017). Connexins and pannexins in bone and skeletal muscle. *Current Osteoporosis Reports*, *15*, 326–334.

Plotkin, L. I., Manolagas, S. C., & Bellido, T. (2002). Transduction of cell survival signals by connexin-43 hemichannels. *The Journal of Biological Chemistry*, *277*, 8648–8657.

Plotkin, L. I., & Stains, J. P. (2015). Connexins and pannexins in the skeleton: Gap junctions, hemichannels and more. *Cellular and Molecular Life Sciences*, *72*, 2853–2867.

Rhee, Y., Bivi, N., Farrow, E. G., Lezcano, V., Plotkin, L. I., White, K. E., et al. (2011). Parathyroid hormone receptor signaling in osteocytes increases the expression of fibroblast growth factor-23 in vitro and in vivo. *Bone*, *49*, 636–643.

Ruppert, R., Hoffmann, E., & Sebald, W. (1996). Human bone morphogenetic protein 2 contains a heparin-binding site which modifies its biological activity. *European Journal of Biochemistry*, *237*, 295–302.

Sanjay, A., Houghton, A., Neff, L., Didomenico, E., Bardelay, C., Antoine, E., et al. (2001). Cbl associates with Pyk2 and Src to regulate Src kinase activity, alpha(v)beta(3) integrin-mediated signaling, cell adhesion, and osteoclast motility. *The Journal of Cell Biology*, *152*, 181–195.

Sato, T., Del Carmen, O. M., Hou, P., Heegaard, A. M., Kumegawa, M., Foged, N. T., et al. (1997). Identification of the membrane-type matrix metalloproteinase MT1-MMP in osteoclasts. *Journal of Cell Science*, *110*(Pt 5), 589–596.

Schilling, A. F., Filke, S., Lange, T., Gebauer, M., Brink, S., Baranowsky, A., et al. (2008). Gap junctional communication in human osteoclasts in vitro and in vivo. *Journal of Cellular and Molecular Medicine*, *12*, 2497–2504.

Schluter, K. D. (1999). PTH and PTHrP: Similar structures but different functions. *News in Physiological Sciences*, *14*, 243–249.

Schulte, G. (2015). Frizzleds and WNT/beta-catenin signaling—The black box of ligand-receptor selectivity, complex stoichiometry and activation kinetics. *European Journal of Pharmacology*, *763*, 191–195.

Scimeca, J. C., Franchi, A., Trojani, C., Parrinello, H., Grosgeorge, J., Robert, C., et al. (2000). The gene encoding the mouse homologue of the human osteoclast-specific 116-kDa V-ATPase subunit bears a deletion in osteosclerotic (oc/oc) mutants. *Bone*, *26*, 207–213.

Shankar, G., Gadek, T. R., Burdick, D. J., Davison, I., Mason, W. T., & Horton, M. A. (1995). Structural determinants of calcium signaling by RGD peptides in rat osteoclasts: Integrin-dependent and -independent actions. *Experimental Cell Research*, *219*, 364–371.

Shao, J., Zhou, Y., & Xiao, Y. (2018). The regulatory roles of notch in osteocyte differentiation via the crosstalk with canonical Wnt pathways during the transition of osteoblasts to osteocytes. *Bone*, *108*, 165–178.

Sheikh, M. S., & Fornace, A. J., Jr. (2000). Death and decoy receptors and p53-mediated apoptosis. *Leukemia*, *14*, 1509–1513.

Shinohara, M., Koga, T., Okamoto, K., Sakaguchi, S., Arai, K., Yasuda, H., et al. (2008). Tyrosine kinases Btk and Tec regulate osteoclast differentiation by linking RANK and ITAM signals. *Cell*, *132*, 794–806.

Siller-Jackson, A. J., Burra, S., Gu, S., Xia, X., Bonewald, L. F., Sprague, E., et al. (2008). Adaptation of connexin 43-hemichannel prostaglandin release to mechanical loading. *The Journal of Biological Chemistry*, *283*, 26374–26382.

Sims, N. A. (2016). Cell-specific paracrine actions of IL-6 family cytokines from bone, marrow and muscle that control bone formation and resorption. *The International Journal of Biochemistry & Cell Biology*, *79*, 14–23.

Sims, N. A., & Martin, T. J. (2014). Coupling the activities of bone formation and resorption: A multitude of signals within the basic multicellular unit. *Bonekey Reports*, *3*, 481.

Sims, N. A., & Vrahnas, C. (2014). Regulation of cortical and trabecular bone mass by communication between osteoblasts, osteocytes and osteoclasts. *Archives of Biochemistry and Biophysics*, *561*, 22–28.

Spinardi, L., Rietdorf, J., Nitsch, L., Bono, M., Tacchetti, C., Way, M., et al. (2004). A dynamic podosome-like structure of epithelial cells. *Experimental Cell Research*, *295*, 360–374.

Stains, J. P., & Civitelli, R. (2005). Gap junctions regulate extracellular signal-regulated kinase signaling to affect gene transcription. *Molecular Biology of the Cell*, *16*, 64–72.

Stains, J. P., Lecanda, F., Screen, J., Towler, D. A., & Civitelli, R. (2003). Gap junctional communication modulates gene transcription by altering the recruitment of Sp1 and Sp3 to connexin—Response elements in osteoblast promoters. *The Journal of Biological Chemistry*, *278*, 24377–24387.

Stepan, J. J., Alenfeld, F., Boivin, G., Feyen, J. H., & Lakatos, P. (2003). Mechanisms of action of antiresorptive therapies of postmenopausal osteoporosis. *Endocrine Regulations, 37*, 225–238.

Stevens, J. R., Miranda-Carboni, G. A., Singer, M. A., Brugger, S. M., Lyons, K. M., & Lane, T. F. (2010). Wnt10b deficiency results in age-dependent loss of bone mass and progressive reduction of mesenchymal progenitor cells. *Journal of Bone and Mineral Research, 25*, 2138–2147.

Strazzulla, L. C., & Cronstein, B. N. (2016). Regulation of bone and cartilage by adenosine signaling. *Purinergic Signal, 12*, 583–593.

Styrkarsdottir, U., Thorleifsson, G., Sulem, P., Gudbjartsson, D. F., Sigurdsson, A., Jonasdottir, A., et al. (2013). Nonsense mutation in the LGR4 gene is associated with several human diseases and other traits. *Nature, 497*, 517–520.

Sugatani, T., Alvarez, U., & Hruska, K. A. (2003). PTEN regulates RANKL- and osteopontin-stimulated signal transduction during osteoclast differentiation and cell motility. *The Journal of Biological Chemistry, 278*, 5001–5008.

Sun, Y., Buki, K. G., Ettala, O., Vaaraniemi, J. P., & Vaananen, H. K. (2005). Possible role of direct Rac1-Rab7 interaction in ruffled border formation of osteoclasts. *The Journal of Biological Chemistry, 280*, 32356–32361.

Sundaram, K., Nishimura, R., Senn, J., Youssef, R. F., London, S. D., & Reddy, S. V. (2007). RANK ligand signaling modulates the matrix metalloproteinase-9 gene expression during osteoclast differentiation. *Experimental Cell Research, 313*, 168–178.

Takada, T., Katagiri, T., Ifuku, M., Morimura, N., Kobayashi, M., Hasegawa, K., et al. (2003). Sulfated polysaccharides enhance the biological activities of bone morphogenetic proteins. *The Journal of Biological Chemistry, 278*, 43229–43235.

Takegahara, N., Takamatsu, H., Toyofuku, T., Tsujimura, T., Okuno, T., Yukawa, K., et al. (2006). Plexin-A1 and its interaction with DAP12 in immune responses and bone homeostasis. *Nature Cell Biology, 8*, 615–622.

Takeyama, S., Yoshimura, Y., Shirai, Y., Deyama, Y., Hasegawa, T., Yawaka, Y., et al. (2000). Low calcium environment effects osteoprotegerin ligand/osteoclast differentiation factor. *Biochemical and Biophysical Research Communications, 276*, 524–529.

Tanaka, S., Wakeyama, H., Akiyama, T., Takahashi, K., Amano, H., Nakayama, K. I., et al. (2010). Regulation of osteoclast apoptosis by Bcl-2 family protein bim and caspase-3. *Advances in Experimental Medicine and Biology, 658*, 111–116.

Tella, S. H., & Gallagher, J. C. (2014). Prevention and treatment of postmenopausal osteoporosis. *The Journal of Steroid Biochemistry and Molecular Biology, 142*, 155–170.

Thompson, W. R., Rubin, C. T., & Rubin, J. (2012). Mechanical regulation of signaling pathways in bone. *Gene, 503*, 179–193.

Tonna, S., & Sims, N. A. (2014). Talking among ourselves: Paracrine control of bone formation within the osteoblast lineage. *Calcified Tissue International, 94*, 35–45.

Troen, B. (2004). The role of cathepsin K in normal bone resorption. *Drug News & Perspectives, 17*, 19–28.

Tsubaki, M., Komai, M., Itoh, T., Imano, M., Sakamoto, K., Shimaoka, H., et al. (2014). Nitrogen-containing bisphosphonates inhibit RANKL- and M-CSF-induced osteoclast formation through the inhibition of ERK1/2 and Akt activation. *Journal of Biomedical Science, 21*, 10.

Tu, X., Delgado-Calle, J., Condon, K. W., Maycas, M., Zhang, H., Carlesso, N., et al. (2015). Osteocytes mediate the anabolic actions of canonical Wnt/b-catenin signaling in bone. *Proceedings of the National Academy of Sciences of the United States of America, 112*, E478–E486.

Vaananen, H. K., Zhao, H., Mulari, M., & Halleen, J. M. (2000). The cell biology of osteoclast function. *Journal of Cell Science, 113*(Pt 3), 377–381.

Vaira, S., Alhawagri, M., Anwisye, I., Kitaura, H., Faccio, R., & Novack, D. V. (2008). RelA/p65 promotes osteoclast differentiation by blocking a RANKL-induced apoptotic JNK pathway in mice. *The Journal of Clinical Investigation*, *118*, 2088–2097.

Verlinden, L., Vanderschueren, D., & Verstuyf, A. (2016). Semaphorin signaling in bone. *Molecular and Cellular Endocrinology*, *432*, 66–74.

Vignery, A. (2005). Macrophage fusion: The making of osteoclasts and giant cells. *The Journal of Experimental Medicine*, *202*, 337–340.

Waguespack, S. G., Koller, D. L., White, K. E., Fishburn, T., Carn, G., Buckwalter, K. A., et al. (2003). Chloride channel 7 (ClCN7) gene mutations and autosomal dominant osteopetrosis, type II. *Journal of Bone and Mineral Research*, *18*, 1513–1518.

Wakeyama, H., Akiyama, T., Takahashi, K., Amano, H., Kadono, Y., Nakamura, M., et al. (2007). Negative feedback loop in the bim-caspase-3 axis regulating apoptosis and activity of osteoclasts. *Journal of Bone and Mineral Research*, *22*, 1631–1639.

Walsh, M. C., & Choi, Y. (2003). Biology of the TRANCE axis. *Cytokine & Growth Factor Reviews*, *14*, 251–263.

Wang, L., Liu, S., Zhao, Y., Liu, D., Liu, Y., Chen, C., et al. (2015). Osteoblast-induced osteoclast apoptosis by fas ligand/FAS pathway is required for maintenance of bone mass. *Cell Death and Differentiation*, *22*, 1654–1664.

Wang, Y., Menendez, A., Fong, C., Elalieh, H. Z., Chang, W., & Bikle, D. D. (2014). Ephrin B2/EphB4 mediates the actions of IGF-I signaling in regulating endochondral bone formation. *Journal of Bone and Mineral Research*, *29*, 1900–1913.

Watkins, M., Grimston, S. K., Norris, J. Y., Guillotin, B., Shaw, A., Beniash, E., et al. (2011). Osteoblast connexin43 modulates skeletal architecture by regulating both arms of bone remodeling. *Molecular Biology of the Cell*, *22*, 1240–1251.

Wein, M. N., & Kronenberg, H. M. (2018). Regulation of bone remodeling by parathyroid hormone. *Cold Spring Harbor Perspectives in Medicine*, *8*, pii: a031237.

Wijenayaka, A. R., Kogawa, M., Lim, H. P., Bonewald, L. F., Findlay, D. M., & Atkins, G. J. (2011). Sclerostin stimulates osteocyte support of osteoclast activity by a RANKL-dependent pathway. *PLoS One*, *6*, e25900.

Wiktor-Jedrzejczak, W., Bartocci, A., Ferrante, A. W., Jr., Ahmed-Ansari, A., Sell, K. W., Pollard, J. W., et al. (1990). Total absence of colony-stimulating factor 1 in the macrophage-deficient osteopetrotic (op/op) mouse. *Proceedings of the National Academy of Sciences of the United States of America*, *87*, 4828–4832.

Williams, B. O. (2017). LRP5: From bedside to bench to bone. *Bone*, *102*, 26–30.

Wong, B. R., Besser, D., Kim, N., Arron, J. R., Vologodskaia, M., Hanafusa, H., et al. (1999). TRANCE, a TNF family member, activates Akt/PKB through a signaling complex involving TRAF6 and c-Src. *Molecular Cell*, *4*, 1041–1049.

Wu, M., Chen, G., & Li, Y. P. (2016). TGF-beta and BMP signaling in osteoblast, skeletal development, and bone formation, homeostasis and disease. *Bone Research*, *4*, 16009.

Wu, X., Mckenna, M. A., Feng, X., Nagy, T. R., & Mcdonald, J. M. (2003). Osteoclast apoptosis: The role of Fas in vivo and in vitro. *Endocrinology*, *144*, 5545–5555.

Wu, H., Xu, G., & Li, Y. P. (2009). Atp6v0d2 is an essential component of the osteoclast-specific proton pump that mediates extracellular acidification in bone resorption. *Journal of Bone and Mineral Research*, *24*, 871–885.

Xiao, L., Esliger, A., & Hurley, M. M. (2012). Nuclear FGF2 isoforms inhibit bone marrow stromal cell mineralization through FGF23/FGFR/MAPK in vitro. *Journal of Bone and Mineral Research*, *28*, 35–45.

Xiao, L., Liu, P., Li, X., Doetschman, T., Coffin, J. D., Drissi, H., et al. (2009). Exported 18-kDa isoform of fibroblast growth factor-2 is a critical determinant of bone mass in mice. *The Journal of Biological Chemistry*, *284*, 3170–3182.

Xiong, Y., Song, D., Cai, Y., Yu, W., Yeung, Y. G., & Stanley, E. R. (2011). A CSF-1 receptor phosphotyrosine 559 signaling pathway regulates receptor ubiquitination and tyrosine phosphorylation. *The Journal of Biological Chemistry, 286*, 952–960.
Xu, X., Han, J., Ito, Y., Bringas, P., Jr., Deng, C., & Chai, Y. (2008). Ectodermal Smad4 and p38 MAPK are functionally redundant in mediating TGF-beta/BMP signaling during tooth and palate development. *Developmental Cell, 15*, 322–329.
Yagi, M., Miyamoto, T., Sawatani, Y., Iwamoto, K., Hosogane, N., Fujita, N., et al. (2005). DC-STAMP is essential for cell-cell fusion in osteoclasts and foreign body giant cells. *The Journal of Experimental Medicine, 202*, 345–351.
Yagi, M., Ninomiya, K., Fujita, N., Suzuki, T., Iwasaki, R., Morita, K., et al. (2007). Induction of DC-STAMP by alternative activation and downstream signaling mechanisms. *Journal of Bone and Mineral Research, 22*, 992–1001.
Yakar, S., Werner, H., & Rosen, C. J. (2018). Insulin-like growth factors: Actions on the skeleton. *Journal of Molecular Endocrinology, 61*, T115–T137.
Yamashita, T., Takahashi, N., & Udagawa, N. (2012). New roles of osteoblasts involved in osteoclast differentiation. *World Journal of Orthopedics, 3*, 175–181.
Yang, M., Birnbaum, M. J., Mackay, C. A., Mason-Savas, A., Thompson, B., & Odgren, P. R. (2008). Osteoclast stimulatory transmembrane protein (OC-STAMP), a novel protein induced by RANKL that promotes osteoclast differentiation. *Journal of Cellular Physiology, 215*, 497–505.
Yang, D. Q., Feng, S., Chen, W., Zhao, H., Paulson, C., & Li, Y. P. (2012). V-ATPase subunit ATP6AP1 (Ac45) regulates osteoclast differentiation, extracellular acidification, lysosomal trafficking, and protease exocytosis in osteoclast-mediated bone resorption. *Journal of Bone and Mineral Research, 27*, 1695–1707.
Yang, C. R., Wang, J. H., Hsieh, S. L., Wang, S. M., Hsu, T. L., & Lin, W. W. (2004). Decoy receptor 3 (DcR3) induces osteoclast formation from monocyte/macrophage lineage precursor cells. *Cell Death and Differentiation, 11*(Suppl 1), S97–107.
Yi, Z., Lin, W. W., Stunz, L. L., & Bishop, G. A. (2014). Roles for TNF-receptor associated factor 3 (TRAF3) in lymphocyte functions. *Cytokine & Growth Factor Reviews, 25*, 147–156.
Yi, J., Xiong, W., Gong, X., Bellister, S., Ellis, L. M., & Liu, Q. (2013). Analysis of LGR4 receptor distribution in human and mouse tissues. *PLoS One, 8*, e78144.
Yokouchi, M., Kondo, T., Sanjay, A., Houghton, A., Yoshimura, A., Komiya, S., et al. (2001). Src-catalyzed phosphorylation of c-Cbl leads to the interdependent ubiquitination of both proteins. *Journal of Biological Chemistry, 276*, 35185–35193.
Yoshiko, Y., Wang, H., Minamizaki, T., Ijuin, C., Yamamoto, R., Suemune, S., et al. (2007). Mineralized tissue cells are a principal source of FGF23. *Bone, 40*, 1565–1573.
Yu, K., Xu, J., Liu, Z., Sosic, D., Shao, J., Olson, E. N., et al. (2003). Conditional inactivation of FGF receptor 2 reveals an essential role for FGF signaling in the regulation of osteoblast function and bone growth. *Development, 130*, 3063–3074.
Yun, Y. R., Won, J. E., Jeon, E., Lee, S., Kang, W., Jo, H., et al. (2010). Fibroblast growth factors: Biology, function, and application for tissue regeneration. *Journal of Tissue Engineering, 2010*, 218142.
Zanotti, S., & Canalis, E. (2012). Notch regulation of bone development and remodeling and related skeletal disorders. *Calcified Tissue International, 90*, 69–75.
Zanotti, S., Smerdel-Ramoya, A., Stadmeyer, L., Durant, D., Radtke, F., & Canalis, E. (2008). Notch inhibits osteoblast differentiation and causes osteopenia. *Endocrinology, 149*, 3890–3899.
Zhang, Y., Paul, E. M., Sathyendra, V., Davidson, A., Bronson, S., Srinivasan, S., et al. (2011). Enhanced osteoclastic resorption and responsiveness to mechanical load in gap junction deficient bone. *PLoS One, 6* e23516.

Zhao, C., Irie, N., Takada, Y., Shimoda, K., Miyamoto, T., Nishiwaki, T., et al. (2006). Bidirectional ephrinB2-EphB4 signaling controls bone homeostasis. *Cell Metabolism*, *4*, 111–121.

Zhou, M., Guo, S., Yuan, L., Zhang, Y., Zhang, M., Chen, H., et al. (2017). Blockade of LGR4 inhibits proliferation and odonto/osteogenic differentiation of stem cells from apical papillae. *Journal of Molecular Histology*, *48*, 389–401.

Zhu, C., Zheng, X. F., Yang, Y. H., Li, B., Wang, Y. R., Jiang, S. D., et al. (2016). LGR4 acts as a key receptor for R-spondin 2 to promote osteogenesis through Wnt signaling pathway. *Cellular Signalling*, *28*, 989–1000.

Zmuda, J. M., Yerges, L. M., Kammerer, C. M., Cauley, J. A., Wang, X., Nestlerode, C. S., et al. (2009). Association analysis of WNT10B with bone mass and structure among individuals of African ancestry. *Journal of Bone and Mineral Research*, *24*, 437–447.

Zuo, H., & Wan, Y. (2017). Nuclear receptors in skeletal homeostasis. *Current Topics in Developmental Biology*, *125*, 71–107.

CHAPTER SEVEN

Activating mutations of the gp130/JAK/STAT pathway in human diseases

Juliane Lokau, Christoph Garbers*
Department of Pathology, Otto-von-Guericke-University Magdeburg, Medical Faculty, Magdeburg, Germany
*Corresponding author: e-mail address: christoph.garbers@med.ovgu.de

Contents

Abstract

Cytokines of the interleukin-6 (IL-6) family are involved in numerous physiological and pathophysiological processes. Dysregulated and increased activities of its members can be found in practically all human inflammatory diseases including cancer. All cytokines activate several intracellular signaling cascades, including the Jak/STAT, MAPK, PI3K, and Src/YAP signaling pathways. Additionally, several mutations in proteins involved in these signaling cascades have been identified in human patients, which render these proteins constitutively active and result in a hyperactivation of the signaling pathway. Interestingly, some of these mutations are associated with or even causative for distinct human diseases, making them interesting targets for therapy. This chapter describes the basic biology of the gp130/Jak/STAT pathway, summarizes what is known about the molecular mechanisms of the activating mutations, and gives an outlook how this knowledge can be exploited for targeted therapy in human diseases.

Advances in Protein Chemistry and Structural Biology, Volume 116
ISSN 1876-1623
https://doi.org/10.1016/bs.apcsb.2018.11.007

1. Introduction: The IL-6 family of cytokines

The interleukin-6 (IL-6) family of cytokines consists of its members IL-6, IL-11, cardiotrophin-like cytokine (CLC), ciliary neurotrophic factor (CNTF), cardiotrophin-1 (CT-1), oncostatin M (OSM), leukemia inhibitory factor (LIF), IL-27 (composed of EBI3 and p28/IL-30), IL-30, and IL-31 (Fig. 1 and Garbers et al., 2012). They all belong to the four-helical bundle family and share the same protein fold, in which the four helices are arranged in an up-up-down-down topology (Bazan, 1989, 1990).

However, this fold was also found in a variety of other proteins that do not belong to the IL-6 cytokine family. Therefore, the hallmark of all family members is their ability to use the β-receptor glycoprotein 130 (gp130) for signaling. Due to the fact that gp130 is expressed ubiquitously throughout the body, other cytokine receptors have to participate in the signal transduction and thus signaling pathway initiation, because otherwise all IL-6-type

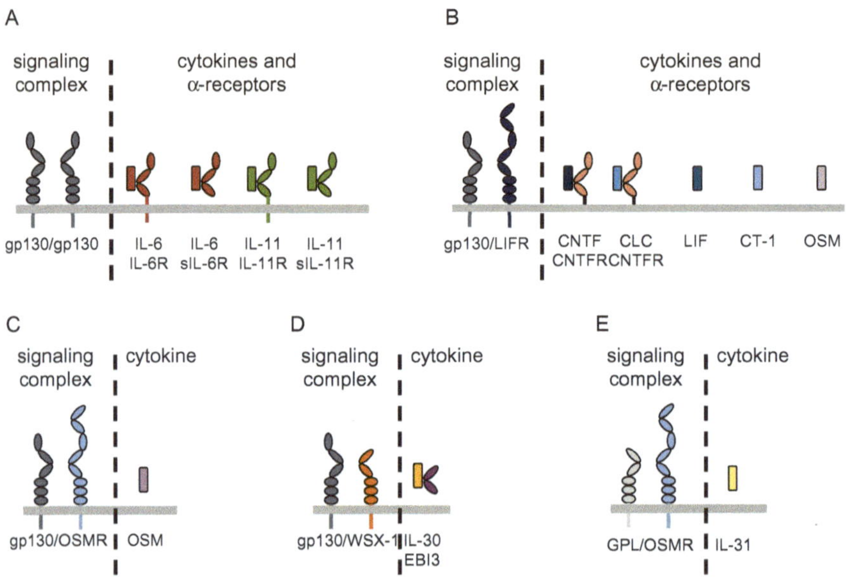

Fig. 1 The IL-6 family of cytokines. (A) IL-6 and IL-11 can signal either via their membrane-bound IL-6R/IL-11R (termed classic signaling) or via their soluble IL-6R/IL-11R (trans-signaling). In both cases, a gp130 homodimer is engaged for signaling. (B) CNTF and CLC (via the CNTF receptor) and LIF, CT-1 and OSM signal via a heterodimer of gp130 and the LIFR. (C) Besides signaling via gp130/LIFR, OSM can also signal via a heterodimer of gp130 and the OSMR. (D) IL-27, the only heterodimeric cytokine of the IL-6 family that consists of p28/IL-30 and EBI3, signals via WSX-1 and gp130. (E) IL-31 induces intracellular signaling via GPL and OSMR.

cytokines would be able to activate all cells simultaneously. Consequently, none of these cytokines can directly activate gp130. Instead, all cytokines need at least one additional receptor for signaling.

The name-giving member IL-6 itself has to bind first to a non-signaling α-receptor (IL-6 receptor, IL-6R), and the resulting IL-6/IL-6R complex then recruits and thereby activates a gp130 homodimer. A similar molecular mechanism exists for the biological activity of IL-11, which first binds to the IL-11 receptor (IL-11R), and the IL-11/IL-11R complex afterward transmits its signals through a gp130/gp130 homodimer. Importantly, both cytokines can also bind to and signal via soluble forms of their α-receptors (trans-signaling), which significantly widens the spectrum of cells that can be activated by IL-11 and IL-6, because cells that are the target of trans-signaling do not need to express membrane-bound IL-6R and IL-11R, respectively (Lokau, Agthe, Flynn, & Garbers, 2017).

The third non-signaling α-receptor is the CNTF receptor (CNTFR). CNTF and CLC can bind to the CNTFR, and both cytokines activate intracellular signaling cascades via a heterodimer of gp130 in combination with the LIF receptor (LIFR, Gearing et al., 1991). CT-1, OSM, and LIF also activate gp130/LIFR, but these three cytokines do not need the help of an α-receptor, but can rather directly bind to and activate the two β-receptors (Pennica et al., 1995). Furthermore, OSM can activate a second heterodimer, namely gp130/OSM receptor (OSMR), an unique feature that no other member of the family possesses (Hermanns, 2015). IL-27, the only heterodimeric cytokine among the IL-6-type cytokines, activates cells via a heterodimer consisting of gp130 and WSX-1 (Pflanz et al., 2004). IL-30/p28, the cytokine subunit of IL-27, has also properties independent of EBI3, and can bind to and activate the IL-6R, which then leads to the formation of a gp130 homodimer (Crabé et al., 2009; Garbers et al., 2013). Additionally, IL-31 is part of the IL-6 family, although it does not engage gp130 for signaling, but uses GPL, the gp130-like cytokine receptor, in combination with the OSMR (Dreuw et al., 2004; Ghilardi et al., 2002).

All cytokines activate (with varying efficiency, strength, and duration) the same intracellular signaling cascades: the Janus kinase/signal transducer and activator of transcription (Jak/STAT) pathway, the mitogen-activated protein kinase/extracellular-regulated kinase pathway (MAPK/ERK), and the phosphatidylinositol-4,5-bisphosphate 3-kinase (PI3K) pathway. Most recently, activation of the protein YAP (yes-associated protein, usually part of the Hippo/YAP signaling pathway; Moroishi, Hansen, & Guan, 2015) by Src kinases has been described to be dependent on gp130 activation (Taniguchi et al., 2017, 2015). More details on this topic are given in Section 2.

2. Intracellular signaling cascades activated by IL-6-type cytokines

Signal transduction via gp130 is initiated by binding of an IL-6-type cytokine to its specific receptors, which results in the formation of active signaling complexes containing gp130 homo- or heterodimers (Fig. 2). Signaling of IL-6-type cytokines activates four different signaling pathways: the

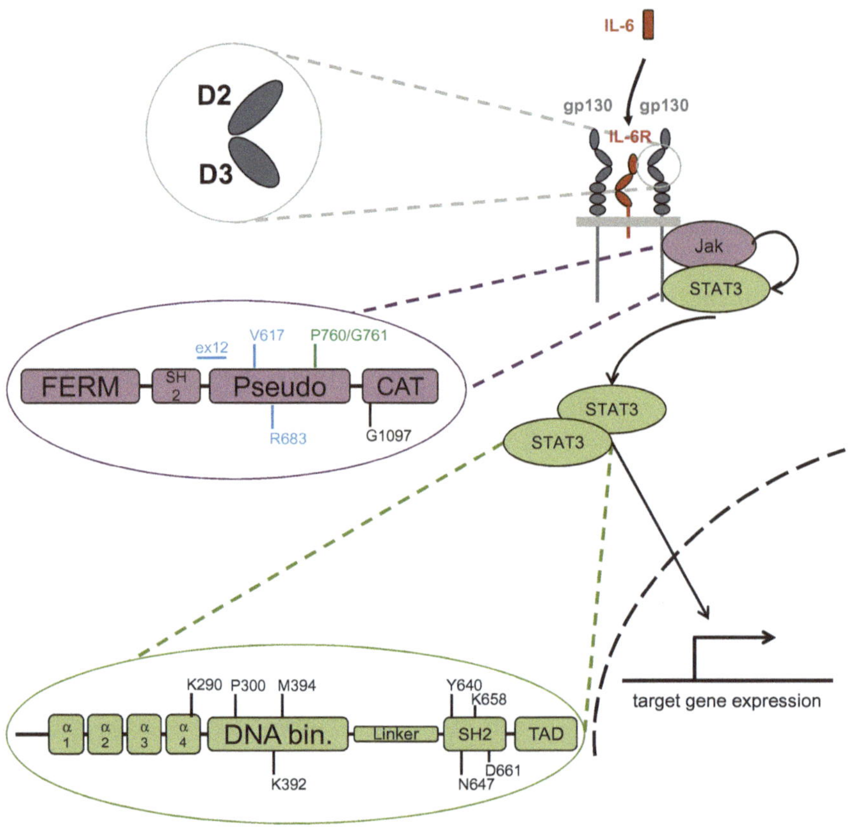

Fig. 2 Signal transduction of IL-6-family cytokines and activating mutations. IL-6 binds to the IL-6R (both shown in red) and recruits a gp130 homodimer (gray). Afterwards, Jak kinases (purple) are activated, which phosphorylate tyrosine residues within the gp130 intracellular region and on STAT3 transcription factors (green, phosphorylation not depicted). STAT3 homodimers translocate into the nucleus to activate target gene expression. The domain organization of Jak and STAT are shown in insets, and common mutations mentioned in the text are indicated. Mutations in Jak2 are shown in blue, in Jak1 are shown in black, and mutations in Tyk2 are shown in green.

Janus kinase (Jak)/signal transducer and activator of transcription (STAT), phosphatidyl-inositol-3-kinase (PI3K), mitogen-activated protein kinase (MAPK), and Src/yes-associated protein 1 (YAP1) cascades (Eulenfeld et al., 2012; Garbers et al., 2012; Taniguchi et al., 2015).

Through conformational changes of the β-receptors upon complex formation, intracellularly associated Jaks are brought in close proximity to each other, resulting in their autophosphorylation (Lütticken et al., 1994; Matadeen, Hon, Heath, Jones, & Fuller, 2007; Stahl et al., 1994). Among the members of the Jak family, Jak1, Jak2, and Tyk2 have been described to be associated with the β-receptors of the IL-6 family (Stahl et al., 1994). However, Jak1 seems to be crucial for signal transduction, since loss of Jak1 drastically reduces signaling of IL-6-type cytokines while loss of Jak2 or Tyk2 has less impact (Guschin et al., 1995; Rodig et al., 1998; Schaper et al., 1998; Sohn et al., 2013; Takahashi et al., 2008). In addition to autophosphorylation, the Jaks also phosphorylate specific tyrosine residues within the intracellular region of the β-receptors (Garbers et al., 2012; Heinrich et al., 2003).

Between the β-receptors of the IL-6 family, the mechanisms of intracellular signal transduction are rather similar. Therefore, when describing specific motifs relevant for signaling, we will focus on residues identified in the common β-receptor gp130 (Fig. 2).

Among the tyrosine residues phosphorylated upon gp130 activation, pY767, pY814, pY905, and pY915 interact with STAT3, and pY905 and pY915 additionally with STAT1 (Gerhartz et al., 1996). Upon binding to the receptors, STATs are also phosphorylated. Phosphorylation of STAT1 and STAT3 at Y701 or Y705, respectively (Kaptein, Paillard, & Saunders, 1996; Shuai, Stark, Kerr, & Darnell, 1993), leads to the formation of active STAT dimers, which translocate into the nucleus where they regulate the transcription of target genes (Heinrich et al., 2003; Shuai et al., 1993; Zhong, Wen, & Darnell, 1994). In addition to the tyrosine phosphorylation, STAT1/3 are also phosphorylated at S727 (Decker & Kovarik, 2000; Zhang, Blenis, Li, Schindler, & Chen-Kiang, 1995). While the serine phosphorylation is required for full activity, the impact appears to be specific for individual target genes (Eulenfeld et al., 2012; Wen, Zhong, & Darnell, 1995). Most of the IL-6-type cytokines activate mainly STAT3, and to a lesser extent STAT1 (Garbers et al., 2012; Heinrich et al., 2003). The exception is IL-27, which primarily activates STAT1 since the intracellular region of WSX-1 does not contain a binding motif for STAT3 (Owaki et al., 2008; Pflanz et al., 2004).

One well-studied STAT target gene is suppressor of cytokine signaling 3 (SOCS3), the most prominent negative regulator of gp130 signaling (Babon, Varghese, & Nicola, 2014; Yoshimura, Naka, & Kubo, 2007). SOCS3 binds to pY759 in activated gp130 (Schmitz, Weissenbach, Haan, Heinrich, & Schaper, 2000) and furthermore binds to the kinase domain of Jaks, thereby inhibiting further phosphorylation of STATs (Sasaki et al., 1999). Additionally, SOCS3 promotes proteosomal degradation of gp130 and Jaks (Kershaw, Laktyushin, Nicola, & Babon, 2014), offering a second mechanism to restrict signaling. Consequently, loss of SOCS3 leads to prolonged STAT phosphorylation upon IL-6 stimulation (Croker et al., 2003).

While many gp130 functions are attributed to STAT activity, also other Jak-dependent pathways are activated upon gp130 activation. In addition to the above-mentioned STAT binding interface, Y759 in the gp130 intracellular region presents a binding site for Src homology-2 domain containing phosphatase (SHP2) (Schmitz et al., 2000; Tebbutt et al., 2002). Upon ligand-induced activation, Jak1 phosphorylates SHP2 at the amino acid residues Y304 and Y327. Activated SHP2 afterward associates with the multisite docking protein gab1 (Eulenfeld et al., 2012; Schaper et al., 1998; Yin, Shen, Feng, & Yang, 1997). Gab1 links the activated receptor complex to PI3K and Ras, and the latter then induces activation of Erk and MAPK (Cunnick et al., 2002; Eulenfeld et al., 2012; Takahashi-Tezuka et al., 1998). Interestingly, IL-31 is the only IL-6-type cytokine that does not activate Erk signaling, most probably because GPL does not recruit SHP2 to its cytoplasmic portion (Dreuw et al., 2004).

In addition to activation of Erk/MAPK and PI3K signaling, SHP2 has an inhibitory function in STAT activation (Lehmann et al., 2003). In contrast to SOCS3, the inhibitory effect of SHP2 is not on the induced phosphorylation of STAT3, but rather limits basal STAT3 activation in the absence of cytokines (Dittrich et al., 2012). Therefore, as binding site for SOCS3 and SHP2, Y759 in the gp130 intracellular region represents an important residue for restriction of STAT-mediated gene expression. Furthermore, it has been reported that an imbalance between MAPK/PI3K and STAT3 signals is associated with different diseases, highlighting the importance of the regulatory residue Y759 (Ernst & Jenkins, 2004; Eulenfeld et al., 2012; Judd et al., 2004).

Most research concerning the signal transduction of IL-6-type cytokines has focused on Jak-induced mechanisms, but gp130 also associates with kinases of the Src-family (Ernst, Gearing, & Dunn, 1994; Hallek et al., 1997; Schaeffer et al., 2001). Recently, it was found that the Src-family kinases Src and Yes interact with gp130 and, upon cytokine binding, phosphorylate yes-associated protein 1 (YAP1) (Tamm, Böwer, & Anneren, 2011; Taniguchi et al., 2015). The tyrosine-phosphorylated YAP1 then translocates into the nucleus and induces the transcription of target genes (Tamm et al., 2011). Interestingly, while STAT3 as well as YAP1 are required for mucosal regeneration, the two transcription factors control different target genes (Taniguchi et al., 2015).

3. Mutations in IL-6 family cytokines

In human inflammatory diseases, the amount of IL-6-type cytokines is usually increased. The prime example for this is IL-6 itself, which is largely absent or in the range of a few picogram per mL in the sera of healthy humans, but can rise dramatically during disease, e.g., in meningococcal septic shock (Waage, Brandtzaeg, Halstensen, Kierulf, & Espevik, 1989). Thus, the strength and duration of intracellular signaling provoked by cytokines are determined by the amount of extracellular ligands. Therapeutic strategies which exploit this fact are therefore directed against either the cytokines itself or their receptors, as evident for IL-6 and the IL-6R (Garbers, Heink, Korn, & Rose-John, 2018).

One example for a natural occurring variant is the coding SNP rs4252548 of the *Il11* gene, which results in the exchange of the amino acid residue Arg112 into His112. This polymorphism is associated with adult human height in individuals of European ancestry and resulted in a reduction of 8 mm (Lanktree et al., 2011; Marouli et al., 2017). Mechanistically, the mutation did not alter the biological activity of IL-11 or its ability to interact with IL-11R and gp130, but rather compromised the stability of the protein (Lokau, Göttert, et al., 2017). Given the pivotal role of IL-11 in bone homeostasis (Sims, 2016), reduced stability might be the cause for the human phenotype.

No natural occurring mutations in IL-6-type cytokines are known which lead to a stronger activation of the intracellular signaling cascades and might therefore be of interest in terms of disease or therapy. However, in vitro studies have shown that cytokines can be modified in ways that lead to muteins with enhanced biological activity. The generation of such variants has been described for IL-6 (Cabibbo et al., 1995; Toniatti et al., 1996) and IL-11 (Harmegnies et al., 2003; Sano, Takei, Ueda, & Tsumoto, 2017; Yanaka et al., 2011), and is therefore possible that such variants exist in humans and have just not been discovered yet.

4. Mutations in the α-receptors IL-6R, IL-11R, and CNTFR

Not all members of the IL-6 family of cytokines are capable of binding to their signal-transducing β-receptors directly. Instead, they rely on individual α-receptors, which bind the cytokine in a first step, and the resulting cytokine/receptor complex engages a dimer of two β-receptors in a second step, resulting in the formation of the final signaling complex. Among all family members, only IL-6, IL-11, CLC, and CNTF require an α-receptor, and while IL-6 and IL-11 initiate the intracellular signaling pathways via gp130 homodimers, CNTF activates a heterodimer consisting of gp130 and LIFR.

The three α-receptors are not signaling competent on their own. Although IL-6R and IL-11R have intracellular regions, these lack any motifs associated with signal initiation like tyrosine residues or the so-called box 1 and box 2 motifs (Heinrich et al., 2003). This is further underlined by the fact that both receptors exist as soluble forms that are able to activate gp130-dependent signaling in combination with their ligands like their membrane-bound counterparts, showing that the intracellular regions are indeed dispensable for signaling of IL-6 and IL-11 (Lokau, Agthe, et al., 2017). The intracellular parts are instead crucial for sorting of both receptors to the basolateral and/or apical side of polarized cells, as they contain certain amino acid-based sorting motifs that are recognized by cargo proteins and ensure correct intracellular trafficking of the receptors (Monhasery et al., 2016). In contrast, the CNTFR is a GPI-anchored protein that lacks an intracellular region entirely (Davis et al., 1993).

Due to the lack of signaling capacity of the three receptors, it is clear that mutations occurring within them will not directly activate intracellular signaling pathways. Nevertheless, several human mutations are known that at least indirectly affect intracellular signaling both qualitatively and quantitatively. The best-studied mutation of the IL-6R is a single-nucleotide polymorphism (SNP) that results in the exchange of the amino acid residue Asp358 into Ala358 (rs2228145, Galicia et al., 2004; Rafiq et al., 2007). This SNP increases sIL-6R serum levels in the affected individuals (Galicia et al., 2004; Garbers et al., 2014; Rafiq et al., 2007). This is accompanied with increased IL-6 serum levels, although the molecular reason for this is unclear. Furthermore, people homozygous for this SNP have less membrane-bound IL-6R on CD4+ T cells and monocytes (Ferreira et al., 2013) and display reduced C-reactive protein levels (Sarwar et al., 2012;

Swerdlow et al., 2012). Interestingly, this SNP has a great impact on patients with coronary heart disease. For every copy of the minor allele inherited, not only sIL-6R was increased by ~34% but also the risk of coronary heart disease was reduced by 3.4% (Sarwar et al., 2012; Swerdlow et al., 2012). The reason for this rather protective effect of the sIL-6R is currently under debate. The reduced IL-6R levels on the hepatocytes make them less responsive for stimulation by IL-6, which is also indicated by the reduced CRP levels in the serum. Reduced activation of STAT1 and STAT3 signaling in humans homozygous for the Asp358Ala mutation has also been demonstrated in T cells and monocytes (Ferreira et al., 2013). However, the increased sIL-6R levels in combination with sgp130 are also believed to increase the capacity of the endogenous buffer system to block systemic low-grade IL-6 effects (Garbers, Aparicio-Siegmund, & Rose-John, 2015; Scheller & Rose-John, 2012). Whether one of these mechanisms or a combination of both is responsible for the observed effects warrants further investigation.

The Asp358Ala mutation is located within the so-called stalk region of the IL-6R, which connects the three extracellular domains with the plasma membrane and is believed to have no own structure. Although more differential splicing of the IL-6R mRNA has been described in humans homozygous for this SNP (Stephens et al., 2012), the majority of sIL-6R is not generated by splicing, but rather by proteolytic cleavage of the membrane-bound IL-6R (Riethmueller et al., 2017). Indeed, the Asp358Ala mutation is located close to the cleavage site used by the metalloproteases ADAM10 and ADAM17 (Riethmueller et al., 2017), and this mutation renders the IL-6R more susceptible for proteolytic cleavage (Garbers et al., 2014; Riethmueller et al., 2017), which might be the molecular reason for the increased sIL-6R serum levels in humans.

Mutations within the IL-11R are not directly associated with inflammatory diseases, but appear to rather affect developmental processes. Several recent publication described patients with craniosynostosis, which is a malformation of the skull provoked by premature closure of the sutures (Brischoux-Boucher et al., 2018; Clarke et al., 2018; Keupp et al., 2013; Nieminen et al., 2011). The further clinical phenotype appears to be variable, as some patients show tooth abnormalities and have conductive hearing loss, but this is not the case for all patients. The mutations are mostly coding mutations that exchange a single amino acid residue and are distributed all over the IL-11R ectodomain. Initial experiments suggest that they inactivate the IL-11R, which is corroborated by findings with *Il11ra*$^{-/-}$ mice that have a comparable phenotype. A common feature of the patient mutations is the absence of intracellular signaling when cells overexpressing corresponding IL-11R mutants are stimulated with IL-11 (Keupp et al., 2013; Nieminen et al., 2011). The exact molecular reasons are still unclear, but first data show retention of mutant receptors within the endoplasmic reticulum (ER), lack of IL-11R maturation and impaired cell-surface appearance, suggesting that folding of the protein is impaired (Agthe et al., 2018). Whether these findings hold true for all mutations has to be analyzed. To our knowledge, no functional mutations in the CNTFR have been described yet.

In summary, while mutations in the three α-receptors are not directly activating intracellular signaling cascades, they are crucial for the proper activation of signaling cascades by their respective ligands. Thus, mutations in IL-6R and IL-11R reduce or completely prevent activation of signaling pathways and are therefore important components of the signaling cascade.

5. Mutations in gp130 and related receptors

In contrast to the α-receptors (see Section 4), mutations in the common β-receptor of the IL-6 family, gp130, have been described that result in the constitutive activation of intracellular signaling pathways (Fig. 2). The landmark publication by Rebouissou et al. identified several in-frame deletion variants of gp130 in inflammatory hepatocellular tumors, which were constitutively active, meaning that they did not need to be extracellularly activated by IL-6/IL-6R or IL-11/IL-11R in order to provoke an intracellular response (Rebouissou et al., 2009). Expression of the most common variant in Ba/F3-gp130 cells resulted in ligand-independent proliferation of the cell line, underlining its constitutive activity (Sommer et al., 2012). Interestingly, an anti-gp130 antibody that specifically blocks IL-11 signaling has been shown to block activity of this gp130 variant (Sommer et al., 2012). However, other studies using other cell lines as Ba/F3 cells showed convincingly that the majority of the mutated gp130 proteins signal from intracellular compartments, particularly the endosome, making an antibody-based approach to target such malignancies difficult, if not impossible (Rinis, Kuster, Schmitz-Van de Leur, Mohr, & Müller-Newen, 2014; Schmidt-Arras et al., 2013). Instead, targeting of the downstream kinase Jak1 might be a potential therapeutic target (Poussin et al., 2013). Mechanistically, activating gp130 mutations increase the interdomain angle of the D2 and D3 domain and disrupt the hydrophobic D2–D3 connectivity, resulting in activation of gp130 (Schütt et al., 2013). Activating gp130 mutations have furthermore been described in benign hepatocellular neoplasms developed on cirrhotic livers (Calderaro et al., 2016) and in hepatocellular adenomas from patients with glycogen storage disease type I (Calderaro et al., 2013).

Coding mutations are also known in other signal-transducing β-receptors used by the IL-6 family. Mutations in OSMR cause a human disease called familial primary localized cutaneous amyloidosis (FPLCA). FPLCA is characterized by the deposition of amyloids in the upper dermis of the skin, and the patients suffer from chronic skin itching (Arita et al., 2008). It is an autosomal dominant disorder that has until now been described in several families (Arita et al., 2009; Lin et al., 2010; Tanaka et al., 2010; Wang et al., 2012). Interestingly, only point mutations located in the extracellular part of OSMR have been found, and no deletions as in gp130. Accordingly, the mutations diminish the activity of the receptor, resulting in reduced activation of the intracellular signaling pathways, and do not lead to a constitutive activation.

Mutations in the LIFR cause the autosomal-recessive Stüve-Wiedemann syndrome, which is characterized by, e.g., bone abnormalities and often results in infant mortality (Mikelonis, Jorcyk, Tawara, & Oxford, 2014). However, like for the OSMR, all known mutations appear to result in loss of function. Recently, mutations in the LIFR have also been associated with congenital abnormalities of the kidneys and urinary tract (Kosfeld et al., 2017).

It is currently unclear whether small deletions would also activate OSMR and/or LIFR, and whether findings from gp130 can be easily transferred from one receptor to the other. Given the different expression pattern of gp130, LIFR, and OSMR, it is also not possible to speculate what the disease or phenotype would be if a constitutively active variant of LIFR or OSMR was found. No coding mutations in GPL and WSX-1 have been described to date.

In summary, gp130 appears to be the only receptor of the IL-6 family that can be constitutively activated by mutations.

6. Mutations in Janus kinases

The receptors of the IL-6 family lack intrinsic tyrosine kinase activity and thus require the activity of constitutively associated kinases (Heinrich et al., 2003). Different kinases are linked to IL-6-type cytokine signaling, but most research focused on the members of the Janus kinase family (Jak1, Jak2, Tyk2, Fig. 2).

Jaks consist of four different domains: an N-terminal FERM (four-point-one, ezrin, radixin, moesin) domain, followed by a SH2 domain, a pseudokinase domain, and the tyrosine kinase domain (Heinrich et al., 2003; Morris, Kershaw, & Babon, 2018). The kinase domain is responsible for the catalytic activity of Jaks (Morris et al., 2018). The pseudokinase domain is devoid of catalytic activity but is considered to act as auto-inhibitor for the adjacent kinase domain, although the exact mechanism is not yet known (Lupardus et al., 2014). At least part of the inhibitory activity is, however, assigned to a low-level kinase activity of the so-called pseudokinase domain, which phosphorylates negative regulatory sites in the kinase domain (Ungureanu et al., 2011). The FERM and SH2 domains interact with specific intracellular regions of the cognate receptors and might be involved in the specificity of receptor-Jak interactions (Ferrao & Lupardus, 2017; Haan, Behrmann, & Haan, 2010; Wallweber, Tam, Franke, Starovasnik, & Lupardus, 2014). The domains of Jaks are connected by flexible linker regions, which allow for a high number of different conformations. This flexibility might be involved in the regulation of Jaks (Babon, Lucet, Murphy, Nicola, & Varghese, 2014; Lupardus et al., 2011).

Mutations resulting in constitutively active Jaks have been described in different diseases, mainly in myeloid malignancies but also in cancers of non-hematopoietic origin (Chen, Staudt, & Green, 2012; Flex et al., 2008; Godfrey & Green, 2012; Jeong et al., 2008; Pilati et al., 2014; Pilati & Zucman-Rossi, 2015). Notably, while these mutations lead to cytokine-independent Jak phosphorylation, they require the presence of a receptor dimer (Bercovich et al., 2008; Hornakova et al., 2009; Lu et al., 2005). Supposedly, the receptors function as scaffolds, positioning the Jaks in a distance that allows phosphorylation (Lu et al., 2005). Interestingly, while mutations are distributed among the whole protein, the majority of described activating mutations is found in the pseudokinase domain (Babon, Lucet, et al., 2014; Chen et al., 2012; Haan et al., 2010; Lupardus et al., 2014).

The first pathological Jak variations where identified in acute leukemias, where the kinase domain of Jak2 is fused to the oligomerization domain of TEL (Lacronique et al., 1997). In contrast to point mutations, TEL-Jak2 dimerizes independent of cytokine receptors, resulting in activation of Jak2 (Lacronique et al., 1997). Since then, other constitutive active Jak2 fusion proteins have been identified, which are thought to function similar, although they have not yet been analyzed in detail (Chen et al., 2012; Griesinger et al., 2005; Nebral et al., 2009; Poitras, Dal Cin, Aster, Deangelo, & Morton, 2008; Reiter et al., 2005).

The most frequent and best-studied Jak mutation is Jak2-V617F, which is prevalent in myeloproliferative disorders, especially in polycythemia vera (Godfrey & Green, 2012; James et al., 2005; Kralovics et al., 2005). Notably, the corresponding Jak1 mutation was identified in patients suffering from acute myeloid leukemia (AML) or acute lymphoblastic leukemia (ALL) (Jeong et al., 2008; Mullighan et al., 2009). V617F is located in the pseudokinase domain of Jak2 and predicted to be part of the kinase/pseudokinase domain interface (Giordanetto & Kroemer, 2002). The mutation has been reported to abrogate the kinase activity of the pseudokinase domain that is required to maintain an inactive Jak2, thus resulting in a dysregulated kinase domain (Ungureanu et al., 2011). Likewise, other Jak2 mutations found in exon 12, which result in distinct myeloproliferative neoplasms (MPN) (Ma et al., 2009; Scott, 2011; Scott et al., 2007), are also involved in the interaction between kinase and pseudokinase domain, and are thought to result in a constitutive active kinase via a similar mechanism (Lupardus et al., 2014).

Additionally, other Jak2 mutations are found in B-cell lineage ALL near the residue R683, most frequently R683G (Bercovich et al., 2008; Kearney et al., 2009; Mullighan et al., 2009). While R683 is also located in the pseudokinase domain, it is not predicted to be involved in the kinase/pseudokinase interaction. Instead, it is considered to alter the properties of a binding pocket, potentially altering intra- or intermolecular interactions (Bercovich et al., 2008; Chen et al., 2012; Haan et al., 2010).

Furthermore, also activating mutations within the kinase domain of Jak2 have been described (Lupardus et al., 2014; Mercher et al., 2006; Mullighan et al., 2009; Vainchenker & Constantinescu, 2013), although less common than in the pseudokinase domain. Notably, these are also located at the kinase/pseudokinase interface, thus indicating a mechanism similar to that of V617F (Lupardus et al., 2014; Mercher et al., 2006).

In summary, mutations within Jak2 where described in distinct hotspots which, while all resulting in ligand-independent activation of the kinase, appear to function via different mechanisms. Furthermore, they are associated with distinct diseases (Haan et al., 2010; Lupardus et al., 2014). The different diseases following various Jak2 mutations were recapitulated in mice, where it was also shown that they differently couple to receptors and thus affect different cells (Yao et al., 2017).

In comparison to Jak2, far less is known about mutations in the other Jaks. Activating Jak1 mutations are most commonly found in ALL but also, albeit less frequently, in other cancers, including tumors of non-hematopoietic origin (Blombery et al., 2016; Chen et al., 2012; Flex et al., 2008; Jeong et al., 2008; Pilati et al., 2014). Also for Jak1, the majority of activating mutations is found in the pseudokinase domain, more specifically in the interaction site between the kinase and the pseudokinase domain. Consequently, the mechanisms that lead to the constitutive active Jak1 are considered to be similar to those described for Jak2 (Flex et al., 2008; Lupardus et al., 2014).

Activating Jak1 mutations at residue G1097 have been identified in anaplastic large cell lymphomas (ALCL) (Crescenzo et al., 2015). G1097 is located at the surface of the kinase domain and does not interact with the pseudokinase domain. Instead, it might be involved in protein-protein interactions (Crescenzo et al., 2015). Interestingly, some ALCL patients harbored a STAT3 mutation in addition to the Jak1 mutation, and these mutations were shown to act synergistically in vitro (Crescenzo et al., 2015). Notably, in a cell culture assay, Jak1-G1097V did not result in constitutive phosphorylation of STAT3, but rather an enhanced phosphorylation upon IL-6 stimulation (Chen et al., 2017).

Additionally, mutations in the linker between FERM domain and SH2 domain have been described in ALL patients (Flex et al., 2008). While the FERM and SH2 domains have been mainly associated with receptor interaction (Morris et al., 2018), there were also indications for a role in regulation of the kinase activity (Zhou et al., 2001). However, it is not yet understood how the mutations in this linker sequence result in an active Jak1 (Flex et al., 2008).

The first activating Tyk2 mutations have only recently been found in myeloproliferating disorders. The mechanisms leading to Tyk2 hyperactivation are in most cases not yet known. However, mutations are found in the FERM domain, the pseudokinase domain, and the kinase domain, and the molecular consequences are most likely similar to those described for Jak1 and Jak2 mutations (Leitner, Witalisz-Siepracka, Strobl, & Müller, 2017; Sanda et al., 2013; Waanders et al., 2017). Two Tyk2 mutations found in the pseudokinase domain, namely P760L and G761V, are located in the aspartic acid-proline-glycine (DPG) motif (Waanders et al., 2017), which is a conserved motif required for ATP binding (Bandaranayake et al., 2012; Lupardus et al., 2014). Modeling of these variants indicated that both alter the structure of this motif, potentially influencing ATP binding to the pseudokinase domain (Waanders et al., 2017), and loss of ATP binding has previously been reported to suppress pathogenic Jak activation (Hammarén et al., 2015).

An nucleophosmin (Npm1)-Tyk2 fusion protein, that results in constitutive active Tyk2, has been described in lymphoproliferative disorders

(Velusamy et al., 2014). Mechanistically, similar to observations made with Jak2 fusion proteins, the oligomerization domain of Npm1 likely orientates catalytic domains of Tyk2 in a manner that allows for autophosphorylation (Velusamy et al., 2014). Afterwards, other constitutively active Tyk2 fusion proteins have been described, which presumably function in the same manner (Crescenzo et al., 2015; Leitner et al., 2017; Roberts et al., 2014).

In addition to the mutations mentioned here, various other point mutations in Jaks have been described to be associated with different diseases (Chen et al., 2012; Hammarén, Virtanen, Raivola, & Silvennoinen, 2018; Lupardus et al., 2014; Pilati & Zucman-Rossi, 2015). For most of them, however, the detailed molecular mechanism that results in dysregulated Jak activity is not yet known. Nevertheless, as hyperactive Jaks are involved in various pathological conditions, different Jak inhibitors are currently in clinical trials for different diseases (Babon, Lucet, et al., 2014; Garbers et al., 2018). Furthermore, a study with patient-derived xenograft models indicated that an activating Jak1 mutation predicts the efficacy of a Jak inhibitor (Yang et al., 2016) pointing toward a potential use in personalized medicine. Others, however, reported in an in vitro study that mutations in Jak1 do not determine the sensitivity toward Jak inhibition (Chen et al., 2017).

7. Mutations in STAT transcription factors

IL-6 family cytokines activate the transcription factors STAT1 and STAT3 (see Section 2 for more details). Mutations in both STAT proteins have been described, although mutations that constitutively activate the transcription factor and thus do not need an extracellular stimulus that induces the activation of the Jak/STAT signaling cascade are only present in STAT3. In contrast, loss-of-function mutations are described for STAT1, which has important implications for mycobacterial and viral infections (see, e.g., Chapgier et al., 2006). We will therefore focus in the following on mutations described in STAT3 (Fig. 2).

One of the major papers by Koskela et al. described somatic STAT3 mutations in 31 patients with large granular lymphocytic (LGL) leukemia, a lymphoproliferative disorder characterized by expansion of cytotoxic T cells (Koskela et al., 2012). The authors used next-generation exome sequencing and found that mutations (Y640F, D661V, D661Y, and N647I) were all clustered within the Src homology 2 (SH2) domain of STAT3, which is crucial for the ability of the protein to form dimers. The mutations increased the hydrophobicity of the protein, resulting in enhanced dimerization, phosphorylation, and an increased nuclear localization compared to STAT3 wild type (Koskela et al., 2012). Activating STAT3 mutations have also been found in chronic natural killer-cell lymphoproliferative disorder, an LGL subtype (Jerez et al., 2012) and have been later confirmed to be highly specific for large granular lymphocytic leukemia (Fasan et al., 2013). Interestingly, activating STAT3 mutations have also been found outside of the SH2 domain in patients with LGL leukemia (Andersson et al., 2016). Analogous mutations have also been described for STAT5b, although the prevalence appears to be much smaller compared to STAT3 (Rajala, Porkka, Maciejewski, Loughran, & Mustjoki, 2014). Recent research revealed that patients are able to carry multiple STAT3 mutations in different lymphocyte clones, making LGL leukemia a complex and challenging disease (Rajala et al., 2015).

Besides its role in LGL leukemia, STAT mutations have also been described in other diseases involving T cells. Activating mutations of STAT5B and STAT3 have for example been described in NK/T-cell lymphomas and γδ-T-cell lymphomas (Kücük et al., 2015). A single patient with a STAT3-Y640F mutation was also detected among 63 patients with extranodal NK/T-cell lymphoma, although activating mutations in Jak3 were here much more frequent (Sim et al., 2017). An earlier report, which described dominant negative mutations in STAT3, highlighted the importance of STAT3 signaling for the development of Th17 cells (de Beaucoudrey et al., 2008).

Two recent reports consistently show a connection between activating STAT3 mutations and neonatal diabetes (Saarimaki-Vire et al., 2017; Velayos et al., 2017). However, the proposed mechanisms of action appear to differ. Velayos et al. found the point mutation STAT3-P300S within the DNA-binding domain of transcription factor. The mutation led to constitutive activation of STAT3 and decreased insulin expression (Velayos et al., 2017). The study by Saarimaki-Vire et al. analyzed the mutation STAT3-K392R. Using induced pluripotent stem cells from the patient, the authors concluded that the STAT3 mutation causes premature endocrine differentiation, which is the underlying cause for the phenotype (Saarimaki-Vire et al., 2017). The latter patient was earlier described in a study which reported activating germline STAT3 mutations in five patients with early-onset autoimmune disease, which included diabetes type 1 (Flanagan et al., 2014). In a different study, three patients were presented with mutations, which are located in the DNA-binding (K392R and M394T) or dimerization (K658N) domains of STAT3. The patients had autoimmunity, hypogammaglobulinemia, and lymphoproliferation (Haapaniemi et al., 2015).

The authors additionally reported T-cell large granular lymphocytic leukemia in the patient with the K392R mutation (Haapaniemi et al., 2015). Most recently, a novel activating mutation within the coiled-coil domain of STAT3 (K290N) was identified in a patient with common variable immunodeficiency (Russell et al., 2018).

One year before the discovery of STAT3 mutations in LGL leukemia, somatic mutations that activate STAT3 were described in human inflammatory hepatocellular adenomas, a form of benign liver tumors (Pilati et al., 2011). The majority of mutations found were substitutions of single amino acid residues within the SH2 domain of STAT3, suggesting that the mechanism proposed by Koskela et al. would also apply here.

8. Concluding remarks

Cytokines of the IL-6 family have important physiological roles. The probably best known function in this regard is the ability of IL-6 to activate the hepatic acute phase response (Andus et al., 1988), a mechanism that is essential for the defense against pathological bacteria. Interference with IL-6 signaling on hepatocytes is therefore undesired, and recurrent bacterial infections are indeed a major adverse effect when rheumatoid arthritis patients are treated with the IL-6R-blocking antibody tocilizumab (Garbers et al., 2018; Lang et al., 2012).

However, the biological activities of IL-6 and the other cytokines of the family have to be tightly controlled. Uncontrolled and overshooting activities are associated with and can be found in practically all inflammatory human diseases, ranging from the above-mentioned rheumatoid arthritis to several cancer entities. While increased cytokine levels can be targeted with specific antibodies, targeting activating mutations in the key proteins of the gp130/Jak/STAT signaling pathway is quite challenging. This chapter comprehensively summarizes the current knowledge in this field of research.

Interestingly, mutations are known in all involved proteins that are signaling competent, i.e., gp130, the Janus kinases, and the STAT transcription factors. Several Janus kinase inhibitors with different selectivity profiles among the family members have been developed and are already approved for clinical use. This is not the case for therapeutics directed against gp130 or STATs, and whether and how such drugs would be effective for gp130 and STAT variants that contain activating mutations is unclear. Therefore, further research in this direction is warranted.

References

Agthe, M., Brügge, J., Garbers, Y., Wandel, M., Kespohl, B., Arnold, P., et al. (2018). Mutations in craniosynostosis patients cause defective interleukin-11 receptor maturation and drive craniosynostosis-like disease in mice. *Cell Reports*, *25*(1), 10–18. e15.

Andersson, E., Kuusanmaki, H., Bortoluzzi, S., Lagstrom, S., Parsons, A., Rajala, H., et al. (2016). Activating somatic mutations outside the SH2-domain of STAT3 in LGL leukemia. *Leukemia*, *30*(5), 1204–1208.

Andus, T., Geiger, T., Hirano, T., Kishimoto, T., Tran-Thi, T. A., Decker, K., et al. (1988). Regulation of synthesis and secretion of major rat acute-phase proteins by recombinant human interleukin-6 (BSF-2/IL-6) in hepatocyte primary cultures. *European Journal of Biochemistry*, *173*(2), 287–293.

Arita, K., Abe, R., Baba, K., McGrath, J. A., Akiyama, M., & Shimizu, H. (2009). A novel OSMR mutation in familial primary localized cutaneous amyloidosis in a Japanese family. *Journal of Dermatological Science*, *55*(1), 64–65.

Arita, K., South, A. P., Hans-Filho, G., Sakuma, T. H., Lai-Cheong, J., Clements, S., et al. (2008). Oncostatin M receptor-beta mutations underlie familial primary localized cutaneous amyloidosis. *American Journal of Human Genetics*, *82*(1), 73–80.

Babon, J. J., Lucet, I. S., Murphy, J. M., Nicola, N. A., & Varghese, L. N. (2014). The molecular regulation of Janus kinase (JAK) activation. *The Biochemical Journal*, *462*(1), 1–13.

Babon, J. J., Varghese, L. N., & Nicola, N. A. (2014). Inhibition of IL-6 family cytokines by SOCS3. *Seminars in Immunology*, *26*(1), 13–19.

Bandaranayake, R. M., Ungureanu, D., Shan, Y., Shaw, D. E., Silvennoinen, O., & Hubbard, S. R. (2012). Crystal structures of the JAK2 pseudokinase domain and the pathogenic mutant V617F. *Nature Structural & Molecular Biology*, *19*(8), 754–759.

Bazan, J. F. (1989). A novel family of growth factor receptors: A common binding domain in the growth hormone, prolactin, erythropoietin and IL-6 receptors, and the p75 IL-2 receptor beta-chain. *Biochemical and Biophysical Research Communications*, *164*(2), 788–795.

Bazan, J. F. (1990). Haemopoietic receptors and helical cytokines. *Immunology Today*, *11*(10), 350–354.

Bercovich, D., Ganmore, I., Scott, L. M., Wainreb, G., Birger, Y., Elimelech, A., et al. (2008). Mutations of JAK2 in acute lymphoblastic leukaemias associated with Down's syndrome. *Lancet*, *372*(9648), 1484–1492.

Blombery, P., Thompson, E. R., Jones, K., Arnau, G. M., Lade, S., Markham, J. F., et al. (2016). Whole exome sequencing reveals activating JAK1 and STAT3 mutations in breast implant-associated anaplastic large cell lymphoma anaplastic large cell lymphoma. *Haematologica*, *101*(9), e387–e390.

Brischoux-Boucher, E., Trimouille, A., Baujat, G., Goldenberg, A., Schaefer, E., Guichard, B., et al. (2018). IL11RA-related Crouzon-like autosomal recessive craniosynostosis in 10 new patients: Resemblances and differences. *Clinical Genetics*, *94*(3–4), 373–380.

Cabibbo, A., Sporeno, E., Toniatti, C., Altamura, S., Savino, R., Paonessa, G., et al. (1995). Monovalent phage display of human interleukin (hIL)-6: Selection of superbinder variants from a complex molecular repertoire in the hIL-6 D-helix. *Gene*, *167*(1–2), 41–47.

Calderaro, J., Labrune, P., Morcrette, G., Rebouissou, S., Franco, D., Prevot, S., et al. (2013). Molecular characterization of hepatocellular adenomas developed in patients with glycogen storage disease type I. *Journal of Hepatology*, *58*(2), 350–357.

Calderaro, J., Nault, J. C., Balabaud, C., Couchy, G., Saint-Paul, M. C., Azoulay, D., et al. (2016). Inflammatory hepatocellular adenomas developed in the setting of chronic liver disease and cirrhosis. *Modern Pathology*, *29*(1), 43–50.

Chapgier, A., Boisson-Dupuis, S., Jouanguy, E., Vogt, G., Feinberg, J., Prochnicka-Chalufour, A., et al. (2006). Novel STAT1 alleles in otherwise healthy patients with mycobacterial disease. *PLoS Genetics*, *2*(8), e131.

Chen, E., Staudt, L. M., & Green, A. R. (2012). Janus kinase deregulation in leukemia and lymphoma. *Immunity*, *36*(4), 529–541.
Chen, J., Zhang, Y., Petrus, M. N., Xiao, W., Nicolae, A., Raffeld, M., et al. (2017). Cytokine receptor signaling is required for the survival of ALK-anaplastic large cell lymphoma, even in the presence of JAK1/STAT3 mutations. *Proceedings of the National Academy of Sciences of the United States of America*, *114*(15), 3975–3980.
Clarke, C. M., Fok, V. T., Gustafson, J. A., Smyth, M. D., Timms, A. E., Frazar, C. D., et al. (2018). Single suture craniosynostosis: Identification of rare variants in genes associated with syndromic forms. *American Journal of Medical Genetics Part A*, *176*(2), 290–300.
Crabé, S., Guay-Giroux, A., Tormo, A. J., Duluc, D., Lissilaa, R., Guilhot, F., et al. (2009). The IL-27 p 28 subunit binds cytokine-like factor 1 to form a cytokine regulating NK and T cell activities requiring IL-6R for signaling. *Journal of Immunology*, *183*(12), 7692–7702.
Crescenzo, R., Abate, F., Lasorsa, E., Tabbo, F., Gaudiano, M., Chiesa, N., et al. (2015). Convergent mutations and kinase fusions lead to oncogenic STAT3 activation in anaplastic large cell lymphoma. *Cancer Cell*, *27*(4), 516–532.
Croker, B. A., Krebs, D. L., Zhang, J. G., Wormald, S., Willson, T. A., Stanley, E. G., et al. (2003). SOCS3 negatively regulates IL-6 signaling in vivo. *Nature Immunology*, *4*(6), 540–545.
Cunnick, J. M., Meng, S., Ren, Y., Desponts, C., Wang, H. G., Djeu, J. Y., et al. (2002). Regulation of the mitogen-activated protein kinase signaling pathway by SHP2. *The Journal of Biological Chemistry*, *277*(11), 9498–9504.
Davis, S., Aldrich, T. H., Stahl, N., Pan, L., Taga, T., Kishimoto, T., et al. (1993). LIFR beta and gp130 as heterodimerizing signal transducers of the tripartite CNTF receptor. *Science*, *260*(5115), 1805–1808.
de Beaucoudrey, L., Puel, A., Filipe-Santos, O., Cobat, A., Ghandil, P., Chrabieh, M., et al. (2008). Mutations in STAT3 and IL12RB1 impair the development of human IL-17-producing T cells. *The Journal of Experimental Medicine*, *205*(7), 1543–1550.
Decker, T., & Kovarik, P. (2000). Serine phosphorylation of STATs. *Oncogene*, *19*(21), 2628–2637.
Dittrich, A., Quaiser, T., Khouri, C., Gortz, D., Monnigmann, M., & Schaper, F. (2012). Model-driven experimental analysis of the function of SHP-2 in IL-6-induced Jak/STAT signaling. *Molecular BioSystems*, *8*(8), 2119–2134.
Dreuw, A., Radtke, S., Pflanz, S., Lippok, B. E., Heinrich, P. C., & Hermanns, H. M. (2004). Characterization of the signaling capacities of the novel gp130-like cytokine receptor. *The Journal of Biological Chemistry*, *279*(34), 36112–36120.
Ernst, M., Gearing, D. P., & Dunn, A. R. (1994). Functional and biochemical association of Hck with the LIF/IL-6 receptor signal transducing subunit gp130 in embryonic stem cells. *The EMBO Journal*, *13*(7), 1574–1584.
Ernst, M., & Jenkins, B. J. (2004). Acquiring signalling specificity from the cytokine receptor gp130. *Trends in Genetics*, *20*(1), 23–32.
Eulenfeld, R., Dittrich, A., Khouri, C., Muller, P. J., Mutze, B., Wolf, A., et al. (2012). Interleukin-6 signalling: More than Jaks and STATs. *European Journal of Cell Biology*, *91*(6–7), 486–495.
Fasan, A., Kern, W., Grossmann, V., Haferlach, C., Haferlach, T., & Schnittger, S. (2013). STAT3 mutations are highly specific for large granular lymphocytic leukemia. *Leukemia*, *27*(7), 1598–1600.
Ferrao, R., & Lupardus, P. J. (2017). The Janus kinase (JAK) FERM and SH2 domains: Bringing specificity to JAK-receptor interactions. *Frontiers in Endocrinology*, *8*, 71.
Ferreira, R., Freitag, D., Cutler, A., Howson, J. M., Rainbow, D., Smyth, D., et al. (2013). Functional IL6R 358Ala allele impairs classical IL-6 receptor signaling and influences risk of diverse inflammatory diseases. *PLoS Genetics*, *9*(4), e1003444.

Flanagan, S. E., Haapaniemi, E., Russell, M. A., Caswell, R., Allen, H. L., De Franco, E., et al. (2014). Activating germline mutations in STAT3 cause early-onset multi-organ autoimmune disease. *Nature Genetics*, *46*(8), 812–814.

Flex, E., Petrangeli, V., Stella, L., Chiaretti, S., Hornakova, T., Knoops, L., et al. (2008). Somatically acquired JAK1 mutations in adult acute lymphoblastic leukemia. *The Journal of Experimental Medicine*, *205*(4), 751–758.

Galicia, J., Tai, H., Komatsu, Y., Shimada, Y., Akazawa, K., & Yoshie, H. (2004). Polymorphisms in the IL-6 receptor (IL-6R) gene: Strong evidence that serum levels of soluble IL-6R are genetically influenced. *Genes and Immunity*, *5*(6), 513–516.

Garbers, C., Aparicio-Siegmund, S., & Rose-John, S. (2015). The IL-6/gp130/STAT3 signaling axis: Recent advances towards specific inhibition. *Current Opinion in Immunology*, *34*, 75–82.

Garbers, C., Heink, S., Korn, T., & Rose-John, S. (2018). Interleukin-6: Designing specific therapeutics for a complex cytokine. *Nature Reviews. Drug Discovery*, *17*(6), 395–412.

Garbers, C., Hermanns, H., Schaper, F., Müller-Newen, G., Grötzinger, J., Rose-John, S., et al. (2012). Plasticity and cross-talk of interleukin 6-type cytokines. *Cytokine & Growth Factor Reviews*, *23*(3), 85–97.

Garbers, C., Monhasery, N., Aparicio-Siegmund, S., Lokau, J., Baran, P., Nowell, M. A., et al. (2014). The Interleukin-6 receptor Asp358Ala single nucleotide polymorphism rs2228145 confers increased proteolytic conversion rates by ADAM proteases. *Biochimica et Biophysica Acta*, *1842*(9), 1485–1494.

Garbers, C., Spudy, B., Aparicio-Siegmund, S., Waetzig, G. H., Sommer, J., Hölscher, C., et al. (2013). An interleukin-6 receptor-dependent molecular switch mediates signal transduction of the IL-27 cytokine subunit p28 (IL-30) via a gp130 protein receptor homodimer. *The Journal of Biological Chemistry*, *288*(6), 4346–4354.

Gearing, D. P., Thut, C. J., VandeBos, T., Gimpel, S. D., Delaney, P. B., King, J., et al. (1991). Leukemia inhibitory factor receptor is structurally related to the IL-6 signal transducer, gp130. *The EMBO Journal*, *10*(10), 2839–2848.

Gerhartz, C., Heesel, B., Sasse, J., Hemmann, U., Landgraf, C., Schneider-Mergener, J., et al. (1996). Differential activation of acute phase response factor/STAT3 and STAT1 via the cytoplasmic domain of the interleukin 6 signal transducer gp130. I. Definition of a novel phosphotyrosine motif mediating STAT1 activation. *The Journal of Biological Chemistry*, *271*(22), 12991–12998.

Ghilardi, N., Li, J., Hongo, J. A., Yi, S., Gurney, A., & de Sauvage, F. J. (2002). A novel type I cytokine receptor is expressed on monocytes, signals proliferation, and activates STAT-3 and STAT-5. *The Journal of Biological Chemistry*, *277*(19), 16831–16836.

Giordanetto, F., & Kroemer, R. T. (2002). Prediction of the structure of human Janus kinase 2 (JAK2) comprising JAK homology domains 1 through 7. *Protein Engineering*, *15*(9), 727–737.

Godfrey, A. L., & Green, A. R. (2012). Genotype-phenotype interactions in the myeloproliferative neoplasms. *Hematology/Oncology Clinics of North America*, *26*(5), 993–1015.

Griesinger, F., Hennig, H., Hillmer, F., Podleschny, M., Steffens, R., Pies, A., et al. (2005). A BCR-JAK2 fusion gene as the result of a t(9;22)(p24;q11.2) translocation in a patient with a clinically typical chronic myeloid leukemia. *Genes, Chromosomes & Cancer*, *44*(3), 329–333.

Guschin, D., Rogers, N., Briscoe, J., Witthuhn, B., Watling, D., Horn, F., et al. (1995). A major role for the protein tyrosine kinase JAK1 in the JAK/STAT signal transduction pathway in response to interleukin-6. *The EMBO Journal*, *14*(7), 1421–1429.

Haan, C., Behrmann, I., & Haan, S. (2010). Perspectives for the use of structural information and chemical genetics to develop inhibitors of Janus kinases. *Journal of Cellular and Molecular Medicine*, *14*(3), 504–527.

Haapaniemi, E. M., Kaustio, M., Rajala, H. L., van Adrichem, A. J., Kainulainen, L., Glumoff, V., et al. (2015). Autoimmunity, hypogammaglobulinemia, lymphoproliferation, and mycobacterial disease in patients with activating mutations in STAT3. *Blood*, *125*(4), 639–648.

Hallek, M., Neumann, C., Schaffer, M., Danhauser-Riedl, S., von Bubnoff, N., de Vos, G., et al. (1997). Signal transduction of interleukin-6 involves tyrosine phosphorylation of multiple cytosolic proteins and activation of Src-family kinases Fyn, Hck, and Lyn in multiple myeloma cell lines. *Experimental Hematology*, *25*(13), 1367–1377.

Hammarén, H. M., Ungureanu, D., Grisouard, J., Skoda, R. C., Hubbard, S. R., & Silvennoinen, O. (2015). ATP binding to the pseudokinase domain of JAK2 is critical for pathogenic activation. *Proceedings of the National Academy of Sciences of the United States of America*, *112*(15), 4642–4647.

Hammarén, H. M., Virtanen, A. T., Raivola, J., & Silvennoinen, O. (2018). The regulation of JAKs in cytokine signaling and its breakdown in disease. *Cytokine*. https://doi.org/10.1016/j.cyto.2018.03.041. in press.

Harmegnies, D., Wang, X. M., Vandenbussche, P., Leon, A., Vusio, P., Grötzinger, J., et al. (2003). Characterization of a potent human interleukin-11 agonist. *The Biochemical Journal*, *375*(Pt. 1), 23–32.

Heinrich, P. C., Behrmann, I., Haan, S., Hermanns, H. M., Müller-Newen, G., & Schaper, F. (2003). Principles of interleukin (IL)-6-type cytokine signalling and its regulation. *The Biochemical Journal*, *374*(Pt. 1), 1–20.

Hermanns, H. M. (2015). Oncostatin M and interleukin-31: Cytokines, receptors, signal transduction and physiology. *Cytokine & Growth Factor Reviews*, *26*(5), 545–558.

Hornakova, T., Staerk, J., Royer, Y., Flex, E., Tartaglia, M., Constantinescu, S. N., et al. (2009). Acute lymphoblastic leukemia-associated JAK1 mutants activate the Janus kinase/STAT pathway via interleukin-9 receptor alpha homodimers. *The Journal of Biological Chemistry*, *284*(11), 6773–6781.

James, C., Ugo, V., Le Couédic, J. P., Staerk, J., Delhommeau, F., Lacout, C., et al. (2005). A unique clonal JAK2 mutation leading to constitutive signalling causes polycythaemia vera. *Nature*, *434*(7037), 1144–1148.

Jeong, E. G., Kim, M. S., Nam, H. K., Min, C. K., Lee, S., Chung, Y. J., et al. (2008). Somatic mutations of JAK1 and JAK3 in acute leukemias and solid cancers. *Clinical Cancer Research*, *14*(12), 3716–3721.

Jerez, A., Clemente, M. J., Makishima, H., Koskela, H., Leblanc, F., Peng Ng, K., et al. (2012). STAT3 mutations unify the pathogenesis of chronic lymphoproliferative disorders of NK cells and T-cell large granular lymphocyte leukemia. *Blood*, *120*(15), 3048–3057.

Judd, L. M., Alderman, B. M., Howlett, M., Shulkes, A., Dow, C., Moverley, J., et al. (2004). Gastric cancer development in mice lacking the SHP2 binding site on the IL-6 family co-receptor gp130. *Gastroenterology*, *126*(1), 196–207.

Kaptein, A., Paillard, V., & Saunders, M. (1996). Dominant negative stat3 mutant inhibits interleukin-6-induced Jak-STAT signal transduction. *The Journal of Biological Chemistry*, *271*(11), 5961–5964.

Kearney, L., Gonzalez De Castro, D., Yeung, J., Procter, J., Horsley, S. W., Eguchi-Ishimae, M., et al. (2009). Specific JAK2 mutation (JAK2R683) and multiple gene deletions in Down syndrome acute lymphoblastic leukemia. *Blood*, *113*(3), 646–648.

Kershaw, N. J., Laktyushin, A., Nicola, N. A., & Babon, J. J. (2014). Reconstruction of an active SOCS3-based E3 ubiquitin ligase complex in vitro: Identification of the active components and JAK2 and gp130 as substrates. *Growth Factors*, *32*(1), 1–10.

Keupp, K., Li, Y., Vargel, I., Hoischen, A., Richardson, R., Neveling, K., et al. (2013). Mutations in the interleukin receptor IL11RA cause autosomal recessive Crouzon-like craniosynostosis. *Molecular Genetics & Genomic Medicine*, *1*(4), 223–237.

Kosfeld, A., Brand, F., Weiss, A. C., Kreuzer, M., Goerk, M., Martens, H., et al. (2017). Mutations in the leukemia inhibitory factor receptor (LIFR) gene and Lifr deficiency cause urinary tract malformations. *Human Molecular Genetics, 26*(9), 1716–1731.

Koskela, H., Eldfors, S., Ellonen, P., van Adrichem, A., Kuusanmäki, H., Andersson, E., et al. (2012). Somatic STAT3 mutations in large granular lymphocytic leukemia. *The New England Journal of Medicine, 366*(20), 1905–1913.

Kralovics, R., Passamonti, F., Buser, A. S., Teo, S. S., Tiedt, R., Passweg, J. R., et al. (2005). A gain-of-function mutation of JAK2 in myeloproliferative disorders. *The New England Journal of Medicine, 352*(17), 1779–1790.

Kücük, C., Jiang, B., Hu, X., Zhang, W., Chan, J. K., Xiao, W., et al. (2015). Activating mutations of STAT5B and STAT3 in lymphomas derived from gammadelta-T or NK cells. *Nature Communications, 6*, 6025.

Lacronique, V., Boureux, A., Valle, V. D., Poirel, H., Quang, C. T., Mauchauffé, M., et al. (1997). A TEL-JAK2 fusion protein with constitutive kinase activity in human leukemia. *Science, 278*(5341), 1309–1312.

Lang, V. R., Englbrecht, M., Rech, J., Nusslein, H., Manger, K., Schuch, F., et al. (2012). Risk of infections in rheumatoid arthritis patients treated with tocilizumab. *Rheumatology, 51*(5), 852–857.

Lanktree, M. B., Guo, Y., Murtaza, M., Glessner, J. T., Bailey, S. D., Onland-Moret, C. N., et al. (2011). Meta-analysis of Dense Genecentric Association studies reveals common and uncommon variants associated with height. *American Journal of Human Genetics, 88*(1), 6–18.

Lehmann, U., Schmitz, J., Weissenbach, M., Sobota, R. M., Hortner, M., Friederichs, K., et al. (2003). SHP2 and SOCS3 contribute to Tyr-759-dependent attenuation of interleukin-6 signaling through gp130. *The Journal of Biological Chemistry, 278*(1), 661–671.

Leitner, N. R., Witalisz-Siepracka, A., Strobl, B., & Müller, M. (2017). Tyrosine kinase 2—Surveillant of tumours and bona fide oncogene. *Cytokine, 89*, 209–218.

Lin, M. W., Lee, D. D., Liu, T. T., Lin, Y. F., Chen, S. Y., Huang, C. C., et al. (2010). Novel IL31RA gene mutation and ancestral OSMR mutant allele in familial primary cutaneous amyloidosis. *European Journal of Human Genetics, 18*(1), 26–32.

Lokau, J., Agthe, M., Flynn, C. M., & Garbers, C. (2017). Proteolytic control of interleukin-11 and interleukin-6 biology. *Biochimica et Biophysica Acta, 1864*(11 Pt. B), 2105–2117.

Lokau, J., Göttert, S., Arnold, P., Düsterhöft, S., Massa López, D., Grötzinger, J., et al. (2017). The SNP rs4252548 (R112H) which is associated with reduced human height compromises the stability of IL-11. *Biochimica et Biophysica Acta, 1865*(3), 496–506.

Lu, X., Levine, R., Tong, W., Wernig, G., Pikman, Y., Zarnegar, S., et al. (2005). Expression of a homodimeric type I cytokine receptor is required for JAK2V617F-mediated transformation. *Proceedings of the National Academy of Sciences of the United States of America, 102*(52), 18962–18967.

Lupardus, P. J., Skiniotis, G., Rice, A. J., Thomas, C., Fischer, S., Walz, T., et al. (2011). Structural snapshots of full-length Jak1, a transmembrane gp130/IL-6/IL-6Ralpha cytokine receptor complex, and the receptor-Jak1 holocomplex. *Structure, 19*(1), 45–55.

Lupardus, P. J., Ultsch, M., Wallweber, H., Bir Kohli, P., Johnson, A. R., & Eigenbrot, C. (2014). Structure of the pseudokinase–kinase domains from protein kinase TYK2 reveals a mechanism for Janus kinase (JAK) autoinhibition. *Proceedings of the National Academy of Sciences, 111*(22), 8025–8030.

Lütticken, C., Wegenka, U. M., Yuan, J., Buschmann, J., Schindler, C., Ziemiecki, A., et al. (1994). Association of transcription factor APRF and protein kinase Jak1 with the interleukin-6 signal transducer gp130. *Science, 263*(5143), 89–92.

Ma, W., Kantarjian, H., Zhang, X., Yeh, C. H., Zhang, Z. J., Verstovsek, S., et al. (2009). Mutation profile of JAK2 transcripts in patients with chronic myeloproliferative neoplasias. *The Journal of Molecular Diagnostics, 11*(1), 49–53.

Marouli, E., Graff, M., Medina-Gomez, C., Lo, K. S., Wood, A. R., Kjaer, T. R., et al. (2017). Rare and low-frequency coding variants alter human adult height. *Nature*, *542*(7640), 186–190.
Matadeen, R., Hon, W.-C., Heath, J. K., Jones, Y. E., & Fuller, S. (2007). The dynamics of signal triggering in a gp130-receptor complex. *Structure*, *15*(4), 441–448.
Mercher, T., Wernig, G., Moore, S. A., Levine, R. L., Gu, T. L., Frohling, S., et al. (2006). JAK2T875N is a novel activating mutation that results in myeloproliferative disease with features of megakaryoblastic leukemia in a murine bone marrow transplantation model. *Blood*, *108*(8), 2770–2779.
Mikelonis, D., Jorcyk, C. L., Tawara, K., & Oxford, J. T. (2014). Stüve-Wiedemann syndrome: LIFR and associated cytokines in clinical course and etiology. *Orphanet Journal of Rare Diseases*, *9*, 34.
Monhasery, N., Moll, J., Cuman, C., Franke, M., Lamertz, L., Nitz, R., et al. (2016). Transcytosis of IL-11 and apical redirection of gp130 is mediated by IL-11alpha receptor. *Cell Reports*, *16*(4), 1067–1081.
Moroishi, T., Hansen, C. G., & Guan, K. L. (2015). The emerging roles of YAP and TAZ in cancer. *Nature Reviews. Cancer*, *15*(2), 73–79.
Morris, R., Kershaw, N. J., & Babon, J. J. (2018). The molecular details of cytokine signalling via the JAK/STAT pathway. *Protein Science*, *27*(12), 1984–2009.
Mullighan, C. G., Zhang, J., Harvey, R. C., Collins-Underwood, J. R., Schulman, B. A., Phillips, L. A., et al. (2009). JAK mutations in high-risk childhood acute lymphoblastic leukemia. *Proceedings of the National Academy of Sciences of the United States of America*, *106*(23), 9414–9418.
Nebral, K., Denk, D., Attarbaschi, A., Konig, M., Mann, G., Haas, O. A., et al. (2009). Incidence and diversity of PAX5 fusion genes in childhood acute lymphoblastic leukemia. *Leukemia*, *23*(1), 134–143.
Nieminen, P., Morgan, N., Fenwick, A., Parmanen, S., Veistinen, L., Mikkola, M., et al. (2011). Inactivation of IL11 signaling causes craniosynostosis, delayed tooth eruption, and supernumerary teeth. *American Journal of Human Genetics*, *89*(1), 67–81.
Owaki, T., Asakawa, M., Morishima, N., Mizoguchi, I., Fukai, F., Takeda, K., et al. (2008). STAT3 is indispensable to IL-27-mediated cell proliferation but not to IL-27-induced Th1 differentiation and suppression of proinflammatory cytokine production. *Journal of Immunology*, *180*(5), 2903–2911.
Pennica, D., Shaw, K. J., Swanson, T. A., Moore, M. W., Shelton, D. L., Zioncheck, K. A., et al. (1995). Cardiotrophin-1. Biological activities and binding to the leukemia inhibitory factor receptor/gp130 signaling complex. *The Journal of Biological Chemistry*, *270*(18), 10915–10922.
Pflanz, S., Hibbert, L., Mattson, J., Rosales, R., Vaisberg, E., Bazan, J. F., et al. (2004). WSX-1 and glycoprotein 130 constitute a signal-transducing receptor for IL-27. *Journal of Immunology*, *172*(4), 2225–2231.
Pilati, C., Amessou, M., Bihl, M., Balabaud, C., Nhieu, J., Paradis, V., et al. (2011). Somatic mutations activating STAT3 in human inflammatory hepatocellular adenomas. *The Journal of Experimental Medicine*, *208*(7), 1359–1366.
Pilati, C., Letouzé, E., Nault, J. C., Imbeaud, S., Boulai, A., Calderaro, J., et al. (2014). Genomic profiling of hepatocellular adenomas reveals recurrent FRK-activating mutations and the mechanisms of malignant transformation. *Cancer Cell*, *25*(4), 428–441.
Pilati, C., & Zucman-Rossi, J. (2015). Mutations leading to constitutive active gp130/JAK1/STAT3 pathway. *Cytokine & Growth Factor Reviews*, *26*(5), 499–506.
Poitras, J. L., Dal Cin, P., Aster, J. C., Deangelo, D. J., & Morton, C. C. (2008). Novel SSBP2-JAK2 fusion gene resulting from a t(5;9)(q14.1;p24.1) in pre-B acute lymphocytic leukemia. *Genes, Chromosomes & Cancer*, *47*(10), 884–889.

Poussin, K., Pilati, C., Couchy, G., Calderaro, J., Bioulac-Sage, P., Bacq, Y., et al. (2013). Biochemical and functional analyses of gp130 mutants unveil JAK1 as a novel therapeutic target in human inflammatory hepatocellular adenoma. *Oncoimmunology*, *2*(12), e27090.

Rafiq, S., Frayling, T., Murray, A., Hurst, A., Stevens, K., Weedon, M., et al. (2007). A common variant of the interleukin 6 receptor (IL-6r) gene increases IL-6r and IL-6 levels, without other inflammatory effects. *Genes and Immunity*, *8*(7), 552–559.

Rajala, H. L., Olson, T., Clemente, M. J., Lagstrom, S., Ellonen, P., Lundan, T., et al. (2015). The analysis of clonal diversity and therapy responses using STAT3 mutations as a molecular marker in large granular lymphocytic leukemia. *Haematologica*, *100*(1), 91–99.

Rajala, H. L., Porkka, K., Maciejewski, J. P., Loughran, T. P., Jr., & Mustjoki, S. (2014). Uncovering the pathogenesis of large granular lymphocytic leukemia-novel STAT3 and STAT5b mutations. *Annals of Medicine*, *46*(3), 114–122.

Rebouissou, S., Amessou, M., Couchy, G., Poussin, K., Imbeaud, S., Pilati, C., et al. (2009). Frequent in-frame somatic deletions activate gp130 in inflammatory hepatocellular tumours. *Nature*, *457*(7226), 200–204.

Reiter, A., Walz, C., Watmore, A., Schoch, C., Blau, I., Schlegelberger, B., et al. (2005). The t(8;9)(p22;p24) is a recurrent abnormality in chronic and acute leukemia that fuses PCM1 to JAK2. *Cancer Research*, *65*(7), 2662–2667.

Riethmueller, S., Somasundaram, P., Ehlers, J. C., Hung, C.-W. W., Flynn, C. M., Lokau, J., et al. (2017). Proteolytic origin of the soluble human IL-6R in vivo and a decisive role of N-glycosylation. *PLoS Biology*, *15*(1), e2000080.

Rinis, N., Kuster, A., Schmitz-Van de Leur, H., Mohr, A., & Müller-Newen, G. (2014). Intracellular signaling prevents effective blockade of oncogenic gp130 mutants by neutralizing antibodies. *Cell Communication and Signaling*, *12*, 14.

Roberts, K. G., Li, Y., Payne-Turner, D., Harvey, R. C., Yang, Y. L., Pei, D., et al. (2014). Targetable kinase-activating lesions in pH-like acute lymphoblastic leukemia. *The New England Journal of Medicine*, *371*(11), 1005–1015.

Rodig, S. J., Meraz, M. A., White, M. J., Lampe, P. A., Riley, J. K., Arthur, C. D., et al. (1998). Disruption of the Jak1 gene demonstrates obligatory and nonredundant roles of the Jaks in cytokine-induced biologic responses. *Cell*, *93*(3), 373–383.

Russell, M. A., Pigors, M., Houssen, M. E., Manson, A., Kelsell, D., Longhurst, H., et al. (2018). A novel de novo activating mutation in STAT3 identified in a patient with common variable immunodeficiency (CVID). *Clinical Immunology*, *187*, 132–136.

Saarimaki-Vire, J., Balboa, D., Russell, M. A., Saarikettu, J., Kinnunen, M., Keskitalo, S., et al. (2017). An activating STAT3 mutation causes neonatal diabetes through premature induction of pancreatic differentiation. *Cell Reports*, *19*(2), 281–294.

Sanda, T., Tyner, J. W., Gutierrez, A., Ngo, V. N., Glover, J., Chang, B. H., et al. (2013). TYK2-STAT1-BCL2 pathway dependence in T-cell acute lymphoblastic leukemia. *Cancer Discovery*, *3*(5), 564–577.

Sano, E., Takei, T., Ueda, T., & Tsumoto, K. (2017). Production and characterization of genetically modified human IL-11 variants. *Biochimica et Biophysica Acta—General Subjects*, *1861*(2), 205–217.

Sarwar, N., Butterworth, A. S., Freitag, D. F., Gregson, J., Willeit, P., Gorman, D. N., et al. (2012). Interleukin-6 receptor pathways in coronary heart disease: A collaborative meta-analysis of 82 studies. *Lancet*, *379*(9822), 1205–1213.

Sasaki, A., Yasukawa, H., Suzuki, A., Kamizono, S., Syoda, T., Kinjyo, I., et al. (1999). Cytokine-inducible SH2 protein-3 (CIS3/SOCS3) inhibits Janus tyrosine kinase by binding through the N-terminal kinase inhibitory region as well as SH2 domain. *Genes to Cells*, *4*(6), 339–351.

Schaeffer, M., Schneiderbauer, M., Weidler, S., Tavares, R., Warmuth, M., de Vos, G., et al. (2001). Signaling through a novel domain of gp130 mediates cell proliferation and activation of Hck and Erk kinases. *Molecular and Cellular Biology*, *21*(23), 8068–8081.

Schaper, F., Gendo, C., Eck, M., Schmitz, J., Grimm, C., Anhuf, D., et al. (1998). Activation of the protein tyrosine phosphatase SHP2 via the interleukin-6 signal transducing receptor protein gp130 requires tyrosine kinase Jak1 and limits acute-phase protein expression. *The Biochemical Journal*, *335*(Pt. 3), 557–565.

Scheller, J., & Rose-John, S. (2012). The interleukin 6 pathway and atherosclerosis. *Lancet*, *380*(9839), 338.

Schmidt-Arras, D., Müller, M., Stevanovic, M., Horn, S., Schütt, A., Bergmann, J., et al. (2013). Oncogenic deletion mutants of gp130 signal from intracellular compartments. *Journal of Cell Science*, *127*(15), 341–353.

Schmitz, J., Weissenbach, M., Haan, S., Heinrich, P. C., & Schaper, F. (2000). SOCS3 exerts its inhibitory function on interleukin-6 signal transduction through the SHP2 recruitment site of gp130. *The Journal of Biological Chemistry*, *275*(17), 12848–12856.

Schütt, A., Zacharias, M., Schneider, N., Horn, S., Grötzinger, J., Rose-John, S., et al. (2013). gp130 activation is regulated by D2-D3 interdomain connectivity. *The Biochemical Journal*, *450*(3), 487–496.

Scott, L. M. (2011). The JAK2 exon 12 mutations: A comprehensive review. *American Journal of Hematology*, *86*(8), 668–676.

Scott, L. M., Tong, W., Levine, R. L., Scott, M. A., Beer, P. A., Stratton, M. R., et al. (2007). JAK2 exon 12 mutations in polycythemia vera and idiopathic erythrocytosis. *The New England Journal of Medicine*, *356*(5), 459–468.

Shuai, K., Stark, G. R., Kerr, I. M., & Darnell, J. E., Jr. (1993). A single phosphotyrosine residue of Stat91 required for gene activation by interferon-gamma. *Science*, *261*(5129), 1744–1746.

Sim, S. H., Kim, S., Kim, T. M., Jeon, Y. K., Nam, S. J., Ahn, Y. O., et al. (2017). Novel JAK3-activating mutations in extranodal NK/T-cell lymphoma, nasal type. *The American Journal of Pathology*, *187*(5), 980–986.

Sims, N. A. (2016). Cell-specific paracrine actions of IL-6 family cytokines from bone, marrow and muscle that control bone formation and resorption. *The International Journal of Biochemistry & Cell Biology*, *79*, 14–23.

Sohn, S. J., Barrett, K., Van Abbema, A., Chang, C., Kohli, P. B., Kanda, H., et al. (2013). A restricted role for TYK2 catalytic activity in human cytokine responses revealed by novel TYK2-selective inhibitors. *The Journal of Immunology*, *191*(5), 2205–2216.

Sommer, J., Effenberger, T., Volpi, E., Waetzig, G., Bernhardt, M., Suthaus, J., et al. (2012). Constitutively active mutant gp130 receptor protein from inflammatory hepatocellular adenoma is inhibited by an anti-gp130 antibody that specifically neutralizes interleukin 11 signaling. *The Journal of Biological Chemistry*, *287*(17), 13743–13751.

Stahl, N., Boulton, T. G., Farruggella, T., Ip, N. Y., Davis, S., Witthuhn, B. A., et al. (1994). Association and activation of Jak-Tyk kinases by CNTF-LIF-OSM-IL-6 beta receptor components. *Science*, *263*(5143), 92–95.

Stephens, O. W., Zhang, Q., Qu, P., Zhou, Y., Chavan, S., Tian, E., et al. (2012). An intermediate-risk multiple myeloma subgroup is defined by sIL-6r: Levels synergistically increase with incidence of SNP rs2228145 and 1q21 amplification. *Blood*, *119*(2), 503–512.

Swerdlow, D., Holmes, M., Kuchenbaecker, K., Engmann, J., Shah, T., Sofat, R., et al. (2012). The interleukin-6 receptor as a target for prevention of coronary heart disease: A mendelian randomisation analysis. *Lancet*, *379*(9822), 1214–1224.

Takahashi, Y., Takahashi, M., Carpino, N., Jou, S. T., Chao, J. R., Tanaka, S., et al. (2008). Leukemia inhibitory factor regulates trophoblast giant cell differentiation via Janus kinase 1-signal transducer and activator of transcription 3-suppressor of cytokine signaling 3 pathway. *Molecular Endocrinology*, *22*(7), 1673–1681.

Takahashi-Tezuka, M., Yoshida, Y., Fukada, T., Ohtani, T., Yamanaka, Y., Nishida, K., et al. (1998). Gab1 acts as an adapter molecule linking the cytokine receptor gp130 to ERK mitogen-activated protein kinase. *Molecular and Cellular Biology*, *18*(7), 4109–4117.

Tamm, C., Böwer, N., & Anneren, C. (2011). Regulation of mouse embryonic stem cell self-renewal by a Yes-YAP-TEAD2 signaling pathway downstream of LIF. *Journal of Cell Science*, *124*, 1136–1144. Pt. 7.

Tanaka, A., Lai-Cheong, J. E., van den Akker, P. C., Nagy, N., Millington, G., Diercks, G. F., et al. (2010). The molecular skin pathology of familial primary localized cutaneous amyloidosis. *Experimental Dermatology*, *19*(5), 416–423.

Taniguchi, K., Moroishi, T., de Jong, P. R., Krawczyk, M., Grebbin, B. M., Luo, H., et al. (2017). YAP-IL-6ST autoregulatory loop activated on APC loss controls colonic tumorigenesis. *Proceedings of the National Academy of Sciences of the United States of America*, *114*(7), 1643–1648.

Taniguchi, K., Wu, L.-W., Grivennikov, S. I., de Jong, P. R., Lian, I., Yu, F.-X., et al. (2015). A gp130-Src-YAP module links inflammation to epithelial regeneration. *Nature*, *519*(7541), 57–62.

Tebbutt, N., Giraud, A., Inglese, M., Jenkins, B., Waring, P., Clay, F., et al. (2002). Reciprocal regulation of gastrointestinal homeostasis by SHP2 and STAT-mediated trefoil gene activation in gp130 mutant mice. *Nature Medicine*, *8*(10), 1089–1097.

Toniatti, C., Cabibbo, A., Sporena, E., Salvati, A. L., Cerretani, M., Serafini, S., et al. (1996). Engineering human interleukin-6 to obtain variants with strongly enhanced bioactivity. *The EMBO Journal*, *15*(11), 2726–2737.

Ungureanu, D., Wu, J., Pekkala, T., Niranjan, Y., Young, C., Jensen, O. N., et al. (2011). The pseudokinase domain of JAK2 is a dual-specificity protein kinase that negatively regulates cytokine signaling. *Nature Structural & Molecular Biology*, *18*(9), 971–976.

Vainchenker, W., & Constantinescu, S. N. (2013). JAK/STAT signaling in hematological malignancies. *Oncogene*, *32*(21), 2601–2613.

Velayos, T., Martinez, R., Alonso, M., Garcia-Etxebarria, K., Aguayo, A., Camarero, C., et al. (2017). An activating mutation in STAT3 results in neonatal diabetes through reduced insulin synthesis. *Diabetes*, *66*(4), 1022–1029.

Velusamy, T., Kiel, M. J., Sahasrabuddhe, A. A., Rolland, D., Dixon, C. A., Bailey, N. G., et al. (2014). A novel recurrent NPM1-TYK2 gene fusion in cutaneous CD30-positive lymphoproliferative disorders. *Blood*, *124*(25), 3768–3771.

Waage, A., Brandtzaeg, P., Halstensen, A., Kierulf, P., & Espevik, T. (1989). The complex pattern of cytokines in serum from patients with meningococcal septic shock. Association between interleukin 6, interleukin 1, and fatal outcome. *The Journal of Experimental Medicine*, *169*(1), 333–338.

Waanders, E., Scheijen, B., Jongmans, M. C., Venselaar, H., van Reijmersdal, S. V., van Dijk, A. H., et al. (2017). Germline activating TYK2 mutations in pediatric patients with two primary acute lymphoblastic leukemia occurrences. *Leukemia*, *31*(4), 821–828.

Wallweber, H. J., Tam, C., Franke, Y., Starovasnik, M. A., & Lupardus, P. J. (2014). Structural basis of recognition of interferon-alpha receptor by tyrosine kinase 2. *Nature Structural & Molecular Biology*, *21*(5), 443–448.

Wang, W. H., Li, L. F., Huang, E. S., Zhang, Q., Sun, T. T., Song, Q. H., et al. (2012). A new c.1845A->T of oncostatin M receptor-beta mutation and slightly enhanced oncostatin M receptor-beta expression in a Chinese family with primary localized cutaneous amyloidosis. *European Journal of Dermatology*, *22*(1), 29–33.

Wen, Z., Zhong, Z., & Darnell, J. E., Jr. (1995). Maximal activation of transcription by Stat1 and Stat3 requires both tyrosine and serine phosphorylation. *Cell*, *82*(2), 241–250.

Yanaka, S., Sano, E., Naruse, N., Miura, K., Futatsumori-Sugai, M., Caaveiro, J. M., et al. (2011). Non-core region modulates interleukin-11 signaling activity: Generation of agonist and antagonist variants. *The Journal of Biological Chemistry*, *286*(10), 8085–8093.

Yang, S., Luo, C., Gu, Q., Xu, Q., Wang, G., Sun, H., et al. (2016). Activating JAK1 mutation may predict the sensitivity of JAK-STAT inhibition in hepatocellular carcinoma. *Oncotarget*, 7(5), 5461–5469.

Yao, H., Ma, Y., Hong, Z., Zhao, L., Monaghan, S. A., Hu, M. C., et al. (2017). Activating JAK2 mutants reveal cytokine receptor coupling differences that impact outcomes in myeloproliferative neoplasm. *Leukemia*, *31*(10), 2122–2131.

Yin, T., Shen, R., Feng, G. S., & Yang, Y. C. (1997). Molecular characterization of specific interactions between SHP-2 phosphatase and JAK tyrosine kinases. *The Journal of Biological Chemistry*, *272*(2), 1032–1037.

Yoshimura, A., Naka, T., & Kubo, M. (2007). SOCS proteins, cytokine signalling and immune regulation. *Nature Reviews. Immunology*, 7(6), 454–465.

Zhang, X., Blenis, J., Li, H. C., Schindler, C., & Chen-Kiang, S. (1995). Requirement of serine phosphorylation for formation of STAT-promoter complexes. *Science*, *267*(5206), 1990–1994.

Zhong, Z., Wen, Z., & Darnell, J. E. (1994). Stat3: A STAT family member activated by tyrosine phosphorylation in response to epidermal growth factor and interleukin-6. *Science*, *264*(5155), 95–98.

Zhou, Y. J., Chen, M., Cusack, N. A., Kimmel, L. H., Magnuson, K. S., Boyd, J. G., et al. (2001). Unexpected effects of FERM domain mutations on catalytic activity of Jak3: Structural implication for Janus kinases. *Molecular Cell*, *8*(5), 959–969.

CHAPTER EIGHT

Aquaporin water channels: New perspectives on the potential role in inflammation

Margherita Sisto*, Domenico Ribatti, Sabrina Lisi
Department of Basic Medical Sciences, Neurosciences and Sensory Organs (SMBNOS), Section of Human Anatomy and Histology, University of Bari "Aldo Moro", Bari, Italy
*Corresponding author: e-mail address: margherita.sisto@uniba.it

Contents

Abstract

Aquaporins (AQPs) are a family of membrane water channel proteins that osmotically modulate water fluid homeostasis in several tissues; some of them also transport small solutes such as glycerol. At the cellular level, the AQPs regulate not only cell migration and transepithelial fluid transport across membranes, but also common events that are crucial for the inflammatory response. Emerging data reveal a new function of AQPs in

Advances in Protein Chemistry and Structural Biology, Volume 116
ISSN 1876-1623
https://doi.org/10.1016/bs.apcsb.2018.11.010

the inflammatory process, as demonstrated by their dysregulation in a wide range of inflammatory diseases including edematous states, cancer, obesity, wound healing and several autoimmune diseases. This chapter summarizes the discoveries made so far about the structure and functions of the AQPs and provides updated information on the underlying mechanisms of AQPs in several human inflammatory diseases. The discovery of new functions for AQPs opens new vistas offering promise for the discovery of mechanisms and therapeutic opportunities in inflammatory disorders.

1. Introduction

The inflammatory process and the complex mechanisms governing its regulation have attracted strong scientific interest in the last years. Although the pathological aspects of many types of inflammation have been well described, their physiological functions are mostly unknown. Inflammation is characterized by a wide variety of physiological and pathological processes involving the body's physiological response to injury of tissues and cells. A large range of adverse conditions orchestrates all inflammatory responses, in which a pivotal role is played by the immune system that triggers the recruitment of leukocytes and plasma proteins to the involved tissue site. By contrast, when the inflammatory process becomes prolonged and persistent over time it leads to maladaptive responses and attempts to repair the tissue target (Wynn & Vannella, 2016). Indeed, failure to resolve this condition leads to chronic inflammation, extensive tissue destruction and progressive fibrosis. Such persistent inflammation is linked to many chronic human conditions and diseases, including allergy, atherosclerosis, cancer, arthritis and autoimmune diseases (Chen et al., 2017; Medzhitov, 2010). Inflammatory mediators and regulatory pathways influence the start of inflammatory process beginning from the dysregulation of cellular and tissue homeostasis: a critical role is played by the membranes that contribute to maintain the microenvironment in a state of equilibrium. In fact, the cells regulate their internal microenvironment in response to external states, and the loss of their ability to modulate the movement of fluid across the membrane leads to intense alterations of cell physiology (Brocker, Thompson, & Vasiliou, 2012; Medzhitov, 2008; Verbalis, 2003). In addition, local or systemic inflammation could be characterized by perturbed homeostasis; this process was shown to be linked to irregular ion transport (Kotas & Medzhitov, 2015). Many inflammatory mediators, such as cytokines and chemokines, can change homeostasis of the cells by varying their susceptibility to homeostatic signals or modulating the channel access

(Medzhitov, 2008). A clear sign of homeostasis disturbance and inflammation is the swelling of cells and tissues together with a remarkable increase of extracellular fluids (edema) in the inflammation site (Meli, Pirozzi, & Pelagalli, 2018).

Aquaporins (AQPs) are a family of water channel molecules that contribute to water transport through cell membranes in response to osmotic gradients. They are thus convenient biomarkers, since they regulate not only cell migration, but also common events that are essential to the inflammatory response (Agre, 2006; Day et al., 2014; Verkman, 2013; Verkman, Anderson, & Papadopoulos, 2014). The involvement of AQPs in several inflammatory conditions, as discovered by their dysregulation in many human diseases, reveals a new, interesting role in the protection and response to different harmful stimuli, including bacterial infection. Moreover, recent data have addressed their possible involvement in inflammatory processes, participating as regulators of the innate host defense at cell membrane level (Rump & Adamzik, 2018; Verkman, 2013).

This chapter discusses the background, discovery, and function of AQPs in various bodily processes, especially focusing on the potential role of AQPs in the inflammatory process, as an adaptive response to the loss of cellular homeostasis or tissue damage. The function or alteration of AQPs in different inflammatory diseases is also considered.

2. AQPs structure, function and regulation

AQPs belong to a highly conserved group of membrane proteins that form a large family comprising a great number of integral membrane proteins found in a wide range of organisms (Abascal, Irisarri, & Zardoya, 2014). Three families of AQPs have so far been identified, called classical aquaporins, aquaglyceroporins, that are permeable to water and glycerol, and other small solutes in addition to water, respectively, and superaquaporins (S-aquaporins) or subcellular aquaporins, that are only detectable in animals (Ishibashi, Tanaka, & Morishita, 2014) and have an uncertain permeability. In 1992, the molecular-structure property of the first water channel, AQP1, was firstly described in the erythrocyte plasma membrane (Preston, Smith, Zeidel, Moulds, & Agre, 1994). AQP1 Sequencing showed its homology to the major intrinsic protein (MIP) in the plasma membrane of eye lens fiber cells, which was subsequently renamed AQP0. The study of AQP1 structure rapidly led to the identification of several other AQPs in animals, plants, yeast, and bacteria, revealing at least 13 AQPs in humans.

2.1 Characteristics of AQPs structure

Cell membranes are fairly waterproof, forming a barrier that resists the crossing of water molecules. Some cells, however, need to allow more water to flow through the membrane. For instance, the concentration of waste in the kidneys and the internal pressure of the eye demand accurate control of water. In these cases, cells use AQPs to control the flow of water in and out of the cell. AQPs have a characteristic basic structure, in which AQPs monomers consist of six transmembrane helical segments and two short helical segments that surround cytoplasmic and extracellular vestibules connected by a narrow aqueous pore (Fig. 1) (Kruse, Uehlein, & Kaldenhoff, 2006). AQPs are composed of four identical chains, each with its own channel at the center of the subunit; each 28-kDa subunit in the AQPs homotetramer contains an individual aqueous pore, which allows single-file water transport in which steric factors ensure water selectivity (Khalili-Araghi et al., 2009). The larger hole formed at the center of the four chains is lined with carbon-rich amino acids and is presumably plugged by membrane lipids in the cell. A recent, very high resolution analysis of the structure of AQPs revealed that two charged amino acids at the entry to the water channel are important to ensure that only water, and not ions,

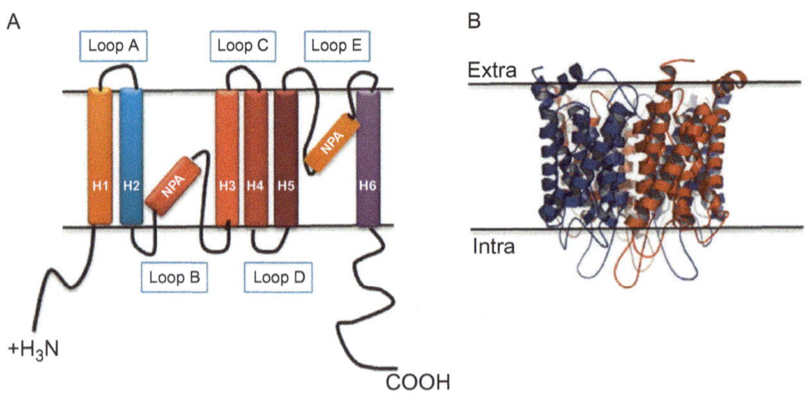

Fig. 1 (A) Schematic diagram of the shared structural architecture of AQPs. Six transmembrane a-helices (H1–H6) are connected by alternating extracellular and intracellular loops with the carboxyl and amino termini orientated toward the cytoplasm. Two of these connecting loops (HB and HE) fold into the transmembrane pore so that the protein resembles an hour-glass. Loops B and E each form a helical region containing highly conserved Asn-Pro-Ala (NPA) motifs that facilitate selectivity of the pore. (B) Three-dimensional structure of an AQP subunit monomer.

passes through the channel. The AQPs monomers contain several conserved motifs, including the Asn-Pro-Ala (NPA) motif, in their short helical segments. The NPA boxes and adjacent residues are thought to be essential for water transport activity (Yousif & Bhave, 2011) (Fig. 1). In the tetramer, AQPs monomers work independently. Therefore, the monomer is the functional unit, transporting water independently of the other subunits in the complex (Verkman & Mitra, 2000). Some AQPs, such as mammalian AQP4, can further aggregate in cell membranes to form supramolecular crystalline assemblies—called orthogonal arrays of particles—which are maintained by inter-tetrameric N-terminal interactions involving specific residues (Crane & Verkman, 2009).

2.2 Molecular characterization of AQPs function

The primary function of most AQPs is to transport water across cell membranes in response to osmotic gradients created by active solute transport. As reported above, X-ray of the crystal structures existing in all AQPs revealed that the AQPs tetramers in membranes are composed of four individual monomers (Verkman & Mitra, 2000); because the water transport capacity of AQPs monomers is low, membranes often contain a high density of AQPs to increase water permeability substantially above the level found in the absence of AQPs. A subset of AQPs, called aquaglyceroporins, also transports glycerol. The aquaglyceroporins have a less constricted pore with a slightly greater diameter than that of the water selective AQPs (a diameter of 3.4 Å as compared to 2.8 Å, respectively), with a larger proportion of hydrophobic pore-lining residues (Wu & Beitz, 2007). In addition to water and glycerol, there is evidence, some of which is still controversial, that gases, various small solutes and ions pass through some AQPs. Non transporting functions for some AQPs have also been suggested, such as cell-cell adhesion, membrane polarization and regulation of interacting proteins, such as ion channels, suggesting that some cellular and tissue functions of AQPs water transport are not obvious *a priori*. In fact, some studies have suggested that AQP9 can also transport amino acids, sugars and arsenite (Carbrey et al., 2009; Tsukaguchi et al., 1998; Wu & Beitz, 2007), while AQP6 has been proposed to function as an intracellular chloride channel (Yasui et al., 1999) and several AQPs have also been thought to transport other small molecules and gases, including carbon dioxide, ammonia, nitric oxide and hydrogen peroxide (Miller, Dickinson, & Chang, 2010; Musa-Aziz, Chen, Pelletier, & Boron, 2009; Wang & Tajkhorshid, 2010).

2.3 AQPs regulation

Because most AQPs are constitutively expressed at the cell plasma membrane, it has become clear that many signaling pathways regulate the expression of AQPs genes through complex transcriptional and post-transcriptional controls. Many reports have concluded that in mammals, AQPs are up- and downregulated in response to various physiological situations or stresses; the stimulation of vasopressin (also called ADH)-V2-receptor leads to an increased AQP2 gene transcription, suggesting that AQP2 transcription is controlled by an ADH-dependent mechanism (Hasler, Nielsen, Feraille, & Martin, 2006). Oliveira and colleagues demonstrated that AQP1 and AQP9 expression is regulated by oestrogens (Oliveira, Carnes, Franca, Hermo, & Hess, 2005), and pregnancy-related hormones determine an increase in AQP2 mRNA and protein expression in the renal collecting ducts of rats, causing an expansion of extracellular fluid volume and water retention, as normally occurs during pregnancy (Schrier, Cadnapaphornchai, & Ohara, 2001). AQP4 expression undergoes modifications during brain trauma and inflammation (Aoki-Yoshino et al., 2005; Tomura et al., 2011). The recently identified function of AQP8 as a two-way transporter of water and hydrogen peroxide across biological membranes leads to the demonstration that diverse cellular stress conditions, such as heat, hypoxia, and ER stress, reversibly inhibit the permeability of AQP8 to H2O2 and water (Medraño-Fernandez et al., 2016). Nevertheless, with the exception of AQP2 and a few other AQPs, the biological significance of the transcriptional regulation of AQPs expression is unclear. Ongoing works on the study of the expression of AQP genes and regulation will hopefully elucidate in more detail the involvement of AQPs in pathophysiological processes.

3. Physiological role of mammalian AQPs

3.1 AQPs in epithelial fluid transport

A major role of epithelial cells is the transport of fluid across tissue barriers. Therefore, a principal role of the AQPs, as water or glycerol transporting proteins expressed at plasma membranes in epithelial and endothelial cells, is in transepithelial fluid transport.

In kidney, AQPs-facilitated transepithelial water transport across the tubules is necessary for the formation of concentrated urine (Ma et al., 1998). About 60–70% of the water in the glomerular filtrate is absorbed by the highly permeable proximal tubule, facilitated by the presence of

AQP1 water channels. The remaining part of the glomerular filtrate enters the loop of Henle where further reabsorption of water without solute occurs, leading to urine concentration. The descending thin limb of the loop of Henle is highly permeable to water thanks to the containing AQP1 water channels that allow the re-absorbance of about 15% (Schnermann et al., 1998). The thin ascending loop and the thick ascending loop of Henle are, in contrast, impermeable to water because of the absence of AQPs channels (Tradtrantip, Tajima, Li, & Verkman, 2009). As expected, AQP1-deficient mice show a defective fluid absorption (Schnermann et al., 1998). Recent works reported that water transport across the collecting duct epithelium involves vasopressin (ADH)-regulated AQP2 trafficking to the apical plasma membrane; the cortical collecting duct has low water permeability in its basal state, but in the presence of ADH it can absorb water in the filtrate, allowing an increase of the urine volume depending upon whole-body water composition and requirements. In fact, the urine that enters the medullary collecting duct is hypotonic (Lee & Kwon, 2007; Wilson, Miranda, & Knepper, 2013). The permeability of the collecting ducts varies depending on the need for body water; in fact, if conditions warrant the loss of water from the body, the medullary collecting duct remains impermeable to water, and the hypotonic urine is excreted. On the contrary, if it is necessary to conserve water, the collecting duct absorbs considerable amounts of water from the urine, thereby leading to the formation of hypertonic urine (Sands & Layton, 2009). This antidiuretic mechanism is ADH-dependent. The hypothalamus-secreted ADH binds to a V2 receptor present on the basolateral surface of the principal cells of the collecting duct, activating intracellular cAMP production; the increased level of intracellular cAMP leads to the phosphorylation of residues in the carboxyl-terminal domain of AQP2, allowing the insertion of AQP2 into the luminal membrane (Christensen, Zelenina, Aperia, & Nielsen, 2000; Hoffert et al., 2007). The inserted AQP2 allows water flow across a tubule lumen-interstitium osmotic gradient; it has been clearly demonstrated that tubule water enters the cell through AQP2, traverses the cytosol, and the basolateral AQP3, and AQP4 channels transfer water within the collecting duct cells to the renal interstitium and subsequently into the circulation, thus concentrating the urine (Ishibashi et al., 1997; Kim et al., 2005; Noda, Sohara, Ohta, & Sasaki, 2010; Verkman, 2008). In the absence of ADH, AQP2 water channels retrieval occurs from the apical plasma membrane into vesicles (Nielsen et al., 1995) for potential reuse by recycling, and the collecting duct becomes once more impermeable to water. Based on these considerations, in humans, loss-of-function mutations in AQP2 result in the rare genetic disease called nephrogenic diabetes insipidus (NDI) (Deen et al., 1994; Fujiwara & Bichet, 2005), which is characterized by polyuria and urinary hypo-osmolality. Abnormal kidney expression of the other AQPs involved in water movement across the tubules determines defects in the urinary concentration. Specifically, the phenotype of AQP1 and AQP3 null mice was characterized by marked polyuria with a substantially lower urine osmolality than in wild-type mice (Ma, Fukuda, Song, Matthay, & Verkman, 2000; Ma, Song, et al., 2000). AQP1 water channels are present in the proximal tubule, thin descending limb of Henle, and outer medullary vasa recta. In AQP1 null mice a urinary-concentrating defect was evident, associated with a defective proximal tubule fluid absorption and defective counter-current multiplication (Ma et al., 1998). Ma and colleagues demonstrated that transgenic mice lacking AQP4 were not polyuric but had a deficient mechanism of urinary-concentrating ability in response to water deprivation (Ma et al., 1997). Their studies suggested that, given the location of AQP4 in the basolateral membrane of the inner medullary collecting duct, and because most of the fluid in the antidiuretic kidney is extracted from the lumen in the cortical collecting duct, AQP3 might be more critical than AQP4 for the formation of concentrated urine (Chou, Ma, Yang, Knepper, & Verkman, 1998).

Recently, a hypothetical role for AQP11 in urine production was suggested, because mice lacking AQP11 manifest an interesting polycystic kidney disease phenotype (Morishita et al., 2005), although how the loss of AQP11 produces renal cysts remains unclear.

The high transepithelial water permeability conferred by AQPs has been demonstrated in other organs; knockout mice lacking AQP5 show defective salivary and airway submucosal glands function and a decreased or absent secretion of airway mucus (Shen et al., 2011) and saliva, that results hypertonic and more viscous (Ma et al., 1999). Moreover, AQPs water channels are expressed in many cell types in the eye. At the ocular surface, AQP1 is expressed in corneal endothelium, AQP3 and AQP5 in corneal epithelium, and AQP3 in conjunctival epithelium. AQPs are also expressed in lens fiber cells, lens epithelium (AQP1), ciliary epithelium (AQP1, AQP4) and retinal Müller cells (AQP4). Analysis of knockout mice lacking individual AQPs suggests their involvement in the maintenance of corneal and lens transparency, corneal epithelial repair, intraocular pressure regulation, retinal signal transduction and retinal swelling following injury (Verkman, Ruiz-Ederra, & Levin, 2008; Zhang, Vetrivel, & Verkman, 2002). AQP 1 and 4 are abundant in the brain and are related to the regulation of water physiology (Tait, Saadoun, Bell, & Papadopoulos, 2008). AQP1 is abundant in the choroid plexus, where it regulates cerebrospinal fluid physiology, and AQP4 is distributed in the brain parenchyma (Filippidis, Kalani, & Rekate, 2012; Kalani, Filippidis, & Rekate, 2012; Nagelhus & Ottersen, 2013). Experiments performed on animal models suggest that AQPs appear to be required for active transepithelial fluid transport only when the fluid transport rate, normalized to the epithelial surface area, is very high, as occurs in the proximal tubule and salivary gland. By contrast, when the area-normalized rate of fluid absorption is low, as it is in lung alveolar epithelium, the AQPs independent-basal water permeability is sufficient to support a relatively slow fluid absorption; in fact, deletion of AQP5 in mice does not impair active fluid absorption (Ma, Song, et al., 2000). In addition to their role in active water transport, AQPs also facilitate passive, osmotically driven water transport, as in sustained large osmotic gradients extraction of water in the thin descending limb of Henle and the collecting duct in the kidney (Verkman, 2006). Fig. 2 represents a schematic diagram showing the localization of AQPs in various epithelia.

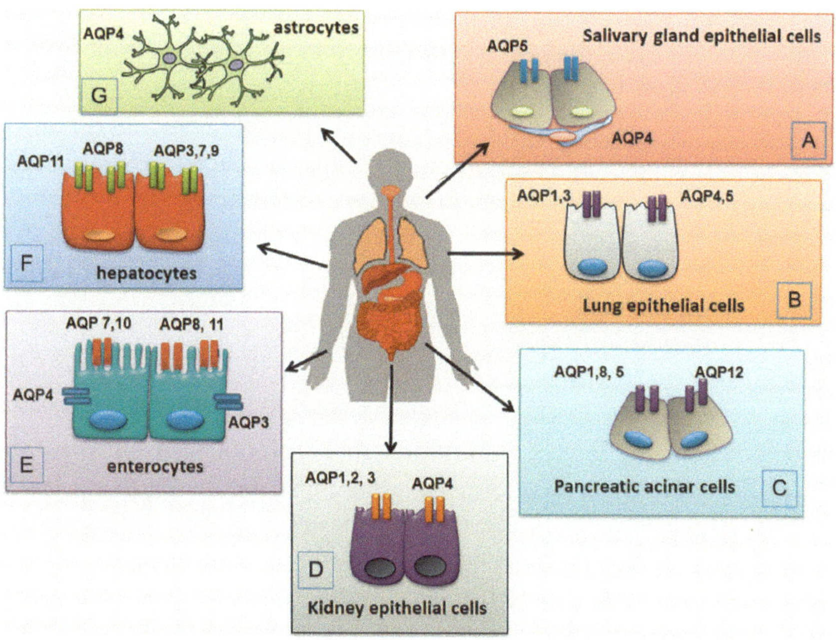

Fig. 2 Mammalian AQPs expression pattern among tissues. The schematic diagram depicts the localization of AQPs in various epithelia (A–F) and glial cells (G).

3.2 AQPs distribution in nervous system

AQPs are widely expressed in the central nervous system (CNS) and data in the literature show that the glial cells are involved rather than epithelial cells. Among the subtypes of AQPs, AQP1 and 4 are the most abundant in the brain according to the literature. AQP4 is the predominant water channel in the mammalian brain and is mainly expressed in the perivascular astrocyte endfeet at the brain-blood interface (Assentoft, Larsen, & MacAulay, 2015) (Fig. 2G). Specifically, AQP4 is expressed in the endfeet of astrocytes and these endfeet are in contact with blood vessels associated with the blood-brain and brain-fluid interfaces (Assentoft et al., 2015). The importance of AQP4 in the water exchange mechanisms that involve the astrocytic endfeet has been established through experiments based on the deletion of AQP4 in mice, which resulted in a marked decrease in brain water uptake (Haj-Yasein et al., 2011). AQP1 is expressed in the choroid plexus, where it regulates cerebrospinal fluid (CSF) secretion; in fact, AQP1 is found in the ventricular-facing cell plasma membrane of choroid plexus epithelial cells (Jeon, Park, Park, Hwang, & Hwang, 2017) and Oshio and colleagues reported that deletion of AQP 1 in mice reduced intracranial pressure by slowing CSF secretion (Oshio, Watanabe, Song, Verkman, & Manley, 2005).

AQP9, a cloned water channel permeable to monocarboxylate, glycerol and urea (Tsukaguchi et al., 1998), is weakly expressed in the brain, being almost exclusively observed in catecholaminergic cells including the adrenergic, noradrenergic and dopaminergic groups and therefore with a fairly preferential localization in catecholaminergic nuclei known to be involved in many cerebral functions. Its expression in neuronal cells not directly involved in osmoregulation suggests that brain AQP9 could also be used as a metabolite channel, since lactate and glycerol can be energy substrates for neurons (Lindskog, Asplund, Catrina, Nielsen, & Rützler, 2016).

As regards the spinal cord, AQP4 is primarily localized in the subpial *astrocyte processes* that form the *glial-limiting membrane*, AQP1 immunolabeling was detected mostly in processes of non-myelinated neurons in the dorsal horns, and AQP9 may be expressed predominantly in spinal cord radial astrocytes (Oklinski et al., 2016). A pattern of AQPs expression was observed also in the peripheral nervous system; in ganglia, AQP1 was expressed in a subpopulation of neuronal C-fibers with a specific colocalization with substance P and the capsaicin receptor TRPV1 (transient receptor potential cation channel subfamily V member 1) (Papadopoulos

& Verkman, 2013). AQP1 is also expressed in neurons of the trigeminal ganglion where it mediates pain perception from the head. To support the involvement of AQP1 in nociception, recently, Oshio and colleagues demonstrated the presence of AQP1 in the dorsal laminae of the horns of the spinal cord, associated with pain perception (Oshio et al., 2004). Few isolated reports have described AQP1 expression in other parts of the peripheral nervous system, including the sciatic nerve (Gao et al., 2006; Oshio et al., 2004) and the periodontal Ruffini mechanoreceptors (Nandasena et al., 2007).

Studies conducted on sensory organs demonstrated AQP4 expression in retinal glia (Li, Patil, & Verkman, 2002), and in supporting cells in the olfactory epithelium (Lu, Zhang, Zador, & Verkman, 2008); similarly, AQP4 was expressed in supporting cells of the organ of Corti in the inner ear (Li & Verkman, 2001); interestingly, however, it is absent from excitable cells in these tissues. This is an analogous situation to that observed in the CNS, where AQP4 is found in astrocytes but not in neurons. In addition, several AQPs can be detected in the eye, where they regulate intraocular pressure, the transparency of the corneal and lens, and the barrier function of the cornea and conjunctiva (Verkman, 2008). As shown by functional studies on AQPs knockout mice, AQP1 is expressed in the corneal endothelium, keratocytes and the ciliary epithelium. AQP4 is expressed in the ciliary epithelium. AQP3 is expressed in the conjunctival epithelium and, in association with AQP5, in the corneal epithelium (Verkman, 2008).

Also in the enteric nervous system, various studies have reported the expression of several AQPs regulating motor functions, local blood flow, mucosal transport and secretions, as well as in neurons and glial cells localized in the submucosal and myenteric plexuses. Gao et al. first reported the expression pattern of AQP1 in the human enteric nervous system by immunolocalization in 2006 (Gao et al., 2006). AQP1 has been detected in glia in human oesophageal neurons (Gao et al., 2006) and rat ileal neurons (Nagahama, Ma, Semba, & Naruse, 2006) of the enteric system, indicating a glial cell-specific expression. AQP4 has been detected in rat and mouse myenteric and submucosal neurons (Thi, Spray, & Hanani, 2008); the AQP4 expression in myenteric and submucosal neurons suggests that AQP4 may be involved in intestinal motility but no functional data have yet been reported in the literature as regards AQP1 or AQP4 involvement in mucosal solute transport or gastrointestinal motility.

3.3 AQPs in cellular migration

AQPs-dependent cell migration has been found in a variety of cell types *in vitro* and in mice *in vivo*. Saadoun and colleagues observed an impaired tumor angiogenesis and a reduction of growth and vascularity of implanted tumors in AQP1-null mice (Saadoun, Papadopoulos, Hara-Chikuma, & Verkman, 2005). The proposed mechanism is based on modifications of AQP-dependent polarization to the *leading edge* of *migrating cells*, driving water influx through the plasma membrane (Saadoun, Papadopoulos, Hara-Chikuma, et al., 2005) (Fig. 3A). In fact, it was observed that migration can be modulated by changes in extracellular osmolality and transcellular osmotic gradients (Saadoun, Papadopoulos, Watanabe, et al., 2005). The water influx, in turn, causes expansion of the adjacent plasma membrane, followed by actin repolymerization to stabilize the cell membrane protrusion (Charras, Yarrow, Horton, Mahadevan, & Mitchison, 2005). *In vitro* experiments on aortic endothelial cells from Aqp1-null mice showed that these cells migrated more slowly than wild-type cells in response to a chemotactic stimulus, and that cell migration was increased in various cells after AQPs transfection (Papadopoulos, Saadoun, & Verkman, 2008). However, alternative mechanisms, such as AQP-dependent changes in cell volume during the migration and interaction of AQPs with other proteins, are currently under study. AQP-facilitated cell

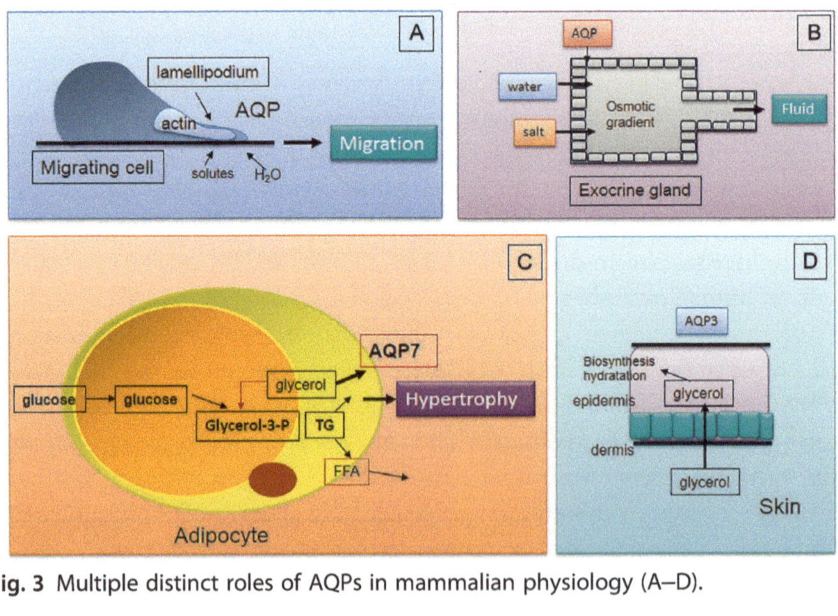

Fig. 3 Multiple distinct roles of AQPs in mammalian physiology (A–D).

migration appears to be a general mechanism applicable also to tumor spread, wound healing and immune cell chemotaxis. AQP4 deletion slows the migration of reactive astrocytes toward a chemotaxic stimulus and increases glial scarring (Auguste et al., 2007; Saadoun, Papadopoulos, Watanabe, et al., 2005). AQP1-expressing tumor cells have an increased ability to extravasate across blood vessels and to invade locally (Hu & Verkman, 2006). Moreover, a high level of AQPs expression was detected in many tumor types, and correlated with tumor grade; this was observed, for example, in glioblastomas (Verkman et al., 2008). An impaired cell migration has also been seen in AQP1-deficient proximal tubule epithelial cells, and AQP3-deficient corneal epithelial cells, enterocytes, and skin keratinocytes. On the other hand, the expression of AQP3 in skin and cornea facilitates wound healing (Hara-Chikuma & Verkman, 2008a, 2008b; Levin & Verkman, 2006). Furthermore, although the evidence remains preliminary, AQPs might also be involved in intracellular vesicular transport, such as secretory granule exocytosis in pancreatic acinar cells and astrocyte cytokine vesicular secretion (Cho et al., 2002; Li, Zhang, Varrin-Doyer, Zamvil, & Verkman, 2011).

3.4 AQPs distribution in the skin and their role in hydration

Several AQPs are expressed in mammalian skin. Skin is composed of three regions: the deepest, the hypodermis, contains adipocytes. Above the hypodermis is the dermis, a complex matrix of vascularized collagen and elastin; finally, the most superficial region of the skin is the epidermis, which is supported by the dermis. The epidermis is an epithelium that is continually renewed by the proliferation of keratinocytes located in the basal layer and their differentiation in the stratum spinosum and stratum granulosum. Terminally differentiated keratinocytes form the stratum corneum (SC), composed of corneocytes. The lipid matrix and corneocytes of the SC pose the primary barrier of the epidermis against body fluids and water evaporation. For this reason, AQPs were not expected to be present in mammalian epidermis, which is impermeable to water. Experiments performed on animal models and human cells demonstrated that AQP3 was found in the epidermis in mice (Frigeri et al., 1995) and human keratinocytes (Sougrat et al., 2002), AQP1 was detected in endothelial cells of the human dermis (Mobasheri et al., 2004). In addition, AQP5 is expressed in sweat glands, although its role remains somewhat controversial (Song, Sonawane, & Verkman, 2002) (a possible generic mechanism of AQPs involvement in exocrine gland function is reported in Fig. 3B); finally, mice lacking AQP7 manifest progressive adipocyte fat accumulation and hypertrophy, and in human adipocytes AQP7 has been shown to play an important role in glycerol transport (Hara-Chikuma et al., 2005). Reduced plasma membrane glycerol permeability in AQP7 deficiency results in an increased glycerol concentration in the adipocyte cytoplasm, leading to triglycerides biosynthesis (Hibuse et al., 2005). According to the literature, among the AQPs, AQP3 is the most abundantly-expressed in the plasma membrane of human keratinocytes and acts as a pH-sensitive water channel (Sougrat et al., 2002), but the role of AQP3 as aquaglyceroporin, permeable both to water and glycerol, has now been demonstrated. Mice lacking AQP3 have dry skin and a decreased water and glycerol epidermis permeability (Ma, Hara, Sougrat, Verbavatz, & Verkman, 2002). Glycerol replacement was found to correct the hydration, elasticity and barrier function (Hara & Verkman, 2003), suggesting that defective glycerol transport is sufficient to account for the dry skin. Thus, the principal function of AQP3 in mouse skin epidermis seems to be glycerol transport, not water transport (Hara-Chikuma & Verkman, 2005) (Fig. 3D). Topical or systemic administration of glycerol normalized the stratum corneum glycerol content and corrected the skin hydration defect caused by AQP3 deficiency (Hara & Verkman, 2003), providing a rationale for the use of glycerol in cosmetics and various skin medical formulations.

3.5 Aquaglyceroporin AQP7 in fat metabolism

In the last years, a key role in obesity has been proposed for AQP7. AQP7 is expressed in the plasma membrane of adipocytes and facilitates the transport of glycerol across cell membranes (Hara-Chikuma & Verkman, 2005; Hibuse et al., 2005) (Fig. 3C). AQP7 gene null mice show a remarkable progressive increase in glycerol and triglycerides storage in their adipocytes, provoking adipocyte hypertrophy that subsequently leads them to develop adult-onset obesity (Hibuse et al., 2005). Overall, these results provide evidence that an increase of the adipocyte volume may be due to a reduced adipocyte glycerol permeability and consequent accumulation of glycerol and triglycerides. Nevertheless, further studies on AQP9 knockout mice have showed markedly increased plasma levels of glycerol and triglycerides and a deficient glycerol metabolism, and these results have suggested that AQP9 may be a metabolic regulator for hepatic glycerol internalization

(Rojek et al., 2007). Thereby, AQP7 and AQP9 are emerging as important facilitators of glycerol transport across the cell membrane and modulators in controlling the metabolism of glycerol, contributing to the pathophysiology of obesity and diabetes (Carbrey et al., 2003; Maeda, Hibuse, & Funahashi, 2009).

3.6 AQP3 in cellular proliferation

A significant role of AQP3 in cell proliferation was observed in various AQP3-expressing tissues, including skin, colon and cornea. AQP3-null mice show an altered cutaneous and corneal wound healing (Hara-Chikuma & Verkman, 2008a, 2008b; Levin & Verkman, 2006) and impairment enterocyte proliferation (Thiagarajah, Zhao, & Verkman, 2007). An enhanced expression of AQP3 was reported in colorectal carcinogenesis (Moon et al., 2003), human lung cancer (Liu, Matsuzaki, et al., 2007; Liu, Promeneur, et al., 2007), gastric adenocarcinomas (Huang et al., 2010) and human skin squamous cell carcinomas, revealing a novel role for AQP3 in cancer using transgenic mice (Kusayama et al., 2011). Indeed, AQP3 knockout mice exhibit total resistance to skin cancers development in response to a tumor starting-promoter that triggers multiple tumors in wild-type mice (Hara-Chikuma & Verkman, 2008c). Furthermore, AQP3-deficient mice show an altered cellular glycerol metabolism and biosynthesis, suggesting an involvement of AQP3-facilitated glycerol transport in epidermal cell proliferation and carcinogenesis through a mechanism implicating cellular glycerol in the production of ATP energy. Results obtained supported a mechanism in which the reduced epidermal cell "energy" in AQP3-deficient mice leads to a reduced cell proliferation (Hara-Chikuma & Verkman, 2008c). In fact, in these mice a reduction of the cellular glycerol concentration and ATP levels was observed, and an impaired lipid biosynthesis, MAPK signaling and, finally, as a consequence, a reduced cell proliferation (Hara-Chikuma & Verkman, 2008c). The same mechanism was observed in breast cancer, in which it is possible that an increased AQP3 expression in breast cancer cells may facilitate glycerol transport into the cell, contributing to the formation of ATP, and providing cells with sufficient energy for cancer proliferation (Arif et al., 2018). Nevertheless, additional studies are needed to support a potential use of AQP3 inhibitors. This field of research is strongly attractive for prevention and therapy in skin cancer and other tumors linked to AQP3 overexpression.

4. AQPs in inflammation

The latest interesting findings reveal a new function of AQPs in several inflammatory processes, suggesting they could be used in the clinical field as a novel therapeutic target for anti-inflammatory therapy. The cells involved in inflammatory processes lose variations of the osmotic microenvironment, determining a marked increase in cell water permeability, cellular volume and alterations in cytoskeletal structure (Maidhof, Jacobsen, Papatheodorou, & Chahine, 2014). Active resolution of the inflammatory process is essential as it facilitates tissue repair after injury; failure to resolve the process leads to chronic inflammation, causing continuous tissue damage and progressive fibrosis (Ortega-Gómez, Perretti, & Soehnlein, 2013). Many articles have highlighted the possible involvement of AQPs in the development of these inflammatory processes, providing new evidence on the importance of water homeostasis regulation through AQPs during infections and inflammation, and paving the way for further investigations of roles for AQPs in inflammatory processes. Accordingly, many data demonstrate that AQPs show a role in maintaining the homeostasis of many physiological processes related to secretive and absorptive activities of several tissues, such as kidney, salivary gland, lung, skin, sweat glands, and intestine (Laforenza, 2012; Pelagalli, Squillacioti, Mirabella, & Meli, 2016) (Fig. 4).

5. Potential role of AQPs in different models of inflammation

5.1 AQPs in inflammatory kidney disorders

Water handling is abnormal in a variety of kidney disorders, and the AQPs play a highly important role in the reabsorption of water through the renal tubular fluid (Nielsen et al., 2002; Verkman, 2013; Wang, Zhang, Yu, Liu, & Jin, 2018). Some of these pathological conditions are primary renal disorders, whereas others reflect changes in other organs or systems. The role of changes in the expression and/or function of AQPs has been studied in a range of these conditions, including genetic defects, acquired defects of renal responsiveness, and conditions in which there is a marked inflammatory process following chronic kidney disease (CKD) (Verkman, 2012) (Fig. 4A). CKD is a common, progressive disease, characterized by fibrotic changes in the kidney and by persistent high-grade inflammation that causes irreversible loss of function of this organ (Wang et al., 2008). It leads to rapidly

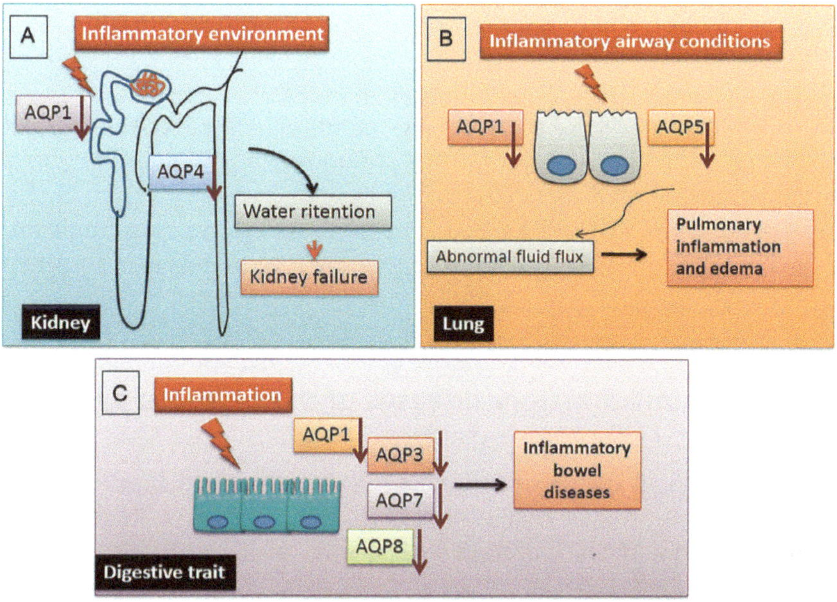

Fig. 4 The involvement of AQPs in several inflammatory processes in Kidney (A), Lung (B) and Digestive trait (C). Under inflammatory conditions, the cells can undergo alteration of electrolytes and water transport mediated by AQPs causing an increase in cell hydraulic permeability and size and thus, a dysregulation in cytoskeletal structure occurs that induce tissue impairment and organ injury.

progressive renal failure and cardiovascular system diseases, including heart failure (Gajjala, Sanati, & Jankowski, 2015; Thomas, Kanso, & Sedor, 2008).

Therefore, on the basis of the current literature, several AQPs are involved in differential pathological mechanisms in sepsis, and acute kidney injury is a common, serious complication of sepsis in which renal inflammation is one of the main pathological changes observed (Rump & Adamzik, 2018). Recently, Wang et al. have used lipopolysaccharide (LPS), an important inflammatory factor in *in vitro* models, to investigate inflammation-associated renal disease and AQPs expression (Wang et al., 2018), demonstrating that AQP1 had the lowest expression in the human proximal tubule cell line treated with LPS. Overall, these results suggest that AQP1 has a cytoprotective role and may be an effective treatment for acute kidney injury caused by sepsis. In addition, dysregulation of AQP2 has been shown to be associated to acute and chronic renal failure. Indeed, interestingly, the downregulation of AQP2 observed in sepsis models causes an intense inflammatory response and thus renal failure (Chagnon et al., 2008; Rump & Adamzik, 2018). Recently,

AQP4 expression was investigated, and its regulation in a mouse model of human immunodeficiency virus-associated nephropathy that mimics this pathology featuring an intense inflammatory response and rapid progressive kidney failure (D'Agati, Suh, Carbone, Cheng, & Appel, 1989). A decrease of AQP4 expression was demonstrated, associated with an enhanced inflammatory response that contributed to renal injury in these mice (Bryant et al., 2018). Indeed, renal dysfunction in Aqp4 null mice was associated with an enhanced oxidative stress and inflammatory process (Chang et al., 2013).

5.2 AQPs in inflammatory diseases of the digestive system

The digestive system is a major site of fluid movement and the wide distribution of AQPs in the gastrointestinal (GI) tract suggests that they likely have an important role in channel-mediated water transport, intestinal permeability, and fluid secretion/absorption (Zhu, Chen, & Jiang, 2016). Interestingly, this is consistent with a potential role for AQPs in the pathophysiology of gastrointestinal system diseases linked to an intense inflammatory response within the organ network (Day et al., 2014; Gregoire et al., 2015; Laforenza, 2012; Zhu et al., 2016) (Fig. 4C). Inflammatory bowel disease (IBD) is a chronic disease of the GI system, characterized by persistent activation of the immune system that leads to chronic, marked inflammation of the intestinal tract (Abraham & Cho, 2009; Matricon, Barnich, & Ardid, 2010). The expression of AQP1, 3, 7 and 8 is significantly reduced in patients with IBD, as well as in patients with Crohn's disease and ulcerative colitis (UC) (Hardin et al., 2004; Petr et al., 2015; Zahn et al., 2007), thus inducing speculation that this decrease is involved in the development of these diseases. Hepatic cirrhosis is another serious chronic inflammatory condition linked to pathologic water retention, being the final common endpoint in a variety of toxic, metabolic, infectious, and autoimmune forms of chronic liver disease (Schuppan & Afdhal, 2008; Zhou, Zhang, & Qiao, 2014). Several reports have demonstrated a marked overexpression of AQP1 in both human and rodent chronic liver diseases (Huebert et al., 2011, 2010; Lehmann, Larocca, Soria, & Marinelli, 2008). Therefore, an increased AQP1 expression during cirrhosis promotes angiogenesis, enhancing endothelial invasion through the dense extracellular microenvironment linked with the disease (Huebert et al., 2010). Indeed, current studies show dramatic *in vivo* AQP1 gene knockout consequences on the angiogenesis, fibrosis, and portal hypertension that follows bile duct ligation in mice, further validating this water channel as a convincing treatment target in chronic liver disease (Day et al., 2014; Huebert et al., 2011; Rump & Adamzik, 2018).

5.3 AQPs in inflammatory lung diseases

Several AQPs are expressed in the lung and in the bronchopulmonary tissues, where they act as regulators of transcellular water transport (Borok & Verkman, 2002; Verkman, 2002); their dysregulated expression alters the lung physiology, provoking marked airway inflammation (Verkman, 2002, 2007, 2009) (Fig. 4B). The localization of AQP5 in the apical membrane of the alveolar epithelial cells of the lung and acinar glandular cells provides the main route for the osmotically driven flow of water within the whole airway system, thus regulating the hydration of the airway surfaces (Folkesson, Matthay, Hasegawa, Kheradmand, & Verkman, 1994; Nielsen, King, Christensen, & Agre, 1997; Nielsen, Smith, Christensen, Knepper, & Agre, 1993). Indeed, studies with Aqp5 null mice have demonstrated that AQP5 enhances fluid secretion representing a rate-limiting barrier in gland fluid secretion (Song & Verkman, 2001). Interestingly, Towne et al. showed that AQP5 is significantly downregulated during pulmonary inflammation and edema (Towne, Harrod, Krane, & Menon, 2000) and, in addition, AQP5 expression is reduced in inflammatory airway conditions, such as chronic obstructive pulmonary disease (Shen et al., 2011). Validating important roles of AQPs in multiple human cancers and in different and distinct aspects of the inflammatory process linked to tumor development and progression, recently, the overexpression of AQP1 has been linked to many types of lung cancer and considered as a distinctive clinical prognostic factor (Hoque et al., 2006). Liu et al. evidenced AQP3 expression in patients affected by cancer lung, demonstrating that this AQP seems to be involved in tumor differentiation and processes related to the clinical stage in lung adenocarcinomas (Liu, Matsuzaki, et al., 2007; Liu, Promeneur, et al., 2007; Wang et al., 2015).

Several reports have demonstrated that AQPs are regulated by growth factors and inflammatory mediators in respiratory physiology (King, Nielsen, & Agre, 1997; Towne et al., 2000). In this inflammatory context, AQP5 and AQP1 show a decreased expression following adenoviral infection, resulting in pulmonary inflammation and edema in mouse lung (Towne et al., 2000). In particular, autocrine and/or paracrine mediators cause several pathophysiological modifications of the alveolar capillary and among these mediators, pro-inflammatory cytokines are released after infection by different bacteria or viruses (Al-Samir et al., 2016; Musa-Aziz et al., 2009). Interestingly, a study conducted by Towne et al. using an *in vivo* mouse model of pulmonary inflammation resulting from adenovirus infection showed a marked and significant decrease of AQP1 and AQP5 expression. In addition, the reduction in AQP1 and AQP5 expression was demonstrated at sites at a distance from that of the infection, suggesting a humoral and cell-mediated modulation of these AQPs (King, Kozono, & Agre, 2004; Towne et al., 2000).

Further data support these findings, showing that Th2 cytokines and IL-4 are downregulated in AQP5-deficient mice and so revealing a new function of AQPs in the inflammatory process (Karras et al., 2007). Indeed, AQPs were individuated as potential targets of altered gene expression in various mice models of induced asthma. Recent work has shown that AQP5 function is crucial in modulating bronchoconstrictive responses. AQP5-deficient mice are hyper-responsive to cholinergic-stimulated bronchoconstriction, a clinical feature of asthma that is a common, complex genetic disease caused by a combination of inherited and environmental factors (Krane et al., 2009). In fact, preclinical data showed that AQP3 and AQP5 are involved in allergic airway inflammation in induced asthma (Ikezoe et al., 2016).

5.4 AQPs in neuroinflammation

The study of AQPs in the CNS has provided insight into water distribution in the various brain compartments, suggesting a possible link with neurological disorders associated to disturbed water movements like cerebral edema (Agre, 2004) (Fig. 5). Therefore, the potential contribution of these channel proteins in neuroinflammation has been widely analyzed in neuromyelitis optica (NMO) and multiple sclerosis (Oklinski et al., 2016). Since physiological brain function is intricately linked to water homeostasis (Amiry-Moghaddam & Ottersen, 2003), AQP4, the main AQP expressed in the CNS, is expected to play a key role in the regulation of the brain water flux (Medici, Frassoni, Tassi, Spreafico, & Garbelli, 2011; Nagelhus & Ottersen, 2013; Oshio et al., 2004). Indeed, in the brain of Aqp4-deficient mice, measurement of water movement within CNS tissues showed a decrease by 50% (Manley et al., 2000), confirming the key role of AQP4 in water flow regulation.

Many of the CNS disorders are associated with neuroinflammatory processes and recent studies have started to demonstrate interrelations between AQP4 and neuroinflammation in chronic and acute brain diseases. AQP4 is

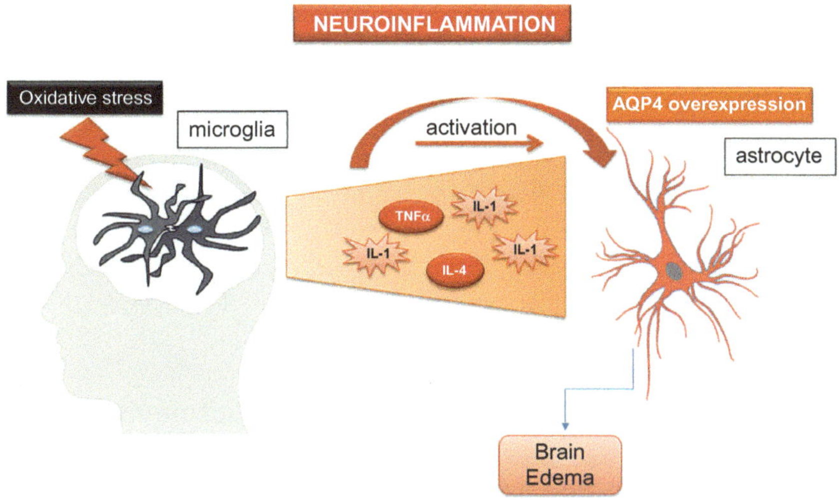

Fig. 5 AQP4 mediates the communication between astrocytes and microglia during neuroinflammation with consequent cell swelling.

commonly upregulated in neuroinflammation (Fukuda & Badaut, 2012) and is considered a key player in edema formation. Furthermore, its expression is increased after cerebral injury. Edema is observed after different brain injuries and is associated with blood-brain barrier (BBB) disruption. BBB disruption could lead to plasma protein leakage and extravascular fluid accumulation, which could be a part of the neuroinflammatory response (Bowman et al., 2007; Erickson, Dohi, & Banks, 2012). Current knowledge of AQP4 functions in the nervous system is largely derived from transgenic experiments on AQP4 knockout mice in several pathological conditions (Verkman, 2013). Interestingly, the presence of AQP4 has also been reported in reactive microglia after intra-nigral injection of LPS in rats, and activated astrocytes trigger the secretion of chemokines and cytokines, causing BBB disruption and vasogenic edema formation (Tomás-Camardiel et al., 2004).

Recent data have confirmed the pro-inflammatory role of AQP4 in a model of experimental autoimmune encephalomyelitis (EAE). This harmful role of AQP4 in EAE patients is supported by a less severe clinical and tissue inflammation than in wild type animals following intracerebral LPS injection in AQP4 null mice (Fukuda & Badaut, 2012). This is most likely the cause of the reduced production of the pro-inflammatory cytokines TNFα and IL-6, observed in Aqp4 null mice astrocyte cultures (Fukuda & Badaut, 2012).

A possible link between neuroinflammation and AQP4 was also observed in NMO, a demyelinating disease characterized by an inflammatory process of the optic nerve and lesions in the spinal cord. Several studies have suggested the presence of specific autoantibodies targeting AQP4 expressed on the astrocytic membrane in NMO patients (Bennett et al., 2009; Bradl et al., 2009; Fukuda & Badaut, 2012; Hinson et al., 2012). Moreover, an involvement of APQ4 in brain edema has been clearly demonstrated, since AQP4-deficient mice show a clear reduction of cerebral edema and swelling of pericapillary astrocytic foot processes (Manley et al., 2000).

5.5 Potential role of AQPs in inflammatory autoimmune disorders

AQPs are involved in different inflammatory diseases and participate as regulators of innate host defense at cell membrane level. Several autoimmune diseases are characterized by a marked inflammatory response, as is well-known in rheumatoid arthritis (RA) characterized by enormously elevated levels of inflammatory cytokines, causing damage to the cartilage and bone. Indeed, several AQPs have been localized in chondrocytes, where they modulate the traffic of ions and molecules (Meli et al., 2018; Mobasheri et al., 2004) and, therefore, regulate the cartilage physiology.

Recently, it has been demonstrated that human chondrocytes and synoviocytes express AQP1, and it is upregulated in RA cartilage. Interestingly, AQP1 may be involved in joint swelling and synovial edema formation. According to this mechanism, AQP1 could act as a regulator of metabolic or extracellular matrix water on chondrocyte membranes (Trujillo et al., 2004). Several authors have demonstrated that in synovial tissues derived from patients affected by osteoarthritis and RA, TNF-α modulates both *Aqp9* gene and protein expression, indicating the cytokines as regulators of the AQPs function (Mobasheri et al., 2004; Nagahara et al., 2010; Richardson et al., 2003). Recently, Cai et al. have revealed AQP4 expression in articular cartilage in an adjuvant-induced arthritis (AIA) rat model, evidencing a potential involvement of AQP4 activation in the development of rat AIA (Cai et al., 2017).

Since the transcellular water flow is involved in specialized secretory tissues, and the main function of secretory glands is fluid secretion, AQPs could contribute to and modulate fluid production in exocrine glands (Delporte & Steinfeld, 2006) (Fig. 3B). Numerous studies have evaluated the expression of AQPs in secretory glands, such as the salivary and lacrimal glands, and the pancreas. For example, AQP1 expression was significantly increased in plasma membranes of pancreatic ducts in patients suffering from autoimmune pancreatitis, in which fluid secretion from the pancreas is severely impaired (Ko et al., 2009). Furthermore, AQP1 gene expression has been demonstrated in human parotid, submandibular and labial glands, while the protein was located in myoepithelial and endothelial cells rather than in glandular tissues (Gresz et al., 1999, 2001; Mobasheri & Marples, 2004). In the salivary glands, AQP5 regulates transcellular water movement in both acinar and parotid salivary cells (Day et al., 2014; Delporte, Bryla, & Perret, 2016); the salivary cells isolated from AQP5-deficient mice have a decreased membrane water permeability and altered cell volume regulation, indicating an involvement of AQP5 in saliva secretion (Krane et al., 2001). Furthermore, AQP3 has been demonstrated, located on the basolateral membranes of mucous and serous acinar cells of salivary glands, illustrating a critical role for AQP3 as well as AQP5 in transcellular water flow during saliva formation (Delporte et al., 2016; Gresz et al., 2001). AQP8 was shown to be expressed in rat salivary gland acinar cells (Hoque et al., 2002; Wellner, Hoque, Goldsmith, & Baum, 2000) and located in myoepithelial cells (Elkjaer et al., 2001; Wellner, Redman, Swaim, & Baum, 2006).

Since water movement is involved in saliva secretion, the expression, localization, and function of the AQPs have been extensively studied in salivary glands of patients affected by Sjögren's syndrome (SS). SS is one of the most common autoimmune diseases. Early clinical manifestations of SS are primarily a dramatically decreased tear and saliva secretion, leading to dry eye and dry mouth syndrome and atrophy of the lacrimal and salivary glands. In labial salivary gland biopsies of SS patients, AQP1 expression is considerably decreased in myoepithelial cells (MECs) as compared to controls, while no changes were observed in endothelial cells of non-fenestrated capillaries (Beroukas et al., 2002). This decreased AQP1 expression in MECs raised the hypothesis that an alteration of these cells plays a role in the disease pathogenesis (Beroukas et al., 2002). Interestingly, the treatment of a patient affected by SS with rituximab was demonstrated to lead to a marked increase in AQP1 expression in MECs, as well as a great improvement of xerostomia (Beroukas et al., 2002; Ring, Kallenbach, Praetorius, Nielsen, & Melgaard, 2006; Shanahan et al., 1999). Additionally, an abnormal localization of AQP5 has been observed in SS patients, in whom the localization is predominantly at the basolateral membrane rather than at the apical membrane of

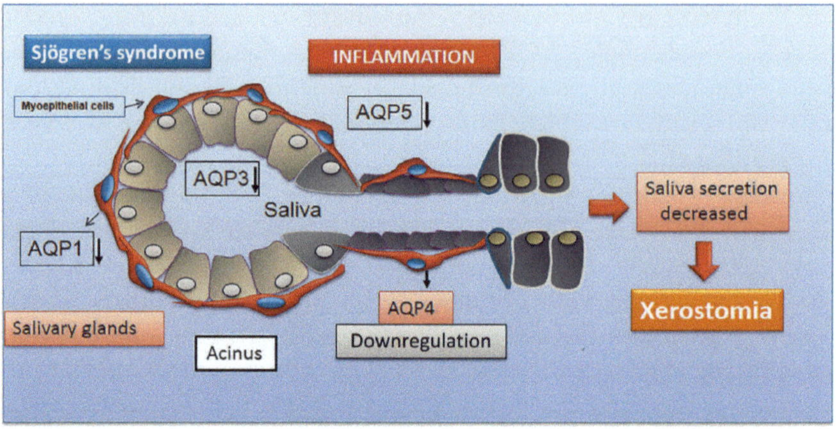

Fig. 6 Altered molecular mechanisms of saliva secretion in Sjögren's syndrome (SS) salivary gland. Decreased AQP3 and AQP5 proteins expression in acinar and ductal cells and AQP1 and AQP4 in myoepithelial cells causes the impairment of salivary gland function and thus xerostomia and hyposalivation.

acinar cells. An altered AQP5 expression has been proposed to contribute to the deficient fluid secretion observed in SS patients (Steinfeld et al., 2001). Recently, the presence of the AQP4 gene and the distribution of AQP4 protein in healthy controls and SS salivary gland biopsies was demonstrated, confined to the basal region of acini, to the lateral and apical membrane of intercalated and striated ducts; AQP4 was also present in myoepithelial cells (MECs) (Sisto et al., 2017). The importance of this study was the observation that AQP4 membrane expression was decreased in SS MECs compared to healthy controls. Thus, the marked downregulation of AQP4 expression in SS MECs suggests that these cells do not function normally in SS glands, and therefore some phases of salivary secretion may be impaired and could be responsible for the loss of saliva secretion in these patients (Sisto et al., 2017) (Fig. 6).

6. Conclusions

In the last few decades, new emerging findings have suggested that the AQPs may be regulators of water transport across membranes; consistently with their expression in most tissues, AQPs are linked to various physiological and pathophysiological processes in response to infectious and inflammatory stimuli. Information largely acquired from recent analyses of AQP-knockout mice has revealed unexpected cellular roles of AQPs.

The discovery of AQPs involvement in inflammatory conditions occurring in a wide variety of diseases has certainly contributed to a better knowledge of the complex mechanisms regulating host-pathogen communications and other types of inflammation. Future studies are needed to better understand the molecular mechanisms driving the inflammatory response linked to dysregulations of AQPs expression, and to clarify the signaling pathways involved in the regulation of AQPs functions. Additional research into the regulatory mechanisms of AQPs may identify the AQPs as new potential therapeutic targets and will undoubtedly contribute to improved prevention measures and treatment of inflammatory diseases.

Acknowledgment

We are grateful to M.V.C. Pragnell, B.A., a professional scientific text editor, for critical reading of the manuscript.

References

Abascal, F., Irisarri, I., & Zardoya, R. (2014). Diversity and evolution of membrane intrinsic proteins. *Biochimica et Biophysica Acta, 1840*, 1468–1481.

Abraham, C., & Cho, J. H. (2009). Inflammatory bowel disease. *The New England Journal of Medicine, 361*, 2066–2078.

Agre, P. (2004). Aquaporin water channels (Nobel Lecture). *Angewandte Chemie (International Ed. in English), 43*, 4278–4290.

Agre, P. (2006). The aquaporin water channels. *Proceedings of the American Thoracic Society, 3*, 5–13.

Al-Samir, S., Goossens, D., Cartron, J. P., Nielsen, S., Scherbarth, F., Steinlechner, S., et al. (2016). Maximal oxygen consumption is reduced in aquaporin-1 knockout mice. *Frontiers in Physiology, 7*, 347.

Amiry-Moghaddam, M., & Ottersen, O. P. (2003). The molecular basis of water transport in the brain. *Nature Reviews. Neuroscience, 4*, 991–1001.

Aoki-Yoshino, K., Uchihara, T., Duyckaerts, C., Nakamura, A., Hauw, J. J., & Wakayama, Y. (2005). Enhanced expression of aquaporin 4 in human brain with inflammatory diseases. *Acta Neuropathologica, 110*, 281–288.

Arif, M., Kitchen, P., Conner, M. T., Hill, E. J., Nagel, D., Bill, R. M., et al. (2018). Down-regulation of aquaporin 3 inhibits cellular proliferation, migration and invasion in the MDA-MB-231 breast cancer cell line. *Oncology Letters, 16*, 713–720.

Assentoft, M., Larsen, B. R., & MacAulay, N. (2015). Regulation and function of AQP4 in the central nervous system. *Neurochemical Research, 40*, 2615–2627.

Auguste, K. I., Jin, S., Uchida, K., Yan, D., Manley, G. T., Papadopoulos, M. C., et al. (2007). Greatly impaired migration of implanted aquaporin-4-deficient astroglial cells in mouse brain toward a site of injury. *The FASEB Journal, 21*, 108–116.

Bennett, J. L., Lam, C., Kalluri, S. R., Saikali, P., Bautista, K., Dupree, C., et al. (2009). Intrathecal pathogenic anti-aquaporin-4 antibodies in early neuromyelitis optica. *Annals of Neurology, 66*, 617–629.

Beroukas, D., Hiscock, J., Gannon, B. J., Jonsson, R., Gordon, T. P., & Waterman, S. A. (2002). Selective down-regulation of aquaporin-1 in salivary glands in primary Sjögren's syndrome. *Laboratory Investigation, 82*, 1547–1552.

Borok, Z., & Verkman, A. S. (2002). Lung edema clearance: 20 years of progress: Invited review: Role of aquaporin water channels in fluid transport in lung and airways. *Journal of Applied Physiology*, *93*, 2199–2206.

Bowman, G. L., Kaye, J. A., Moore, M., Waichunas, D., Carlson, N. E., & Quinn, J. F. (2007). Blood-brain barrier impairment in Alzheimer disease: Stability and functional significance. *Neurology*, *68*, 1809–1814.

Bradl, M., Misu, T., Takahashi, T., Watanabe, M., Mader, S., Reindl, M., et al. (2009). Neuromyelitis optica: Pathogenicity of patient immunoglobulin in vivo. *Annals of Neurology*, *66*, 630–643.

Brocker, C., Thompson, D. C., & Vasiliou, V. (2012). The role of hyperosmotic stress in inflammation and disease. *Biomolecular Concepts*, *3*, 345–364.

Bryant, J. L., Guda, P. R., Ray, S., Asemu, G., Sagi, A. R., Mubariz, F., et al. (2018). Renal aquaporin-4 associated pathology in TG-26 mice. *Experimental and Molecular Pathology*, *104*, 239–249.

Cai, L., Lei, C., Li, R., Chen, W. N., Hu, C. M., Chen, X. Y., et al. (2017). Overexpression of aquaporin 4 in articular chondrocytes exacerbates the severity of adjuvant-induced arthritis in rats: An in vivo and in vitro study. *Journal of Inflammation (London, England)*, *14*, 6.

Carbrey, J. M., Gorelick-Feldman, D. A., Kozono, D., Praetorius, J., Nielsen, S., & Agre, P. (2003). Aquaglyceroporin AQP9: Solute permeation and metabolic control of expression in liver. *Proceedings of the National Academy of Sciences of the United States of America*, *100*, 2945–2950.

Carbrey, J. M., Song, L., Zhou, Y., Yoshinaga, M., Rojek, A., Wang, Y., et al. (2009). Reduced arsenic clearance and increased toxicity in aquaglyceroporin-9-null mice. *Proceedings of the National Academy of Sciences of the United States of America*, *106*, 15956–15960.

Chagnon, F., Vaidya, V. S., Plante, G. E., Bonventre, J. V., Bernard, A., Guindi, C., et al. (2008). Modulation of aquaporin-2/vasopressin2 receptor kidney expression and tubular injury after endotoxin (lipopolysaccharide) challenge. *Critical Care Medicine*, *36*, 3054–3056.

Chang, K. H., Lyu, R. K., Chen, C. M., Wu, Y. R., Chang, H. S., Huang, C. C., et al. (2013). Distinct features between longitudinally extensive transverse myelitis presenting with and without anti-aquaporin 4 antibodies. *Multiple Sclerosis*, *19*, 299–307.

Charras, G. T., Yarrow, J. C., Horton, M. A., Mahadevan, L., & Mitchison, T. J. (2005). Non-equilibration of hydrostatic pressure in blebbing cells. *Nature*, *435*, 365–369.

Chen, L., Deng, H., Cui, H., Fang, J., Zuo, Z., Deng, J., et al. (2017). Inflammatory responses and inflammation-associated diseases in organs. *Oncotarget*, *9*, 7204–7218.

Cho, S. J., Sattar, A. K., Jeong, E. H., Satchi, M., Cho, J. A., Dash, S., et al. (2002). Aquaporin 1 regulates GTP-induced rapid gating of water in secretory vesicles. *Proceedings of the National Academy of Sciences of the United States of America*, *99*, 4720–4724.

Chou, C. L., Ma, T., Yang, B., Knepper, M. A., & Verkman, A. S. (1998). Fourfold reduction of water permeability in inner medullary collecting duct of aquaporin-4 knockout mice. *The American Journal of Physiology*, *274*, C549–C554.

Christensen, B. M., Zelenina, M., Aperia, A., & Nielsen, S. (2000). Localization and regulation of PKA-phosphorylated AQP2 in response to V(2)-receptor agonist/antagonist treatment. *American Journal of Physiology. Renal Physiology*, *278*, F29–F42.

Crane, J. M., & Verkman, A. S. (2009). Reversible, temperature-dependent supramolecular assembly of aquaporin-4 orthogonal arrays in live cell membranes. *Biophysical Journal*, *97*, 3010–3018.

D'Agati, V., Suh, J. I., Carbone, L., Cheng, J. T., & Appel, G. (1989). Pathology of HIV-associated nephropathy: A detailed morphologic and comparative study. *Kidney International*, *35*, 1358–1370.

Day, R. E., Kitchen, P., Owen, D. S., Bland, C., Marshall, L., Conner, A. C., et al. (2014). Human aquaporins: Regulators of transcellular water flow. *Biochimica et Biophysica Acta, 1840*, 1492–1506.

Deen, P. M., Weghuis, D. O., Sinke, R. J., Geurts van Kessel, A., Wieringa, B., & van Os, C. H. (1994). Assignment of the human gene for the water channel of renal collecting duct Aquaporin 2 (AQP2) to chromosome 12 region q12→q13. *Cytogenetics and Cell Genetics, 66*, 260–262.

Delporte, C., Bryla, A., & Perret, J. (2016). Aquaporins in salivary glands: From basic research to clinical applications. *International Journal of Molecular Sciences, 17*, 166.

Delporte, C., & Steinfeld, S. (2006). Distribution and roles of aquaporins in salivary glands. *Biochimica et Biophysica Acta, 1758*, 1061–1070.

Elkjaer, M. L., Nejsum, L. N., Gresz, V., Kwon, T. H., Jensen, U. B., Frøkiaer, J., et al. (2001). Immunolocalization of aquaporin-8 in rat kidney, gastrointestinal tract, testis, and airways. *American Journal of Physiology. Renal Physiology, 281*, F1047–F1057.

Erickson, M. A., Dohi, K., & Banks, W. (2012). Neuroinflammation: A common pathway in CNS diseases as mediated at the blood-brain barrier. *Neuroimmunomodulation, 19*, 121–130.

Filippidis, A. S., Kalani, M. Y., & Rekate, H. L. (2012). Hydrocephalus and aquaporins: The role of aquaporin-4. *Acta Neurochirurgica. Supplement, 113*, 55–58.

Folkesson, H. G., Matthay, M. A., Hasegawa, H., Kheradmand, F., & Verkman, A. S. (1994). Transcellular water transport in lung alveolar epithelium through mercury-sensitive water channels. *Proceedings of the National Academy of Sciences of the United States of America, 91*, 4970–4974.

Frigeri, A., Gropper, M. A., Umenishi, F., Kawashima, M., Brown, D., & Verkman, A. S. (1995). Localization of MIWC and GLIP water channel homologs in neuromuscular, epithelial and glandular tissues. *Journal of Cell Science, 108*, 2993–3002.

Fujiwara, T. M., & Bichet, D. G. (2005). Molecular biology of hereditary diabetes insipidus. *Journal of the American Society of Nephrology, 16*, 2836–2846.

Fukuda, A. M., & Badaut, J. (2012). Aquaporin 4: A player in cerebral edema and neuroinflammation. *Journal of Neuroinflammation, 9*, 279.

Gajjala, P. R., Sanati, M., & Jankowski, J. (2015). Cellular and molecular mechanisms of chronic kidney disease with diabetes mellitus and cardiovascular diseases as its comorbidities. *Frontiers in Immunology, 6*, 340.

Gao, H., He, C., Fang, X., Hou, X., Feng, X., Yang, H., et al. (2006). Localization of aquaporin-1 water channel in glial cells of the human peripheral nervous system. *Glia, 53*, 783–787.

Gregoire, F., Lucidi, V., Zerrad-Saadi, A., Virreira, M., Bolaky, N., Delforge, V., et al. (2015). Analysis of aquaporin expression in liver with a focus on hepatocytes. *Histochemistry and Cell Biology, 144*, 347–363.

Gresz, V., Burghardt, B., Ferguson, C. J., Hurley, P. T., Takács, M., Nielsen, S., et al. (1999). Expression of aquaporin 1 (AQP1) water channels in human labial salivary glands. *Archives of Oral Biology, 44*(Suppl. 1), S53–S57.

Gresz, V., Kwon, T. H., Hurley, P. T., Varga, G., Zelles, T., Nielsen, S., et al. (2001). Identification and localization of aquaporin water channels in human salivary glands. *American Journal of Physiology. Gastrointestinal and Liver Physiology, 281*, G247–G254.

Haj-Yasein, N. N., Vindedal, G. F., Eilert-Olsen, M., Gundersen, G. A., Skare, O., Laake, P., et al. (2011). Glial-conditional deletion of aquaporin-4 (Aqp4) reduces blood–brain water uptake and confers barrier function on perivascular astrocyte endfeet. *Proceedings of the National Academy of Sciences of the United States of America, 108*, 17815–17820.

Hara, M., & Verkman, A. S. (2003). Glycerol replacement corrects defective skin hydration, elasticity, and barrier function in aquaporin-3-deficient mice. *Proceedings of the National Academy of Sciences of the United States of America, 100*, 7360–7365.

Hara-Chikuma, M., Sohara, E., Rai, T., Ikawa, M., Okabe, M., Sasaki, S., et al. (2005). Progressive adipocyte hypertrophy in aquaporin-7 deficient mice: Adipocyte glycerol permeability as a novel regulator of fat accumulation. *The Journal of Biological Chemistry*, *280*, 15493–15496.

Hara-Chikuma, M., & Verkman, A. S. (2005). Aquaporin-3 functions as a glycerol transporter in mammalian skin. *Biology of the Cell*, *97*, 479–486.

Hara-Chikuma, M., & Verkman, A. S. (2008a). Roles of aquaporin-3 in the epidermis. *The Journal of Investigative Dermatology*, *128*, 2145–2151.

Hara-Chikuma, M., & Verkman, A. S. (2008b). Aquaporin-3 facilitates epidermal cell migration and proliferation during wound healing. *Journal of Molecular Medicine*, *86*, 221–231.

Hara-Chikuma, M., & Verkman, A. S. (2008c). Prevention of skin tumorigenesis and impairment of epidermal cell proliferation by targeted aquaporin-3 gene disruption. *Molecular and Cellular Biology*, *28*, 328–332.

Hardin, J. A., Wallace, L. E., Wong, J. F., O'Loughlin, E. V., Urbanski, S. J., Gall, D. G., et al. (2004). Aquaporin expression is downregulated in a murine model of colitis and in patients with ulcerative colitis, Crohn's disease and infectious colitis. *Cell and Tissue Research*, *318*, 313–323.

Hasler, U., Nielsen, S., Feraille, E., & Martin, P. Y. (2006). Posttranscriptional control of aquaporin-2 abundance by vasopressin in renal collecting duct principal cells. *American Journal of Physiology. Renal Physiology*, *290*, F177–F187.

Hibuse, T., Maeda, N., Funahashi, T., Yamamoto, K., Nagasawa, A., Mizunoya, W., et al. (2005). Aquaporin 7 deficiency is associated with development of obesity through activation of adipose glycerol kinase. *Proceedings of the National Academy of Sciences of the United States of America*, *102*, 10993–10998.

Hinson, S. R., Romero, M. F., Popescu, B. F., Lucchinetti, C. F., Fryer, J. P., Wolburg, H., et al. (2012). Molecular outcomes of neuromyelitis optica (NMO)-IgG binding to aquaporin-4 in astrocytes. *Proceedings of the National Academy of Sciences of the United States of America*, *109*, 1245–1250.

Hoffert, J. D., Nielsen, J., Yu, M. J., Pisitkun, T., Schleicher, S. M., Nielsen, S., et al. (2007). Dynamics of aquaporin-2 serine-261 phosphorylation in response to short-term vasopressin treatment in collecting duct. *American Journal of Physiology. Renal Physiology*, *292*, F691–F700.

Hoque, M. O., Soria, J. C., Woo, J., Lee, T., Lee, J., Jang, S. J., et al. (2006). Aquaporin 1 is overexpressed in lung cancer and stimulates NIH-3T3 cell proliferation and anchorage-independent growth. *The American Journal of Pathology*, *168*, 1345–1353.

Hoque, A. T., Yamano, S., Liu, X., Swaim, W. D., Goldsmith, C. M., Delporte, C., et al. (2002). Expression of the aquaporin 8 water channel in a rat salivary epithelial cell. *Journal of Cellular Physiology*, *191*, 336–341.

Hu, J., & Verkman, A. S. (2006). Increased migration and metastatic potential of tumor cells expressing aquaporin water channels. *The FASEB Journal*, *20*, 1892–1894.

Huang, Y., Zhu, Z., Sun, M., Wang, J., Guo, R., Shen, L., et al. (2010). Critical role of aquaporin-3 in the human epidermal growth factor-induced migration and proliferation in the human gastric adenocarcinoma cells. *Cancer Biology & Therapy*, *9*, 1000–1007.

Huebert, R. C., Jagavelu, K., Hendrickson, H. I., Vasdev, M. M., Arab, J. P., Splinter, P. L., et al. (2011). Aquaporin-1 promotes angiogenesis, fibrosis, and portal hypertension through mechanisms dependent on osmotically sensitive microRNAs. *The American Journal of Pathology*, *179*, 1851–1860.

Huebert, R. C., Vasdev, M. M., Shergill, U., Das, A., Huang, B. Q., Charlton, M. R., et al. (2010). Aquaporin-1 facilitates angiogenic invasion in the pathological neovasculature that accompanies cirrhosis. *Hepatology (Baltimore, Md.)*, *52*, 238–248.

Ikezoe, K., Oga, T., Honda, T., Hara-Chikuma, M., Ma, X., Tsuruyama, T., et al. (2016). Aquaporin-3 potentiates allergic airway inflammation in ovalbumin-induced murine asthma. *Scientific Reports*, *6*, 25781.

Ishibashi, K., Sasaki, S., Fushimi, K., Yamamoto, T., Kuwahara, M., & Marumo, F. (1997). Immunolocalization and effect of dehydration on AQP3, a basolateral water channel of kidney collecting ducts. *The American Journal of Physiology*, *272*, F235–F241.

Ishibashi, K., Tanaka, Y., & Morishita, Y. (2014). The role of mammalian superaquaporins inside the cell. *Biochimica et Biophysica Acta*, *1840*, 1507–1512.

Jeon, T., Park, K. S., Park, S. H., Hwang, J. H., & Hwang, S. K. (2017). Expression of aquaporin 1 and 4 in the choroid plexus and brain parenchyma of kaolin-induced hydrocephalic rats. *Korean Journal of Neurotrauma*, *13*, 68–75.

Kalani, M. Y., Filippidis, A. S., & Rekate, H. L. (2012). Hydrocephalus and aquaporins: The role of aquaporin-1. *Acta Neurochirurgica. Supplement*, *113*, 51–54.

Karras, J. G., Crosby, J. R., Guha, M., Tung, D., Miller, D. A., Gaarde, W. A., et al. (2007). Anti-inflammatory activity of inhaled IL-4 receptor-alpha antisense oligonucleotide in mice. *American Journal of Respiratory Cell and Molecular Biology*, *36*, 276–285.

Khalili-Araghi, F., Gumbart, J., Wen, P. C., Sotomayor, M., Tajkhorshid, E., & Schulten, K. (2009). Molecular dynamics simulations of membrane channels and transporters. *Current Opinion in Structural Biology*, *19*, 128–137.

Kim, S. W., Gresz, V., Rojek, A., Wang, W., Verkman, A. S., Frøkiaer, J., et al. (2005). Decreased expression of AQP2 and AQP4 water channels and Na,K-ATPase in kidney collecting duct in AQP3 null mice. *Biology of the Cell*, *97*, 765–778.

King, L. S., Kozono, D., & Agre, P. (2004). From structure to disease: The evolving tale of aquaporin biology. *Nature Reviews. Molecular Cell Biology*, *5*, 687–698.

King, L. S., Nielsen, S., & Agre, P. (1997). Aquaporins in complex tissues. I. Developmental patterns in respiratory and glandular tissues of the rat. *American Journal of Physiology*, *273*, C1541–C1548.

Ko, S. B., Mizuno, N., Yatabe, Y., Yoshikawa, T., Ishiguro, H., Yamamoto, A., et al. (2009). Aquaporin 1 water channel is overexpressed in the plasma membranes of pancreatic ducts in patients with autoimmune pancreatitis. *The Journal of Medical Investigation*, *56*, 318–321.

Kotas, M. E., & Medzhitov, R. (2015). Homeostasis, inflammation, and disease susceptibility. *Cell*, *160*, 816–827.

Krane, C. M., Deng, B., Mutyam, V., McDonald, C. A., Pazdziorko, S., Mason, L., et al. (2009). Altered regulation of aquaporin gene expression in allergen and IL-13-induced mouse models of asthma. *Cytokine*, *46*, 111–118.

Krane, C. M., Melvin, J. E., Nguyen, H. V., Richardson, L., Towne, J. E., Doetschman, T., et al. (2001). Salivary acinar cells from aquaporin 5-deficient mice have decreased membrane water permeability and altered cell volume regulation. *The Journal of Biological Chemistry*, *276*, 23413–23420.

Kruse, E., Uehlein, N., & Kaldenhoff, R. (2006). The aquaporins. *Genome Biology*, *7*, 206.

Kusayama, M., Wada, K., Nagata, M., Ishimoto, S., Takahashi, H., Yoneda, M., et al. (2011). Critical role of aquaporin 3 on growth of human esophageal and oral squamous cell carcinoma. *Cancer Science*, *102*, 1128–1136.

Laforenza, U. (2012). Water channel proteins in the gastrointestinal tract. *Molecular Aspects of Medicine*, *33*, 642–650.

Lee, B. H., & Kwon, T. H. (2007). Regulation of AQP2 in collecting duct: An emphasis on the effects of angiotensin II or aldosterone. *Electrolyte & Blood Pressure*, *5*, 15–22.

Lehmann, G. L., Larocca, M. C., Soria, L. R., & Marinelli, R. A. (2008). Aquaporins: Their role in cholestatic liver disease. *World Journal of Gastroenterology*, *14*, 7059–7067.

Levin, M. H., & Verkman, A. S. (2006). Aquaporin-3-dependent cell migration and proliferation during corneal re-epithelialization. *Investigative Ophthalmology & Visual Science, 47*, 4365–4372.

Li, J., Patil, R. V., & Verkman, A. S. (2002). Mildly abnormal retinal function in transgenic mice without Müller cell aquaporin-4 water channels. *Investigative Ophthalmology & Visual Science, 43*, 573–579.

Li, J., & Verkman, A. S. (2001). Impaired hearing in mice lacking aquaporin-4 water channels. *The Journal of Biological Chemistry, 276*, 31233–31237.

Li, L., Zhang, H., Varrin-Doyer, M., Zamvil, S. S., & Verkman, A. S. (2011). Proinflammatory role of aquaporin-4 in autoimmune neuroinflammation. *The FASEB Journal, 25*, 1556–1566.

Lindskog, C., Asplund, A., Catrina, A., Nielsen, S., & Rützler, M. (2016). A systematic characterization of aquaporin-9 expression in human normal and pathological tissues. *The Journal of Histochemistry and Cytochemistry, 64*, 287–300.

Liu, Y. L., Matsuzaki, T., Nakazawa, T., Murata, S. I., Nakamura, N., Kondo, T., et al. (2007). Expression of aquaporin 3 (AQP3) in normal and neoplastic lung tissues. *Human Pathology, 38*, 171–178.

Liu, Y. L., Promeneur, D., Rojek, A., Kumar, N., Frøkiaer, J., Nielsen, S., et al. (2007). Aquaporin 9 is the major pathway for glycerol uptake by mouse erythrocytes, with implications for malarial virulence. *Proceedings of the National Academy of Sciences of the United States of America, 104*, 12560–12564.

Lu, D. C., Zhang, H., Zador, Z., & Verkman, A. S. (2008). Impaired olfaction in mice lacking aquaporin-4 water channels. *The FASEB Journal, 22*, 3216–3223.

Ma, T., Fukuda, N., Song, Y., Matthay, M. A., & Verkman, A. S. (2000). Lung fluid transport in aquaporin-5 knockout mice. *The Journal of Clinical Investigation, 105*, 93–100.

Ma, T., Hara, M., Sougrat, R., Verbavatz, J. M., & Verkman, A. S. (2002). Impaired stratum corneum hydration in mice lacking epidermal water channel aquaporin-3. *The Journal of Biological Chemistry, 277*, 17147–17153.

Ma, T., Song, Y., Gillespie, A., Carlson, E. J., Epstein, C. J., & Verkman, A. S. (1999). Defective secretion of saliva in transgenic mice lacking aquaporin-5 water channels. *The Journal of Biological Chemistry, 274*, 20071–20074.

Ma, T., Song, Y., Yang, B., Gillespie, A., Carlson, E. J., Epstein, C. J., et al. (2000). Nephrogenic diabetes insipidus in mice lacking aquaporin-3 water channels. *Proceedings of the National Academy of Sciences of the United States of America, 97*, 4386–4391.

Ma, T., Yang, B., Gillespie, A., Carlson, E. J., Epstein, C. J., & Verkman, A. S. (1997). Generation and phenotype of a transgenic knockout mouse lacking the mercurial-insensitive water channel aquaporin-4. *The Journal of Clinical Investigation, 100*, 957–962.

Ma, T., Yang, B., Gillespie, A., Carlson, E. J., Epstein, C. J., & Verkman, A. S. (1998). Severely impaired urinary concentrating ability in transgenic mice lacking aquaporin-1 water channels. *The Journal of Biological Chemistry, 273*, 4296–4299.

Maeda, N., Hibuse, T., & Funahashi, T. (2009). Role of aquaporin-7 and aquaporin-9 in glycerol metabolism; involvement in obesity. *Handbook of Experimental Pharmacology, 190*, 233–249.

Maidhof, R., Jacobsen, T., Papatheodorou, A., & Chahine, N. O. (2014). Inflammation induces irreversible biophysical changes in isolated nucleus pulposus cells. *PLoS One, 9*, e99621.

Manley, G. T., Fujimura, M., Ma, T., Noshita, N., Filiz, F., Bollen, A. W., et al. (2000). Aquaporin-4 deletion in mice reduces brain edema after acute water intoxication and ischemic stroke. *Nature Medicine, 6*, 159–163.

Matricon, J., Barnich, N., & Ardid, D. (2010). Immunopathogenesis of inflammatory bowel disease. *Self/Nonself, 1*, 299–309.

Medici, V., Frassoni, C., Tassi, L., Spreafico, R., & Garbelli, R. (2011). Aquaporin 4 expression in control and epileptic human cerebral cortex. *Brain Research*, *1367*, 330–339.
Medraño-Fernandez, I., Bestetti, S., Bertolotti, M., Bienert, G. P., Bottino, C., Laforenza, U., et al. (2016). Stress regulates aquaporin-8 permeability to impact cell growth and survival. *Antioxidants & Redox Signaling*, *24*, 1031–1044.
Medzhitov, R. (2008). Origin and physiological roles of inflammation. *Nature*, *454*, 428–435.
Medzhitov, R. (2010). Inflammation 2010: New adventures of an old flame. *Cell*, *140*, 771–776.
Meli, R., Pirozzi, C., & Pelagalli, A. (2018). New perspectives on the potential role of aquaporins (AQPs) in the physiology of inflammation. *Frontiers in Physiology*, *9*, 101.
Miller, E. W., Dickinson, B. C., & Chang, C. J. (2010). Aquaporin-3 mediates hydrogen peroxide uptake to regulate downstream intracellular signaling. *Proceedings of the National Academy of Sciences of the United States of America*, *107*, 15681–15686.
Mobasheri, A., & Marples, D. (2004). Expression of the AQP-1 water channel in normal human tissues: A semiquantitative study using tissue microarray technology. *American Journal of Physiology. Cell Physiology*, *286*, C529–C537.
Mobasheri, A., Trujillo, E., Bell, S., Carter, S. D., Clegg, P. D., Martin-Vasallo, P., et al. (2004). Aquaporin water channels AQP1 and AQP3, are expressed in equine articular chondrocytes. *Veterinary Journal*, *168*, 143–150.
Moon, C., Soria, J. C., Jang, S. J., Lee, J., Obaidul, H. M., Sibony, M., et al. (2003). Involvement of aquaporins in colorectal carcinogenesis. *Oncogene*, *22*, 6699–6703.
Morishita, Y., Matsuzaki, T., Hara-chikuma, M., Andoo, A., Shimono, M., Matsuki, A., et al. (2005). Disruption of aquaporin-11 produces polycystic kidneys following vacuolization of the proximal tubule. *Molecular and Cellular Biology*, *25*, 7770–7779.
Musa-Aziz, R., Chen, L. M., Pelletier, M. F., & Boron, W. F. (2009). Relative CO2/NH3 selectivities of AQP1, AQP4, AQP5, AmtB, and RhAG. *Proceedings of the National Academy of Sciences of the United States of America*, *106*, 5406–5411.
Nagahama, M., Ma, N., Semba, R., & Naruse, S. (2006). Aquaporin 1 immunoreactive enteric neurons in the rat ileum. *Neuroscience Letters*, *395*, 206–210.
Nagahara, M., Waguri-Nagaya, Y., Yamagami, T., Aoyama, M., Tada, T., Inoue, K., et al. (2010). TNF-alpha-induced aquaporin 9 in synoviocytes from patients with OA and RA. *Rheumatology*, *49*, 898–906.
Nagelhus, E. A., & Ottersen, O. P. (2013). Physiological roles of aquaporin-4 in brain. *Physiological Reviews*, *93*, 1543–1562.
Nandasena, B. G., Suzuki, A., Aita, M., Kawano, Y., Nozawa-Inoue, K., & Maeda, T. (2007). Immunolocalization of aquaporin-1 in the mechano receptive Ruffini endings in the periodontal ligament. *Brain Research*, *1157*, 32–40.
Nielsen, S., Chou, C. L., Marples, D., Christensen, E. I., Kishore, B. K., & Knepper, M. A. (1995). Vasopressin increases water permeability of kidney collecting duct by inducing translocation of aquaporin-CD water channels to plasma membrane. *Proceedings of the National Academy of Sciences of the United States of America*, *92*, 1013–1017.
Nielsen, S., Frøkiaer, J., Marples, D., Kwon, T. H., Agre, P., & Knepper, M. A. (2002). Aquaporins in the kidney: From molecules to medicine physiological reviews. *Physiological Reviews*, *82*, 205–244.
Nielsen, S., King, L. S., Christensen, B. M., & Agre, P. (1997). Aquaporins in complex tissues. II. Subcellular distribution in respiratory and glandular tissues of rat. *The American Journal of Physiology*, *273*, C1549–C1561.
Nielsen, S., Smith, B. L., Christensen, E. I., Knepper, M. A., & Agre, P. (1993). CHIP28 water channels are localized in constitutively water-permeable segments of the nephron. *The Journal of Cell Biology*, *120*, 371–383.

Noda, Y., Sohara, E., Ohta, E., & Sasaki, S. (2010). Aquaporins in kidney pathophysiology. *Nature Reviews Nephrology*, *6*, 168–178.

Oklinski, M. K., Skowronski, M. T., Skowronska, A., Rützler, M., Nørgaard, K., Nieland, J. D., et al. (2016). Aquaporins in the Spinal Cord. *International Journal of Molecular Sciences*, *17*, 2050.

Oliveira, C. A., Carnes, K., Franca, L. R., Hermo, L., & Hess, R. A. (2005). Aquaporin-1 and -9 are differentially regulated by oestrogen in the efferent ductule epithelium and initial segment of the epididymis. *Biology of the Cell*, *97*, 385–395.

Ortega-Gómez, A., Perretti, M., & Soehnlein, O. (2013). Resolution of inflammation: An integrated view. *EMBO Molecular Medicine*, *5*, 661–674.

Oshio, K., Binder, D. K., Yang, B., Schecter, S., Verkman, A. S., & Manley, G. T. (2004). Expression of aquaporin water channels in mouse spinal cord. *Neuroscience*, *127*, 685–693.

Oshio, K., Watanabe, H., Song, Y., Verkman, A. S., & Manley, G. T. (2005). Reduced cerebrospinal fluid production and intracranial pressure in mice lacking choroid plexus water channel Aquaporin-1. *The FASEB Journal*, *19*, 76–78.

Papadopoulos, M. C., Saadoun, S., & Verkman, A. S. (2008). Aquaporins and cell migration. *Pflügers Archiv*, *456*, 693–700.

Papadopoulos, M. C., & Verkman, A. S. (2013). Aquaporin water channels in the nervous system. *Nature Reviews. Neuroscience*, *14*, 265–277.

Pelagalli, A., Squillacioti, C., Mirabella, N., & Meli, R. (2016). Aquaporins in health and disease: An overview focusing on the gut of different species. *International Journal of Molecular Sciences*, *17*, E1213.

Petr, R., Lunde, L. K., Frye, S. A., Støen, M., Nygård, S., Morth, J. P., et al. (2015). Reduced expression of aquaporins in human intestinal mucosa in early stage inflammatory bowel disease. *Clinical and Experimental Gastroenterology*, *8*, 49–67.

Preston, G. M., Smith, B. L., Zeidel, M. L., Moulds, J. J., & Agre, P. (1994). Mutations in aquaporin-1 in phenotypically normal humans without functional CHIP water channels. *Science*, *265*, 1585–1587.

Richardson, S., Neama, G., Phillips, T., Bell, S., Carter, S. D., Moley, K. H., et al. (2003). Molecular characterization and partial cDNA cloning of facilitative glucose transporters expressed in human articular chondrocytes; stimulation of 2-deoxyglucose uptake by IGF-I and elevated MMP-2 secretion by glucose deprivation. *Osteoarthritis and Cartilage*, *11*, 92–101.

Ring, T., Kallenbach, M., Praetorius, J., Nielsen, S., & Melgaard, B. (2006). Successful treatment of a patient with primary Sjögren's syndrome with Rituximab. *Clinical Rheumatology*, *25*, 891–894.

Rojek, A. M., Skowronski, M. T., Füchtbauer, E. M., Füchtbauer, A. C., Fenton, R. A., Agre, P., et al. (2007). Defective glycerol metabolism in aquaporin 9 (AQP9) knockout mice. *Proceedings of the National Academy of Sciences of the United States of America*, *104*, 3609–3614.

Rump, K., & Adamzik, M. (2018). Function of aquaporins in sepsis: A systematic review. *Cell & Bioscience*, *8*, 10.

Saadoun, S., Papadopoulos, M. C., Hara-Chikuma, M., & Verkman, A. S. (2005). Impairment of angiogenesis and cell migration by targeted of aquaporin-1 gene disruption. *Nature*, *434*, 786–792.

Saadoun, S., Papadopoulos, M. C., Watanabe, H., Yan, D., Manley, G. T., & Verkman, A. S. (2005). Involvement of aquaporin-4 in astroglial cell migration and glial scar formation. *Journal of Cell Science*, *118*, 5691–5698.

Sands, J. M., & Layton, H. E. (2009). The physiology of urinary concentration: An update. *Seminars in Nephrology*, *29*, 178–195.

Schnermann, J., Chou, C. L., Ma, T., Traynor, T., Knepper, M. A., & Verkman, A. S. (1998). Defective proximal tubular fluid reabsorption in transgenic aquaporin-1 null mice. *Proceedings of the National Academy of Sciences of the United States of America*, *95*, 9660–9664.

Schrier, R. W., Cadnapaphornchai, M. A., & Ohara, M. (2001). Water retention and aquaporins in heart failure, liver disease and pregnancy. *Journal of the Royal Society of Medicine*, *94*, 265–269.

Schuppan, D., & Afdhal, N. H. (2008). Liver cirrhosis. *Lancet*, *371*, 838–851.

Shanahan, C. M., Connolly, D. L., Tyson, K. L., Cary, N. R., Osbourn, J. K., Agre, P., et al. (1999). Aquaporin-1 is expressed by vascular smooth muscle cells and mediates rapid water transport across vascular cell membranes. *Journal of Vascular Research*, *36*, 353–362.

Shen, Y., Wang, Y., Chen, Z., Wang, D., Wang, X., Jin, M., et al. (2011). Role of aquaporin 5 in antigen-induced airway inflammation and mucous hyperproduction in mice. *Journal of Cellular and Molecular Medicine*, *15*, 1355–1363.

Sisto, M., Lorusso, L., Ingravallo, G., Nico, B., Ribatti, D., Ruggieri, S., et al. (2017). Abnormal distribution of AQP4 in minor salivary glands of primary Sjögren's syndrome patients. *Autoimmunity*, *50*, 202–210.

Song, Y., Sonawane, N., & Verkman, A. S. (2002). Localization of aquaporin-5 in sweat glands and functional analysis using knockout mice. *The Journal of Physiology*, *541*, 561–568.

Song, Y., & Verkman, A. S. (2001). Aquaporin-5 dependent fluid secretion in airway submucosal glands. *The Journal of Biological Chemistry*, *276*, 41288–41292.

Sougrat, R., Morand, M., Gondran, C., Barré, P., Gobin, R., Bonté, F., et al. (2002). Functional expression of AQP3 in human skin epidermis and reconstructed epidermis. *The Journal of Investigative Dermatology*, *118*, 678–685.

Steinfeld, S., Cogan, E., King, L. S., Agre, P., Kiss, R., & Delporte, C. (2001). Abnormal distribution of aquaporin-5 water channel protein in salivary glands from Sjögren's syndrome patients. *Laboratory Investigation*, *81*, 143–148.

Tait, M. J., Saadoun, S., Bell, B. A., & Papadopoulos, M. C. (2008). Water movements in the brain: Role of aquaporins. *Trends in Neurosciences*, *31*, 37–43.

Thi, M. M., Spray, D. C., & Hanani, M. (2008). Aquaporin-4 water channels in enteric neurons. *Journal of Neuroscience Research*, *86*, 448–456.

Thiagarajah, J. R., Zhao, D., & Verkman, A. S. (2007). Impaired enterocyte proliferation in aquaporin-3 deficiency in mouse models of colitis. *Gut*, *56*, 1529–1535.

Thomas, R., Kanso, A., & Sedor, J. R. (2008). Chronic kidney disease and its complications. *Primary Care*, *35*, 329–344vii.

Tomás-Camardiel, M., Venero, J. L., de Pablos, R. M., Rite, I., Machado, A., & Cano, J. (2004). In vivo expression of aquaporin-4 by reactive microglia. *Journal of Neurochemistry*, *91*, 891–899.

Tomura, S., Nawashiro, H., Otani, N., Uozumi, Y., Toyooka, T., Ohsumi, A., et al. (2011). Effect of decompressivecraniectomy on aquaporin-4 expression after lateral fluid percussion injury in rats. *Journal of Neurotrauma*, *28*, 237–243.

Towne, J. E., Harrod, K. S., Krane, C. M., & Menon, A. G. (2000). Decreased expression of aquaporin (AQP)1 and AQP5 in mouse lung after acute viral infection. *American Journal of Respiratory Cell and Molecular Biology*, *22*, 34–44.

Tradtrantip, L., Tajima, M., Li, L., & Verkman, A. (2009). Aquaporin water channels in transepithelial fluid transport. *The Journal of Medical Investigation*, *56*, 179–184.

Trujillo, E., González, T., Marín, R., Martín-Vasallo, P., Marples, D., & Mobasheri, A. (2004). Human articular chondrocytes, synoviocytes and synovial microvessels express aquaporin water channels; upregulation of AQP1 in rheumatoid arthritis. *Histology and Histopathology*, *19*, 435–444.

Tsukaguchi, H., Shayakul, C., Berger, U. V., Mackenzie, B., Devidas, S., Guggino, W. B., et al. (1998). Molecular characterization of a broad selectivity neutral solute channel. *The Journal of Biological Chemistry*, *273*, 24737–24743.

Verbalis, J. G. (2003). Disorders of body water homeostasis. *Best Practice & Research. Clinical Endocrinology & Metabolism*, *17*, 471–503.

Verkman, A. S. (2002). Aquaporin water channels and endothelial cell function. *Journal of Anatomy*, *200*, 617–627.

Verkman, A. S. (2006). Roles of aquaporins in kidney revealed by transgenic mice. *Seminars in Nephrology*, *26*, 200–208.

Verkman, A. S. (2007). Role of aquaporins in lung fluid physiology. *Respiratory Physiology & Neurobiology*, *159*, 324–330.

Verkman, A. S. (2008). Dissecting the roles of aquaporins in renal pathophysiology using transgenic mice. *Seminars in Nephrology*, *28*, 217–226.

Verkman, A. S. (2009). Knock-out models reveal new aquaporin functions. *Handbook of Experimental Pharmacology*, *190*, 359–381.

Verkman, A. S. (2012). Aquaporins in clinical medicine. *Annual Review of Medicine*, *63*, 303–316.

Verkman, A. S. (2013). Aquaporins. *Current Biology*, *23*, R52–R55.

Verkman, A. S., Anderson, M. O., & Papadopoulos, M. C. (2014). Aquaporins: Important but elusive drug targets. *Nature Reviews. Drug Discovery*, *13*, 259–277.

Verkman, A. S., & Mitra, A. K. (2000). Structure and function of aquaporin water channels. *American Journal of Physiology. Renal Physiology*, *278*, F13–F28.

Verkman, A. S., Ruiz-Ederra, J., & Levin, M. H. (2008). Functions of aquaporins in the eye. *Progress in Retinal and Eye Research*, *27*, 420–433.

Wang, H., Hao, X., Zhang, L., Liu, S., Wang, Q., Liu, N., et al. (2015). Decreased expression of aquaporin-1 in lung tissue of silicotic rats. *Clinical Laboratory*, *61*, 1163–1169.

Wang, W., Li, C., Summer, S. N., Falk, S., Wang, W., Ljubanovic, D., et al. (2008). Role of AQP1 in endotoxemia-induced acute kidney injury. *American Journal of Physiology. Renal Physiology*, *294*, 80.

Wang, Y., & Tajkhorshid, E. (2010). Nitric oxide conduction by the brain aquaporin AQP4. *Proteins*, *78*, 661–670.

Wang, Y., Zhang, W., Yu, G., Liu, Q., & Jin, Y. (2018). Cytoprotective effect of aquaporin 1 against lipopolysaccharide-induced apoptosis and inflammation of renal epithelial HK-2 cells. *Experimental and Therapeutic Medicine*, *15*, 4243–4252.

Wellner, R. B., Hoque, A. T., Goldsmith, C. M., & Baum, B. J. (2000). Evidence that aquaporin-8 is located in the basolateral membrane of rat submandibular gland acinar cells. *Pflügers Archiv*, *441*, 49–56.

Wellner, R. B., Redman, R. S., Swaim, W. D., & Baum, B. J. (2006). Further evidence for AQP8 expression in the myoepithelium of rat submandibular and parotid glands. *Pflügers Archiv*, *451*, 642–645.

Wilson, J. L., Miranda, C. A., & Knepper, M. A. (2013). Vasopressin and the regulation of aquaporin-2. *Clinical and Experimental Nephrology*, *17*, 751–764.

Wu, B., & Beitz, E. (2007). Aquaporins with selectivity for unconventional permeants. *Cellular and Molecular Life Sciences*, *64*, 2413–2421.

Wynn, T. A., & Vannella, K. M. (2016). Macrophages in tissue repair, regeneration, and fibrosis. *Immunity*, *44*, 450–462.

Yasui, M., Hazama, A., Kwon, T. H., Nielsen, S., Guggino, W. B., & Agre, P. (1999). Rapid gating and anion permeability of an intracellular aquaporin. *Nature*, *402*, 184–187.

Yousif, A., & Bhave, M. (2011). Identification of PIP aquaporin genes from wheat. *World Academy of Science, Engineering and Technology*, *59*, 2192–2196.

Zahn, A., Moehle, C., Langmann, T., Ehehalt, R., Autschbach, F., Stremmel, W., et al. (2007). Aquaporin-8 expression is reduced in ileum and induced in colon of patients with ulcerative colitis. *World Journal of Gastroenterology*, *13*, 1687–1695.
Zhang, D., Vetrivel, L., & Verkman, A. S. (2002). Aquaporin deletion in mice reduces intraocular pressure and aqueous fluid production. *The Journal of General Physiology*, *119*, 561–569.
Zhou, W. C., Zhang, Q. B., & Qiao, L. (2014). Pathogenesis of liver cirrhosis. *World Journal of Gastroenterology*, *20*, 7312–7324.
Zhu, C., Chen, Z., & Jiang, Z. (2016). Expression, distribution and role of aquaporin water channels in human and animal stomach and intestines. *International Journal of Molecular Sciences*, *17*, 1399.

CHAPTER NINE

Intracellular protein complexes involved in synapse assembly in presynaptic neurons

Kyung Ah Han, Ji Won Um, Jaewon Ko*
Department of Brain and Cognitive Sciences, Daegu Gyeongbuk Institute of Science and Technology (DGIST), Daegu, Korea
*Corresponding author: e-mail address: jaewonko@dgist.ac.kr

Contents

Abstract

The presynaptic active zone, composed of evolutionarily conserved protein complexes, is a specialized area that serves to orchestrate precise and efficient neurotransmitter release by organizing various presynaptic proteins involved in mediating docking and priming of synaptic vesicles, recruiting voltage-gated calcium channels, and modulating presynaptic nerve terminals with aligned postsynaptic structures. Among membrane proteins localized to active zone, presynaptic neurexins and LAR-RPTPs (leukocyte common antigen-related receptor tyrosine phosphatase) have emerged as hubs that orchestrate both shared and distinct extracellular synaptic adhesion pathways. In this chapter, we discuss intracellular signaling cascades involved in recruiting various intracellular proteins at both excitatory and inhibitory synaptic sites. In particular,

Advances in Protein Chemistry and Structural Biology, Volume 116
ISSN 1876-1623
https://doi.org/10.1016/bs.apcsb.2018.11.008

we highlight recent studies on key active zone proteins that physically and functionally link these cascades with neurexins and LAR-RPTPs in both vertebrate and invertebrate model systems. These studies allow us to build a general, universal view of how presynaptic active zones operate together with postsynaptic structures in neural circuits.

1. Introduction

The brain is composed of an astronomical number of nerve cells that form the neural circuits and synapses that underlie all brain functions, ranging from thinking and feeling to learning and memory. Synapses—specialized intercellular junctions between a presynaptic nerve cell and a postsynaptic nerve cell—mediate the synaptic transmissions that transfer and process neural information throughout the nervous system (Biederer, Kaeser, & Blanpied, 2017; Südhof, 2017a). Decades of extensive research have achieved significant progress in defining key mechanisms underlying the development of the presynaptic nerve terminal, synaptic cleft, and postsynaptic specialization (Biederer et al., 2017). A "nanocolumn" model, integrating diverse molecular machineries involved in synapse organization, was recently proposed to account for the synaptic alignment of neurotransmitter release sites and reception sites (Tang et al., 2016).

A variety of *trans*-synaptic adhesion molecules that form laterally connected net-like structures may pattern the heterogeneity of synaptic clefts (Lucic, Yang, Schweikert, Forster, & Baumeister, 2005). Molecular identities of multiple *trans*-synaptic adhesion complexes have recently been unveiled, partly due to accelerated development of mass spectrometry-based technologies (Aebersold & Mann, 2016; Cravatt, Simon, & Yates, 2007). A selected subset of *trans*-synaptic adhesion molecules has been demonstrated to function as key synaptic organizers (Condomitti & de Wit, 2018; de Wit & Ghosh, 2016; Park, Bae, Yoon, & Ko, 2018; Südhof, 2017b; Um & Ko, 2017). Among them, presynaptic neurexins and LAR-RPTPs (leukocyte common antigen-related receptor protein tyrosine phosphatases) have been well established to function as "hubs" in mammalian brains that orchestrate discrete synaptic signaling pathways by interacting with non-overlapping postsynaptic synaptogenic adhesion molecules (Han, Jeon, Um, & Ko, 2016; Südhof, 2017b; Um & Ko, 2013). Intriguingly, both neurexins and LAR-RPTPs possess evolutionarily conserved paralogs in invertebrates, suggesting that a comprehensive understanding of synapse organization could be obtained by investigating these molecular complexes (Chia, Li, & Shen, 2013; Petzoldt, Lutzkendorf, & Sigrist, 2016).

Recent investigations have been geared toward identification of the extracellular mechanisms underlying neurexin-based and LAR-RPTP-based synapse organizations (Südhof, 2017b; Um & Ko, 2013) (Fig. 1). Moreover, extensive crystallography-based structural analyses have provided key insights into how *trans*-synaptic adhesion pathways are structurally shaped (Coles, Jones, & Aricescu, 2015; Han et al., 2016; Missler, Südhof, & Biederer, 2012). Strikingly, neurexins modulate presynaptic neurotransmitter release—probability through signaling based on neuroligin–PSD-95 (postsynaptic density 95) interactions (Futai et al., 2007), and link various postsynaptogenic signals in a retrograde manner to mediate presynaptic assembly using as yet unidentified co-receptors (Gokce & Südhof, 2013). In contrast, LAR-RPTPs, through their D2 domain, employ liprin-α family scaffold proteins and selected substrates to perform a similar role (Han et al., 2018) (Fig. 2). Although these mechanisms appear to be dissimilar, liprin-α proteins are required in common for recruitment of the presynaptic machineries necessary for the downstream actions of neurexins and LAR-RPTPs (Han et al., 2018). These studies suggest the tantalizing concept that neurexins and LAR-RPTPs crosstalk with each other in a context-dependent manner to encode the identity of associated neural circuits. However, it remains to be determined how neurexins or LAR-RPTPs functionally and physically couple with a variety of presynaptic proteins, particularly those residing in the presynaptic active zone (Biederer et al., 2017; Südhof, 2012).

In this chapter, we highlight recent reports on intracellular signaling mechanisms mediated by various presynaptic proteins that interact with neurexins or LAR-RPTPs. As a complement to studies in vertebrates, we also summarize related progress in invertebrate model organisms. Lastly, we provide future perspectives on how *trans*-cellular nanocolumns, together with active zone functions, contribute to the molecular organization of various synapse types, and by extension, neural circuit types.

2. Intracellular synaptic signaling by presynaptic neurexins

2.1 Neurexin-interacting intracellular proteins

Neurexins, mainly expressed in presynaptic neurons, are type-I transmembrane proteins that function as key neuronal cell-adhesion molecules by interacting with various extracellular postsynaptic membrane proteins, including neuroligins, LRRTMs (leucine-rich repeat transmembrane proteins),

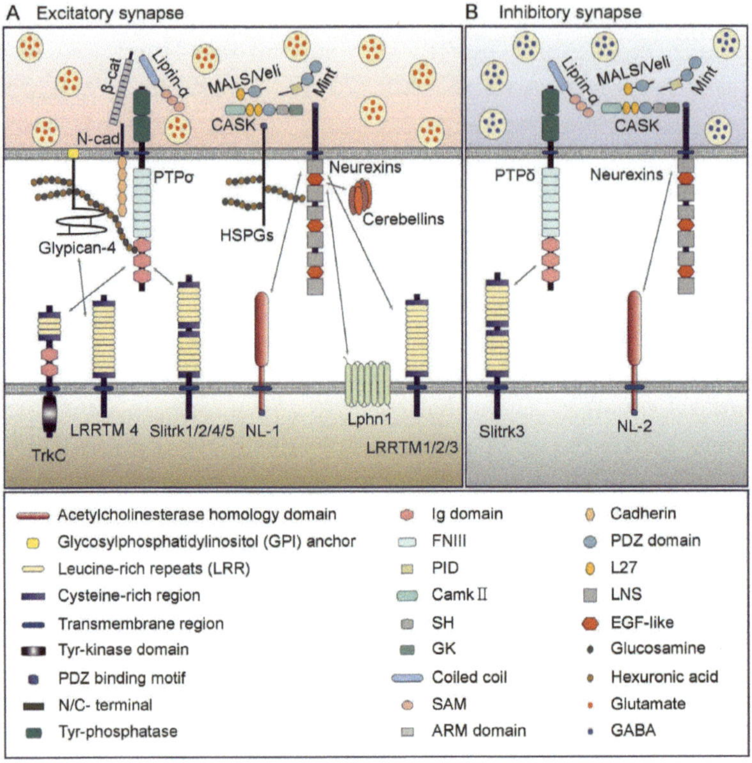

Fig. 1 Overview of presynaptic transmembrane proteins involved in synapse organization. (A) Excitatory synapse development through molecular interactions between key presynaptic cell-adhesion molecules with diverse extracellular and intracellular proteins. Neurexins control excitatory presynapse development through interactions with neuroligin-1 (NL-1) and LRRTMs. LAR-RPTPs, on the other hand, govern excitatory synapse development through interactions with Slitrks (except for Slitrk3) and TrkC. The heparan sulfate chains of heparan sulfate proteoglycans (e.g., glypicans and syndecans) promote excitatory synapse development by enhancing the binding affinity of neurexins and PTPσ for their respective postsynaptic ligands. (B) Inhibitory synapse development through molecular interactions of presynaptic cell-adhesion molecules with diverse extracellular and intracellular proteins. Neurexins control inhibitory presynapse development through interactions with neuroligin-2 (NL-2). PTPδ, on the other hand, induce inhibitory synapses through interaction with Slitrk3. Note that NL-2 and Slitrk3 were reported to directly bind to each other. Abbreviations: *CASK*, calcium/calmodulin-dependent serine protein kinase; *HSPGs*, heparan sulfate proteoglycans; *Lphn1*, latrophilin-1; *LRRTMs*, leucine-rich repeat transmembrane proteins; *MALS/Veli*, mammalian LIN-7/vertebrate homolog of LIN-7; *Mint*, Munc-18-interacting protein; *NL-1*, neuroligin-1; *NL-2*, neuroligin-2; *PTPδ*, receptor tyrosine phosphatase delta; *PTPσ*, receptor tyrosine phosphatase sigma; *Slitrk*, Slit- and Trk-like protein; and *TrkC*, neurotrophin receptor tyrosine kinase C.

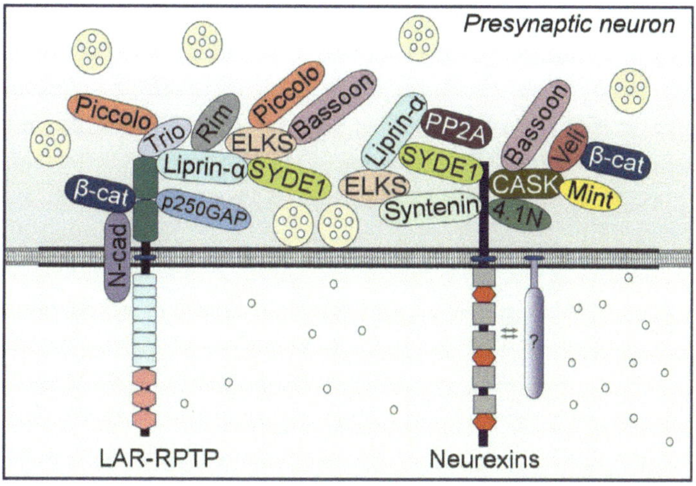

Fig. 2 Model of actions of vertebrate neurexins and LAR-RPTPs in mediating assembly of presynaptic nerve terminals. Neurexins and LAR-RPTPs mediate the assembly of presynaptic nerve terminals by coordinating various intracellular presynaptic proteins. Neurexins and LAR-RPTPs possess distinct sets of intracellular binding proteins. Specifically, neurexins directly bind to CASK, syntenins, and 4.1N protein, whereas LAR-RPTPs directly interact with liprin-α, β-catenin, N-cadherin, p250GAP, and Trio. Note that these intracellular proteins are indirectly associated with each other through multiple interactions, likely forming a large complex (i.e., neurexins and LAR-RPTPs are physically complexed at intracellular parts).

cerebellins, latrophilins, and others (Südhof, 2017b) (Fig. 1). The neurexin family is composed of the longer α-neurexins and shorter β-neurexins, which share most of the same extracellular ligands; both α- and β-neurexins undergo extensive alternative splicing events, notably at splice site #4, that are important in determining ligand binding (Südhof, 2017b). Neurexins bind to a variety of intracellular proteins through distinct cytosolic domains, including an erythrocyte membrane cytoskeletal protein 4.1-binding motif and a C-terminal PDZ-binding motif (Reissner, Runkel, & Missler, 2013). Here, we focus on recent studies on selective subsets of intracellular proteins that associate with neurexins in neurons. We will draw heavily on both vertebrate and invertebrate studies in our account of the roles of these neurexin-interacting proteins in neurexin-mediated presynaptic assembly.

2.1.1 Calcium/calmodulin-dependent serine protein kinase (CASK)

CASK is a scaffold protein ubiquitously expressed in neuronal synapses that plays an important role in neuronal development. CASK was originally discovered in yeast two-hybrid screens as a binding partner of neurexins (Hata, Butz, & Südhof, 1996). It was subsequently demonstrated to be involved in synaptic functions, polarized and regulated synaptic targeting, and surface trafficking of neurexins (Atasoy et al., 2007; Fairless et al., 2008; Gokce & Südhof, 2013). Notably, CASK levels are upregulated in human embryonic stem cells containing heterozygous conditional mutations in neurexin 1 (*NRXN1*) (Pak et al., 2015). Although CASK is expressed in both presynaptic and postsynaptic neurons, only CASK roles in presynaptic neurons are discussed in this review (for postsynaptic functions of CASK, see reviews by Yi-Ping Hsueh on mammalian CASK and Chen & Featherstone, 2011 on Drosophila CASK). CASK is involved in organizing the presynaptic release machinery through assembly of multiprotein complexes involving MALS/Veli (mammalian LIN-7/vertebrate homolog of LIN-7), Mint-1 (munc-18 interacting protein 1), liprin-α, and KIF1A (kinesin family member 1A) (Olsen et al., 2005; Shin et al., 2003; Spangler et al., 2013; Wu, Muthaiyan Shanmugam, Bhan, Huang, & Wagner, 2016). Intriguingly, mutations in the human *CASK* gene are associated with X-linked mental retardation and microcephaly with pontine and cerebellar hypoplasia (Hayashi et al., 2017).

2.1.2 Syntenin-1

Syntenin-1 is a tandem PDZ domain-containing scaffold that binds to various classes of membrane proteins, including syndecan (Grootjans et al., 1997), B class ephrins (Lin, Gish, Songyang, & Pawson, 1999; Torres et al., 1998), AMPA (α-amino-3-hydroxy-5-methyl-4-isoxazolepropionic acid), kainite receptors (Hirbec et al., 2003, 2002), glycine transporter 2 (Ohno et al., 2004), and a subset of synaptic adhesion molecules (Biederer et al., 2002; Koroll, Rathjen, & Volkmer, 2001). Intriguingly, syntenin-1 controls its associations with membrane proteins through high-affinity binding to phosphatidylinositol 4,5-biphosphate-containing lipid layers (Zimmermann et al., 2002). Importantly, syntenin-1 assists ephrin-B1 and -B2 in mediating EphB-dependent presynaptic development (McClelland, Sheffler-Collins, Kayser, & Dalva, 2009). Moreover, syntenin-1 directly binds to members of the ELKS family of active zone proteins, which further associate with RIMs (Rab3-interacting molecules), liprin-α, piccolo, and bassoon (Held & Kaeser, 2018; Ko, Na, Kim, Lee, & Kim, 2003; Ko et al., 2006). Although its precise roles in presynaptic development and underlying mechanisms remain unclear, syntenin-1 is likely involved in organizing the presynaptic machinery together with other interacting proteins. Syntenin-1 may control certain properties of synapses, and by extension, neural circuits. In support of this idea, a recent study using syntenin-1 knockout mice suggested that syntenin-1 regulates cued fear memory and learning involving activation of basolateral amygdala and infralimbic cortical neurons (Talukdar, Inoue, Yoshida, & Mori, 2018). Whether the related syntenin-2 performs similar roles in presynaptic neurons is a question that warrants comprehensive investigation.

2.1.3 4.1N protein

4.1N protein is a neuronal homolog of the erythrocyte membrane cytoskeletal protein 4.1 that directly binds to a membrane-proximal region of the AMPA receptor subunit GluA1 to link AMPA receptor surface expression with actin cytoskeletal association (Shen, Liang, Walensky, & Huganir, 2000). Coracle, a Drosophila homolog of mammalian 4.1N protein, was similarly reported to interact with glutamate receptors (Chen, Merino, Sigrist, & Featherstone, 2005). In addition to interacting with AMPA glutamate receptors, 4.1N protein also interacts through an apparently evolutionarily conserved domain with neurexin family proteins to contribute to the formation of septate or septate-like junctions (Baumgartner et al., 1996; Ward, Lamb, & Fehon, 1998). 4.1N protein also binds to a positively charged region of CASK and recruits actin and spectrin to the cytosolic domain of neurexins (Biederer & Südhof, 2001). These results suggest the possibility that 4.1N protein may assemble actin/spectrin microfilament structures to perform presynaptic functions, although neurexins do not appear to recruit actin to synaptogenic sites in the course of heterologous synapse formation (Gokce & Südhof, 2013).

2.1.4 Synapse-defective-1 (SYDE1)

SYD-1, a *Caenorhabditis elegans* homolog of mammalian SYDE1, was initially characterized as a protein that specifies axon identity in *C. elegans* (Hallam, Goncharov, McEwen, Baran, & Jin, 2002), whereas DSYD-1, a Drosophila homolog of SYDE1, was demonstrated to organize the synchronization of pre- and postsynaptic assembly by directly interacting with neurexins and Bruchpilot (a Drosophila homolog of mammalian ELKS family proteins) (Owald et al., 2010, 2012). DSYD-1 is critical for synaptic clustering and immobilization of Nrx-1 (an invertebrate homolog of canonical neurexins), presynaptic clustering of Bruchpilot and postsynaptic clustering of Nlg-1 (an invertebrate homolog of mammalian neuroligins), in part though collaboration with liprin-α proteins (Owald et al., 2012; Spinner, Walla, & Herman, 2018). Similarly, DSYD-1 operates upstream of liprin-α and protein phosphatase 2A (PP2A) to stabilize functional active zone specification (Li et al., 2014). Another study showed that a short isoform of DSYD-1 (i.e., Syd-1C lacking the PDZ domain) orchestrates UNC-41/DCC and SAX-3 (Robo) signaling to the MIG-2/Rac pathway, which serves to negatively regulate RhoGAP activity-mediated actin polymerization in the growth cone and promote axon guidance (Xu & Quinn, 2016). However, DSYD-1 also positively regulates the Rac pathway by promoting the guanine nucleotide exchange factor (GEF) activity of Trio in selecting synaptic target sites and promoting Bruchpilot clustering (Holbrook, Finley, Lyons, & Herman, 2012; Spinner et al., 2018), although the precise operation of this mechanism remains to be clearly defined. A more recent study provided intriguing evidence that the ratio of DSYD-1 to Bruchpilot varies across synapse types in the Drosophila brain and contributes to clustering of two different UNC-13 isoforms—one that is close to voltage-gated calcium channels (UNC13A) and one that is more remotely localized (UNC13B) (Fulterer et al., 2018). It is possible that similar mechanisms operate in mammalian synapses. In mammals (mice), mSYD1A (synapse-defective 1A) was identified as a regulator of presynaptic functions, including tethering of presynaptic vesicles at synapses (Wentzel et al., 2013). mSYD1A was shown to exhibit GTPase-activating protein (GAP) activity toward RhoA, that is not required for mSYD1A-dependent recruitment of synaptic vesicles (Wentzel et al., 2013). Moreover, mSYD1A forms biochemical complexes with presynaptic neurexin- and LAR-RPTP-dependent synapse formation (Wentzel et al., 2013). mSYD1A-knockout mice exhibit a loss of docked vesicles and impaired synaptic transmission in the hippocampus (Xu & Quinn, 2016).

2.2 Presynaptic assembly by neurexins: Invertebrate studies

The roles of invertebrate neurexins in mediating synapse assembly, synaptic connectivity, and cytoskeletal remodeling have been extensively studied (Calahorro & Izquierdo, 2018; Philbrook et al., 2018). In *C. elegans*, postsynaptic Nrx-1 and presynaptic Nlg-1 organize a retrograde signal that inhibits acetylcholine release at the neuromuscular junction (Hu et al., 2012; Tong et al., 2017). This inhibition is mediated by increased presynaptic levels of tomosyn proteins (Hu et al., 2012) and by inhibition of the function of presynaptic UNC-2/Ca_V2 calcium channels (Tong et al., 2017). From the standpoint of extracellular mechanisms, postsynaptic Nrx-1 is cleaved by the protease SUP-17/ADAM10, which releases the soluble ectodomain of Nrx-1; this domain, in turn, mediates retrograde inhibition by decreasing

UNC-2/Ca_V2 channel activity through interactions with presynaptic UNC-36/α2δ (an auxiliary subunit of UNC-2) (Tong et al., 2017). These observations have been replicated in studies on mammalian synapses, which showed that mammalian neurexin-1α binds selectively to α2δ-3 and decreases $Ca_V2.2$ currents, suggesting a conserved mechanism for neurexins in regulating synaptic assembly and transmission. At GABAergic synaptic sites in *C. elegans*, Nlg-1 localizes specifically to postsynapses (similar to a mammalian counterpart) and is required for UNC-49/$GABA_A$ receptor clustering (Maro et al., 2015; Zhang et al., 2010). Presynaptic punctin/MADD-4 (a secreted molecule resembling members of the ADAMTS family of proteases) binds to both Nlg-1 and Nrx-1, forming a tripartite receptor-ligand complex (Maro et al., 2015). Intriguingly, presynaptic punctin and Nrx-1 are cooperatively required for clustering of postsynaptic Nlg-1 and $GABA_A$ receptors, suggesting a mechanism that clusters neurotransmitter receptors locally to achieve the precise apposition between pre- and postsynaptic specializations (Maro et al., 2015). Like the corresponding *C. elegans* proteins, *Drosophila melanogaster* neurexin (DNrx) and neuroligin-1 (DNlg1) are critical for synaptic structure, function, and assembly, acting through various intracellular proteins, including DSYD-1, WAVE regulatory complex (WRC), Scribble, DPix, Wishful thinking (Wit), and spinophilin, to coordinate actin cytoskeletal organization (Banerjee, Venkatesan, & Bhat, 2017; Muhammad et al., 2015; Owald et al., 2012; Rui et al., 2017; Xing et al., 2018). Specifically, DNlg1 directly interacts with and recruits WRC to the postsynaptic membrane to promote presynaptic F-actin assembly (Xing et al., 2018), whereas DNrx directly interacts with the scaffold protein Scribble and the GEF protein DPix to activate Rac1 and subsequently stimulate the assembly of presynaptic F-actin necessary for synaptic vesicle clustering (Rui et al., 2017). Moreover, DNrx binds to spinophilin, antagonizing DSYD-1 to fine-tune DNrx-DNlg1 signaling for optimized active zone architecture and function (Muhammad et al., 2015). These ongoing studies in *C. elegans* and *D. melanogaster* strongly underscore the significance of DNrx and Nrx-1 and their ligands DNlg1 and Nlg-1 in regulating synaptic assembly, and provide important insights into how mammalian neurexins and their ligands similarly shape synaptic structure and function. Notably, among mammalian neurexin ligands, only neuroligins, latrophilins, and calsyntenin-3 are known to have invertebrate paralogs (Südhof, 2017b). Thus, investigations of the precise roles of lat-1 (a *C. elegans* homolog of mammalian latrophilins), dCirl (a *D. melanogaster* homolog of mammalian latrophilins), and CASY-1 (a *C. elegans* homolog of mammalian calsyntenin-3) are critical for expanding our current understanding of neurexin-mediated synaptic assembly in synaptic structures of invertebrates.

2.3 Presynaptic assembly by neurexins: Vertebrate studies

Mammalian neurexins were proposed to mediate the assembly of presynaptic terminals long before invertebrate neurexins were investigated. Dean et al. showed that oligomerization of neuroligin-1 induces presynaptic clustering of neurexins, which triggers the recruitment of synaptic vesicles through cytoplasmic domain-mediated interactions (Dean et al., 2003). However, Gokce and Südhof recently demonstrated that presynaptic neurexins do not require cytoplasmic domain-mediated interactions to trigger heterologous synapse formation (Gokce & Südhof, 2013) (Fig. 2). Moreover, monomeric neuroligin-1 was shown to retain its ability to bind to neurexins and recruit synaptic vesicles (Ko et al., 2009), suggesting that the interactions of neurexins with cognate postsynaptic ligands are sufficient to create the nascent synaptogenic hubs. The C-terminal PDZ domain-interaction motif of neurexins is required for surface transport, but is dispensable for heterologous synapse formation (Gokce & Südhof, 2013). Strikingly, the membrane-tethered monomeric laminin/neurexin/sex hormone-binding globulin (LNS) domain of β-neurexin was found to be sufficient for restoring the impaired heterologous synapse formation induced by neuroligin-1 or LRRTM2 (leucine-rich repeat transmembrane protein 2), both of which are neurexin ligands, in neurexin-deficient neurons when lentivirally expressed as a glycosylphosphatidylinositol-anchored protein (Gokce & Südhof, 2013). These results suggest that presynaptic neurexins require as yet unidentified co-receptors to propagate the retrograde signal originating from postsynaptic ligands to presynaptic neurons. Intriguingly, a recent study revealed that the presynaptic active zone proteins, liprin-α2 and liprin-α3, may be involved in neurexin-mediated presynaptic assembly (Han et al., 2018), suggesting that neurexins are functionally linked to intracellular machineries through co-receptor(s).

3. Intracellular synaptic signaling by presynaptic LAR-RPTPs

3.1 LAR-RPTP-interacting intracellular proteins

Like neurexins, LAR-RPTPs are evolutionarily conserved type-I transmembrane proteins that possess typical extracellular structural motifs and intracellular domains with tyrosine phosphatase activity (Um & Ko, 2013). LAR-RPTPs bind to an array of divergent postsynaptic ligands, distinct from postsynaptic neurexin ligands, through interactions that are governed by alternative splicing events at mini-exon A and B sites of LAR-RPTPs (Südhof, 2017b) (Fig. 1). LAR-RPTPs bind to diverse sets of intracellular proteins through the distal D2 domain and to a series of substrates through the proximal D1 domain (Um & Ko, 2013). Although most studies have focused on extracellular networks involving LAR-RPTPs and their distinct ligands, here we describe three specific intracellular proteins that bind to LAR-RPTPs in presynaptic neurons. As was the case for neurexins, recent studies on both vertebrates and invertebrates will be discussed, particularly with respect to the presynaptic functions of LAR-RPTPs.

3.1.1 α-Liprins

Liprin family proteins, the first intracellular binding proteins identified for LAR-RPTP family proteins, interact through their C-terminal SAM (sterile α-motif) domains (reviewed in Um & Ko, 2013). In *D. melanogaster*, liprins act in concert with other binding proteins, including DLAR, DTrio, N-cadherin, and kinesin-1, to play crucial roles in synapse formation, synaptic vesicle trafficking, and axon targeting (Astigarraga, Hofmeyer, Farajian, & Treisman, 2010; Kaufmann, DeProto, Ranjan, Wan, & Van Vactor, 2002; Miller et al., 2005; Prakash et al., 2009). In *C. elegans*, Syd-2/liprin regulates synapse formation and axon guidance, differentiation of presynaptic termini, presynaptic assembly, and activation of motor proteins (Ackley et al., 2005; Dai et al., 2006; Kittelmann et al., 2013; Patel et al., 2006; Taru & Jin, 2011; Wagner et al., 2009; Zhen & Jin, 1999). In mammals, members of the liprin-α family are differentially expressed at distinct synapse types and form macromolecular complexes with other active zone components (Han et al., 2016). Liprin-α1 is primarily localized at dendrites, where it regulates dendritic morphogenesis (Hoogenraad et al., 2007). Notably, presynaptic liprin-α2 acts as a key scaffold that organizes presynaptic ultrastructure and neurotransmitter release by regulating presynaptic vesicle pool size through recruitment of RIM1 and CASK proteins (Spangler et al., 2013). In mice, liprin-α2 directly interacts with mSYD1A to regulate tethering of presynaptic vesicles (Wentzel et al., 2013). However, using stimulated emission depletion super-resolution microscopy, a recent study showed that liprin-α3, but not liprin-α2, predominantly colocalizes with other active zone proteins (Wong et al., 2018). Moreover, liprin-α3 knockout mice exhibit reduced synaptic vesicle tethering and docking, and impaired vesicle exocytosis in hippocampal neurons (Wong et al., 2018). Intriguingly, deletion of liprin-α3 induces rearrangement of various active zone components, notably including translocation of liprin-α2 to the active zone. These results suggest that liprin-α3 has a unique scaffolding role in modulating liprin-α2 active zone localization. Future studies should resolve redundant and diverse roles of all four liprin-α proteins and other liprins (liprin-β and liprin-γ) in synaptic assembly.

3.1.2 p250RhoGAP

p250RhoGAP was initially identified as a binding protein for the NR2B subunit of the *N*-methyl-D-aspartate subtype of ionotropic glutamate receptors (NMDARs), which are concentrated in postsynaptic densities of mammalian brains (Nakazawa et al., 2003; Taniguchi et al., 2003). It also regulates dendritic spine morphology in an NMDAR-dependent manner (Nakazawa et al., 2008). Intriguingly, p250RhoGAP is a major neuronal target of microRNA 132 in the regulation of activity-dependent structural and functional plasticity, and leptin- and cAMP-response element-binding protein (CREB)-induced synaptogenesis (Dhar et al., 2014; Wayman et al., 2008). Collectively, these data suggest that the main function of p250RhoGAP is postsynaptic. However, p250RhoGAP regulates axon growth downstream of the Cdh1-APC/Smurf1 pathway (Kannan, Lee, Schwedhelm-Domeyer, Nakazawa, & Stegmuller, 2012) and axon branching downstream of brain-derived neurotrophic factor (BDNF) during retinocollicular/tectal map formation (Marler et al., 2014). In addition, p250RhoGAP was identified as a substrate for protein tyrosine phosphatase-σ (PTPσ), hinting at a potential role in modulating Rac signaling in presynaptic neurons (Chagnon et al., 2010). RRC-1 in *C. elegans* was identified as a homolog of mammalian p250RhoGAP (Delawary et al., 2007), but its functional significance has not yet been investigated. Evidence to support the significance of F-actin in presynaptic assembly has rapidly accumulated; thus, presynaptic active zone assembly downstream of LAR-RPTPs is likely to depend on actin cytoskeletal remodeling, in contrast to assembly downstream of neurexins (Gokce & Südhof, 2013).

3.1.3 β-Catenin

β-Catenin performs diverse roles in both presynaptic and postsynaptic neurons. In presynaptic neurons, β-catenin modulates vesicle localization by recruiting Veli to cadherin sites via its PDZ-binding motif to regulate a reserved pool of synaptic vesicles (Bamji et al., 2003). These functions are mediated by the action of BDNF, which is responsible for mobilizing synaptic vesicles by causing disengagement of cadherin adhesion complexes following tyrosine phosphorylation of β-catenin at tyrosine residue 654 (Y654) by Trk receptor kinases (Bamji, Rico, Kimes, & Reichardt, 2006). Moreover, dephosphorylation of β-catenin at tyrosine residue 654 (Y654) enhances the interaction of β-catenin with N-cadherin, resulting in increased postsynaptic surface levels of N-cadherin and an increase in the density of presynaptic vesicle proteins, number of active presynaptic boutons, and size of total recycling vesicles (Chen, Chen, Wang, Huang, & Tai, 2017). Although the interaction of β-catenin with N-cadherin is dispensable for presynaptic vesicle localization, it is essential for presynaptic vesicle turnover via an N-cadherin-mediated *trans*-synaptic pathway involving tyrosine phosphorylation-dependent signaling (Chen et al., 2017; Vitureira, Letellier, White, & Goda, 2012). In contrast, stimulation with hepatocyte growth factor induces tyrosine phosphorylation of β-catenin at Y142. This leads to dissociation of β-catenin from MET receptor tyrosine kinase accompanied by an increase in β-catenin/N-cadherin complexes, resulting in modulation of excitatory synapse formation (David et al., 2008; Xie, Eagleson, Wu, & Levitt, 2016). These reports suggest that the role of β-catenin in organizing synapse development reflects differential involvement of distinct tyrosine phosphorylation cascades in presynaptic and postsynaptic neurons. β-Catenin has also been identified as a direct substrate of LAR-RPTPs in the control of activity-dependent postsynaptic AMPA receptor targeting in postsynaptic neurons (Dunah et al., 2005; Muise et al., 2007; Murase, Mosser, & Schuman, 2002; Togashi et al., 2002). However, its significance as a substrate for LAR-RPTPs in presynaptic neurons has remained unclear (Han et al., 2018).

In addition to activity-dependent tyrosine phosphorylation signaling pathways, β-catenin also employs other mechanisms in its regulation of various aspects of synapse development, including local translation (Taylor, Wu, Tai, & Schuman, 2013), retrograde signaling involving Slit2 (Li et al., 2008; Wu et al., 2015), and actin cytoskeletal remodeling involving Scribble and β-PIX (a Rac/Cdc42 guanine nucleotide exchange factor) (Sun, Aiga, Yoshida, Humbert, & Bamji, 2009; Sun & Bamji, 2011). Moreover, presynaptic β-catenin, together with cadherin and p140Cap, is required for coordinated assembly of presynaptic transmitter release and stabilization of postsynaptic spines and synapse functions in the neocortex

(Li et al., 2017). Whether all of these mechanisms operate in both pre- and postsynaptic neurons across diverse brain areas is a question that needs to be more fully addressed to understand how β-catenin contributes to synapse assembly and stabilization.

3.1.4 Trio

Trio was originally identified as a Rac- and Rho-specific GEF for LAR-RPTPs (Debant et al., 1996). Intriguingly, Trio and DLar, together with Abl, Pak, and Dock proteins, act as key transducers of presynaptic axon guidance signals in Drosophila (Awasaki et al., 2000; Bateman, Shu, & Van Vactor, 2000; Newsome et al., 2000). In addition, Trio, DLar, and diaphanous (a formin) act in concert to regulate the growth of presynaptic boutons at the *Drosophila* neuromuscular junction (Pawson, Eaton, & Davis, 2008). Moreover, Trio is an essential component of the miniature (but not evoked) neurotransmission signaling mechanism in presynaptic neurons that is involved in regulating synaptic terminal maturation in *Drosophila* (Choi et al., 2014). In *C. elegans*, UNC-73/trio is crucial for neurotransmitter release involving dense core vesicle-mediated neuromodulation of locomotion (Hu, Pawson, & Steven, 2011). In vertebrates, Trio directly interacts with two key active zone proteins, piccolo and bassoon, in addition to LAR-RPTPs (Terry-Lorenzo et al., 2016), although the significance of these interactions is unclear. Postsynaptically, Trio promotes dendritic branching, field coverage, and dendritic outgrowth in various *Drosophila* sensory neurons in a cell-autonomous manner (Iyer et al., 2012). In rats, Trio is functionally redundant with Kalirin (a brain-specific RhoGEF) in acting as a substrate for CamKII to mediate AMPA receptor upregulation during induction of long-term potentiation (Herring & Nicoll, 2016). Intriguingly, a large cluster of *de novo* missense mutations was found in human patients with autism spectrum disorders (Sadybekov, Tian, Arnesano, Katritch, & Herring, 2017) or mild intellectual disability (Ba et al., 2016). These mutations produce either altered glutamatergic neurotransmission or accelerated maturation of neuronal network, thereby influencing both dendritic branching and synapse function.

3.2 Presynaptic assembly by LAR-RPTPs: Focus on vertebrate studies

Both neurexins and LAR-RPTPs are proposed to localize at the presynaptic active zone of mammalian brains (Südhof, 2012). Similar to neurexins, PTPσ forms higher-order assemblies, not only through ligand binding but also through interactions with glycosaminoglycan chains of heparan sulfate proteoglycans, likely leading to promotion of sensory neuron growth (Coles et al., 2011; Won et al., 2017) (Fig. 2). However, until recent reports (Han et al., 2018; Nakamura et al., 2017), it had remained unclear how this oligomerization status of LAR-RPTPs is transduced intracellularly in mammalian neurons (Fig. 1). Using systematic structure-function analyses in cultured hippocampal neurons to probe molecular mechanisms underlying PTPσ-mediated excitatory presynaptic differentiation and synapse development, Han et al. (2018) showed that PTPσ is required for Slitrk1- or TrkC-induced presynaptic differentiation through extracellular interactions involving immunoglobulin domains and heparan sulfates. They further found that, unlike the mechanisms of presynaptic assembly mediated by neurexins (Gokce & Südhof, 2013), presynaptic PTPσ requires cytoplasmic domain (D2 domain)-mediated scaffold interactions (with liprin-α2 and liprin-α3) and tyrosine phosphorylation pathways (with p250RhoGAP and N-cadherin) (Han et al., 2018). Intriguingly, knockdown of liprin-α2 and liprin-α3 proteins in cultured hippocampal neurons impaired presynaptic differentiation induced by both neurexins and LAR-RPTPs, suggesting that these two presynaptic adhesion molecules share common intracellular mechanisms in presynaptic neurons (Han et al., 2018). PTPδ also serves a postsynaptic role in regulating semaphorin-3A-mediated cortical dendritic outgrowth through activation of Fyn tyrosine kinase (Nakamura et al., 2017). These studies collectively suggest that tyrosine phosphorylation signaling pathways are crucial for regulation of various aspects of synapse development in both presynaptic and postsynaptic neurons. Future studies should investigate whether these mechanisms similarly apply in invertebrates.

4. Other presynaptic membrane proteins

4.1 Glypicans

Glypicans are an evolutionarily conserved family of heparan sulfate proteoglycans localized to neuronal membrane via a glycosylphosphatidylinositol (GPI) anchor (Condomitti & de Wit, 2018; Um & Ko, 2017) (Fig. 1). Two perisynaptically localized Drosophila glypicans, Dally and Dally-like, serve to restrict presynaptic active zone morphogenesis by outcompeting Drosophila syndecan (Sdc) through interactions with DLar (Johnson et al., 2006). Similar to the presynaptic role of Drosophila glypicans at neuromuscular junctions, mammalian glypicans, particularly glypican-4, also act as functional receptors for the postsynaptogenic adhesion molecule LRRTM4 (leucine-rich transmembrane protein 4), at excitatory synapses. There, they use PTPσ as a heparan sulfate-dependent co-receptor on the presynaptic membrane to organize presynaptic differentiation and neurotransmission (de Wit et al., 2013; Ko et al., 2015; Siddiqui et al., 2013). A recent study in *C. elegans* showed that heparan sulfate 3-*O*-sulfotransferase and two glypicans, lon-2/glypican and gpn-1/glypican, are critical for presynaptic specialization, particularly vesicle fusion, in B-type ray sensory neurons (Lazaro-Pena, Diaz-Balzac, Bulow, & Emmons, 2018). Moreover, glypican-4 and -6, expressed and secreted from astrocytes, positively regulate excitatory synapse formation in cultured retinal ganglion cells and promote AMPA receptor clustering (Allen et al., 2012). These proteins bind to presynaptic PTPσ and PTPδ and induce the release of neuronal pentraxin 1 from presynaptic neurons, actions that are accompanied by clustering of postsynaptic AMPA receptors (Farhy-Tselnicker et al., 2017). These studies consistently support the significance of presynaptic glypicans in synapse formation. Given that neurexins, as heparan sulfate proteoglycans, have recently been proposed to coordinate normal synapse development and mouse survival (Zhang et al., 2018), it is plausible that a subset of glypicans (and other heparan sulfate proteoglycans) can directly bind to neurexins in presynaptic neurons, acting as key regulatory codes to provide diverse fine-tuning functions.

4.2 Syndecans

Syndecans, composed of four members in mammals (syndecan-1–4), are membrane-bound heparan sulfate proteoglycans, like glypicans (Reizes, Benoit, & Clegg, 2008). Syndecan-2 and -3 exhibit both overlapping and distinct distributions in the forebrain, whereas syndecan-1 is expressed exclusively in the cerebellum, and syndecan-4 is expressed mostly in astrocytes (Hsueh & Sheng, 1999). Intriguingly, immunostaining experiments using syndecan-3-specific antibodies have shown that syndecan-3 exhibits an axonal distribution in the brain and in cultured neurons, consistent with its role as a receptor for heparin-binding growth-associated molecules in promoting neurite outgrowth (Hsueh & Sheng, 1999; Rauvala & Peng, 1997). Drosophila syndecan (Sdc) promotes the growth of presynaptic terminals by binding to DLar, a role distinct from that of Drosophila glypicans (Johnson et al., 2006). Postsynaptic syndecan-2 was recently shown to be required for presynaptic maturation through binding to presynaptic fibroblast growth factor 22 (FGF-22) (Hu, Umemori, & Hsueh, 2016). However, the contribution of syndecans to the assembly of presynaptic nerve terminals has not been extensively investigated in invertebrate or vertebrate models. Since both neurexins and syndecans share common intracellular binding proteins (CASK, syntenins, and 4.1N protein), and the intracellular region of neurexins is dispensable for the induction of presynaptic differentiation (Gokce & Südhof, 2013), it is conceivable that syndecans may act as co-receptors for neurexins to coordinate intracellular signaling cascades in presynaptic neurons. Future studies should directly address this possibility.

4.3 LRP4

Lrp4 (low-density lipoprotein receptor-related protein 4), a co-receptor of agrin (a motor neuron-derived ligand) that forms a complex with MuSK (a receptor tyrosine kinase expressed in skeletal muscle) in a manner that depends on agrin (Kim et al., 2008; Zhang et al., 2008), is critical for neuromuscular junction formation (Weatherbee, Anderson, & Niswander, 2006). Suppression of Lrp4 expression in muscle cells attenuates agrin binding, agrin-induced MuSK tyrosine phosphorylation, and acetylcholine receptor clustering (Zhang et al., 2008). Importantly, Lrp4-mediated neuromuscular junction formation is mediated by its extracellular, but not intracellular, region (Gomez & Burden, 2011). Moreover, Lrp4 directly binds to an unknown receptor in the axons of motoneurons and induces presynaptic differentiation by providing retrograde signals, independent of its role in coordinating MuSK activation-dependent postsynaptic differentiation (Yumoto, Kim, & Burden, 2012). In Drosophila, Lrp4 is localized to nerve terminals at or near active zones of excitatory neurons (Mosca, Luginbuhl, Wang, & Luo, 2017). Disrupting Lrp4 function in excitatory neurons reduces excitatory synapse number in both putative olfactory receptor neurons and projection neurons (Mosca et al., 2017). Intriguingly, presynaptic Lrp4 in olfactory receptor neurons acts through direct interactions with SRPK79D kinase to play a critical role in normal olfactory attraction behavior in Drosophila (Mosca et al., 2017). Future studies should explore whether the presynaptic functions revealed in Drosophila are conserved in the mammalian central nervous system.

5. Conclusions and future directions

Presynaptic neurexins and LAR-RPTPs, arguably the best-characterized synaptic adhesion molecules, *trans*-synaptically interact with non-overlapping postsynaptogenic adhesion molecules, most of which are essential for diverse facets of synapse and neural circuit development across vertebrates and invertebrates. However, investigations of intracellular protein machineries underlying the function of these adhesion molecules in extracellular signal propagation have been largely limited to invertebrate neurexins. Recent studies have just begun to unveil molecular mechanisms responsible for presynaptic assembly. Intriguingly, although neurexins and LAR-RPTPs appear to share similar intracellular molecular components, they employ completely different extracellular molecular components, suggesting that intracellular signaling pathways in presynaptic neurons may prove to be uninstructive with respect to specific functions. One intriguing hypothesis is that postsynaptic partner(s) of neurexins and LAR-RPTPs might specify the connectivity and properties of synapse and neural circuit types, in combination with specific variants of neurexins and LAR-RPTPs; notably, both neurexins and LAR-RPTPs are subject to extensive alternative splicing events. There are numerous unanswered questions surrounding neurexins, LAR-RPTPs, their respective postsynaptic ligands, and intracellular protein components in mediating synapse and circuit development. Efforts to address these questions will contribute to a better understanding of how synapses and neural circuits are molecularly organized in animals.

Acknowledgments

This work was supported by grants from the Korea Healthcare Technology R&D Project, funded by the Ministry for Health and Welfare Affairs, Republic of Korea (HI17C0080 to J.K.), and Basic Science Research Program through the National Research Foundation of Korea (NRF) funded by the Ministry of Education (2018R1A6A3A11045857 to K.A.H.).

References

Ackley, B. D., Harrington, R. J., Hudson, M. L., Williams, L., Kenyon, C. J., Chisholm, A. D., et al. (2005). The two isoforms of the Caenorhabditis elegans leukocyte-common antigen related receptor tyrosine phosphatase PTP-3 function independently in axon guidance and synapse formation. *The Journal of Neuroscience*, *25*(33), 7517–7528. https://doi.org/10.1523/JNEUROSCI.2010-05.2005.

Aebersold, R., & Mann, M. (2016). Mass-spectrometric exploration of proteome structure and function. *Nature*, *537*(7620), 347–355. https://doi.org/10.1038/nature19949.

Allen, N. J., Bennett, M. L., Foo, L. C., Wang, G. X., Chakraborty, C., Smith, S. J., et al. (2012). Astrocyte glypicans 4 and 6 promote formation of excitatory synapses via GluA1 AMPA receptors. *Nature*, *486*(7403), 410–414. https://doi.org/10.1038/nature11059.

Astigarraga, S., Hofmeyer, K., Farajian, R., & Treisman, J. E. (2010). Three Drosophila liprins interact to control synapse formation. *The Journal of Neuroscience*, *30*(46), 15358–15368. https://doi.org/10.1523/JNEUROSCI.1862-10.2010.

Atasoy, D., Schoch, S., Ho, A., Nadasy, K. A., Liu, X., Zhang, W., et al. (2007). Deletion of CASK in mice is lethal and impairs synaptic function. *Proceedings of the National Academy of Sciences of the United States of America, 104*(7), 2525–2530. https://doi.org/10.1073/pnas.0611003104.

Awasaki, T., Saito, M., Sone, M., Suzuki, E., Sakai, R., Ito, K., et al. (2000). The Drosophila trio plays an essential role in patterning of axons by regulating their directional extension. *Neuron, 26*(1), 119–131.

Ba, W., Yan, Y., Reijnders, M. R., Schuurs-Hoeijmakers, J. H., Feenstra, I., Bongers, E. M., et al. (2016). TRIO loss of function is associated with mild intellectual disability and affects dendritic branching and synapse function. *Human Molecular Genetics, 25*(5), 892–902. https://doi.org/10.1093/hmg/ddv618.

Bamji, S. X., Rico, B., Kimes, N., & Reichardt, L. F. (2006). BDNF mobilizes synaptic vesicles and enhances synapse formation by disrupting cadherin-beta-catenin interactions. *The Journal of Cell Biology, 174*(2), 289–299. https://doi.org/10.1083/jcb.200601087.

Bamji, S. X., Shimazu, K., Kimes, N., Huelsken, J., Birchmeier, W., Lu, B., et al. (2003). Role of beta-catenin in synaptic vesicle localization and presynaptic assembly. *Neuron, 40*(4), 719–731.

Banerjee, S., Venkatesan, A., & Bhat, M. A. (2017). Neurexin, neuroligin and wishful thinking coordinate synaptic cytoarchitecture and growth at neuromuscular junctions. *Molecular and Cellular Neurosciences, 78*, 9–24. https://doi.org/10.1016/j.mcn.2016.11.004.

Bateman, J., Shu, H., & Van Vactor, D. (2000). The guanine nucleotide exchange factor trio mediates axonal development in the Drosophila embryo. *Neuron, 26*(1), 93–106.

Baumgartner, S., Littleton, J. T., Broadie, K., Bhat, M. A., Harbecke, R., Lengyel, J. A., et al. (1996). A Drosophila neurexin is required for septate junction and blood-nerve barrier formation and function. *Cell, 87*(6), 1059–1068.

Biederer, T., Kaeser, P. S., & Blanpied, T. A. (2017). Transcellular nanoalignment of synaptic function. *Neuron, 96*(3), 680–696. https://doi.org/10.1016/j.neuron.2017.10.006.

Biederer, T., Sara, Y., Mozhayeva, M., Atasoy, D., Liu, X., Kavalali, E. T., et al. (2002). SynCAM, a synaptic adhesion molecule that drives synapse assembly. *Science (New York, N.Y.), 297*(5586), 1525–1531. https://doi.org/10.1126/science.1072356.

Biederer, T., & Südhof, T. C. (2001). CASK and protein 4.1 support F-actin nucleation on neurexins. *The Journal of Biological Chemistry, 276*(51), 47869–47876. https://doi.org/10.1074/jbc.M105287200.

Calahorro, F., & Izquierdo, P. G. (2018). The presynaptic machinery at the synapse of C. elegans. *Invertebrate Neuroscience, 18*(2), 4. https://doi.org/10.1007/s10158-018-0207-5.

Chagnon, M. J., Wu, C. L., Nakazawa, T., Yamamoto, T., Noda, M., Blanchetot, C., et al. (2010). Receptor tyrosine phosphatase sigma (RPTPsigma) regulates, p250GAP, a novel substrate that attenuates Rac signaling. *Cellular Signalling, 22*(11), 1626–1633. https://doi.org/10.1016/j.cellsig.2010.06.001.

Chen, C. Y., Chen, Y. T., Wang, J. Y., Huang, Y. S., & Tai, C. Y. (2017). Postsynaptic Y654 dephosphorylation of beta-catenin modulates presynaptic vesicle turnover through increased n-cadherin-mediated transsynaptic signaling. *Developmental Neurobiology,* 77(1), 61–74. https://doi.org/10.1002/dneu.22411.

Chen, K., & Featherstone, D. E. (2011). Pre and postsynaptic roles for Drosophila CASK. *Molecular and Cellular Neurosciences, 48*(2), 171–182. https://doi.org/10.1016/j.mcn.2011.07.009.

Chen, K., Merino, C., Sigrist, S. J., & Featherstone, D. E. (2005). The 4.1 protein coracle mediates subunit-selective anchoring of Drosophila glutamate receptors to the postsynaptic actin cytoskeleton. *The Journal of Neuroscience, 25*(28), 6667–6675. https://doi.org/10.1523/JNEUROSCI.1527-05.2005.

Chia, P. H., Li, P., & Shen, K. (2013). Cell biology in neuroscience: Cellular and molecular mechanisms underlying presynapse formation. *The Journal of Cell Biology*, *203*(1), 11–22. https://doi.org/10.1083/jcb.201307020.

Choi, B. J., Imlach, W. L., Jiao, W., Wolfram, V., Wu, Y., Grbic, M., et al. (2014). Miniature neurotransmission regulates drosophila synaptic structural maturation. *Neuron*, *82*(3), 618–634. https://doi.org/10.1016/j.neuron.2014.03.012.

Coles, C. H., Jones, E. Y., & Aricescu, A. R. (2015). Extracellular regulation of type IIa receptor protein tyrosine phosphatases: Mechanistic insights from structural analyses. *Seminars in Cell & Developmental Biology*, *37*, 98–107. https://doi.org/10.1016/j.semcdb.2014.09.007.

Coles, C. H., Shen, Y., Tenney, A. P., Siebold, C., Sutton, G. C., Lu, W., et al. (2011). Proteoglycan-specific molecular switch for RPTPsigma clustering and neuronal extension. *Science (New York, N.Y.)*, *332*(6028), 484–488. https://doi.org/10.1126/science.1200840.

Condomitti, G., & de Wit, J. (2018). Heparan sulfate proteoglycans as emerging players in synaptic specificity. *Frontiers in Molecular Neuroscience*, *11*, 14. https://doi.org/10.3389/fnmol.2018.00014.

Cravatt, B. F., Simon, G. M., & Yates, J. R., 3rd. (2007). The biological impact of mass-spectrometry-based proteomics. *Nature*, *450*(7172), 991–1000. https://doi.org/10.1038/nature06525.

Dai, Y., Taru, H., Deken, S. L., Grill, B., Ackley, B., Nonet, M. L., et al. (2006). SYD-2 Liprin-alpha organizes presynaptic active zone formation through ELKS. *Nature Neuroscience*, *9*(12), 1479–1487. https://doi.org/10.1038/nn1808.

David, M. D., Yeramian, A., Dunach, M., Llovera, M., Canti, C., de Herreros, A. G., et al. (2008). Signalling by neurotrophins and hepatocyte growth factor regulates axon morphogenesis by differential beta-catenin phosphorylation. *Journal of Cell Science*, *121*(Pt. 16), 2718–2730. https://doi.org/10.1242/jcs.029660.

de Wit, J., & Ghosh, A. (2016). Specification of synaptic connectivity by cell surface interactions. *Nature Reviews. Neuroscience*, *17*(1), 22–35. https://doi.org/10.1038/nrn.2015.3.

de Wit, J., O'Sullivan, M. L., Savas, J. N., Condomitti, G., Caccese, M. C., Vennekens, K. M., et al. (2013). Unbiased discovery of glypican as a receptor for LRRTM4 in regulating excitatory synapse development. *Neuron*, *79*(4), 696–711. https://doi.org/10.1016/j.neuron.2013.06.049.

Dean, C., Scholl, F. G., Choih, J., DeMaria, S., Berger, J., Isacoff, E., et al. (2003). Neurexin mediates the assembly of presynaptic terminals. *Nature Neuroscience*, *6*(7), 708–716. https://doi.org/10.1038/nn1074.

Debant, A., Serra-Pages, C., Seipel, K., O'Brien, S., Tang, M., Park, S. H., et al. (1996). The multidomain protein Trio binds the LAR transmembrane tyrosine phosphatase, contains a protein kinase domain, and has separate rac-specific and rho-specific guanine nucleotide exchange factor domains. *Proceedings of the National Academy of Sciences of the United States of America*, *93*(11), 5466–5471.

Delawary, M., Nakazawa, T., Tezuka, T., Sawa, M., Iino, Y., Takenawa, T., et al. (2007). Molecular characterization of a novel RhoGAP, RRC-1 of the nematode Caenorhabditis elegans. *Biochemical and Biophysical Research Communications*, *357*(2), 377–382. https://doi.org/10.1016/j.bbrc.2007.03.192.

Dhar, M., Zhu, M., Impey, S., Lambert, T. J., Bland, T., Karatsoreos, I. N., et al. (2014). Leptin induces hippocampal synaptogenesis via CREB-regulated microRNA-132 suppression of p250GAP. *Molecular Endocrinology*, *28*(7), 1073–1087. https://doi.org/10.1210/me.2013-1332.

Dunah, A. W., Hueske, E., Wyszynski, M., Hoogenraad, C. C., Jaworski, J., Pak, D. T., et al. (2005). LAR receptor protein tyrosine phosphatases in the development and maintenance of excitatory synapses. *Nature Neuroscience*, *8*(4), 458–467. https://doi.org/10.1038/nn1416.

Fairless, R., Masius, H., Rohlmann, A., Heupel, K., Ahmad, M., Reissner, C., et al. (2008). Polarized targeting of neurexins to synapses is regulated by their C-terminal sequences. *The Journal of Neuroscience*, *28*(48), 12969–12981. https://doi.org/10.1523/JNEUROSCI.5294-07.2008.
Farhy-Tselnicker, I., van Casteren, A. C. M., Lee, A., Chang, V. T., Aricescu, A. R., & Allen, N. J. (2017). Astrocyte-secreted glypican 4 regulates release of neuronal pentraxin 1 from axons to induce functional synapse formation. *Neuron*, *96*(2). https://doi.org/10.1016/j.neuron.2017.09.053. 428–445.e413.
Fulterer, A., Andlauer, T. F. M., Ender, A., Maglione, M., Eyring, K., Woitkuhn, J., et al. (2018). Active zone scaffold protein ratios tune functional diversity across brain synapses. *Cell Reports*, *23*(5), 1259–1274. https://doi.org/10.1016/j.celrep.2018.03.126.
Futai, K., Kim, M. J., Hashikawa, T., Scheiffele, P., Sheng, M., & Hayashi, Y. (2007). Retrograde modulation of presynaptic release probability through signaling mediated by PSD-95-neuroligin. *Nature Neuroscience*, *10*(2), 186–195. https://doi.org/10.1038/nn1837.
Gokce, O., & Südhof, T. C. (2013). Membrane-tethered monomeric neurexin LNS-domain triggers synapse formation. *The Journal of Neuroscience*, *33*(36), 14617–14628. https://doi.org/10.1523/JNEUROSCI.1232-13.2013.
Gomez, A. M., & Burden, S. J. (2011). The extracellular region of Lrp4 is sufficient to mediate neuromuscular synapse formation. *Developmental Dynamics*, *240*(12), 2626–2633. https://doi.org/10.1002/dvdy.22772.
Grootjans, J. J., Zimmermann, P., Reekmans, G., Smets, A., Degeest, G., Durr, J., et al. (1997). Syntenin, a PDZ protein that binds syndecan cytoplasmic domains. *Proceedings of the National Academy of Sciences of the United States of America*, *94*(25), 13683–13688.
Hallam, S. J., Goncharov, A., McEwen, J., Baran, R., & Jin, Y. (2002). SYD-1, a presynaptic protein with PDZ, C2 and rhoGAP-like domains, specifies axon identity in C. elegans. *Nature Neuroscience*, *5*(11), 1137–1146. https://doi.org/10.1038/nn959.
Han, K. A., Jeon, S., Um, J. W., & Ko, J. (2016). Emergent synapse organizers: LAR-RPTPs and their companions. *International Review of Cell and Molecular Biology*, *324*, 39–65. https://doi.org/10.1016/bs.ircmb.2016.01.002.
Han, K. A., Ko, J. S., Pramanik, G., Kim, J. Y., Tabuchi, K., Um, J. W., et al. (2018). PTPsigma drives excitatory presynaptic assembly via various extracellular and intracellular mechanisms. *The Journal of Neuroscience*, *38*(30), 6700–6721. https://doi.org/10.1523/JNEUROSCI.0672-18.2018.
Hata, Y., Butz, S., & Südhof, T. C. (1996). CASK: A novel dlg/PSD95 homolog with an N-terminal calmodulin-dependent protein kinase domain identified by interaction with neurexins. *The Journal of Neuroscience*, *16*(8), 2488–2494.
Hayashi, S., Uehara, D. T., Tanimoto, K., Mizuno, S., Chinen, Y., Fukumura, S., et al. (2017). Comprehensive investigation of CASK mutations and other genetic etiologies in 41 patients with intellectual disability and microcephaly with pontine and cerebellar hypoplasia (MICPCH). *PLoS One*, *12*(8). e0181791. https://doi.org/10.1371/journal.pone.0181791.
Held, R. G., & Kaeser, P. S. (2018). ELKS active zone proteins as multitasking scaffolds for secretion. *Open Biology*, *8*(2). https://doi.org/10.1098/rsob.170258. pii: 170258.
Herring, B. E., & Nicoll, R. A. (2016). Kalirin and Trio proteins serve critical roles in excitatory synaptic transmission and LTP. *Proceedings of the National Academy of Sciences of the United States of America*, *113*(8), 2264–2269. https://doi.org/10.1073/pnas.1600179113.
Hirbec, H., Francis, J. C., Lauri, S. E., Braithwaite, S. P., Coussen, F., Mulle, C., et al. (2003). Rapid and differential regulation of AMPA and kainate receptors at hippocampal mossy fibre synapses by PICK1 and GRIP. *Neuron*, *37*(4), 625–638.
Hirbec, H., Perestenko, O., Nishimune, A., Meyer, G., Nakanishi, S., Henley, J. M., et al. (2002). The PDZ proteins PICK1, GRIP, and syntenin bind multiple glutamate receptor subtypes. Analysis of PDZ binding motifs. *The Journal of Biological Chemistry*, *277*(18), 15221–15224. https://doi.org/10.1074/jbc.C200112200.

Holbrook, S., Finley, J. K., Lyons, E. L., & Herman, T. G. (2012). Loss of syd-1 from R7 neurons disrupts two distinct phases of presynaptic development. *The Journal of Neuroscience*, *32*(50), 18101–18111. https://doi.org/10.1523/JNEUROSCI.1350-12.2012.

Hoogenraad, C. C., Feliu-Mojer, M. I., Spangler, S. A., Milstein, A. D., Dunah, A. W., Hung, A. Y., et al. (2007). Liprinalpha1 degradation by calcium/calmodulin-dependent protein kinase II regulates LAR receptor tyrosine phosphatase distribution and dendrite development. *Developmental Cell*, *12*(4), 587–602. https://doi.org/10.1016/j.devcel.2007.02.006.

Hsueh, Y. P., & Sheng, M. (1999). Regulated expression and subcellular localization of syndecan heparan sulfate proteoglycans and the syndecan-binding protein CASK/LIN-2 during rat brain development. *The Journal of Neuroscience*, *19*(17), 7415–7425.

Hu, Z., Hom, S., Kudze, T., Tong, X. J., Choi, S., Aramuni, G., et al. (2012). Neurexin and neuroligin mediate retrograde synaptic inhibition in C. elegans. *Science*, *337*(6097), 980–984. https://doi.org/10.1126/science.1224896.

Hu, S., Pawson, T., & Steven, R. M. (2011). UNC-73/trio RhoGEF-2 activity modulates Caenorhabditis elegans motility through changes in neurotransmitter signaling upstream of the GSA-1/Galphas pathway. *Genetics*, *189*(1), 137–151. https://doi.org/10.1534/genetics.111.131227.

Hu, H. T., Umemori, H., & Hsueh, Y. P. (2016). Postsynaptic SDC2 induces transsynaptic signaling via FGF22 for bidirectional synaptic formation. *Scientific Reports*, *6*, 33592. https://doi.org/10.1038/srep33592.

Iyer, S. C., Wang, D., Iyer, E. P., Trunnell, S. A., Meduri, R., Shinwari, R., et al. (2012). The RhoGEF trio functions in sculpting class specific dendrite morphogenesis in Drosophila sensory neurons. *PLoS One*, 7(3). e33634. https://doi.org/10.1371/journal.pone.0033634.

Johnson, K. G., Tenney, A. P., Ghose, A., Duckworth, A. M., Higashi, M. E., Parfitt, K., et al. (2006). The HSPGs syndecan and dallylike bind the receptor phosphatase LAR and exert distinct effects on synaptic development. *Neuron*, *49*(4), 517–531. https://doi.org/10.1016/j.neuron.2006.01.026.

Kannan, M., Lee, S. J., Schwedhelm-Domeyer, N., Nakazawa, T., & Stegmuller, J. (2012). p250GAP is a novel player in the Cdh1-APC/Smurf1 pathway of axon growth regulation. *PLoS One*, 7(11). e50735. https://doi.org/10.1371/journal.pone.0050735.

Kaufmann, N., DeProto, J., Ranjan, R., Wan, H., & Van Vactor, D. (2002). Drosophila liprin-alpha and the receptor phosphatase Dlar control synapse morphogenesis. *Neuron*, *34*(1), 27–38.

Kim, N., Stiegler, A. L., Cameron, T. O., Hallock, P. T., Gomez, A. M., Huang, J. H., et al. (2008). Lrp4 is a receptor for agrin and forms a complex with MuSK. *Cell*, *135*(2), 334–342. https://doi.org/10.1016/j.cell.2008.10.002.

Kittelmann, M., Hegermann, J., Goncharov, A., Taru, H., Ellisman, M. H., Richmond, J. E., et al. (2013). Liprin-alpha/SYD-2 determines the size of dense projections in presynaptic active zones in C. elegans. *The Journal of Cell Biology*, *203*(5), 849–863. https://doi.org/10.1083/jcb.201302022.

Ko, J., Na, M., Kim, S., Lee, J. R., & Kim, E. (2003). Interaction of the ERC family of RIM-binding proteins with the liprin-alpha family of multidomain proteins. *The Journal of Biological Chemistry*, *278*(43), 42377–42385. https://doi.org/10.1074/jbc.M307561200.

Ko, J. S., Pramanik, G., Um, J. W., Shim, J. S., Lee, D., Kim, K. H., et al. (2015). PTPsigma functions as a presynaptic receptor for the glypican-4/LRRTM4 complex and is essential for excitatory synaptic transmission. *Proceedings of the National Academy of Sciences of the United States of America*, *112*(6), 1874–1879. https://doi.org/10.1073/pnas.1410138112.

Ko, J., Yoon, C., Piccoli, G., Chung, H. S., Kim, K., Lee, J. R., et al. (2006). Organization of the presynaptic active zone by ERC2/CAST1-dependent clustering of the tandem PDZ protein syntenin-1. *The Journal of Neuroscience*, *26*(3), 963–970. https://doi.org/10.1523/JNEUROSCI.4475-05.2006.

Ko, J., Zhang, C., Arac, D., Boucard, A. A., Brunger, A. T., & Südhof, T. C. (2009). Neuroligin-1 performs neurexin-dependent and neurexin-independent functions in synapse validation. *The EMBO Journal*, *28*(20), 3244–3255. https://doi.org/10.1038/emboj.2009.249.

Koroll, M., Rathjen, F. G., & Volkmer, H. (2001). The neural cell recognition molecule neurofascin interacts with syntenin-1 but not with syntenin-2, both of which reveal self-associating activity. *The Journal of Biological Chemistry*, *276*(14), 10646–10654. https://doi.org/10.1074/jbc.M010647200.

Lazaro-Pena, M. I., Diaz-Balzac, C. A., Bulow, H. E., & Emmons, S. W. (2018). Synaptogenesis is modulated by heparan sulfate in Caenorhabditis elegans. *Genetics*, *209*(1), 195–208. https://doi.org/10.1534/genetics.118.300837.

Li, X. M., Dong, X. P., Luo, S. W., Zhang, B., Lee, D. H., Ting, A. K., et al. (2008). Retrograde regulation of motoneuron differentiation by muscle beta-catenin. *Nature Neuroscience*, *11*(3), 262–268. https://doi.org/10.1038/nn2053.

Li, M. Y., Miao, W. Y., Wu, Q. Z., He, S. J., Yan, G., Yang, Y., et al. (2017). A critical role of presynaptic cadherin/catenin/p140Cap complexes in stabilizing spines and functional synapses in the neocortex. *Neuron*, *94*(6). https://doi.org/10.1016/j.neuron.2017.05.022. 1155–1172.e1158.

Li, L., Tian, X., Zhu, M., Bulgari, D., Bohme, M. A., Goettfert, F., et al. (2014). Drosophila Syd-1, liprin-alpha, and protein phosphatase 2A B' subunit Wrd function in a linear pathway to prevent ectopic accumulation of synaptic materials in distal axons. *The Journal of Neuroscience*, *34*(25), 8474–8487. https://doi.org/10.1523/JNEUROSCI.0409-14.2014.

Lin, D., Gish, G. D., Songyang, Z., & Pawson, T. (1999). The carboxyl terminus of B class ephrins constitutes a PDZ domain binding motif. *The Journal of Biological Chemistry*, *274*(6), 3726–3733.

Lucic, V., Yang, T., Schweikert, G., Forster, F., & Baumeister, W. (2005). Morphological characterization of molecular complexes present in the synaptic cleft. *Structure*, *13*(3), 423–434. https://doi.org/10.1016/j.str.2005.02.005.

Marler, K. J., Suetterlin, P., Dopplapudi, A., Rubikaite, A., Adnan, J., Maiorano, N. A., et al. (2014). BDNF promotes axon branching of retinal ganglion cells via miRNA-132 and p250GAP. *The Journal of Neuroscience*, *34*(3), 969–979. https://doi.org/10.1523/JNEUROSCI.1910-13.2014.

Maro, G. S., Gao, S., Olechwier, A. M., Hung, W. L., Liu, M., Ozkan, E., et al. (2015). MADD-4/Punctin and neurexin organize C. elegans GABAergic postsynapses through neuroligin. *Neuron*, *86*(6), 1420–1432. https://doi.org/10.1016/j.neuron.2015.05.015.

McClelland, A. C., Sheffler-Collins, S. I., Kayser, M. S., & Dalva, M. B. (2009). Ephrin-B1 and ephrin-B2 mediate EphB-dependent presynaptic development via syntenin-1. *Proceedings of the National Academy of Sciences of the United States of America*, *106*(48), 20487–20492. https://doi.org/10.1073/pnas.0811862106.

Miller, K. E., DeProto, J., Kaufmann, N., Patel, B. N., Duckworth, A., & Van Vactor, D. (2005). Direct observation demonstrates that liprin-alpha is required for trafficking of synaptic vesicles. *Current Biology*, *15*(7), 684–689. https://doi.org/10.1016/j.cub.2005.02.061.

Missler, M., Südhof, T. C., & Biederer, T. (2012). Synaptic cell adhesion. *Cold Spring Harbor Perspectives in Biology*, *4*(4), a005694. https://doi.org/10.1101/cshperspect.a005694.

Mosca, T. J., Luginbuhl, D. J., Wang, I. E., & Luo, L. (2017). Presynaptic LRP4 promotes synapse number and function of excitatory CNS neurons. *eLife*, *6*. https://doi.org/10.7554/eLife.27347. pii: e27347.

Muhammad, K., Reddy-Alla, S., Driller, J. H., Schreiner, D., Rey, U., Bohme, M. A., et al. (2015). Presynaptic spinophilin tunes neurexin signalling to control active zone architecture and function. *Nature Communications*, *6*, 8362. https://doi.org/10.1038/ncomms9362.

Muise, A. M., Walters, T., Wine, E., Griffiths, A. M., Turner, D., Duerr, R. H., et al. (2007). Protein-tyrosine phosphatase sigma is associated with ulcerative colitis. *Current Biology, 17*(14), 1212–1218. (Epub 2007 Jul 5).

Murase, S., Mosser, E., & Schuman, E. M. (2002). Depolarization drives beta-catenin into neuronal spines promoting changes in synaptic structure and function. *Neuron, 35*(1), 91–105.

Nakamura, F., Okada, T., Shishikura, M., Uetani, N., Taniguchi, M., Yagi, T., et al. (2017). Protein tyrosine phosphatase delta mediates the Sema3A-induced cortical basal dendritic arborization through the activation of fyn tyrosine kinase. *The Journal of Neuroscience, 37*(30), 7125–7139. https://doi.org/10.1523/JNEUROSCI.2519-16.2017.

Nakazawa, T., Kuriu, T., Tezuka, T., Umemori, H., Okabe, S., & Yamamoto, T. (2008). Regulation of dendritic spine morphology by an NMDA receptor-associated Rho GTPase-activating protein, p250GAP. *Journal of Neurochemistry, 105*(4), 1384–1393. https://doi.org/10.1111/j.1471-4159.2008.05335.x.

Nakazawa, T., Watabe, A. M., Tezuka, T., Yoshida, Y., Yokoyama, K., Umemori, H., et al. (2003). p250GAP, a novel brain-enriched GTPase-activating protein for Rho family GTPases, is involved in the N-methyl-D-aspartate receptor signaling. *Molecular Biology of the Cell, 14*(7), 2921–2934. https://doi.org/10.1091/mbc.e02-09-0623.

Newsome, T. P., Schmidt, S., Dietzl, G., Keleman, K., Asling, B., Debant, A., et al. (2000). Trio combines with dock to regulate Pak activity during photoreceptor axon path finding in Drosophila. *Cell, 101*(3), 283–294.

Ohno, K., Koroll, M., El Far, O., Scholze, P., Gomeza, J., & Betz, H. (2004). The neuronal glycine transporter 2 interacts with the PDZ domain protein syntenin-1. *Molecular and Cellular Neurosciences, 26*(4), 518–529. https://doi.org/10.1016/j.mcn.2004.04.007.

Olsen, O., Moore, K. A., Fukata, M., Kazuta, T., Trinidad, J. C., Kauer, F. W., et al. (2005). Neurotransmitter release regulated by a MALS-liprin-alpha presynaptic complex. *The Journal of Cell Biology, 170*(7), 1127–1134. https://doi.org/10.1083/jcb.200503011.

Owald, D., Fouquet, W., Schmidt, M., Wichmann, C., Mertel, S., Depner, H., et al. (2010). A Syd-1 homologue regulates pre- and postsynaptic maturation in Drosophila. *The Journal of Cell Biology, 188*(4), 565–579. https://doi.org/10.1083/jcb.200908055.

Owald, D., Khorramshahi, O., Gupta, V. K., Banovic, D., Depner, H., Fouquet, W., et al. (2012). Cooperation of Syd-1 with neurexin synchronizes pre- with postsynaptic assembly. *Nature Neuroscience, 15*(9), 1219–1226. https://doi.org/10.1038/nn.3183.

Pak, C., Danko, T., Zhang, Y., Aoto, J., Anderson, G., Maxeiner, S., et al. (2015). Human neuropsychiatric disease modeling using conditional deletion reveals synaptic transmission defects caused by heterozygous mutations in NRXN1. *Cell Stem Cell, 17*(3), 316–328. https://doi.org/10.1016/j.stem.2015.07.017.

Park, D., Bae, S., Yoon, T. H., & Ko, J. (2018). Molecular mechanisms of synaptic specificity: Spotlight on hippocampal and cerebellar synapse organizers. *Molecules and Cells, 41*(5), 373–380. https://doi.org/10.14348/molcells.2018.0081.

Patel, M. R., Lehrman, E. K., Poon, V. Y., Crump, J. G., Zhen, M., Bargmann, C. I., et al. (2006). Hierarchical assembly of presynaptic components in defined C. elegans synapses. *Nature Neuroscience, 9*(12), 1488–1498. https://doi.org/10.1038/nn1806.

Pawson, C., Eaton, B. A., & Davis, G. W. (2008). Formin-dependent synaptic growth: Evidence that Dlar signals via diaphanous to modulate synaptic actin and dynamic pioneer microtubules. *The Journal of Neuroscience, 28*(44), 11111–11123. https://doi.org/10.1523/JNEUROSCI.0833-08.2008.

Petzoldt, A. G., Lutzkendorf, J., & Sigrist, S. J. (2016). Mechanisms controlling assembly and plasticity of presynaptic active zone scaffolds. *Current Opinion in Neurobiology, 39*, 69–76. https://doi.org/10.1016/j.conb.2016.04.009.

Philbrook, A., Ramachandran, S., Lambert, C. M., Oliver, D., Florman, J., Alkema, M. J., et al. (2018). Neurexin directs partner-specific synaptic connectivity in C. elegans. *eLife, 7*. https://doi.org/10.7554/eLife.35692. pii: e35692.

Prakash, S., McLendon, H. M., Dubreuil, C. I., Ghose, A., Hwa, J., Dennehy, K. A., et al. (2009). Complex interactions amongst N-cadherin, DLAR, and liprin-alpha regulate Drosophila photoreceptor axon targeting. *Developmental Biology, 336*(1), 10–19. https://doi.org/10.1016/j.ydbio.2009.09.016.

Rauvala, H., & Peng, H. B. (1997). HB-GAM (heparin-binding growth-associated molecule) and heparin-type glycans in the development and plasticity of neuron-target contacts. *Progress in Neurobiology, 52*(2), 127–144.

Reissner, C., Runkel, F., & Missler, M. (2013). Neurexins. *Genome Biology, 14*(9), 213. https://doi.org/10.1186/gb-2013-14-9-213.

Reizes, O., Benoit, S. C., & Clegg, D. J. (2008). The role of syndecans in the regulation of body weight and synaptic plasticity. *The International Journal of Biochemistry & Cell Biology, 40*(1), 28–45. https://doi.org/10.1016/j.biocel.2007.06.011.

Rui, M., Qian, J., Liu, L., Cai, Y., Lv, H., Han, J., et al. (2017). The neuronal protein neurexin directly interacts with the Scribble-Pix complex to stimulate F-actin assembly for synaptic vesicle clustering. *The Journal of Biological Chemistry, 292*(35), 14334–14348. https://doi.org/10.1074/jbc.M117.794040.

Sadybekov, A., Tian, C., Arnesano, C., Katritch, V., & Herring, B. E. (2017). An autism spectrum disorder-related de novo mutation hotspot discovered in the GEF1 domain of Trio. *Nature Communications, 8*(1), 601. https://doi.org/10.1038/s41467-017-00472-0.

Shen, L., Liang, F., Walensky, L. D., & Huganir, R. L. (2000). Regulation of AMPA receptor GluR1 subunit surface expression by a 4. 1N-linked actin cytoskeletal association. *The Journal of Neuroscience, 20*(21), 7932–7940.

Shin, H., Wyszynski, M., Huh, K. H., Valtschanoff, J. G., Lee, J. R., Ko, J., et al. (2003). Association of the kinesin motor KIF1A with the multimodular protein liprin-alpha. *The Journal of Biological Chemistry, 278*(13), 11393–11401. https://doi.org/10.1074/jbc.M211874200.

Siddiqui, T. J., Tari, P. K., Connor, S. A., Zhang, P., Dobie, F. A., She, K., et al. (2013). An LRRTM4-HSPG complex mediates excitatory synapse development on dentate gyrus granule cells. *Neuron, 79*(4), 680–695. https://doi.org/10.1016/j.neuron.2013.06.029.

Spangler, S. A., Schmitz, S. K., Kevenaar, J. T., de Graaff, E., de Wit, H., Demmers, J., et al. (2013). Liprin-alpha2 promotes the presynaptic recruitment and turnover of RIM1/CASK to facilitate synaptic transmission. *The Journal of Cell Biology, 201*(6), 915–928. https://doi.org/10.1083/jcb.201301011.

Spinner, M. A., Walla, D. A., & Herman, T. G. (2018). Drosophila Syd-1 Has RhoGAP activity that is required for presynaptic clustering of bruchpilot/ELKS but Not Neurexin-1. *Genetics, 208*(2), 705–716. https://doi.org/10.1534/genetics.117.300538.

Südhof, T. C. (2012). The presynaptic active zone. *Neuron, 75*(1), 11–25. https://doi.org/10.1016/j.neuron.2012.06.012.

Südhof, T. C. (2017a). Molecular neuroscience in the 21st century: A personal perspective. *Neuron, 96*(3), 536–541. https://doi.org/10.1016/j.neuron.2017.10.005.

Südhof, T. C. (2017b). Synaptic neurexin complexes: A molecular code for the logic of neural circuits. *Cell, 171*(4), 745–769. https://doi.org/10.1016/j.cell.2017.10.024.

Sun, Y., Aiga, M., Yoshida, E., Humbert, P. O., & Bamji, S. X. (2009). Scribble interacts with beta-catenin to localize synaptic vesicles to synapses. *Molecular Biology of the Cell, 20*(14), 3390–3400. https://doi.org/10.1091/mbc.E08-12-1172.

Sun, Y., & Bamji, S. X. (2011). beta-Pix modulates actin-mediated recruitment of synaptic vesicles to synapses. *The Journal of Neuroscience, 31*(47), 17123–17133. https://doi.org/10.1523/JNEUROSCI.2359-11.2011.

Talukdar, G., Inoue, R., Yoshida, T., & Mori, H. (2018). Impairment in extinction of cued fear memory in syntenin-1 knockout mice. *Neurobiology of Learning and Memory, 149*, 58–67. https://doi.org/10.1016/j.nlm.2018.01.006.

Tang, A. H., Chen, H., Li, T. P., Metzbower, S. R., MacGillavry, H. D., & Blanpied, T. A. (2016). A trans-synaptic nanocolumn aligns neurotransmitter release to receptors. *Nature, 536*(7615), 210–214. https://doi.org/10.1038/nature19058.

Taniguchi, S., Liu, H., Nakazawa, T., Yokoyama, K., Tezuka, T., & Yamamoto, T. (2003). p250GAP, a neural RhoGAP protein, is associated with and phosphorylated by Fyn. *Biochemical and Biophysical Research Communications, 306*(1), 151–155.

Taru, H., & Jin, Y. (2011). The liprin homology domain is essential for the homomeric interaction of SYD-2/liprin-alpha protein in presynaptic assembly. *The Journal of Neuroscience, 31*(45), 16261–16268. https://doi.org/10.1523/JNEUROSCI.0002-11.2011.

Taylor, A. M., Wu, J., Tai, H. C., & Schuman, E. M. (2013). Axonal translation of beta-catenin regulates synaptic vesicle dynamics. *The Journal of Neuroscience, 33*(13), 5584–5589. https://doi.org/10.1523/JNEUROSCI.2944-12.2013.

Terry-Lorenzo, R. T., Torres, V. I., Wagh, D., Galaz, J., Swanson, S. K., Florens, L., et al. (2016). Trio, a Rho family GEF, interacts with the presynaptic active zone proteins piccolo and bassoon. *PLoS One, 11*(12). e0167535. https://doi.org/10.1371/journal.pone.0167535.

Togashi, H., Abe, K., Mizoguchi, A., Takaoka, K., Chisaka, O., & Takeichi, M. (2002). Cadherin regulates dendritic spine morphogenesis. *Neuron, 35*(1), 77–89.

Tong, X. J., Lopez-Soto, E. J., Li, L., Liu, H., Nedelcu, D., Lipscombe, D., et al. (2017). Retrograde synaptic inhibition is mediated by alpha-neurexin binding to the alpha2delta subunits of N-Type calcium channels. *Neuron, 95*(2). https://doi.org/10.1016/j.neuron.2017.06.018. 326–340.e325.

Torres, R., Firestein, B. L., Dong, H., Staudinger, J., Olson, E. N., Huganir, R. L., et al. (1998). PDZ proteins bind, cluster, and synaptically colocalize with Eph receptors and their ephrin ligands. *Neuron, 21*(6), 1453–1463.

Um, J. W., & Ko, J. (2013). LAR-RPTPs: Synaptic adhesion molecules that shape synapse development. *Trends in Cell Biology, 23*(10), 465–475. https://doi.org/10.1016/j.tcb.2013.07.004.

Um, J. W., & Ko, J. (2017). Neural glycosylphosphatidylinositol-anchored proteins in synaptic specification. *Trends in Cell Biology, 27*(12), 931–945. https://doi.org/10.1016/j.tcb.2017.06.007.

Vitureira, N., Letellier, M., White, I. J., & Goda, Y. (2012). Differential control of presynaptic efficacy by postsynaptic N-cadherin and beta-catenin. *Nature Neuroscience, 15*(1), 81–89. https://doi.org/10.1038/nn.2995.

Wagner, O. I., Esposito, A., Kohler, B., Chen, C. W., Shen, C. P., Wu, G. H., et al. (2009). Synaptic scaffolding protein SYD-2 clusters and activates kinesin-3 UNC-104 in C. elegans. *Proceedings of the National Academy of Sciences of the United States of America, 106*(46), 19605–19610. https://doi.org/10.1073/pnas.0902949106.

Ward, R. E., Lamb, R. S., & Fehon, R. G. (1998). A conserved functional domain of Drosophila coracle is required for localization at the septate junction and has membrane-organizing activity. *The Journal of Cell Biology, 140*(6), 1463–1473.

Wayman, G. A., Davare, M., Ando, H., Fortin, D., Varlamova, O., Cheng, H. Y., et al. (2008). An activity-regulated microRNA controls dendritic plasticity by down-regulating p250GAP. *Proceedings of the National Academy of Sciences of the United States of America, 105*(26), 9093–9098. https://doi.org/10.1073/pnas.0803072105.

Weatherbee, S. D., Anderson, K. V., & Niswander, L. A. (2006). LDL-receptor-related protein 4 is crucial for formation of the neuromuscular junction. *Development, 133*(24), 4993–5000. https://doi.org/10.1242/dev.02696.

Wentzel, C., Sommer, J. E., Nair, R., Stiefvater, A., Sibarita, J. B., & Scheiffele, P. (2013). mSYD1A, a mammalian synapse-defective-1 protein, regulates synaptogenic signaling and vesicle docking. *Neuron, 78*(6), 1012–1023. https://doi.org/10.1016/j.neuron.2013.05.010.

Won, S. Y., Kim, C. Y., Kim, D., Ko, J., Um, J. W., Lee, S. B., et al. (2017). LAR-RPTP clustering is modulated by competitive binding between synaptic adhesion partners and heparan sulfate. *Frontiers in Molecular Neuroscience*, *10*, 327. https://doi.org/10.3389/fnmol.2017.00327.

Wong, M. Y., Liu, C., Wang, S. S. H., Roquas, A. C. F., Fowler, S. C., & Kaeser, P. S. (2018). Liprin-alpha3 controls vesicle docking and exocytosis at the active zone of hippocampal synapses. *Proceedings of the National Academy of Sciences of the United States of America*, *115*(9), 2234–2239. https://doi.org/10.1073/pnas.1719012115.

Wu, H., Barik, A., Lu, Y., Shen, C., Bowman, A., Li, L., et al. (2015). Slit2 as a beta-catenin/Ctnnb1-dependent retrograde signal for presynaptic differentiation. *eLife*, *4*. https://doi.org/10.7554/eLife.07266. e07266.

Wu, G. H., Muthaiyan Shanmugam, M., Bhan, P., Huang, Y. H., & Wagner, O. I. (2016). Identification and characterization of LIN-2(CASK) as a regulator of kinesin-3 UNC-104(KIF1A) motility and clustering in neurons. *Traffic*, *17*(8), 891–907. https://doi.org/10.1111/tra.12413.

Xie, Z., Eagleson, K. L., Wu, H. H., & Levitt, P. (2016). Hepatocyte growth factor modulates MET receptor tyrosine kinase and beta-catenin functional interactions to enhance synapse formation. *eNeuro*, *3*(4). https://doi.org/10.1523/ENEURO.0074-16.2016. e0074.

Xing, G., Li, M., Sun, Y., Rui, M., Zhuang, Y., Lv, H., et al. (2018). Neurexin-neuroligin 1 regulates synaptic morphology and functions via the WAVE regulatory complex in Drosophila neuromuscular junction. *eLife*, *7*. https://doi.org/10.7554/eLife.30457. pii: e30457.

Xu, Y., & Quinn, C. C. (2016). SYD-1 promotes multiple developmental steps leading to neuronal connectivity. *Molecular Neurobiology*, *53*(10), 6768–6773. https://doi.org/10.1007/s12035-015-9592-5.

Yumoto, N., Kim, N., & Burden, S. J. (2012). Lrp4 is a retrograde signal for presynaptic differentiation at neuromuscular synapses. *Nature*, *489*(7416), 438–442. https://doi.org/10.1038/nature11348.

Zhang, C., Atasoy, D., Arac, D., Yang, X., Fucillo, M. V., Robison, A. J., et al. (2010). Neurexins physically and functionally interact with GABA(A) receptors. *Neuron*, *66*(3), 403–416. https://doi.org/10.1016/j.neuron.2010.04.008.

Zhang, P., Lu, H., Peixoto, R. T., Pines, M. K., Ge, Y., Oku, S., et al. (2018). Heparan sulfate organizes neuronal synapses through neurexin partnerships. *Cell*, *174*(6). https://doi.org/10.1016/j.cell.2018.07.002. 1450–1464.e1423.

Zhang, B., Luo, S., Wang, Q., Suzuki, T., Xiong, W. C., & Mei, L. (2008). LRP4 serves as a coreceptor of agrin. *Neuron*, *60*(2), 285–297. https://doi.org/10.1016/j.neuron.2008.10.006.

Zhen, M., & Jin, Y. (1999). The liprin protein SYD-2 regulates the differentiation of presynaptic termini in C. elegans. *Nature*, *401*(6751), 371–375. https://doi.org/10.1038/43886.

Zimmermann, P., Meerschaert, K., Reekmans, G., Leenaerts, I., Small, J. V., Vandekerckhove, J., et al. (2002). PIP(2)-PDZ domain binding controls the association of syntenin with the plasma membrane. *Molecular Cell*, *9*(6), 1215–1225.

CHAPTER TEN

Dopamine signaling in the striatum

Emmanuel Valjent[a,*], Anne Biever[b], Giuseppe Gangarossa[c], Emma Puighermanal[d]
[a]IGF, CNRS, INSERM, University of Montpellier, Montpellier, France
[b]Max Planck Institute for Brain Research, Frankfurt am Main, Germany
[c]Unité de Biologie Fonctionnelle et Adaptative, CNRS UMR 8251, Université Paris Diderot, Sorbonne Paris Cité, Paris, France
[d]Department of Cell Biology, Physiology and Immunology, Institute of Neuroscience, Autonomous University of Barcelona, Barcelona, Spain
*Corresponding author: e-mail address: emmanuel.valjent@igf.cnrs.fr

Contents

Abstract

The striatum integrates dopamine-mediated reward signals to generate appropriate behavior in response to glutamate-mediated sensory cues. Such associative learning relies on enduring neural plasticity in striatal GABAergic spiny projection neurons which, when altered, can lead to the development of a wide variety of pathological states. Considerable progress has been made in our understanding of the intracellular signaling

Advances in Protein Chemistry and Structural Biology, Volume 116
ISSN 1876-1623
https://doi.org/10.1016/bs.apcsb.2019.01.004

mechanisms in dopamine-related behaviors and pathologies. Through the prism of the regulation of histone H3 and ribosomal protein S6 phosphorylation, we review how dopamine-mediated signaling events regulate gene transcription and mRNA translation. Particularly, we focus on the intracellular cascades controlling these phosphorylations downstream of the modulation of dopamine receptors by psychostimulants, antipsychotics and L-DOPA. Finally, we highlight the importance to precisely determine in which neuronal populations these signaling events occur in order to understand how they participate in remodeling neural circuits and altering dopamine-related behaviors.

1. Introduction

Reinforcement learning is a fundamental process by which animals learn that particular stimuli predict the occurrence of positive or negative events, allowing them to adjust their behavioral responses. Such process critically relies on the activity of midbrain dopamine (DA) neurons. For instance, contextual and/or sensory stimuli associated with the occurrence of an unexpected reward (called primary reward) trigger a transient, or phasic, DA release caused by the increased firing of a subset of DA neurons (Bromberg-Martin, Matsumoto, & Hikosaka, 2010; Schultz, 2016). With time, this activation is transferred from the primary reward to the predictive stimuli (Bromberg-Martin et al., 2010; Schultz, 2016). These DA neurons are also inhibited by the presentation of unexpected aversive stimuli, and thus more generally encode the value of motivation (Bromberg-Martin et al., 2010; Schultz, 2016). A subset of DA neurons can also be activated by stimuli involved in the rapid detection of potentially salient events (Bromberg-Martin et al., 2010; Schultz, 2016). The existence of these distinct populations of DA neurons illustrates the complexity of the neural mechanisms by which DA participates in the learning and control of motivated actions. Therefore the disruption of DA transmission yields to dysfunctions of neural circuits that can ultimately lead to the development of pathological states including Parkinson's disease, schizophrenia, addiction, obsessive-compulsive or anxiety disorders (Gerfen & Surmeier, 2011).

2. Distribution of DA receptors and intracellular signaling in the striatum

Midbrain DA neurons densely innervate the striatum, a subcortical structure involved in motor function and reward-controlled learning

(Gerfen & Surmeier, 2011). Three clusters of DA neurons send topographically organized projections throughout distinct functional striatal territories: the retrorubral field (RrF, or A8 group), the substancia nigra pars compacta (SNc, or A9 group) and the ventral tegmental area (VTA, or A10 group) (Roeper, 2013). Thus, while DA neurons from A9 primarily project to the dorsolateral striatum and control motor actions, those arising from A8 and A10 mainly innervate to the ventral striatum (also known as nucleus accumbens) and regulate motivated behaviors (Gerfen & Surmeier, 2011).

Once locally released, DA's actions are mediated through the activation of five DA receptors (D1R, D2R, D3R, D4R and D5R) unevenly distributed throughout the striatum. D1R and D2R are highly expressed in the entire striatum and after decades of controversy it is now well accepted that GABAergic spiny projection neurons (SPNs) contain either D1R or D2R and that only few coexpress both (Gangarossa, Espallergues, et al., 2013; Matamales et al., 2009). D3R is expressed by a fraction of accumbal D1R- and D2R-SPNs (Gangarossa, Espallergues, et al., 2013), whereas D4R and D5R are barely detected in SPNs (Rivera, Alberti, et al., 2002; Rivera, Cuellar, et al., 2002). The expression of these different DA receptors is, however, not restricted to SPNs. D2R and D5R are highly expressed in cholinergic interneurons (Matamales et al., 2009). D5R is also found in PV-, CR- and NPY/SOM/NOS-expressing interneurons (Rivera, Alberti, et al., 2002; Rivera, Cuellar, et al., 2002). Nevertheless, the distribution of DA receptors should be reconsidered taking into account the diversity of newly discovered striatal GABAergic interneurons (Tepper et al., 2018). Lastly, DA receptors are also found at presynaptic terminals in both the dorsal striatum and the nucleus accumbens where they modulate neurotransmitter release (Bamford, Wightman, & Sulzer, 2018). Presynaptic D2R are the most diverse as they are expressed by SPNs axon collaterals (Burke, Rotstein, & Alvarez, 2017; Taverna, Ilijic, & Surmeier, 2008) as well as by heterogeneous sources of striatal inputs including midbrain DA neurons (Sesack, Aoki, & Pickel, 1994), glutamatergic corticostriatal neurons (Wang & Pickel, 2002) and GABAergic pallidostriatal afferents (Hoover & Marshall, 2004). In contrast, the distribution of presynaptic D1R is more restricted, being confined to SPNs axon collaterals and glutamatergic corticoaccumbal projections (Dumartin, Doudnikoff, Gonon, & Bloch, 2007). By modulating those receptors, DA will tightly regulate the integration of cortical, thalamic and limbic excitatory inputs onto SPNs, thereby strengthening specific striatal subcircuits in order to optimize the use of previously learned information to select and execute the most appropriate behavior (Bamford et al., 2018).

The striatum is certainly one of the brain areas in which the regulation of DA-mediated intracellular signaling mechanisms has been most extensively studied in response to drugs of abuse or therapeutic agents as well as in physiological and pathological contexts. As mentioned above, DA acts through five DA receptors, which are G protein-coupled receptors that, depending of the subtypes (D1R or D2R), stimulate or inhibit the production of intracellular cAMP (Beaulieu & Gainetdinov, 2011). The subsequent modulation of cAMP-dependent protein kinase (PKA), in concert with the regulation of several kinase/phosphatase cascades, results in the short-term regulation of receptors and ion channels, as well as in the phosphorylation of nuclear and cytoplasmic targets that initiate important changes in gene transcription and mRNA translation (Beaulieu & Gainetdinov, 2011).

The main focus of this review is on the intracellular cascades involved in the regulation of histone H3 and rpS6 phosphorylation in the striatum in response to the modulation of D1R and D2R by psychostimulants, haloperidol and L-DOPA. We also discuss how the use of BAC transgenic mice has been instrumental in investigating the regulation of intracellular signaling events taking place in genetically-identified striatal neural circuits.

3. Modulation of histone H3 phosphorylation by dopamine

Histone H3 is one of the four core histones involved in the formation of nucleosomes. Its N-terminal tail undergoes various posttranslational modifications including acetylation, methylation, ADP-ribosylation and phosphorylation, which all contribute to the dynamic regulation of gene expression. To date, seven histone H3 phospho-sites have been identified (Thr3, Thr6, Ser10, Thr11, Ser28, Tyr41 and Thr45). While their phosphorylations involve multiple kinases, their dephosphorylation relies on two phosphatases, protein phophastase-1 (PP1) and protein phophastase-2A (PP2A) (Sawicka & Seiser, 2012). Section 3 summarizes our current knowledge on the mechanisms by which DA triggers rapid changes in the phosphorylation state of histone H3 at S10 and S28 alone or coupled to other posttranslational modifications such as acetylation of Lys14 and trimethylation of Lys27.

3.1 Phosphorylation of histone H3 in D1R-SPNs by psychostimulants and L-DOPA

Enhanced histone H3 phosphorylation (pS10 and pS10acK14) in the dorsal and ventral striatum was first reported following acute *in vivo* administration of cocaine, a potent monoamine reuptake inhibitor (Brami-Cherrier et al., 2005; Kumar et al., 2005; Sanchis-Segura, Lopez-Atalaya, & Barco, 2009). The use of BAC transgenic *Drd1a*-eGFP and *Drd1a*-eGFP-L10a mice has been determinant to reveal that cocaine-induced pS10-H3 and pS10acK14-H3, previously observed in a subset of SPNs (Brami-Cherrier et al., 2005), were confined into the nuclei of D1R-SPNs (Bertran-Gonzalez et al., 2008; Jordi et al., 2013). Since then, several studies have shown that most, if not all, psychostimulant drugs including D-amphetamine, methamphetamine, MDMA and methylphenidate enhance striatal histone H3 phosphorylation at S10 and S28me3K27 selectively in D1R-SPNs (Biever, Puighermanal, et al., 2015; Bonito-Oliva et al., 2016; Rotllant & Armario, 2012) (Fig. 1). Interestingly, unlike other histone H3 posttranslational modifications, such as acetylation or methylation (Jordi et al., 2013; Kumar et al., 2005), the ability of cocaine to promote pS10-H3 and pS10acK14 is barely detectable upon repeated cocaine injection (Bertran-Gonzalez et al., 2008; Kumar et al., 2005).

Histone H3 phosphorylation is also strongly regulated by L-DOPA, the most effective dopamine replacement therapy in Parkinson's disease. Indeed, acute administration of L-DOPA increases pS10-H3, pS10acK14-H3, pS28-H3 and pS28me3K27 in the DA-depleted striatum of parkinsonian rodents (Santini, Alcacer, et al., 2009; Santini et al., 2007; Sodersten et al., 2014). These events occur selectively in D1R-SPNs (Santini, Alcacer, et al., 2009; Santini et al., 2007; Sodersten et al., 2014) and persist in rodents that develop dyskinesia following repeated administration of L-DOPA (Chen et al., 2017; Darmopil, Martin, De Diego, Ares, & Moratalla, 2009; Santini, Alcacer, et al., 2009; Santini et al., 2007; Sodersten et al., 2014) (Fig. 1). Decreased pS10-H3 phosphorylation has been, however, observed in chronically-treated dyskinetic monkeys (Nicholas et al., 2008) supporting the idea that abnormal signaling leading to histone H3 phosphorylation would be preferentially associated with priming which may correspond to the initial phase of L-DOPA treatment (Santini et al., 2010).

D1R activation seems to be primarily involved in the regulation of the level of histone H3 phosphorylation in the striatum. Indeed, direct stimulation of D1R *in vivo* is sufficient to trigger pS10-H3 and pS10acK14-H3 in D1R-SPNs (Gangarossa, Perroy, & Valjent, 2013; Schroeder et al., 2008). In addition, cocaine- and L-DOPA-induced S10-H3 phosphorylation is prevented by pharmacological blockade of D1R or in D1R-deficient mice (Brami-Cherrier et al., 2005; Darmopil et al., 2009; Santini, Alcacer, et al., 2009) (Fig. 2). D1R-mediated pS10-H3 phosphorylation also involves

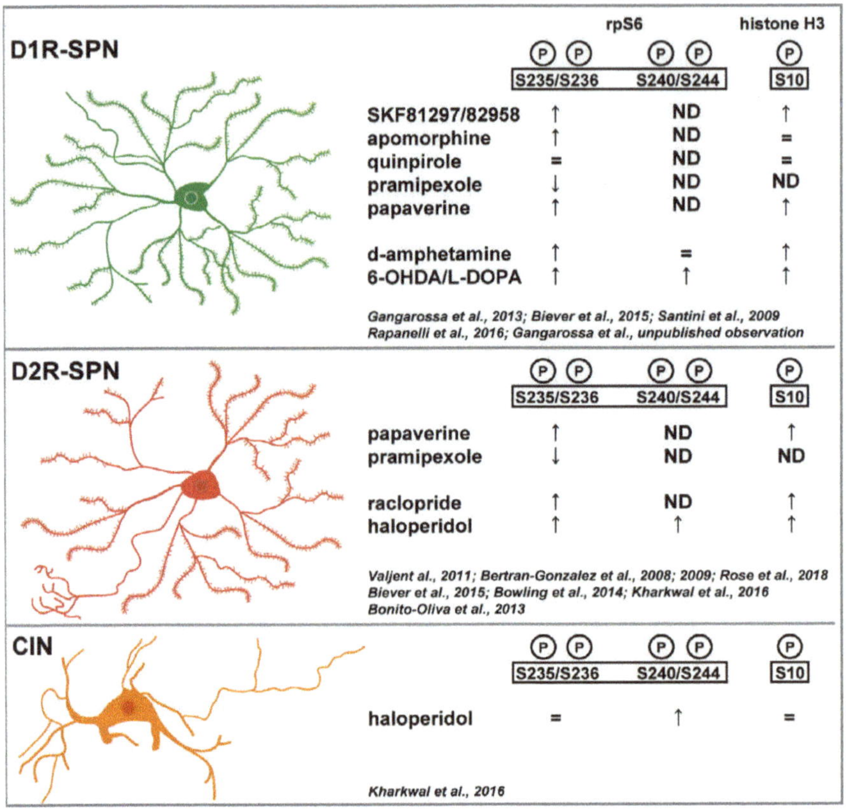

Fig. 1 Summary of the regulation of histone H3 and rpS6 phosphorylation in striatal D1R- and D2R-expressing neurons. Psychostimulants and L-DOPA injected in 6-OHDA-lesioned mice increase histone H3 and rpS6 phosphorylation exclusively in D1R-SPNs while antipsychotics regulate these phosphorylation events in striatal D2R-expressing neurons including SPNs and CINs.

glutamate, as response to cocaine requires NMDA receptor activation (Brami-Cherrier et al., 2005) (Fig. 2). Similar observations have been made in primary striatal neuron culture following D1R stimulation (Gangarossa, Perroy, et al., 2013; Schroeder et al., 2008). On the other hand, metabotropic glutamate receptor type 5 (mGlu5) appears to play a prominent role in the regulation of S10-H3 phosphorylation induced by dopaminergic drugs in Parkinson's disease. Indeed, the ability of the D1R agonist SKF38393 to increase pS10-H3 in the DA-depleted striatum is reduced in 6-OHDA-lesioned mice with selective knockdown of mGlu5 in D1R-SPNs (Fieblinger et al., 2014). Finally, recent evidence reveals that

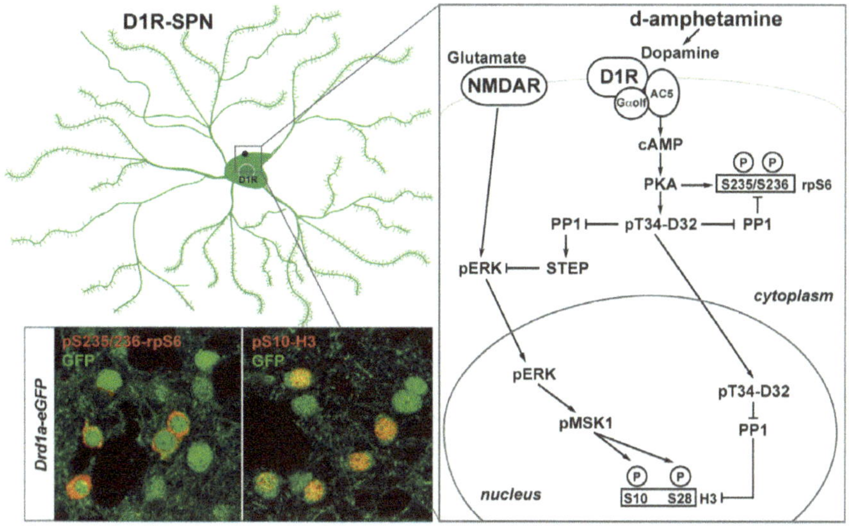

Fig. 2 Intracellular signaling mechanisms involved in D-amphetamine-induced histone H3 and rpS6 phosphorylation in D1R-SPNs. Immunofluorescence of pS235/236-rpS6 and pS10-H3 in combination with eGFP in the striatum of *drd1a*-eGFP mice treated with D-amphetamine (10 mg/kg). D-Amphetamine-induced pS10-H3 and pS28-H3 depend on NMDAR and D1R activation. The cytoplasmic and nuclear inhibition of PP1 by DARPP-32 as well as the activation of the ERK/MSK1 pathway is essential for increased histone H3 phosphorylation in response to D-amphetamine. The phosphorylation of rpS6 at S235/236 relies exclusively on the D1R/Gaolf/cAMP/PKA/DARPP-32 cascade.

L-DOPA-induced pS10-H3 phosphorylation may also requires the D3R, which is over-expressed in D1R-SPNs of dyskinetic mice (Solis, Garcia-Montes, Gonzalez-Granillo, Xu, & Moratalla, 2017).

3.2 Regulation of histone H3 phosphorylation in striatal D2R-SPNs by antipsychotics

Regulation of histone H3 phosphorylation has also been reported following D2R modulation (Fig. 1). In the striatum, tonic stimulation of D2Rs has a strong inhibitory effect on the activation of intracellular signaling pathways in D2R-SPNs (Valjent, Bertran-Gonzalez, Herve, Fisone, & Girault, 2009). This could explain why *in vivo* acute administration of the D2R agonist quinpirole fails to enhance histone H3 phosphorylation (Gangarossa, Perroy, et al., 2013; Li et al., 2004). On the other hand, blockade of D2R by haloperidol or raclopride induces a rapid and sustained increase of pS10-H3, pS10acK14-H3 and pS28me3K27 in D2R-SPNs in the dorsal striatum (Bertran-Gonzalez et al., 2008, 2009; Bonito-Oliva et al., 2016; Li et al., 2004) (Fig. 1). These effects are mediated by the combined stimulation of adenosine A2a and NMDA receptors (Bertran-Gonzalez et al., 2009; Bonito-Oliva et al., 2016; Li et al., 2004) (Fig. 3). Finally, Rose and colleagues unveiled that haloperidol-induced pS10-H3 in D2R-SPNs is mediated by the blockade of D2R/G-protein signaling and is independent of the D2R/β-arrestin coupling (Rose et al., 2018).

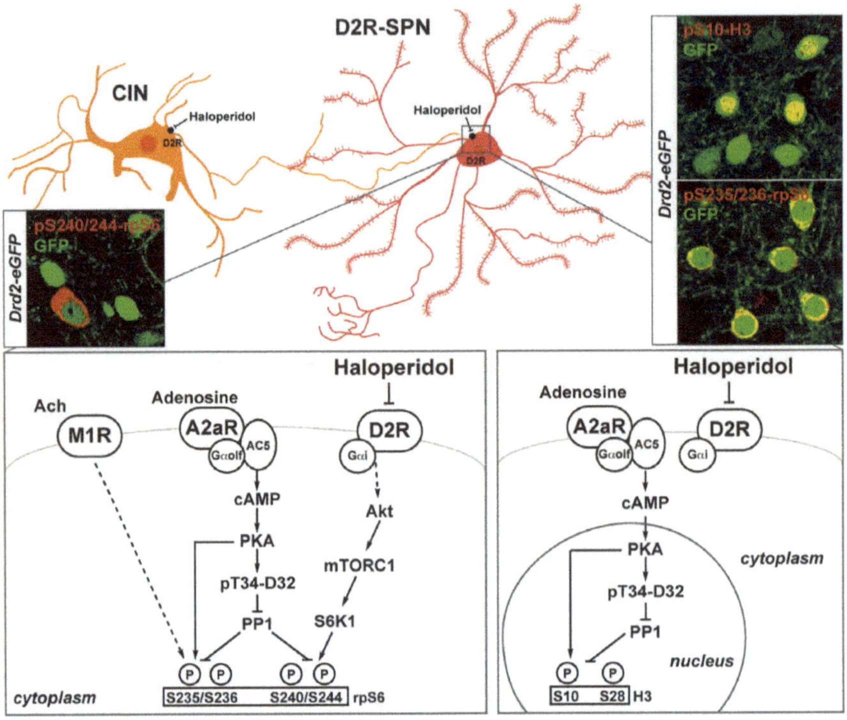

Fig. 3 Regulation of histone H3 and rpS6 phosphorylation by haloperidol in striatal D2R-expressing neurons. Immunofluorescence of pS235/236-rpS6 and pS10-H3 in the striatum of *drd2*-eGFP mice injected with haloperidol (0.5 mg/kg). Note that pS240/244-rpS6 also increases in CINs in response to haloperidol. The activation of the A2aR/Gaolf/cAMP/PKA/DARPP-32 pathway following haloperidol administration triggers S10-H3 and S28-H3 phosphorylation in D2R-SPNs. The intracellular signaling mechanisms involved in the control of rpS6 phosphorylation in D2R-SPNs require the D2R expressed by SPNs and CINs as well as A2aR and M1R expressed by D2R-SPNs. The regulation of rpS6 phosphorylation at S235/236 and at S240/244 depends on the cAMP/PKA/DARPP-32 cascade. Haloperidol-induced S240/244-rpS6 phosphorylation also requires the activation of the mTORC1/p70S6K1 pathway.

3.3 Coordinate control of histone H3 phosphorylation in D1R-SPNs by ERK/MSK1 signaling and cAMP/PKA/DARPP-32 cascade

Over the past years, several studies have pointed out the critical role played by the extracellular signal-regulated kinases (ERKs) in the regulation of histone H3 phosphorylation downstream D1R stimulation (Girault, Valjent, Caboche, & Herve, 2007; Pascoli, Cahill, Bellivier, Caboche, & Vanhoutte, 2014). Indeed, pharmacological blockade of ERK activation prevents increased pS10-H3 and pS10acK14-H3 induced by the D1R agonist SKF81297 (Gangarossa, Perroy, et al., 2013), cocaine (Brami-Cherrier et al., 2005; Papale et al., 2016), D-amphetamine (Biever, Puighermanal, et al., 2015) or L-DOPA (Chen et al., 2017; Santini et al., 2007). Phosphorylation of histone H3 at S10 cannot be achieved directly by ERK. Two targets of ERK, ribosomal subunit protein S6 kinases (RSKs) and mitogen and stress-activated protein kinases (MSKs) have been identified as histone H3 kinases (Brami-Cherrier, Roze, Girault, Betuing, & Caboche, 2009). In the striatum, MSK1 appears to be the kinase responsible for D1R-evoked S10-H3 and S10acK14-H3 phosphorylations. The first evidence came from Brami-Cherrier and colleagues, who reported that acute cocaine administration triggered a transient ERK-dependent increase of pT581-MSK1 and that cocaine-induced pS10-H3 and pS10acK14-H3 were abolished in MSK1-deficient mice (Brami-Cherrier et al., 2005). Since then, increased MSK1 phosphorylation has been systematically observed in D1R-SPNs following cocaine, amphetamine and L-DOPA administration (Bertran-Gonzalez et al., 2008; Santini, Alcacer, et al., 2009; Santini et al., 2007; Westin, Vercammen, Strome, Konradi, & Cenci, 2007) and has been shown to control S10-H3, S10acK14-H3, S28-H3 and S28me3K27-H3 phosphorylations (Alcacer, Charbonnier-Beaupel, Corvol, Girault, & Herve, 2014; Bonito-Oliva et al., 2016; Feyder et al., 2016; Sodersten et al., 2014) (Fig. 2). Interestingly, cocaine-induced pS10-H3 can also occur independently of MSK1, through a mechanism involving the transcription factor Elk-1 (Besnard et al., 2011).

The regulation of histone H3 phosphorylation in response to D1R stimulation also critically depends on cAMP/PKA activation (Schroeder et al., 2008) (Fig. 2). In D1R-SPNs, the production of cAMP requires a specific G-protein heterotrimer subunit containing $G\alpha olf/\beta 2/\gamma 7$, which triggers the activation of adenylyl cyclase type 5 (AC5), the SPNs-enriched AC (Herve, 2011). The absence of S10-H3 phosphorylation in response to L-DOPA in

AC5-deficient mice was the first demonstration of the contribution of cAMP production in the regulation of histone H3 phosphorylation (Park et al., 2014). It should be noticed that 30–50% reduction in D1R-activated cAMP production is, however, not sufficient to prevent S10acK14-H3 phosphorylation induced by L-DOPA since it is preserved in heterozygous mice for Gαolf (*Gnal*$^{+/-}$) (Alcacer et al., 2012).

Although histone H3 can be directly phosphorylated at S10 by PKA (Sawicka & Seiser, 2012), the way PKA regulates pS10-H3 relies on a mechanism involving a phosphatase inhibition cascade (Fig. 2). Once activated, PKA-dependent signaling can be amplified through the recruitment of DARPP-32, a potent PKA-regulated inhibitor of protein phosphatase-1 (PP1) (Yger & Girault, 2011). Phosphorylation of DARPP-32 at T34 is largely increased in D1R-SPNs following psychostimulant and L-DOPA administration (Bateup et al., 2008; Santini et al., 2012; Valjent et al., 2005). Thus, once phosphorylated on T34 by PKA and dephosphorylated on S97 by PP2A, DARPP-32 inhibits nuclear PP1 therefore preventing histone H3 dephosphorylation at S10 in D1R-SPNs (Stipanovich et al., 2008) (Fig. 2). PP1 inhibition also contributes to the regulation of both pS10-H3 induced by L-DOPA and pS28-H3 induced by L-DOPA or amphetamine as demonstrated by the use of mice lacking DARPP-32 in D1R-SPNs and phospho-deficient DARPP-32 knockin mice, in which T34 was mutated to alanine residue (Bonito-Oliva et al., 2016; Santini et al., 2012; Sodersten et al., 2014).

3.4 Phosphorylation of histone H3 in D2R-SPNs exclusively relies on the cAMP/PKA/DARPP-32 cascade

The intracellular mechanisms leading to increased S10-H3 and S10acK14-H3 phosphorylations induced by haloperidol contrast with those of cocaine. Indeed, although haloperidol activates ERK and MSK1 in the dorsal striatum (Bertran-Gonzalez et al., 2008; Gerfen, Miyachi, Paletzki, & Brown, 2002; Pozzi et al., 2003; Valjent, Pages, Herve, Girault, & Caboche, 2004), the concomitant increase in pS10-H3, pS10acK14-H3 and pS28me3K27-H3 observed in D2R-SPNs occur independently of the ERK/MSK1 cascade (Bertran-Gonzalez et al., 2009; Bonito-Oliva et al., 2016) (Fig. 3).

In contrast, the involvement of PKA signaling in the regulation of histone H3 phosphorylation by haloperidol was early suggested. Indeed, Li and colleagues were the first to report that the inactivation of PKA prevented haloperidol-induced pS10-H3 and pS10acK14-H3 (Li et al., 2004). Since then, the mechanism by which haloperidol leads to PKA activation has been identified. It requires the activation of A2aR and its coupling to Gαolf (Bertran-Gonzalez et al., 2009) (Fig. 3). The direct contribution of cAMP production was recently confirmed since both, forskolin, an activator of adenylyl cyclase, or papaverine, an inhibitor of PDE10a phosphodiesterase, are sufficient to enhance pS10-H3 in the striatum (Nishi et al., 2017; Polito et al., 2015). As for psychostimulants and L-DOPA, increased histone H3 phosphorylation induced by haloperidol or papaverine requires the PKA-dependent phosphorylation of the DARPP-32 (Bertran-Gonzalez et al., 2009; Bonito-Oliva et al., 2016; Polito et al., 2015) (Fig. 3).

3.5 Histone H3 phosphorylation and regulation of transcription in the striatum

Histone H3 phosphorylation is highly dynamic and participates to the regulation of chromatin remodeling necessary to trigger initiation and elongation of transcription of specific transcripts (Sawicka & Seiser, 2012). In most of the aforementioned studies, histone H3 phosphorylation at S10, which occurs independently on preacetylated histone H3 has been associated with increased expression of immediate early genes including c-Fos (Bertran-Gonzalez et al., 2009; Brami-Cherrier et al., 2005; Kumar et al., 2005; Santini, Alcacer, et al., 2009; Santini et al., 2007). Importantly, chromatin immunoprecipitation assays demonstrated that pS10acK14-H3 occurs directly at the *c-fos* and *fosB* promoters in striatal neurons in response to cocaine and L-DOPA, respectively (Feyder et al., 2016; Jordi et al., 2013; Kumar et al., 2005). Although the demonstration *per se* of a direct causal link between these two events will be difficult to establish, evidence indicates that blockade of pS10-H3 phosphorylation is associated with altered nucleosomal responses. Thus, the decrease of pS10-H3 and pS10acK14-H3 observed in MSK1-deficient mice or following inhibition of Elk-1 phosphorylation in response to cocaine or L-DOPA is accompanied by reduced expression of several genes including c-fos, zif268, ΔFosB and Arc/Arg3.1 in striatal SPNs (Alcacer et al., 2014; Besnard et al., 2011; Brami-Cherrier et al., 2005; Feyder et al., 2016). Lastly, the increased S28me3K27-H3 phosphorylation induced by L-DOPA, D-amphetamine or haloperidol has been shown to correlate with reduced binding of polycomb group complexes leading to derepression and eventually increased transcription of a subset of genes in D1R- and D2R-SPNs including Atf3, Npas4, Klf4 and Lipg (Bonito-Oliva et al., 2016; Sodersten et al., 2014). It is likely that chromatin immunoprecipitation followed by RNA sequencing will allow the identification of new subsets of genes whose expression or repression are tightly linked to the state of histone H3 phosphorylation at S10 and S28, respectively.

4. Regulation of the ribosomal protein S6 by dopamine

The ribosomal protein rpS6 is a component of the small 40S ribosomal subunit, which undergoes phosphorylation at five Ser residues (S235/236, S240/244 and S247) located in its carboxy-terminal domain. S235/236 phosphorylation sites are controlled by multiple kinases including p70S6K, p90RSK and PKA among others. In contrast, S240/244 sites are solely regulated following p70S6K activation. Dephosphorylation of the five phospho-sites is achieved by a single phosphatase, PP1 (Biever, Valjent, & Puighermanal, 2015). Section 4 summarizes our current knowledge of the molecular mechanisms underlying rpS6 phosphorylation in the striatum following the modulation of dopamine receptors.

4.1 Regulation of rpS6 phosphorylation by D1R and D2R modulation

Most, if not all, pharmacological agents leading to D1R activation promote rpS6 phosphorylation in the striatum and/or the nucleus accumbens (Fig. 1). The administration of apomorphine (Gangarossa, Perroy, et al., 2013), D1R agonists (SKF81297 or SKF82958) (Beckley et al., 2016; Gangarossa, Perroy, et al., 2013; Rapanelli et al., 2016), cocaine (Papale et al., 2016; Sutton & Caron, 2015; Wu, McCallum, Glick, & Huang, 2011) and D-amphetamine (Biever, Puighermanal, et al., 2015; Rapanelli et al., 2014) triggers an increase in pS235/236-rpS6 in D1R-SPNs that requires D1R (Biever, Puighermanal, et al., 2015; Sutton & Caron, 2015). Unlike D-amphetamine, cocaine also regulates rpS6 phosphorylation at S240/244 (Biever, Puighermanal, et al., 2015; Knight et al., 2012). Finally, acute and repeated L-DOPA administration enhances the phosphorylation of rpS6 at the four sites in D1R-SPNs (Gangarossa et al., 2016; Santini, Heiman, Greengard, Valjent, & Fisone, 2009; Santini et al., 2012, 2010) (Fig. 1).

The regulation of rpS6 phosphorylation in the striatum following D2R/D3R stimulation is less consensual and depends on the nature of the agonists tested. While pS235/236-rpS6 are unchanged in the striatum and nucleus accumbens in response to quinpirole or quinelorane (Beckley et al., 2016;

Gangarossa, Perroy, et al., 2013; Salles et al., 2013), these phospho-sites are decreased in both D1R- and D2R-SPNs after the administration of the D3R-preferring agonist, pramipexole (unpublished observation) (Fig. 1). On the other hand, quinelorane induces a transient phosphorylation of rpS6 at S240/244 in both dorsal and ventral striatum. This effect requires both D2R and D3R, since it is totally prevented by raclopride or by the pharmacological blockade and genetic deletion of D3R (Beckley et al., 2016; Gangarossa, Perroy, et al., 2013; Salles et al., 2013).

The effects on rpS6 phosphorylation in response to D2R blockade are more consistent. Enhanced phosphorylation of rpS6 at S240/244 and S235/236 is observed in striatal culture and in the dorsal striatum in response to haloperidol or amisulpiride (Bonito-Oliva et al., 2013; Bowling et al., 2014; Kharkwal et al., 2016; Rose et al., 2018; Valjent et al., 2011) (Fig. 1). Haloperidol-induced pS240/244-rpS6 and pS235/236-rpS6 occur in D2R-SPNs through a mechanism requiring D2R/G-protein signaling (Rose et al., 2018) and the stimulation of A2aR (Bonito-Oliva et al., 2013) (Fig. 1). Recently, one study has provided new insights into the regulation of rpS6 phosphorylation by haloperidol. This involves D2Rs expressed by CINs and acetylcholine (ACh) signaling. By blocking D2Rs located on CINs, haloperidol increases their activity leading to increased rpS6 phosphorylation at S240/244, considered a reliable index of CINs activity (Bertran-Gonzalez, Chieng, Laurent, Valjent, & Balleine, 2012; Kharkwal et al., 2016). The release of ACh activates in return the muscarinic M1R expressed by D2R-SPNs promoting rpS6 phosphorylation at S235/236 (Bertran-Gonzalez et al., 2012; Kharkwal et al., 2016). Thus, haloperidol-induced pS235/236-rpS6 in D2R-SPNs may result from the concomitant blockade of D2R expressed in SPNs and CINs as well as the simultaneous activation of A2aR and M1R in D2R-SPNs (Fig. 3).

4.2 The central role of the cAMP/PKA/DARPP-32 pathway in the regulation of rpS6 phosphorylation in SPNs

The state of rpS6 phosphorylation at S235/236 and S240/244 in both D1R- and D2R-SPNs critically depends on cAMP/PKA activation (Fig. 3). Application of forskolin in striatal slices triggers a PKA-dependent phosphorylation of rpS6 at S235/326 (Biever, Puighermanal, et al., 2015). Similar results are obtained by direct stimulation of PKA in striatal cultures (Valjent et al., 2011). Moreover, D-amphetamine- and haloperidol-induced pS235/236-rpS6 in D1R- and D2R-SPNs, respectively, are strongly reduced in Gαolf heterozygous mice (Biever, Puighermanal, et al., 2015;

Valjent et al., 2011). Finally, the inhibition of PP1 by DARPP-32 is implicated in the regulation of these phospho-sites by haloperidol, D-amphetamine or L-DOPA (Biever, Puighermanal, et al., 2015; Bonito-Oliva et al., 2013; Santini et al., 2012; Valjent et al., 2011) (Fig. 3).

ERK and p70S6K1/mTORC1-dependent mechanisms also participate in the regulation of rpS6 phosphorylation in the striatum. The respective contribution of these two pathways working alone or synergistically might depend on the identity of the phospho-sites studied. This has been well exemplified in the case of the regulation of rpS6 phosphorylation in D2R-SPNs by haloperidol. Indeed, enhanced pS235/236-rpS6 relies solely on the inhibition of PP1 by DARPP-32 (Valjent et al., 2011), whereas the regulation at S240/244 also requires p70S6K1/mTORC1 activation (Bonito-Oliva et al., 2013; Bowling et al., 2014). Another factor includes the nature of the pharmacological stimuli used to trigger rpS6 phosphorylation. Unlike cocaine (Papale et al., 2016), S235/236-rpS6 phosphorylation induced by D-amphetamine occurs independently of ERK (Biever, Puighermanal, et al., 2015; but see Rapanelli et al., 2014). Finally, the physiological state of the animals (pathological *vs* nonpathological) is also important. Indeed, in 6-OHDA-lesioned mice treated with L-DOPA, D1R-mediated rpS6 phosphorylation at S235/236 and S240/244-rpS6 not only requires PP1 inhibition by DARPP-32 but also involves the synergistic activation of the ERK and Rhes/mTORC1 pathways (Santini, Heiman, et al., 2009; Subramaniam et al., 2011).

4.3 Role of rpS6 phosphorylation in the striatum

In several brain areas including the striatum, enhanced phosphorylation of rpS6 is often concomitant to the increased global mRNA translation and synthesis of TOP-mRNA encoded proteins (Biever, Valjent, et al., 2015). Primary mouse striatal neurons treated with haloperidol display enhanced rpS6 phosphorylation at S240/244, which is accompanied by a transient increase in protein synthesis (Bowling et al., 2014). Similarly, the systemic administration of the D1R agonist SKF81927 that promotes pS235/236-rpS6 in the nucleus accumbens also triggers the translation of GluA1, PSD95 and Homer, three important synaptic proteins (Beckley et al., 2016). Such positive correlations are, however, not systematically observed. Indeed, in the dorsal striatum, the increased rpS6 phosphorylation induced by a single administration of D-amphetamine, haloperidol or papaverine is not associated with enhanced global and/or TOP mRNA translation

(Biever, Puighermanal, et al., 2015). Strikingly, the increased rpS6 phosphorylation at S235/236 observed upon repeated D-amphetamine administration in the striatum parallels a transient decrease of *de novo* protein synthesis, illustrating that the two events can be inversely correlated (Biever et al., 2016; Biever, Puighermanal, et al., 2015). The use of rpS6 knockin mice, in which S235, S236, S240, S244 and S247 were replaced by alanine to prevent any phosphorylation (Ruvinsky et al., 2005), has been critical to determine a possible causal link between rpS6 phosphorylation and mRNA translation. While the phosphorylation of rpS6 appears to be dispensable for global protein synthesis, it regulates the translation of a specific subset of mitochondria-related mRNAs that could contribute to the altered synaptic plasticity present at excitatory synapses onto SPNs in the nucleus accumbens of rpS6 knockin mice (Puighermanal et al., 2017). Approaches combining the capture of phosphorylated ribosomes and mRNA profiling should help to determine precisely the identity of the transcripts bound to phosphorylated rpS6 in D1R- and D2R-SPNs in response to psychostimulants, L-DOPA or haloperidol treatment (Knight et al., 2012).

5. Future directions

As illustrated above, selective intracellular mechanisms have been identified through which DA triggers histone H3 and rpS6 phosphorylation in the striatum. The use of transgenic mice expressing fluorescent proteins or Cre-recombinase has been instrumental to precisely identify how and in which striatal populations these posttranslational modifications occurred in response to a wide range of pharmacological compounds. However, several important challenges emerge and it is likely that many of the mechanisms involved in the regulation of striatal intracellular signaling will have to be reconsidered in light of the molecular diversity of neurons defined by their cognate neurotransmitters. For instance, multiple midbrain DA neurons subtypes with distinct molecular and functional properties have been identified (Brichta et al., 2015; Poulin et al., 2014). Among them, some corelease glutamate and GABA (Stuber, Hnasko, Britt, Edwards, & Bonci, 2010; Tecuapetla et al., 2010; Tritsch, Ding, & Sabatini, 2012). How DA/glutamate and DA/GABA corelease influence DA-mediated intracellular signaling is an issue that should be carefully considered. Similarly, the regulation of these intracellular events will need to be reexamined in light of the molecular diversity existing within D1R- and D2R-SPNs as well as in CINs and GABAergic interneurons (Gokce et al., 2016;

Munoz-Manchado et al., 2018; Saunders et al., 2018). Determining whether distinct D1R- and D2R-SPNs subtypes possess a unique combination of intracellular signaling proteins could be key to understand why upon D1R and D2R modulation, topographic patterns of histone H3 and rpS6 phosphorylation are preferentially observed in SPNs throughout the striatum (Gangarossa, Perroy, et al., 2013). Lastly, DA receptors are also present in both astrocytes (Bal et al., 1994; Zanassi, Paolillo, Montecucco, Avvedimento, & Schinelli, 1999) and microglia (Huck et al., 2015), and a role of DA signaling *via* astrocytic D2R in the control of neuro-immune responses has been recently unveiled (Shao et al., 2013). Determining which intracellular mechanisms are activated downstream to the modulation of D1R and D2R expressed by these other cell types and how they influence striatal-dependent behaviors will certainly be part of the future challenge.

Acknowledgments

This work was supported by Inserm, Fondation pour la Recherche Médicale (FRM) and The French National Research Agency (ANR-EPITRACES) (E.V.). G.G. is supported by Allen Foundation Inc. (2016.326) and Nutricia Research Foundation (2017–20) grants. E.P. is a recipient of Beatriu de Pinós fellowship (# 2017BP00132) from University and Research Grants Management Agency (Government of Catalonia).

References

Alcacer, C., Charbonnier-Beaupel, F., Corvol, J. C., Girault, J. A., & Herve, D. (2014). Mitogen- and stress-activated protein kinase 1 is required for specific signaling responses in dopamine-denervated mouse striatum, but is not necessary for L-DOPA-induced dyskinesia. *Neuroscience Letters*, *583*, 76–80.

Alcacer, C., Santini, E., Valjent, E., Gaven, F., Girault, J. A., & Herve, D. (2012). Galpha (olf) mutation allows parsing the role of cAMP-dependent and extracellular signal-regulated kinase-dependent signaling in L-3,4-dihydroxyphenylalanine-induced dyskinesia. *The Journal of Neuroscience*, *32*(17), 5900–5910.

Bal, A., Bachelot, T., Savasta, M., Manier, M., Verna, J. M., Benabid, A. L., et al. (1994). Evidence for dopamine D2 receptor mRNA expression by striatal astrocytes in culture: In situ hybridization and polymerase chain reaction studies. *Brain Research. Molecular Brain Research*, *23*(3), 204–212.

Bamford, N. S., Wightman, R. M., & Sulzer, D. (2018). Dopamine's effects on corticostriatal synapses during reward-based behaviors. *Neuron*, *97*(3), 494–510.

Bateup, H. S., Svenningsson, P., Kuroiwa, M., Gong, S., Nishi, A., Heintz, N., et al. (2008). Cell type-specific regulation of DARPP-32 phosphorylation by psychostimulant and antipsychotic drugs. *Nature Neuroscience*, *11*(8), 932–939.

Beaulieu, J. M., & Gainetdinov, R. R. (2011). The physiology, signaling, and pharmacology of dopamine receptors. *Pharmacological Reviews*, *63*(1), 182–217.

Beckley, J. T., Laguesse, S., Phamluong, K., Morisot, N., Wegner, S. A., & Ron, D. (2016). The first alcohol drink triggers mTORC1-dependent synaptic plasticity in nucleus accumbens dopamine D1 receptor neurons. *The Journal of Neuroscience*, *36*(3), 701–713.

Bertran-Gonzalez, J., Bosch, C., Maroteaux, M., Matamales, M., Herve, D., Valjent, E., et al. (2008). Opposing patterns of signaling activation in dopamine D1 and D2 receptor-expressing striatal neurons in response to cocaine and haloperidol. *The Journal of Neuroscience*, *28*(22), 5671–5685.

Bertran-Gonzalez, J., Chieng, B. C., Laurent, V., Valjent, E., & Balleine, B. W. (2012). Striatal cholinergic interneurons display activity-related phosphorylation of ribosomal protein S6. *PLoS One*, 7(12), e53195.

Bertran-Gonzalez, J., Hakansson, K., Borgkvist, A., Irinopoulou, T., Brami-Cherrier, K., Usiello, A., et al. (2009). Histone H3 phosphorylation is under the opposite tonic control of dopamine D2 and adenosine A2A receptors in striatopallidal neurons. *Neuropsychopharmacology*, *34*(7), 1710–1720.

Besnard, A., Bouveyron, N., Kappes, V., Pascoli, V., Pages, C., Heck, N., et al. (2011). Alterations of molecular and behavioral responses to cocaine by selective inhibition of Elk-1 phosphorylation. *The Journal of Neuroscience*, *31*(40), 14296–14307.

Biever, A., Boubaker-Vitre, J., Cutando, L., Gracia-Rubio, I., Costa-Mattioli, M., Puighermanal, E., et al. (2016). Repeated exposure to D-amphetamine decreases global protein synthesis and regulates the translation of a subset of mRNAs in the striatum. *Frontiers in Molecular Neuroscience*, *9*, 165.

Biever, A., Puighermanal, E., Nishi, A., David, A., Panciatici, C., Longueville, S., et al. (2015). PKA-dependent phosphorylation of ribosomal protein S6 does not correlate with translation efficiency in striatonigral and striatopallidal medium-sized spiny neurons. *The Journal of Neuroscience*, *35*(10), 4113–4130.

Biever, A., Valjent, E., & Puighermanal, E. (2015). Ribosomal protein S6 phosphorylation in the nervous system: From regulation to function. *Frontiers in Molecular Neuroscience*, *8*, 75.

Bonito-Oliva, A., Pallottino, S., Bertran-Gonzalez, J., Girault, J. A., Valjent, E., & Fisone, G. (2013). Haloperidol promotes mTORC1-dependent phosphorylation of ribosomal protein S6 via dopamine- and cAMP-regulated phosphoprotein of 32 kDa and inhibition of protein phosphatase-1. *Neuropharmacology*, *72*, 197–203.

Bonito-Oliva, A., Sodersten, E., Spigolon, G., Hu, X., Hellysaz, A., Falconi, A., et al. (2016). Differential regulation of the phosphorylation of Trimethyl-lysine27 histone H3 at serine 28 in distinct populations of striatal projection neurons. *Neuropharmacology*, *107*, 89–99.

Bowling, H., Zhang, G., Bhattacharya, A., Perez-Cuesta, L. M., Deinhardt, K., Hoeffer, C. A., et al. (2014). Antipsychotics activate mTORC1-dependent translation to enhance neuronal morphological complexity. *Science Signaling*, 7(308), ra4.

Brami-Cherrier, K., Roze, E., Girault, J. A., Betuing, S., & Caboche, J. (2009). Role of the ERK/MSK1 signalling pathway in chromatin remodelling and brain responses to drugs of abuse. *Journal of Neurochemistry*, *108*(6), 1323–1335.

Brami-Cherrier, K., Valjent, E., Herve, D., Darragh, J., Corvol, J. C., Pages, C., et al. (2005). Parsing molecular and behavioral effects of cocaine in mitogen- and stress-activated protein kinase-1-deficient mice. *The Journal of Neuroscience*, *25*(49), 11444–11454.

Brichta, L., Shin, W., Jackson-Lewis, V., Blesa, J., Yap, E. L., Walker, Z., et al. (2015). Identification of neurodegenerative factors using translatome-regulatory network analysis. *Nature Neuroscience*, *18*(9), 1325–1333.

Bromberg-Martin, E. S., Matsumoto, M., & Hikosaka, O. (2010). Dopamine in motivational control: Rewarding, aversive, and alerting. *Neuron*, *68*(5), 815–834.

Burke, D. A., Rotstein, H. G., & Alvarez, V. A. (2017). Striatal local circuitry: A new framework for lateral inhibition. *Neuron*, *96*(2), 267–284.

Chen, G., Nie, S., Han, C., Ma, K., Xu, Y., Zhang, Z., et al. (2017). Antidyskinetic effects of MEK inhibitor are associated with multiple neurochemical alterations in the striatum of hemiparkinsonian rats. *Frontiers in Neuroscience*, *11*, 112.

Darmopil, S., Martin, A. B., De Diego, I. R., Ares, S., & Moratalla, R. (2009). Genetic inactivation of dopamine D1 but not D2 receptors inhibits L-DOPA-induced dyskinesia and histone activation. *Biological Psychiatry*, *66*(6), 603–613.

Dumartin, B., Doudnikoff, E., Gonon, F., & Bloch, B. (2007). Differences in ultrastructural localization of dopaminergic D1 receptors between dorsal striatum and nucleus accumbens in the rat. *Neuroscience Letters*, *419*(3), 273–277.

Feyder, M., Sodersten, E., Santini, E., Vialou, V., LaPlant, Q., Watts, E. L., et al. (2016). A role for mitogen- and stress-activated kinase 1 in L-DOPA-induced dyskinesia and FosB expression. *Biological Psychiatry*, *79*(5), 362–371.

Fieblinger, T., Sebastianutto, I., Alcacer, C., Bimpisidis, Z., Maslava, N., Sandberg, S., et al. (2014). Mechanisms of dopamine D1 receptor-mediated ERK1/2 activation in the parkinsonian striatum and their modulation by metabotropic glutamate receptor type 5. *The Journal of Neuroscience*, *34*(13), 4728–4740.

Gangarossa, G., Espallergues, J., de Kerchove d'Exaerde, A., El Mestikawy, S., Gerfen, C. R., Herve, D., et al. (2013). Distribution and compartmental organization of GABAergic medium-sized spiny neurons in the mouse nucleus accumbens. *Frontiers in Neural Circuits*, 7, 22.

Gangarossa, G., Guzman, M., Prado, V. F., Prado, M. A., Daumas, S., El Mestikawy, S., et al. (2016). Role of the atypical vesicular glutamate transporter VGLUT3 in l-DOPA-induced dyskinesia. *Neurobiology of Disease*, *87*, 69–79.

Gangarossa, G., Perroy, J., & Valjent, E. (2013). Combinatorial topography and cell-type specific regulation of the ERK pathway by dopaminergic agonists in the mouse striatum. *Brain Structure & Function*, *218*(2), 405–419.

Gerfen, C. R., Miyachi, S., Paletzki, R., & Brown, P. (2002). D1 dopamine receptor supersensitivity in the dopamine-depleted striatum results from a switch in the regulation of ERK1/2/MAP kinase. *The Journal of Neuroscience*, *22*(12), 5042–5054.

Gerfen, C. R., & Surmeier, D. J. (2011). Modulation of striatal projection systems by dopamine. *Annual Review of Neuroscience*, *34*, 441–466.

Girault, J. A., Valjent, E., Caboche, J., & Herve, D. (2007). ERK2: A logical AND gate critical for drug-induced plasticity? *Current Opinion in Pharmacology*, 7(1), 77–85.

Gokce, O., Stanley, G. M., Treutlein, B., Neff, N. F., Camp, J. G., Malenka, R. C., et al. (2016). Cellular taxonomy of the mouse striatum as revealed by single-cell RNA-seq. *Cell Reports*, *16*(4), 1126–1137.

Herve, D. (2011). Identification of a specific assembly of the g protein golf as a critical and regulated module of dopamine and adenosine-activated cAMP pathways in the striatum. *Frontiers in Neuroanatomy*, *5*, 48.

Hoover, B. R., & Marshall, J. F. (2004). Molecular, chemical, and anatomical characterization of globus pallidus dopamine D2 receptor mRNA-containing neurons. *Synapse*, *52*(2), 100–113.

Huck, J. H., Freyer, D., Bottcher, C., Mladinov, M., Muselmann-Genschow, C., Thielke, M., et al. (2015). De novo expression of dopamine D2 receptors on microglia after stroke. *Journal of Cerebral Blood Flow and Metabolism*, *35*(11), 1804–1811.

Jordi, E., Heiman, M., Marion-Poll, L., Guermonprez, P., Cheng, S. K., Nairn, A. C., et al. (2013). Differential effects of cocaine on histone posttranslational modifications in identified populations of striatal neurons. *Proceedings of the National Academy of Sciences of the United States of America*, *110*(23), 9511–9516.

Kharkwal, G., Brami-Cherrier, K., Lizardi-Ortiz, J. E., Nelson, A. B., Ramos, M., Del Barrio, D., et al. (2016). Parkinsonism driven by antipsychotics originates from dopaminergic control of striatal cholinergic interneurons. *Neuron*, *91*(1), 67–78.

Knight, Z. A., Tan, K., Birsoy, K., Schmidt, S., Garrison, J. L., Wysocki, R. W., et al. (2012). Molecular profiling of activated neurons by phosphorylated ribosome capture. *Cell*, *151*(5), 1126–1137.

Kumar, A., Choi, K. H., Renthal, W., Tsankova, N. M., Theobald, D. E., Truong, H. T., et al. (2005). Chromatin remodeling is a key mechanism underlying cocaine-induced plasticity in striatum. *Neuron, 48*(2), 303–314.

Li, J., Guo, Y., Schroeder, F. A., Youngs, R. M., Schmidt, T. W., Ferris, C., et al. (2004). Dopamine D2-like antagonists induce chromatin remodeling in striatal neurons through cyclic AMP-protein kinase A and NMDA receptor signaling. *Journal of Neurochemistry, 90*(5), 1117–1131.

Matamales, M., Bertran-Gonzalez, J., Salomon, L., Degos, B., Deniau, J. M., Valjent, E., et al. (2009). Striatal medium-sized spiny neurons: Identification by nuclear staining and study of neuronal subpopulations in BAC transgenic mice. *PLoS One, 4*(3), e4770.

Munoz-Manchado, A. B., Bengtsson Gonzales, C., Zeisel, A., Munguba, H., Bekkouche, B., Skene, N. G., et al. (2018). Diversity of interneurons in the dorsal striatum revealed by single-cell RNA sequencing and PatchSeq. *Cell Reports, 24*(8), 2179-2190.e7.

Nicholas, A. P., Lubin, F. D., Hallett, P. J., Vattem, P., Ravenscroft, P., Bezard, E., et al. (2008). Striatal histone modifications in models of levodopa-induced dyskinesia. *Journal of Neurochemistry, 106*(1), 486–494.

Nishi, A., Matamales, M., Musante, V., Valjent, E., Kuroiwa, M., Kitahara, Y., et al. (2017). Glutamate counteracts dopamine/PKA signaling via dephosphorylation of DARPP-32 Ser-97 and alteration of its cytonuclear distribution. *The Journal of Biological Chemistry, 292*(4), 1462–1476.

Papale, A., Morella, I. M., Indrigo, M. T., Bernardi, R. E., Marrone, L., Marchisella, F., et al. (2016). Impairment of cocaine-mediated behaviours in mice by clinically relevant Ras-ERK inhibitors. *eLife, 5*, pii: e17111.

Park, H. Y., Kang, Y. M., Kang, Y., Park, T. S., Ryu, Y. K., Hwang, J. H., et al. (2014). Inhibition of adenylyl cyclase type 5 prevents L-DOPA-induced dyskinesia in an animal model of Parkinson's disease. *The Journal of Neuroscience, 34*(35), 11744–11753.

Pascoli, V., Cahill, E., Bellivier, F., Caboche, J., & Vanhoutte, P. (2014). Extracellular signal-regulated protein kinases 1 and 2 activation by addictive drugs: A signal toward pathological adaptation. *Biological Psychiatry, 76*(12), 917–926.

Polito, M., Guiot, E., Gangarossa, G., Longueville, S., Doulazmi, M., Valjent, E., et al. (2015). Selective effects of PDE10A inhibitors on striatopallidal neurons require phosphatase inhibition by DARPP-32. *eNeuro, 2*(4), pii: ENEURO.0060-15.2015.

Poulin, J. F., Zou, J., Drouin-Ouellet, J., Kim, K. Y., Cicchetti, F., & Awatramani, R. B. (2014). Defining midbrain dopaminergic neuron diversity by single-cell gene expression profiling. *Cell Reports, 9*(3), 930–943.

Pozzi, L., Hakansson, K., Usiello, A., Borgkvist, A., Lindskog, M., Greengard, P., et al. (2003). Opposite regulation by typical and atypical anti-psychotics of ERK1/2, CREB and Elk-1 phosphorylation in mouse dorsal striatum. *Journal of Neurochemistry, 86*(2), 451–459.

Puighermanal, E., Biever, A., Pascoli, V., Melser, S., Pratlong, M., Cutando, L., et al. (2017). Ribosomal protein S6 phosphorylation is involved in novelty-induced locomotion, synaptic plasticity and mRNA translation. *Frontiers in Molecular Neuroscience, 10*, 419.

Rapanelli, M., Frick, L. R., Horn, K. D., Schwarcz, R. C., Pogorelov, V., Nairn, A. C., et al. (2016). The histamine H3 receptor differentially modulates mitogen-activated protein kinase (MAPK) and Akt signaling in striatonigral and striatopallidal neurons. *The Journal of Biological Chemistry, 291*(40), 21042–21052.

Rapanelli, M., Frick, L. R., Pogorelov, V., Ota, K. T., Abbasi, E., Ohtsu, H., et al. (2014). Dysregulated intracellular signaling in the striatum in a pathophysiologically grounded model of Tourette syndrome. *European Neuropsychopharmacology, 24*(12), 1896–1906.

Rivera, A., Alberti, I., Martin, A. B., Narvaez, J. A., de la Calle, A., & Moratalla, R. (2002). Molecular phenotype of rat striatal neurons expressing the dopamine D5 receptor subtype. *The European Journal of Neuroscience*, *16*(11), 2049–2058.

Rivera, A., Cuellar, B., Giron, F. J., Grandy, D. K., de la Calle, A., & Moratalla, R. (2002). Dopamine D4 receptors are heterogeneously distributed in the striosomes/matrix compartments of the striatum. *Journal of Neurochemistry*, *80*(2), 219–229.

Roeper, J. (2013). Dissecting the diversity of midbrain dopamine neurons. *Trends in Neurosciences*, *36*(6), 336–342.

Rose, S. J., Pack, T. F., Peterson, S. M., Payne, K., Borrelli, E., & Caron, M. G. (2018). Engineered D2R variants reveal the balanced and biased contributions of G-protein and beta-Arrestin to dopamine-dependent functions. *Neuropsychopharmacology*, *43*(5), 1164–1173.

Rotllant, D., & Armario, A. (2012). Brain pattern of histone H3 phosphorylation after acute amphetamine administration: Its relationship to brain c-fos induction is strongly dependent on the particular brain area. *Neuropharmacology*, *62*(2), 1073–1081.

Ruvinsky, I., Sharon, N., Lerer, T., Cohen, H., Stolovich-Rain, M., Nir, T., et al. (2005). Ribosomal protein S6 phosphorylation is a determinant of cell size and glucose homeostasis. *Genes & Development*, *19*(18), 2199–2211.

Salles, M. J., Herve, D., Rivet, J. M., Longueville, S., Millan, M. J., Girault, J. A., et al. (2013). Transient and rapid activation of Akt/GSK-3beta and mTORC1 signaling by D3 dopamine receptor stimulation in dorsal striatum and nucleus accumbens. *Journal of Neurochemistry*, *125*(4), 532–544.

Sanchis-Segura, C., Lopez-Atalaya, J. P., & Barco, A. (2009). Selective boosting of transcriptional and behavioral responses to drugs of abuse by histone deacetylase inhibition. *Neuropsychopharmacology*, *34*(13), 2642–2654.

Santini, E., Alcacer, C., Cacciatore, S., Heiman, M., Herve, D., Greengard, P., et al. (2009). L-DOPA activates ERK signaling and phosphorylates histone H3 in the striatonigral medium spiny neurons of hemiparkinsonian mice. *Journal of Neurochemistry*, *108*(3), 621–633.

Santini, E., Feyder, M., Gangarossa, G., Bateup, H. S., Greengard, P., & Fisone, G. (2012). Dopamine- and cAMP-regulated phosphoprotein of 32-kDa (DARPP-32)-dependent activation of extracellular signal-regulated kinase (ERK) and mammalian target of rapamycin complex 1 (mTORC1) signaling in experimental parkinsonism. *The Journal of Biological Chemistry*, *287*(33), 27806–27812.

Santini, E., Heiman, M., Greengard, P., Valjent, E., & Fisone, G. (2009). Inhibition of mTOR signaling in Parkinson's disease prevents L-DOPA-induced dyskinesia. *Science Signaling*, *2*(80), ra36.

Santini, E., Sgambato-Faure, V., Li, Q., Savasta, M., Dovero, S., Fisone, G., et al. (2010). Distinct changes in cAMP and extracellular signal-regulated protein kinase signalling in L-DOPA-induced dyskinesia. *PLoS One*, *5*(8), e12322.

Santini, E., Valjent, E., Usiello, A., Carta, M., Borgkvist, A., Girault, J. A., et al. (2007). Critical involvement of cAMP/DARPP-32 and extracellular signal-regulated protein kinase signaling in L-DOPA-induced dyskinesia. *The Journal of Neuroscience*, *27*(26), 6995–7005.

Saunders, A., Macosko, E. Z., Wysoker, A., Goldman, M., Krienen, F. M., de Rivera, H., et al. (2018). Molecular diversity and specializations among the cells of the adult mouse brain. *Cell*, *174*(4), 1015-1030.e16.

Sawicka, A., & Seiser, C. (2012). Histone H3 phosphorylation—A versatile chromatin modification for different occasions. *Biochimie*, *94*(11), 2193–2201.

Schroeder, F. A., Penta, K. L., Matevossian, A., Jones, S. R., Konradi, C., Tapper, A. R., et al. (2008). Drug-induced activation of dopamine D(1) receptor signaling and inhibition of class I/II histone deacetylase induce chromatin remodeling in reward circuitry and modulate cocaine-related behaviors. *Neuropsychopharmacology*, *33*(12), 2981–2992.

Schultz, W. (2016). Dopamine reward prediction-error signalling: A two-component response. *Nature Reviews. Neuroscience*, *17*(3), 183–195.

Sesack, S. R., Aoki, C., & Pickel, V. M. (1994). Ultrastructural localization of D2 receptor-like immunoreactivity in midbrain dopamine neurons and their striatal targets. *The Journal of Neuroscience*, *14*(1), 88–106.

Shao, W., Zhang, S. Z., Tang, M., Zhang, X. H., Zhou, Z., Yin, Y. Q., et al. (2013). Suppression of neuroinflammation by astrocytic dopamine D2 receptors via alphaB-crystallin. *Nature*, *494*(7435), 90–94.

Sodersten, E., Feyder, M., Lerdrup, M., Gomes, A. L., Kryh, H., Spigolon, G., et al. (2014). Dopamine signaling leads to loss of polycomb repression and aberrant gene activation in experimental parkinsonism. *PLoS Genetics*, *10*(9), e1004574.

Solis, O., Garcia-Montes, J. R., Gonzalez-Granillo, A., Xu, M., & Moratalla, R. (2017). Dopamine D3 receptor modulates l-DOPA-induced dyskinesia by targeting D1 receptor-mediated striatal signaling. *Cerebral Cortex*, *27*(1), 435–446.

Stipanovich, A., Valjent, E., Matamales, M., Nishi, A., Ahn, J. H., Maroteaux, M., et al. (2008). A phosphatase cascade by which rewarding stimuli control nucleosomal response. *Nature*, *453*(7197), 879–884.

Stuber, G. D., Hnasko, T. S., Britt, J. P., Edwards, R. H., & Bonci, A. (2010). Dopaminergic terminals in the nucleus accumbens but not the dorsal striatum corelease glutamate. *The Journal of Neuroscience*, *30*(24), 8229–8233.

Subramaniam, S., Napolitano, F., Mealer, R. G., Kim, S., Errico, F., Barrow, R., et al. (2011). Rhes, a striatal-enriched small G protein, mediates mTOR signaling and L-DOPA-induced dyskinesia. *Nature Neuroscience*, *15*(2), 191–193.

Sutton, L. P., & Caron, M. G. (2015). Essential role of D1R in the regulation of mTOR complex1 signaling induced by cocaine. *Neuropharmacology*, *99*, 610–619.

Taverna, S., Ilijic, E., & Surmeier, D. J. (2008). Recurrent collateral connections of striatal medium spiny neurons are disrupted in models of Parkinson's disease. *The Journal of Neuroscience*, *28*(21), 5504–5512.

Tecuapetla, F., Patel, J. C., Xenias, H., English, D., Tadros, I., Shah, F., et al. (2010). Glutamatergic signaling by mesolimbic dopamine neurons in the nucleus accumbens. *The Journal of Neuroscience*, *30*(20), 7105–7110.

Tepper, J. M., Koos, T., Ibanez-Sandoval, O., Tecuapetla, F., Faust, T. W., & Assous, M. (2018). Heterogeneity and diversity of striatal GABAergic interneurons: Update 2018. *Frontiers in Neuroanatomy*, *12*, 91.

Tritsch, N. X., Ding, J. B., & Sabatini, B. L. (2012). Dopaminergic neurons inhibit striatal output through non-canonical release of GABA. *Nature*, *490*(7419), 262–266.

Valjent, E., Bertran-Gonzalez, J., Bowling, H., Lopez, S., Santini, E., Matamales, M., et al. (2011). Haloperidol regulates the state of phosphorylation of ribosomal protein S6 via activation of PKA and phosphorylation of DARPP-32. *Neuropsychopharmacology*, *36*(12), 2561–2570.

Valjent, E., Bertran-Gonzalez, J., Herve, D., Fisone, G., & Girault, J. A. (2009). Looking BAC at striatal signaling: Cell-specific analysis in new transgenic mice. *Trends in Neurosciences*, *32*(10), 538–547.

Valjent, E., Pages, C., Herve, D., Girault, J. A., & Caboche, J. (2004). Addictive and non-addictive drugs induce distinct and specific patterns of ERK activation in mouse brain. *The European Journal of Neuroscience*, *19*(7), 1826–1836.

Valjent, E., Pascoli, V., Svenningsson, P., Paul, S., Enslen, H., Corvol, J. C., et al. (2005). Regulation of a protein phosphatase cascade allows convergent dopamine and glutamate signals to activate ERK in the striatum. *Proceedings of the National Academy of Sciences of the United States of America*, *102*(2), 491–496.

Wang, H., & Pickel, V. M. (2002). Dopamine D2 receptors are present in prefrontal cortical afferents and their targets in patches of the rat caudate-putamen nucleus. *The Journal of Comparative Neurology*, *442*(4), 392–404.

Westin, J. E., Vercammen, L., Strome, E. M., Konradi, C., & Cenci, M. A. (2007). Spatiotemporal pattern of striatal ERK1/2 phosphorylation in a rat model of L-DOPA-induced dyskinesia and the role of dopamine D1 receptors. *Biological Psychiatry*, *62*(7), 800–810.

Wu, J., McCallum, S. E., Glick, S. D., & Huang, Y. (2011). Inhibition of the mammalian target of rapamycin pathway by rapamycin blocks cocaine-induced locomotor sensitization. *Neuroscience*, *172*, 104–109.

Yger, M., & Girault, J. A. (2011). DARPP-32, Jack of all trades... master of which? *Frontiers in Behavioral Neuroscience*, *5*, 56.

Zanassi, P., Paolillo, M., Montecucco, A., Avvedimento, E. V., & Schinelli, S. (1999). Pharmacological and molecular evidence for dopamine D(1) receptor expression by striatal astrocytes in culture. *Journal of Neuroscience Research*, *58*(4), 544–552.

CHAPTER ELEVEN

Recent advances in computational studies of GPCR-G protein interactions

Jinan Wang, Yinglong Miao*
Center for Computational Biology and Department of Molecular Biosciences, University of Kansas, Lawrence, KS, United States
*Corresponding author: e-mail address: miao@ku.edu

Contents

Abstract

Protein-protein interactions are key in cellular signaling. G protein-coupled receptors (GPCRs), the largest superfamily of human membrane proteins, are able to transduce extracellular signals (e.g., hormones and neurotransmitters) to intracellular proteins, in particular the G proteins. Since GPCRs serve as primary targets of ~1/3 of currently marketed drugs, it is important to understand mechanisms of GPCR signaling in order to design selective and potent drug molecules. This chapter focuses on recent advances in computational studies of the GPCR-G protein interactions using bioinformatics, protein-protein docking and molecular dynamics simulation approaches.

1. Introduction

Protein-protein interactions (PPIs) are central to many biological processes, including human immune responses and cellular signaling. PPIs have been targeted for developing small-molecule modulators as therapeutic

Advances in Protein Chemistry and Structural Biology, Volume 116
ISSN 1876-1623
https://doi.org/10.1016/bs.apcsb.2018.11.011

drugs (Andreani & Guerois, 2014; Arkin & Wells, 2004). In particular, the interactions between G protein-coupled receptors (GPCRs) and heterotrimeric guanine nucleotide-binding proteins (G proteins) are one of the most important cellular signaling events. Due to critical roles, GPCRs represent primary targets of ~1/3 of currently marketed drugs (Hopkins & Groom, 2002). The classical function of GPCRs is to transmit extracellular signals across the plasma membrane and activate intracellular proteins, e.g., the G proteins, which leads to further signaling of downstream effector proteins. The G protein (Moreira, 2014; Simon, Strathmann, & Gautam, 1991) consists of three structural subunits (G_α, G_β and G_γ), for which 21, 6 and 12 different subtypes have been identified, respectively. There are ~700 unique heterotrimeric G proteins in the human genome. Moreover, GPCRs have ~800 different members in the superfamily. Interactions of GPCRs and the G proteins could thus involve hundreds of thousands of possibilities. However, GPCRs are known to selectively couple with the G proteins (Flock et al., 2017). It is important to understand the mechanism of GPCR-G protein coupling specificity, which will greatly facilitate effective drug design (Pardon et al., 2018; Weiss et al., 2013).

Recent breakthroughs in structural biology including X-ray crystallography and cryo-electron microscopy (cryo-EM) have enabled determination of more than 10 GPCR-G protein complex structures (Carpenter, Nehmé, Warne, Leslie, & Tate, 2016; Chung et al., 2011; DeVree et al., 2016; Draper-Joyce et al., 2018; García-Nafría, Lee, Bai, Carpenter, & Tate, 2018; Huang, Karpiak, et al., 2015; Huang, Manglik, et al., 2015; Liang et al., 2017; Rasmussen, Choi, et al., 2011; Ring et al., 2013; Scheerer et al., 2008). As summarized in Table 1, 17 GPCR structures are complexed with the G proteins or G protein mimics, including opsin coupled with the C-terminal peptide of the G_α subunit (Scheerer et al., 2008), the β_2 adrenergic receptor (β_2AR) with the G_s protein (Rasmussen, DeVree, et al., 2011) or the G-protein mimetic nanobody (Rasmussen, Choi, et al., 2011; Ring et al., 2013), rhodopsin coupled with arrestin (Zhou et al., 2017) or the G_i protein (Kang et al., 2018), the adenosine A_1 receptor (A_1AR) bound by the G_i protein (Draper-Joyce et al., 2018), the adenosine A_{2A} receptor (A_{2A}AR) bound by the "mini-G_s" (Carpenter et al., 2016) or G_s protein (García-Nafría, Lee, et al., 2018), the μ-opioid receptor (μOR) bound by the G-protein mimetic nanobody (Huang, Karpiak, et al., 2015; Huang, Manglik, et al., 2015) or G_i protein (Koehl et al., 2018), the calcitonin receptor coupled with the G_s protein (Liang et al., 2017) and the serotonin 5-HT_{1B} receptor coupled with the G_o protein (García-Nafría, Nehmé, et al., 2018). These structures

Table 1 Structures of GPCRs complexed with the G proteins or G protein mimics.

GPCR	G protein or mimetic nanobody	PDB ID (resolution)	Method	References
Opsin	C-terminal peptide of G_α transducin	3DQB (3.2 Å)	X-ray	Scheerer et al. (2008)
β_2AR	G_s and nanobody Nb35	3SN6 (3.2 Å)	X-ray	Rasmussen, DeVree, et al. (2011)
β_2AR	Nanobody Nb80	3P0G (3.5 Å)	X-ray	Rasmussen, Choi, et al. (2011)
M2	Nanobody Nb9-8	4MQS (3.5 Å)	X-ray	Kruse et al. (2013)
β_2AR	Nanobody Nb6B9	4LDE (2.7 Å), 4LDL (3.1 Å), 4LDO (3.2 Å)	X-ray	Ring et al. (2013)
μOR	Nanobody Nb39	5C1M (2.1 Å)	X-ray	Huang, Karpiak, et al. (2015) and Huang, Manglik, et al. (2015)
A_{2A}AR	Mini-G_s	5G53 (3.4 Å)	X-ray	Carpenter et al. (2016)
GLP-1R	G_s and nanobody Nb35	5VAI (4.1 Å)	Cryo-EM	Zhang et al. (2017)
Calcitonin receptor	G_s and nanobody Nb35	5UZ7 (4.1 Å)	Cryo-EM	Liang et al. (2017)
Rhodopsin	Arrestin	5W0P (3.0 Å)	X-ray	Zhou et al. (2017)
GLP1	G_s and nanobody Nb35	6B3J (3.3 Å)	Cryo-EM	Liang et al. (2018)
KOR	Nanobody Nb39	6B73 (3.1 Å)	X-ray	Che et al. (2018)
μOR	G_i	6DDE (3.5 Å), 6DDF (3.5 Å)	Cryo-EM	Koehl et al. (2018)
A_1AR	G_i	6D9H (3.6 Å)	Cryo-EM	Draper-Joyce et al. (2018)
A_{2A}AR	G_s and nanobody Nb35	6GDG (4.1 Å)	Cryo-EM	García-Nafría, Lee, et al. (2018)
Rhodopsin	G_i	6CMO (4.5 Å)	Cryo-EM	Kang et al. (2018)
5-HT_{1B}	G_o	6G79 (3.8 Å)	Cryo-EM	García-Nafría, Nehmé, Edwards, and Tate (2018)

provide important insights into active conformations of GPCRs and atomic GPCR-G protein interactions. However, the X-ray and cryo-EM structures are rather static images. It remains largely unknown how GPCRs dynamically recognize specific G proteins.

In addition to structural biology, experimental techniques including mutagenesis (Blin, Yun, & Wess, 1995; Burstein, Spalding, & Brann, 1998; Chen et al., 2010; Conklin, Farfel, Lustig, Julius, & Bourne, 1993; Erlenbach et al., 2001; Kostenis, Conklin, & Wess, 1997; Liu, Conklin, Blin, Yun, & Wess, 1995; Marin, Krishna, & Sakmar, 2001; Moro, Lameh, Hogger, & Sadee, 1993; Preininger et al., 2009; Schoneberg, Kostenis, Liu, Gudermann, & Wess, 1998; Slessareva & Graber, 2003; Valiquette, Parent, Loisel, & Bouvier, 1995; Wacker et al., 2008; Xiao et al., 1999), nuclear magnetic resonance (NMR) (Kim et al., 2013), hydrogen-deuterium exchange mass spectrometry (HDXMS) (Chung et al., 2011; Orban et al., 2012), and double electron-electron resonance spectroscopy (DEER) (Van Eps et al., 2018) have been utilized to investigate the GPCR-G protein interactions (Mahoney & Sunahara, 2016; Moreira, 2014; Preininger, Meiler, & Hamm, 2013). While the C-terminal α5 helix in the G_α subunit has been suggested as the primary driver for specific receptor recognition (Blin et al., 1995), the G_α αN helix and receptor intracellular loop (ICL) 2 and transmembrane (TM) helix 6 further contribute to the GPCR-G protein coupling specificity (Burstein et al., 1998; Chen et al., 2010; Neumann, Krause, Claus, & Paschke, 2005; Preininger et al., 2013; Timossi et al., 2002; Zhou, Yan, Yamamoto, & Tai, 1999). Furthermore, dynamic regions in the complex can be crucial for the coupling through allosteric conformational changes (Mahoney & Sunahara, 2016; Preininger et al., 2013). The precise conformation of active GPCRs also depends on chemical properties of the binding agonist. For example, agonist binding often leads to a change in the receptor conformation such as opening of the intracellular G protein binding pocket for coupling to the G proteins (Zocher, Fung, Kobilka, & Muller, 2012). The experimental studies have greatly advanced our knowledge in the field, but the exact determinants of specific GPCR-G protein interactions remain unclear.

On the other hand, computational modeling has proven useful in studying PPIs (Janin et al., 2003; Shoemaker & Panchenko, 2007). Here, we review computational studies of GPCR-G protein interactions using various techniques, including bioinformatics, protein-protein docking and molecular dynamics (MD) simulation.

2. Bioinformatics of GPCR-G protein interactions

Significantly increasing information about the sequences, structures and signaling networks of GPCRs and the G proteins has become available in recent years. A number of bioinformatics and software tools as listed in Table 2 are useful for exploring GPCR-G protein interactions, including the protein data bank (PDB) (Berman et al., 2000), the GPCRdb (Munk et al., 2016), gpDB (Theodoropoulou et al., 2008) and human gpDB (Satagopam et al., 2010).

Table 2 Databases and software tools for modeling GPCR-G protein interactions.

Database/ software	Description	Reference and website
PDB	A database contains biological macromolecular structures determined by experiments	Berman et al. (2000), https://www.rcsb.org/
GPCRdb	A database contains structures, diagrams and web tools of GPCRs	Munk et al. (2016), http://gpcrdb.org/
gpPDB	A database contains information about GPCRs, effectors of GPCRs and their known interactions	Theodoropoulou, Bagos, Spyropoulos, and Hamodrakas (2008), http://bioinformatics.biol.uoa.gr/gpDB/
Human gpDB	A database contains information about 713 human GPCRs, 36 human G proteins and 99 human effectors	Satagopam et al. (2010), http://bioinformatics.biol.uoa.gr/human_gpdb/
OMP	A database provides information about structural classification of membrane proteins, topology, spatial positions in the lipid bilayer, and intracellular localization	Lomize, Pogozheva, Joo, Mosberg, and Lomize (2012), https://opm.phar.umich.edu/
CHARMM-GUI	A web-based graphical user interface that helps preparation of biomolecular systems (including GPCRs and the G proteins) for molecular dynamics simulations	Jo et al. (2017), http://www.charmm-gui.org/

The GPCRdb is a widely used database for studying GPCRs (Munk et al., 2016). It contains valuable information about the structures, known mutations, homologs, ligands and phylogenetic relationships of GPCRs. Besides, the GPCRdb provides useful functions, such as generation of GPCR models and identification of ligand-binding sites for virtual screening. Systematic analysis of data from the GPCRdb could deepen our understanding of GPCRs and their interactions with the G proteins. For example, Suku et al. performed systematic analysis of ligand-binding pockets for GPCRs collected in the GPCRdb (Suku & Giorgetti, 2017). Ten residues including 3.32, 3.33, 3.36, 6.48, 6.51, 6.52, 6.55, 7.35, 7.39 and 7.43 (Ballesteros-Weinstein numbering of GPCRs, Ballesteros & Weinstein, 1995) were identified to interact with ligands. In addition, these residues were found to be conserved and share a common evolutionary history. More recently, a bioinformatics approach has been applied to determine a selectivity barcode (patterns of amino acids) of GPCR-G protein coupling based on the data obtained from the GPCRdb (Flock et al., 2017). While universally conserved residues in the barcode allow GPCRs to bind and activate G protein in a similar manner, different receptors recognize unique positions of the G protein barcode through distinct residues. In summary, bioinformatics has become highly useful in extracting valuable information about GPCR-G protein interactions across the entire family of GPCRs and the G proteins.

3. Protein-protein docking on GPCR-G protein interactions

Since experimental structures of GPCR-G protein complexes are still very limited, protein-protein docking is an efficient computational approach to generate the complex models. It has been successfully applied to construct structures of GPCR-G protein complexes (Alexander et al., 2014; Pawlowski, Saraswathi, Motawea, Chotani, & Kloczkowski, 2014; Shim, Ahn, & Kendall, 2013), in addition to GPCR oligomers (Borroto-Escuela et al., 2018). Pawlowski et al. performed protein-protein docking to investigate the binding specificity between the human α2C-adrenoreceptor (ADRA2C) and the filamin-2 (FLN2) actin binding protein (Pawlowski et al., 2014). There was no experimental structure of the ADRA2C or FLN2. Homology modeling was first performed to obtain their separate structures, which were used to build the complex structure with the HADDOCK server (de Vries, van Dijk, & Bonvin, 2010). Combining multiple sequence alignments and phylogenetic analysis, the authors found that electrostatic interactions between residues R454 and R456 in the ADRA2C and negatively charged residues in the FLN2 play an important role in the protein coupling. In order to investigate the mechanism of GDP release from the G protein, Alexander et al. utilized homology modeling, protein-protein docking and DEER experiments to construct a model of the active state of rhodopsin complexed with a heterotrimeric $G_{\alpha i\beta\gamma}$ protein (Alexander et al., 2014). With the template X-ray structure of the β_2AR-G_s complex, a homology model was first built using the Rosetta software. Then 1000 independent protein-protein docking calculations were performed, resulting in a pool of 739 nonclashing models. Nine structures that could reproduce the DEER distances and signal shapes were identified. These structures suggested that the C-terminus of the G_α α5 helix triggers conformational changes in the helical domain, which lead to GDP release. Based on the resulting models, energetic analysis was performed to identify residues that showed marked changes between the receptor-bound and free forms of the G protein. In another study by the same group, the important role of the α5 helix of G_α in the activation of the G protein was demonstrated through mutagenesis experiments (Kaya et al., 2014). Therefore, protein-protein docking has facilitated building structures and understanding protein interactions of the GPCR-G protein complexes. However, due to the flexible nature of GPCRs and the G proteins, insufficient accuracy of protein-protein docking has largely limited its applications in modeling GPCR-G protein complex (Kaczor, Selent, Sanz, & Pastor, 2013).

4. Molecular dynamics simulations of GPCR-G protein interactions

MD is a powerful computational technique for simulating biomolecular dynamics at an atomistic level (Karplus & McCammon, 2002). MD is able to provide dynamic information about the interactions between GPCRs and the G proteins, which is missing in static experimental structures and protein-protein docking. Thus, MD has been applied to investigate the GPCR-G protein interactions. Although there have been many MD applications on GPCRs or G proteins alone (Grossfield, 2011; Johnston & Filizola, 2011; Miao & McCammon, 2016a; Vanni & Rothlisberger, 2012; Yao et al., 2016), here we will focus on the GPCR-G protein interactions.

Since GPCRs are membrane proteins, their structural dynamics and function (including interactions with the G proteins) could be strongly affected by lipids (Yen et al., 2018). The orientation and position of GPCRs in the lipid membrane need to be carefully modeled. In this regard, the Orientations of Proteins in Membranes (OPM) database (Lomize et al., 2012) is useful in the modeling of membrane proteins, including GPCRs. In addition, CHARMM-GUI is an online webserver (http://www.charmm-gui.org/) (Jo et al., 2017; Jo, Kim, Iyer, & Im, 2008), which can be used to generate a simulation-ready system for a membrane-embedded protein and input files for various MD software packages, including AMBER, NAMD, GROMACS, and so on. It has significantly reduced the effort of system preparation for MD simulations.

Overall, MD simulations have greatly helped us understanding the GPCR-G protein interactions. However, due to limited timescales, direct MD simulations often suffer from insufficient sampling of the GPCR-G protein interactions. To overcome MD limitations, many enhanced sampling methods have been developed during the last several decades (Abrams & Bussi, 2014; Christen & van Gunsteren, 2007; Dellago & Bolhuis, 2009; Gao, Yang, Fan, & Shao, 2008; Liwo, Czaplewski, Ołdziej, & Scheraga, 2008; Miao & McCammon, 2016c; Spiwok, Sucur, & Hosek, 2015). Several enhanced MD methods have been successfully applied to study GPCR-G protein interactions, including umbrella sampling (Kästner, 2011; Rose et al., 2014; Torrie & Valleau, 1977), metadynamics (Alessandro & Francesco, 2008; Laio & Parrinello, 2002; Saleh, Ibrahim, & Clark, 2017; Saleh, Ibrahim, Saladino, Gervasio, & Clark, 2017; Saleh, Saladino, Gervasio, & Clark, 2017) and Gaussian accelerated molecular dynamics (GaMD) (Miao, Feher, & McCammon, 2015; Miao & McCammon, 2018; Pang, Miao, Wang, & McCammon, 2017). As summarized in Table 3, determinants of coupling selectivity between the GPCR and G protein (Kling et al., 2013; Mnpotra et al., 2014; Rose et al., 2014; Shim et al., 2013), effects of different ligand binding on the stability of GPCR-G protein complexes (Bai et al., 2013; Feng et al., 2012; Goetz et al., 2011; Miao & McCammon, 2016b; Saleh, Ibrahim, & Clark, 2017; Saleh, Saladino, et al., 2017; Shirvanyants, Ding, Tsao, Ramachandran, & Dokholyan, 2012), the G protein activation upon binding of a GPCR (Dror et al., 2015) and spontaneous binding of the G-protein mimetic nanobody to a GPCR (Miao & McCammon, 2018) have been investigated through MD simulations and will be discussed in the following.

Table 3 A summary of MD simulation studies on GPCR-G protein interactions.

System	Method	Major findings	References
β_2AR-G_αCT	MD	Interactions between the β_2AR and G_αCT are ligand dependent	Goetz, Lanig, Gmeiner, and Clark (2011)
β_2AR-G_s	MD	Nanobody plays an important role in stabilizing the β_2AR-G_s complex	Feng, Hou, and Li (2012)
β_2AR-G_s	MD	Binding of different ligands affects stability of the β_2AR-G_s complex	Bai, Zhang, Ban, Liu, and Yao (2013)
β_2AR-G_s D_2R-G_i	MD	Receptor ICL3 and the α5-helix of G_α play an important role in GPCR-G protein coupling	Kling, Lanig, Clark, and Gmeiner (2013)
CB1-G_i	MD, Ala mutation	The G_α α5 helix of the G protein plays an important role in the CB1-G_i coupling	Shim et al. (2013)
CB2-G_i	MD, cross-linking	The ICL2 in CB2 and the G_α α5 helix of the G protein play an important role in the CB2-G_i coupling	Mnpotra et al. (2014)
β_2AR-G_i/G_s	MD	The TM6 helix in the β_2AR plays an important role in binding selectivity of G_i and G_s proteins	Rose et al. (2014)
β_2AR-G_s	MD, DEER spectroscopy	Separation of the Ras and helical domain of the G_α subunit is necessary but not sufficient for rapid nucleotide release	Dror et al. (2015)
β_2AR-G_s	Kino-Geometric Sampling	Interaction between the αN helix of the G protein and the receptor ICL2 is important for nucleotide release	Pachov, Fonseca, Arnol, Bernauer, and van den Bedem (2016)

Continued

Table 3 A summary of MD simulation studies on GPCR-G protein interactions.—cont'd

System	Method	Major findings	References
M2-nanobody	GaMD	Nanobody is important in stabilizing the active conformational state of the M2 receptor	Miao and McCammon (2016b)
β_2AR-G_s M2R-nanobody μOR-G_s/nanobody	Metadynamic	The binding of intracellular binding partners alters agonist binding modes	Saleh, Ibrahim, and Clark (2017)
β_2AR-G_s β_2AR-arrestin	Metadynamics	The structure and dynamics of GPCR-G protein complexes depend strongly on the nature of small-molecule ligands	Saleh, Saladino, et al. (2017)
Rhodopsin-G_i	MD, DEER	A model of rhodopsin-G_i is presented	Van Eps et al. (2018)
M2-nanobody	GaMD	GaMD captured spontaneously binding the G protein mimic nanobody to a muscarinic GPCR	Miao and McCammon (2018)
Rhodopsin-arrestin	MD, Fluorescence spectroscopy	GPCRs could stimulate arrestin through interactions mediated by the receptor phosphorylated cytoplasmic tail (RP tail) only, the receptor core only, or both the receptor core and RP tail	Latorraca et al. (2018)

4.1 Determents of GPCR-G protein coupling specificity

MD simulations have been carried out to identify determinants of the GPCR-G protein coupling specificity. Kling et al. (2013) reported microsecond MD simulations of ternary GPCR complexes, including the experimentally determined agonist-bound β_2AR-G_s and two homology models of the dopaminergic D_2 receptor (D2R) bound by the G_i protein. Important residues were located at the receptor intracellular end of the TM5 helix and the N-terminal region of the ICL3, which interacted with the $\alpha5$ helix and $\alpha4$-$\beta6$ loop in the G_α protein subunit.

The TM6 helix of GPCRs was identified as an important domain in determining the coupling selectivity between the receptors and G proteins (Kang et al., 2018; Rose et al., 2014; Shim et al., 2013; Van Eps et al., 2018). MD and umbrella sampling simulations were performed on the β_2AR bound by the C-terminal peptide of the G_α (G_αCT) that was used as a surrogate of the G protein (Rose et al., 2014). The simulations suggested that distinct conformations of the β_2AR induced by binding of different G proteins co-existed in the G-protein free (*apo*) state of the receptor. Conformational heterogeneity of the TM6 emerged when the β_2AR was bound by the G_i or G_s protein. The important role of TM6 in the coupling selectivity was also demonstrated by Xu et al. (Kang et al., 2018) through structural biology and MD simulations. MD simulations were performed on four systems, including the G_i protein complexed with the rhodopsin and μOR and the G_s protein bound by the β_2AR and A_{2A}AR. Results showed that the outward movement of TM6 was less pronounced in the G_i-coupled than in the G_s-coupled receptors. This was consistent with model of the rhodopsin-G_i protein complex, which was tested by MD simulation using the distance constraints from DEER experiments (Van Eps et al., 2018).

The G_α $\alpha5$ helix of the G protein has been also shown to be important for specific GPCR-G protein coupling. Shim et al. combined MD simulation and mutagenesis experiments to identify critical regions for coupling of the CB1 receptor with the G_i protein (Shim et al., 2013). Guided by the X-ray structure of the β_2AR-G_s, a model was built for the CB1-G_i ternary complex. Through an 824 ns MD simulation, they found that tight interactions between the CB1 and the G_α $\alpha5$ helix of the G_i protein were crucial for the receptor-G protein binding. Mnpotra et al. applied MD simulations to explore the interactions between the CB2 and G_i protein (Mnpotra et al., 2014). Results showed that the G_i protein could reorient to a different binding mode in comparison with orientation of the G_s protein in the β_2AR-G_s complex. During reorientation of the G_i protein, two major conformational changes occurred. First, the G_α $\alpha5$ helix of the G_i protein tilted due to outward movement of the TM5 helix in CB2. Second, a 25° clockwise rotation of the G_i protein took place, leading to interaction of the receptor ICL2 with a hydrophobic pocket formed by residues Val34, Leu194, Phe196, Phe336, Thr340, Ile343 and Ile344 in the $G_{\alpha i}$. This structural model was highly consistent with the data obtained from cross-linking studies (Mnpotra et al., 2014).

In summary, the above MD studies have greatly advanced our knowledge of GPCR-G protein interactions. Several important structural motifs that contribute to the GPCR-G protein coupling specificity were identified, including the G_α $\alpha5$ helix of the G protein (Blin et al., 1995; Mnpotra et al., 2014; Shim et al., 2013) and the receptor ICL2/ICL3 and TM6 helix (Kang et al., 2018; Rose et al., 2014; Shim et al., 2013; Van Eps et al., 2018).

4.2 Effects of ligand binding on GPCR-G protein interactions

GPCR signaling occurs via ternary complexes formed under cooperative binding between the receptor, ligand and an intracellular binding partner (IBP). Ligand binding could lead to a conformational change of the receptor (e.g., opening of the intracellular pocket) for coupling to the G protein (DeVree et al., 2016). Conversely, binding of the G protein in the intracellular binding site could allosterically influence ligand binding in the receptor orthosteric site. DeVree et al. demonstrated that binding of the G protein in the β_2AR could allosterically close the receptor extracellular ligand-binding pocket (DeVree et al., 2016). The allosteric interaction between the orthosteric site and the G protein binding pocket is thus involved in the GPCR-G protein interactions. MD simulations were performed on the agonist-β_2AR-$G_{\alpha s}$ complex system (Feng et al., 2012). Interaction between the β_2AR and G_s protein was found to be stable when the complex was bound by the Nb35 nanobody. Without Nb35, the agonist could trigger conformational changes of β_2AR from the extracellular to the intracellular domains. The importance of nanobody in stabilizing the GPCR-agonist-IBP ternary complex was also demonstrated using GaMD in the simulations of the M2 muscarinic receptor (Miao & McCammon, 2016b). The intracellular domain of TM6 could remain in its active state when the receptor was bound with the Nb9-8 nanobody, while removal of Nb9-8 led to inward movement of the TM6 and deactivation of the M2 receptor. These simulation findings were consistent with experimental data obtained from NMR (Nygaard et al., 2013) and DEER studies (Manglik et al., 2015), which indicated that binding of a G protein or G protein mimic is required to stabilize the active conformational state of GPCRs in addition to agonist binding. Metadynamics simulations were also performed to investigate the similarities and differences between β_2AR-agonists bound by different IBPs, including the G_s protein and the G-protein mimetic nanobody (Saleh, Ibrahim, & Clark, 2017). Important intermediate states were identified for the GPCR upon binding of different agonists and IBPs.

Goetz et al. applied MD simulations to determine the effects of different agonist and inverse agonist binding on stability of the β_2AR complexed with the C-terminus of the $G_{\alpha s}$ subunit ($G_{\alpha s}$CT) (Goetz et al., 2011). The simulations showed that the ligand-binding pocket conformation and interaction between the $G_{\alpha s}$CT and β_2AR were different upon binding of the isoprenaline agonist and carazolol inverse agonist. Isoprenaline induced an inward movement of the TM5 in the orthosteric binding site of the β_2AR, whereas carazolol blocked rearrangement of the extracellular domains of the receptor. Moreover, the β_2AR and $G_{\alpha s}$CT formed stable interaction in the presence of isoprenaline, while the complex was destabilized by binding of carazolol. In another study, MD simulations of the β_2AR complexed with the entire G_s protein were performed by Bai et al. to investigate the binding effects of three different ligands (e.g., agonist BI-67107, inverse agonist ICI 118,551 and antagonist alprenolol) (Bai et al., 2013). Their results suggested that BI-67107 formed three more stable hydrogen bonds with the receptor (residues Ser203$^{5.42}$, Ser207$^{5.46}$ and Asn293$^{6.55}$) than ICI 118,551. Thus, BI-67107 was able to stabilize the β_2AR in the active state. Binding of the ICI 118,551 inverse agonist could change β_2AR from the active to the inactive state, as well as inducing dissociation of the G_α and $G_{\beta\gamma}$ subunits.

More recently, Saleh et al. applied metadynamics simulations to investigate structural dynamics and free energy profiles of the β_2AR-arrestin and β_2AR-G_s complexes, in the absence or presence of different ligands (Saleh, Saladino, et al., 2017). The ligands included the full G_s/arrestin agonist isoprenaline, the G_s/arrestin unselective antagonist alprenolol, the G_s inverse agonist/arrestin antagonist ICI 118,551 and the G_s inverse agonist/arrestin partial agonist carvedilol. The simulations suggested that agonists and partial agonists increased the binding affinity of the G protein or arrestin to the β_2AR. Antagonists left the binding affinity largely unaffected or decreased it slightly. Inverse agonists decreased it significantly. An extended ternary complex model was then proposed, in which the ligand bias toward either the G-protein or arrestin pathway is regulated by cooperative binding of the receptor, ligand and IBP. The free energy changes could be used to characterize the ligand signaling bias, which was suggested to be a promising approach for rational design of GPCR biased agonists.

4.3 Activation of the G protein upon binding of GPCRs

The G proteins are molecular switches that turn on intracellular signaling cascades in response to the activation of GPCRs by extracellular stimuli. Their switching function depends on the ability of the G_α subunit to cycle between an inactive GDP-bound state and an active GTP-bound state. Thus, mechanisms about how the GPCR catalyzes GDP release on cognate G protein and how the G protein transits between its different states are significantly important in understanding the signal transduction within the GPCR-G protein complex. Both experimental and computational techniques have been utilized to address the above questions (Duc, Kim, & Chung, 2015; Mahoney & Sunahara, 2016; Nguyen Minh, Hee Ryung, & Ka Young, 2017). Extensive MD simulations performed by Dror et al. demonstrated that separation of the Ras and helical domains was necessary but not sufficient for GDP release from the G protein (Dror et al., 2015). Conformational changes in the G_α $\alpha5$ helix were concomitant with opening of the helical domain. The repositioned $\alpha5$ helix weakened binding of the GDP, facilitating its release from the G_s protein. These predictions were validated by the DEER spectroscopic experiments. Conformational changes in the G_α $\alpha5$ helix and opening of the helical domain in the β_2AR-G_s complex were also observed in computational modeling by Pachov et al. using a Kino-Geometric Sampling (KGS) method (Pachov et al., 2016). This study demonstrated that interactions between the αN helix of the G_s protein and the receptor ICL2 facilitate nucleotide exchange by weakening a salt bridge between the P-loop and switch 1 in the G_s protein. Despite these advances, we still lack a detailed understanding of the mechanisms of the G protein binding to GPCRs, the G protein catalyzed hydrolysis of GTP to GDP and allosteric modulation of nucleotide binding in the G proteins by GPCRs (Duc et al., 2015; Mahoney & Sunahara, 2016; Nguyen Minh et al., 2017).

4.4 Mechanism of GPCR-G protein binding

MD simulations of protein-protein binding are challenge, due to the limited simulation timescales while slowly evolving protein dynamics. Nevertheless, remarkable advances in supercomputing have enabled the D.E. Shaw research group to successfully simulate binding of five different protein-protein systems through exceptionally long-timescale MD simulations (Pan et al., 2018). Hundreds-of-microseconds conventional MD simulations captured spontaneous protein-protein binding events. Furthermore, repeated protein association and dissociation were observed in enhanced MD simulations using a "tempered binding" approach (Pan et al., 2018).

In the context of GPCR-G protein interactions, powerful enhanced MD simulations successfully captured spontaneous binding of a G-protein mimetic nanobody to a muscarinic GPCR using the GaMD method

(Miao & McCammon, 2018). With X-ray structure of the agonist-nanobody-M2 receptor complex, the agonist and nanobody were initially displaced to >20 Å far away from the active M2 receptor. Five 4.5 μs independent GaMD simulations were performed. Although the agonist could not reach the X-ray binding pose in the receptor orthosteric binding pocket, the nanobody could successfully bind to the receptor G-protein coupling site in one GaMD simulation with a minimum RMSD of 2.48 Å in the nanobody core domain compared with the X-ray structure. The GaMD simulations showed significant conformational changes in both the orthosteric ligand-binding pocket and intracellular domains of the M2 receptor upon nanobody binding. Binding of the nanobody switched the orthosteric pocket from the "open" to "closed" conformation and led to activation of the M2 receptor with an increase in the intracellular TM3 – TM6 distance. Moreover, two important low-energy intermediate conformational states were identified during binding of the G-protein mimetic nanobody. The nanobody formed transient electrostatic, hydrogen bonding and hydrophobic interactions with the receptor through the binding process. The flexible receptor ICLs played a key role in the recognition and binding of the nanobody (Miao & McCammon, 2018). Therefore, GaMD simulations provided important insights into the mechanism of the G protein mimic binding to a GPCR.

5. Discussions and outlook

Interactions with the intracellular G proteins represent a canonical signaling pathway of GPCRs, key membrane proteins that serve as primary targets of ~1/3 currently marketed drugs. Structural determination of GPCR-G protein complex has exploded in very recent years, due to breakthroughs in X-ray crystallography and cryo-EM (Table 1). Extensive research studies have been focused on GPCR-G protein interactions using various experimental techniques (e.g., mutagenesis, NMR, HDXMS and DEER) in addition to the structural biology. These studies have greatly facilitated our understanding of GPCR-G protein interactions. However, the experimental techniques often suffer from limited spatial and temporal resolutions, as well as high cost. In this regard, computational modeling has proven useful and efficient in studies of GPCR-G protein interactions. Complementary experimental and computational techniques have been combined in numerous studies in order to obtain a more detailed picture of GPCR-G protein interactions. Here, we have focused on reviewing recent studies of GPCR-G protein interactions with computational approaches, including bioinformatics, protein-protein docking and MD simulations.

With dramatically increasing information that is collected about GPCR-G protein interactions, bioinformatics has been applied to determine major determinants of coupling selectivity between GPCRs and the G proteins (Flock et al., 2017). Bioinformatics is useful in providing an overview of protein-protein interactions for the entire family of GPCRs and the G proteins. On the other hand, docking and MD simulation are able to generate a more detailed picture of target GPCR-G protein interactions of interest. There are several advantages for protein-protein docking. It is highly efficient. The docking software tools and webservers are mostly user friendly. Docking calculations are usually fast without the need of expensive computational resources. They are able to generate computational models of protein complex, e.g., GPCR-G protein complex structures. These models could provide valuable information about overall conformations of the GPCR-G protein complexes. However, applications of protein-protein docking in modeling GPCR-G protein interactions are still limited due to low accuracy. Limited capability to account for protein flexibility and inaccuracy of docking scores often require the use of protein-protein docking in combination with experiments (Alexander et al., 2014) and/or MD simulations for further validation (Shim et al., 2013).

Applications of MD simulations in molecular biology and drug discovery have dramatically increased in recent years especially in the research field of GPCRs (Hollingsworth & Dror, 2018; Latorraca, Venkatakrishnan, & Dror, 2017; Miao & McCammon, 2016a). Remarkable developments in both the computing hardware (e.g., the Anton specialized super computer and fast GPUs) and software tools have enabled long-timescale MD simulations. MD simulations have been performed over microseconds to milliseconds. The MD simulations have provided important insights into the dynamic mechanism of GPCR-G protein interactions at an atomistic level (Latorraca et al., 2017). Nevertheless, binding of intracellular G proteins to GPCRs is challenging for conventional MD simulations. In this regard, enhanced MD simulations are useful to help address the challenge. Notably, enhanced simulations using the GaMD method captured spontaneous binding of the G-protein mimetic nanobody to a GPCR (Miao & McCammon, 2018). Pan et al. simulated both binding and unbinding of five different protein-protein systems using a "tempered binding" approach (Pan et al., 2018), although the GPCR-G protein was not included in their simulated systems. Therefore, innovations in both computing hardware and enhanced sampling methods have opened a new era in MD simulations of protein-protein binding. Continued developments are expected to enable simulations of binding processes between GPCRs and the cognate G proteins in the near future. Such studies will potentially reveal mechanisms of the GPCR-G protein interactions and the cooperative activation of GPCRs and G proteins at an atomistic level.

In summary, bioinformatics, protein-protein docking and MD simulation have proven useful for exploring the GPCR-G protein interactions. Combination of computational modeling and complementary experiments will help us to obtain a detailed understanding of the GPCR-G protein interactions and GPCR signaling mechanism. This will greatly facilitate more effective computer-aided drug design of GPCRs (Huang, Karpiak, et al., 2015; Huang, Manglik, et al., 2015; Korczynska et al., 2018; Miao et al., 2016; Miao & McCammon, 2016a).

Acknowledgments

This work was in part supported by the American Heart Association (Award 17SDG33370094) and the startup funding in the College of Liberal Arts and Sciences at the University of Kansas. Computing time was provided on the Comet, Bridges and Stanford XStream supercomputers through the Extreme Science and Engineering Discovery Environment award TG-MCB170129 and TG-MCB180049 and the Edison and Cori supercomputers through the National Energy Research Scientific Computing Center project M2874.

References

Abrams, C., & Bussi, G. (2014). Enhanced sampling in molecular dynamics using metadynamics, replica-exchange, and temperature-acceleration. *Entropy*, *16*(1), 163–199.

Alessandro, L., & Francesco, L. G. (2008). Metadynamics: A method to simulate rare events and reconstruct the free energy in biophysics, chemistry and material science. *Reports on Progress in Physics*, *71*(12), 126601.

Alexander, N. S., Preininger, A. M., Kaya, A. I., Stein, R. A., Hamm, H. E., & Meiler, J. (2014). Energetic analysis of the rhodopsin—G-protein complex links the α5 helix to GDP release. *Nature Structural & Molecular Biology*, *21*, 56–63.

Andreani, J., & Guerois, R. (2014). Evolution of protein interactions: From interactomes to interfaces. *Archives of Biochemistry and Biophysics*, *554*, 65–75.

Arkin, M. R., & Wells, J. A. (2004). Small-molecule inhibitors of protein–protein interactions: Progressing towards the dream. *Nature Reviews Drug Discovery*, *3*, 301–317.

Bai, Q., Zhang, Y., Ban, Y., Liu, H., & Yao, X. (2013). Computational study on the different ligands induced conformation change of β2 adrenergic receptor-Gs protein complex. *PLoS One*, *8*(7), e68138.

Ballesteros, J. A., & Weinstein, H. (1995). Integrated methods for the construction of three-dimensional models and computational probing of structure-function relations in G protein-coupled receptors. In S. C. Sealfon (Ed.), *Methods in neurosciences: Vol. 25.* (pp. 366–428). Academic Press.

Berman, H. M., Westbrook, J., Feng, Z., Gilliland, G., Bhat, T. N., Weissig, H., et al. (2000). The protein data bank. *Nucleic Acids Research*, *28*(1), 235–242.

Blin, N., Yun, J., & Wess, J. (1995). Mapping of single amino acid residues required for selective activation of Gq/11 by the m3 muscarinic acetylcholine receptor. *Journal of Biological Chemistry*, *270*(30), 17741–17748.

Borroto-Escuela, D. O., Rodriguez, D., Romero-Fernandez, W., Kapla, J., Jaiteh, M., Ranganathan, A., et al. (2018). Mapping the interface of a GPCR dimer: A structural model of the A2A adenosine and D2 dopamine receptor heteromer. *Frontiers in Pharmacology*, *9*, 829.

Burstein, E. S., Spalding, T. A., & Brann, M. R. (1998). The second intracellular loop of the m5 muscarinic receptor is the switch which enables G-protein coupling. *Journal of Biological Chemistry*, *273*(38), 24322–24327.

Carpenter, B., Nehmé, R., Warne, T., Leslie, A. G. W., & Tate, C. G. (2016). Structure of the adenosine A2A receptor bound to an engineered G protein. *Nature*, *536*, 104–106.

Che, T., Majumdar, S., Zaidi, S. A., Ondachi, P., McCorvy, J. D., Wang, S., et al. (2018). Structure of the nanobody-stabilized active state of the kappa opioid receptor. *Cell*, *172*(1), 55–67.

Chen, X. P., Yang, W., Fan, Y., Luo, J. S., Hong, K., Wang, Z., et al. (2010). Structural determinants in the second intracellular loop of the human cannabinoid CB1 receptor mediate selective coupling to G(s) and G(i). *British Journal of Pharmacology*, *161*(8), 1817–1834.

Christen, M., & van Gunsteren, W. F. (2007). On searching in, sampling of, and dynamically moving through conformational space of biomolecular systems: A review. *Journal of Computational Chemistry*, *29*(2), 157–166.

Chung, K. Y., Rasmussen, S. G. F., Liu, T., Li, S., DeVree, B. T., Chae, P. S., et al. (2011). Conformational changes in the G protein Gs induced by the β2 adrenergic receptor. *Nature*, *477*, 611–615.

Conklin, B. R., Farfel, Z., Lustig, K. D., Julius, D., & Bourne, H. R. (1993). Substitution of three amino acids switches receptor specificity of Gqα to that of Giα. *Nature*, *363*, 274–276.

de Vries, S. J., van Dijk, M., & Bonvin, A. M. (2010). The HADDOCK web server for data-driven biomolecular docking. *Nature Protocols*, *5*, 883–897.

Dellago, C., & Bolhuis, P. G. (2009). Transition path sampling and other advanced simulation techniques for rare events. In C. Holm & K. Kremer (Eds.), *Advanced computer simulation approaches for soft matter sciences III* (pp. 167–233). Berlin, Heidelberg: Springer Berlin Heidelberg.

DeVree, B. T., Mahoney, J. P., Vélez-Ruiz, G. A., Rasmussen, S. G. F., Kuszak, A. J., Edwald, E., et al. (2016). Allosteric coupling from G protein to the agonist-binding pocket in GPCRs. *Nature*, *535*, 182–186.

Draper-Joyce, C. J., Khoshouei, M., Thal, D. M., Liang, Y.-L., Nguyen, A. T. N., Furness, S. G. B., et al. (2018). Structure of the adenosine-bound human adenosine A1 receptor–Gi complex. *Nature*, *558*, 559–563.

Dror, R. O., Mildorf, T. J., Hilger, D., Manglik, A., Borhani, D. W., Arlow, D. H., et al. (2015). Structural basis for nucleotide exchange in heterotrimeric G proteins. *Science (New York, N.Y.)*, *348*(6241), 1361–1365.

Duc, N. M., Kim, H. R., & Chung, K. Y. (2015). Structural mechanism of G protein activation by G protein-coupled receptor. *European Journal of Pharmacology*, *763*, 214–222.

Erlenbach, I., Kostenis, E., Schmidt, C., Serradeil-Le Gal, C., Raufaste, D., Dumont, M. E., et al. (2001). Single amino acid substitutions and deletions that alter the G protein coupling properties of the V2 vasopressin receptor identified in yeast by receptor random mutagenesis. *Journal of Biological Chemistry*, *276*, 29382–29392.

Feng, Z., Hou, T., & Li, Y. (2012). Studies on the interactions between β2 adrenergic receptor and Gs protein by molecular dynamics simulations. *Journal of Chemical Information and Modeling*, *52*(4), 1005–1014.

Flock, T., Hauser, A. S., Lund, N., Gloriam, D. E., Balaji, S., & Babu, M. M. (2017). Selectivity determinants of GPCR–G-protein binding. *Nature*, *545*, 317–322.

Gao, Y. Q., Yang, L., Fan, Y., & Shao, Q. (2008). Thermodynamics and kinetics simulations of multi-time-scale processes for complex systems. *International Reviews in Physical Chemistry*, *27*(2), 201–227.

García-Nafría, J., Lee, Y., Bai, X., Carpenter, B., & Tate, C. G. (2018). Cryo-EM structure of the adenosine A2A receptor coupled to an engineered heterotrimeric G protein. *eLife*, *7*, e35946.

García-Nafría, J., Nehmé, R., Edwards, P. C., & Tate, C. G. (2018). Cryo-EM structure of the serotonin 5-HT1B receptor coupled to heterotrimeric Go. *Nature*, *558*(7711), 620–623.

Goetz, A., Lanig, H., Gmeiner, P., & Clark, T. (2011). Molecular dynamics simulations of the effect of the G-protein and diffusible ligands on the β2-adrenergic receptor. *Journal of Molecular Biology*, *414*(4), 611–623.

Grossfield, A. (2011). Recent progress in the study of G protein-coupled receptors with molecular dynamics computer simulations. *Biochimica et Biophysica Acta (BBA)—Biomembranes*, *1808*(7), 1868–1878.

Hollingsworth, S. A., & Dror, R. O. (2018). Molecular dynamics simulation for all. *Neuron*, *99*(6), 1129–1143.

Hopkins, A. L., & Groom, C. R. (2002). The druggable genome. *Nature Reviews Drug Discovery*, *1*, 727–730.

Huang, X., Karpiak, J., Kroeze, W. K., Zhu, H., Chen, X., Moy, S. S., et al. (2015). Allosteric ligands for the pharmacologically dark receptors GPR68 and GPR65. *Nature*, *527*, 477.

Huang, W., Manglik, A., Venkatakrishnan, A. J., Laeremans, T., Feinberg, E. N., Sanborn, A. L., et al. (2015). Structural insights into μ-opioid receptor activation. *Nature*, *524*, 315–321.

Janin, J., Henrick, K., Moult, J., Eyck, L. T., Sternberg, M. J., Vajda, S., et al. (2003). CAPRI: A critical assessment of predicted interactions. *Proteins*, *52*(1), 2–9.

Jo, S., Cheng, X., Lee, J., Kim, S., Park, S.-J., Patel, D. S., et al. (2017). CHARMM-GUI 10 years for biomolecular modeling and simulation. *Journal of Computational Chemistry*, *38*(15), 1114–1124.

Jo, S., Kim, T., Iyer, V. G., & Im, W. (2008). CHARMM-GUI: A web-based graphical user interface for CHARMM. *Journal of Computational Chemistry*, *29*(11), 1859–1865.

Johnston, J. M., & Filizola, M. (2011). Showcasing modern molecular dynamics simulations of membrane proteins through G protein-coupled receptors. *Current Opinion in Structural Biology*, *21*(4), 552–558.

Kaczor, A. A., Selent, J., Sanz, F., & Pastor, M. (2013). Modeling complexes of transmembrane proteins: Systematic analysis of protein-protein docking tools. *Molecular Informatics*, *32*(8), 717–733.

Kang, Y., Kuybeda, O., de Waal, P. W., Mukherjee, S., Van Eps, N., Dutka, P., et al. (2018). Cryo-EM structure of human rhodopsin bound to an inhibitory G protein. *Nature*, *558*, 553–558.

Karplus, M., & McCammon, J. A. (2002). Molecular dynamics simulations of biomolecules. *Nature Structural & Molecular Biology*, *9*(9), 646–652.

Kästner, J. (2011). Umbrella sampling. *Wiley Interdisciplinary Reviews: Computational Molecular Science*, *1*(6), 932–942.

Kaya, A. I., Lokits, A. D., Gilbert, J. A., Iverson, T. M., Meiler, J., & Hamm, H. E. (2014). A conserved phenylalanine as relay between the α5 helix and the GDP binding region of heterotrimeric Gi protein α subunit. *Journal of Biological Chemistry*, *289*, 24475–24487.

Kim, T. H., Chung, K. Y., Manglik, A., Hansen, A. L., Dror, R. O., Mildorf, T. J., et al. (2013). The role of ligands on the equilibria between functional states of a G protein-coupled receptor. *Journal of the American Chemical Society, 135*(25), 9465–9474.

Kling, R. C., Lanig, H., Clark, T., & Gmeiner, P. (2013). Active-state models of ternary GPCR complexes: Determinants of selective receptor-G-protein coupling. *PLoS One, 8*(6), e67244.

Koehl, A., Hu, H., Maeda, S., Zhang, Y., Qu, Q., Paggi, J. M., et al. (2018). Structure of the μ-opioid receptor–Gi protein complex. *Nature, 558*(7711), 547–552.

Korczynska, M., Clark, M. J., Valant, C., Xu, J., Moo, E. V., Albold, S., et al. (2018). Structure-based discovery of selective positive allosteric modulators of antagonists for the M2 muscarinic acetylcholine receptor. *Proceedings of the National Academy of Sciences of the United States of America, 115*(10), E2419–E2428.

Kostenis, E., Conklin, B. R., & Wess, J. (1997). Molecular basis of receptor/G protein coupling selectivity studied by coexpression of wild type and mutant m2 muscarinic receptors with mutant G alpha(q) subunits. *Biochemistry, 36*(6), 1487–1495.

Kruse, A. C., Ring, A. M., Manglik, A., Hu, J., Hu, K., Eitel, K., et al. (2013). Activation and allosteric modulation of a muscarinic acetylcholine receptor. *Nature, 504*, 101–106.

Laio, A., & Parrinello, M. (2002). Escaping free-energy minima. *Proceedings of the National Academy of Sciences of the United States of America, 99*(20), 12562–12566.

Latorraca, N. R., Venkatakrishnan, A. J., & Dror, R. O. (2017). GPCR dynamics: Structures in motion. *Chemical Reviews, 117*(1), 139–155.

Latorraca, N. R., Wang, J. K., Bauer, B., Townshend, R. J. L., Hollingsworth, S. A., Olivieri, J. E., et al. (2018). Molecular mechanism of GPCR-mediated arrestin activation. *Nature, 557*(7705), 452–456.

Liang, Y.-L., Khoshouei, M., Glukhova, A., Furness, S. G. B., Zhao, P., Clydesdale, L., et al. (2018). Phase-plate cryo-EM structure of a biased agonist-bound human GLP-1 receptor–Gs complex. *Nature, 555*, 121.

Liang, Y.-L., Khoshouei, M., Radjainia, M., Zhang, Y., Glukhova, A., Tarrasch, J., et al. (2017). Phase-plate cryo-EM structure of a class B GPCR–G-protein complex. *Nature, 546*, 118–123.

Liu, J., Conklin, B. R., Blin, N., Yun, J., & Wess, J. (1995). Identification of a receptor/G-protein contact site critical for signaling specificity and G-protein activation. *Proceedings of the National Academy of Sciences of the United States of America, 92*(25), 11642–11646.

Liwo, A., Czaplewski, C., Ołdziej, S., & Scheraga, H. A. (2008). Computational techniques for efficient conformational sampling of proteins. *Current Opinion in Structural Biology, 18*(2), 134–139.

Lomize, M. A., Pogozheva, I. D., Joo, H., Mosberg, H. I., & Lomize, A. L. (2012). OPM database and PPM web server: Resources for positioning of proteins in membranes. *Nucleic Acids Research, 40*(D1), D370–D376.

Mahoney, J. P., & Sunahara, R. K. (2016). Mechanistic insights into GPCR–G protein interactions. *Current Opinion in Structural Biology, 41*, 247–254.

Manglik, A., Kim, T. H., Masureel, M., Altenbach, C., Yang, Z., Hilger, D., et al. (2015). Structural insights into the dynamic process of β2-adrenergic receptor signaling. *Cell, 161*(5), 1101–1111.

Marin, E. P., Krishna, A. G., & Sakmar, T. P. (2001). Rapid activation of transducin by mutations distant from the nucleotide-binding site: Evidence for a mechanistic model of receptor-catalyzed nucleotide exchange by G proteins. *Journal of Biological Chemistry, 276*(29), 27400–27405.

Miao, Y., Feher, V. A., & McCammon, J. A. (2015). Gaussian accelerated molecular dynamics: Unconstrained enhanced sampling and free energy calculation. *Journal of Chemical Theory and Computation, 11*(8), 3584–3595.

Miao, Y., Goldfeld, D. A., Moo, E. V., Sexton, P. M., Christopoulos, A., McCammon, J. A., et al. (2016). Accelerated structure-based design of chemically diverse allosteric

modulators of a muscarinic G protein-coupled receptor. *Proceedings of the National Academy of Sciences of the United States of America*, *113*(38), E5675–E5684.
Miao, Y., & McCammon, J. A. (2016a). G-protein coupled receptors: Advances in simulation and drug discovery. *Current Opinion in Structural Biology*, *41*(Suppl. C), 83–89.
Miao, Y., & McCammon, J. A. (2016b). Graded activation and free energy landscapes of a muscarinic G-protein–coupled receptor. *Proceedings of the National Academy of Sciences of the United States of America*, *113*(43), 12162–12167.
Miao, Y., & McCammon, J. A. (2016c). Unconstrained enhanced sampling for free energy calculations of biomolecules: A review. *Molecular Simulation*, *42*(13), 1046–1055.
Miao, Y., & McCammon, J. A. (2018). Mechanism of the G-protein mimetic nanobody binding to a muscarinic G-protein-coupled receptor. *Proceedings of the National Academy of Sciences of the United States of America*, *115*, 3036–3041. https://doi.org/10.1073/pnas.1800756115.
Mnpotra, J. S., Qiao, Z., Cai, J., Lynch, D. L., Grossfield, A., Leioatts, N., et al. (2014). Structural basis of G Protein-coupled receptor-Gi protein interaction: Formation of the cannabinoid CB2 receptor-Gi protein complex. *Journal of Biological Chemistry*, *289*(29), 20259–20272.
Moreira, I. S. (2014). Structural features of the G-protein/GPCR interactions. *Biochimica et Biophysica Acta (BBA)—General Subjects*, *1840*(1), 16–33.
Moro, O., Lameh, J., Hogger, P., & Sadee, W. (1993). Hydrophobic amino acid in the i2 loop plays a key role in receptor-G protein coupling. *Journal of Biological Chemistry*, *268*(30), 22273–22276.
Munk, C., Isberg, V., Mordalski, S., Harpsøe, K., Rataj, K., Hauser, A. S., et al. (2016). GPCRdb: The G protein-coupled receptor database—An introduction. *British Journal of Pharmacology*, *173*(14), 2195–2207.
Neumann, S., Krause, G., Claus, M., & Paschke, R. (2005). Structural determinants for G protein activation and selectivity in the second intracellular loop of the thyrotropin receptor. *Endocrinology*, *146*(1), 477–485.
Nguyen Minh, D., Hee Ryung, K., & Ka Young, C. (2017). Recent progress in understanding the conformational mechanism of heterotrimeric G protein activation. *Biomolecules & Therapeutics*, *25*(1), 4–11.
Nygaard, R., Zou, Y., Dror, R. O., Mildorf, T. J., Arlow, D. H., Manglik, A., et al. (2013). The dynamic process of β2-adrenergic receptor activation. *Cell*, *152*(3), 532–542.
Orban, T., Jastrzebska, B., Gupta, S., Wang, B., Miyagi, M., Chance, M. R., et al. (2012). Conformational dynamics of activation for the pentameric complex of dimeric G protein-coupled receptor and heterotrimeric G protein. *Structure (London, England: 1993)*, *20*(5), 826–840.
Pachov, D. V., Fonseca, R., Arnol, D., Bernauer, J., & van den Bedem, H. (2016). Coupled motions in β2AR:Gαs conformational ensembles. *Journal of Chemical Theory and Computation*, *12*(3), 946–956.
Pan, A. C., Jacobson, D., Borisov, K., Sritharan, D., Weinreich, T. M., & Shaw, D. E. (2018). Atomic-level characterization of protein-protein association. *Biophysical Journal*, *114*(3), 557a.
Pang, Y. T., Miao, Y., Wang, Y., & McCammon, J. A. (2017). Gaussian accelerated molecular dynamics in NAMD. *Journal of Chemical Theory and Computation*, *13*(1), 9–19.
Pardon, E., Betti, C., Laeremans, T., Chevillard, F., Guillemyn, K., Kolb, P., et al. (2018). Nanobody-enabled reverse pharmacology on G-protein-coupled receptors. *Angewandte Chemie (International ed. in English)*, *57*(19), 5292–5295.
Pawlowski, M., Saraswathi, S., Motawea, H. K. B., Chotani, M. A., & Kloczkowski, A. (2014). In silico modeling of human α2C-adrenoreceptor interaction with filamin-2. *PLoS One*, *9*(8), e103099.
Preininger, A. M., Funk, M. A., Oldham, W. M., Meier, S. M., Johnston, C. A., Adhikary, S., et al. (2009). Helix dipole movement and conformational variability contribute to allosteric GDP release in Gαi subunits. *Biochemistry*, *48*(12), 2630–2642.

Preininger, A. M., Meiler, J., & Hamm, H. E. (2013). Conformational flexibility and structural dynamics in GPCR-mediated G protein activation: A perspective. *Journal of Molecular Biology, 425*(13), 2288–2298.

Rasmussen, S. G. F., Choi, H.-J., Fung, J. J., Pardon, E., Casarosa, P., Chae, P. S., et al. (2011). Structure of a nanobody-stabilized active state of the β2 adrenoceptor. *Nature, 469*, 175–180.

Rasmussen, S. G. F., DeVree, B. T., Zou, Y., Kruse, A. C., Chung, K. Y., Kobilka, T. S., et al. (2011). Crystal structure of the β2 adrenergic receptor–Gs protein complex. *Nature, 477*, 549–555.

Ring, A. M., Manglik, A., Kruse, A. C., Enos, M. D., Weis, W. I., Garcia, K. C., et al. (2013). Adrenaline-activated structure of β2-adrenoceptor stabilized by an engineered nanobody. *Nature, 502*, 575–579.

Rose, A. S., Elgeti, M., Zachariae, U., Grubmüller, H., Hofmann, K. P., Scheerer, P., et al. (2014). Position of transmembrane helix 6 determines receptor G protein coupling specificity. *Journal of the American Chemical Society, 136*(32), 11244–11247.

Saleh, N., Ibrahim, P., & Clark, T. (2017). Differences between G-protein-stabilized agonist-GPCR complexes and their nanobody-stabilized equivalents. *Angewandte Chemie (International ed. in English), 56*(31), 9008–9012.

Saleh, N., Ibrahim, P., Saladino, G., Gervasio, F. L., & Clark, T. (2017). An efficient metadynamics-based protocol to model the binding affinity and the transition state ensemble of G-protein-coupled receptor ligands. *Journal of Chemical Information and Modeling, 57*(5), 1210–1217.

Saleh, N., Saladino, G., Gervasio, F. L., & Clark, T. (2017). Investigating allosteric effects on the functional dynamics of β2-adrenergic ternary complexes with enhanced-sampling simulations. *Chemical Science, 8*(5), 4019–4026.

Satagopam, V. P., Theodoropoulou, M. C., Stampolakis, C. K., Pavlopoulos, G. A., Papandreou, N. C., Bagos, P. G., et al. (2010). GPCRs, G-proteins, effectors and their interactions: Human-gpDB, a database employing visualization tools and data integration techniques. *Database: The Journal of Biological Databases and Curation, 2010*, baq019.

Scheerer, P., Park, J. H., Hildebrand, P. W., Kim, Y. J., Krauß, N., Choe, H.-W., et al. (2008). Crystal structure of opsin in its G-protein-interacting conformation. *Nature, 455*, 497–502.

Schoneberg, T., Kostenis, E., Liu, J., Gudermann, T., & Wess, J. (1998). Molecular aspects of vasopressin receptor function. *Advances in Experimental Medicine and Biology, 449*, 347–358.

Shim, J.-Y., Ahn, K. H., & Kendall, D. A. (2013). Molecular basis of cannabinoid CB1 receptor coupling to the G protein heterotrimer Gαiβγ: Identification of key CB1 contacts with the C-terminal helix α5 of Gαi. *Journal of Biological Chemistry, 288*(45), 32449–32465.

Shirvanyants, D., Ding, F., Tsao, D., Ramachandran, S., & Dokholyan, N. V. (2012). Discrete molecular dynamics: An efficient and versatile simulation method for fine protein characterization. *The Journal of Physical Chemistry B, 116*(29), 8375–8382.

Shoemaker, B. A., & Panchenko, A. R. (2007). Deciphering protein-protein interactions. Part II. Computational methods to predict protein and domain interaction partners. *PLoS Computational Biology, 3*(4), e43.

Simon, M., Strathmann, M., & Gautam, N. (1991). Diversity of G proteins in signal transduction. *Science (New York, N.Y.), 252*(5007), 802–808.

Slessareva, J. E., & Graber, S. G. (2003). Reconstitution reveals additional roles for N- and C-terminal domains of g(alpha) in muscarinic receptor coupling. *Biochemistry, 42*(24), 7552–7560.

Spiwok, V., Sucur, Z., & Hosek, P. (2015). Enhanced sampling techniques in biomolecular simulations. *Biotechnology Advances, 33*(6, Pt. 2), 1130–1140.

Suku, E., & Giorgetti, A. (2017). Common evolutionary binding mode of rhodopsin-like GPCRs: Insights from structural bioinformatics. *AIMS Biophysics*, *4*(4), 543–556.
Theodoropoulou, M. C., Bagos, P. G., Spyropoulos, I. C., & Hamodrakas, S. J. (2008). gpDB: A database of GPCRs, G-proteins, effectors and their interactions. *Bioinformatics (Oxford, England)*, *24*(12), 1471–1472.
Timossi, C., Maldonado, D., Vizcaino, A., Lindau-Shepard, B., Conn, P. M., & Ulloa-Aguirre, A. (2002). Structural determinants in the second intracellular loop of the human follicle-stimulating hormone receptor are involved in G(s) protein activation. *Molecular and Cellular Endocrinology*, *189*(1–2), 157–168.
Torrie, G. M., & Valleau, J. P. (1977). Nonphysical sampling distributions in Monte Carlo free-energy estimation: Umbrella sampling. *Journal of Computational Physics*, *23*(2), 187–199.
Valiquette, M., Parent, S., Loisel, T. P., & Bouvier, M. (1995). Mutation of tyrosine-141 inhibits insulin-promoted tyrosine phosphorylation and increased responsiveness of the human beta 2-adrenergic receptor. *EMBO Journal*, *14*(22), 5542–5549.
Van Eps, N., Altenbach, C., Caro, L. N., Latorraca, N. R., Hollingsworth, S. A., Dror, R. O., et al. (2018). Gi- and Gs-coupled GPCRs show different modes of G-protein binding. *Proceedings of the National Academy of Sciences of the United States of America*, *115*(10), 2383–2388.
Vanni, S., & Rothlisberger, U. (2012). A closer look into G protein coupled receptor activation: X-ray crystallography and long-scale molecular dynamics simulations. *Current Medicinal Chemistry*, *19*(8), 1135–1145.
Wacker, J. L., Feller, D. B., Tang, X. B., Defino, M. C., Namkung, Y., Lyssand, J. S., et al. (2008). Disease-causing mutation in GPR54 reveals the importance of the second intracellular loop for class A G-protein-coupled receptor function. *Journal of Biological Chemistry*, *283*(45), 31068–31078.
Weiss, D. R., Ahn, S., Sassano, M. F., Kleist, A., Zhu, X., Strachan, R., et al. (2013). Conformation guides molecular efficacy in docking screens of activated beta-2 adrenergic G protein coupled receptor. *ACS Chemical Biology*, *8*(5), 1018–1026.
Xiao, R.-P., Avdonin, P., Zhou, Y.-Y., Cheng, H., Akhter, S. A., Eschenhagen, T., et al. (1999). Coupling of β2-adrenoceptor to Gi proteins and its physiological relevance in murine cardiac myocytes. *Circulation Research*, *84*(1), 43–52.
Yao, X. Q., Malik, R. U., Griggs, N. W., Skjaerven, L., Traynor, J. R., Sivaramakrishnan, S., et al. (2016). Dynamic coupling and allosteric networks in the alpha subunit of heterotrimeric G proteins. *Journal of Biological Chemistry*, *291*(9), 4742–4753.
Yen, H.-Y., Hoi, K. K., Liko, I., Hedger, G., Horrell, M. R., Song, W., et al. (2018). PtdIns(4,5)P2 stabilizes active states of GPCRs and enhances selectivity of G-protein coupling. *Nature*, *559*(7714), 423–427.
Zhang, Y., Sun, B., Feng, D., Hu, H., Chu, M., Qu, Q., et al. (2017). Cryo-EM structure of the activated GLP-1 receptor in complex with a G protein. *Nature*, *546*, 248–253.
Zhou, X. E., He, Y., de Waal, P. W., Gao, X., Kang, Y., Van Eps, N., et al. (2017). Identification of phosphorylation codes for arrestin recruitment by G protein-coupled receptors. *Cell*, *170*(3), 457–469.
Zhou, H., Yan, F., Yamamoto, S., & Tai, H.-H. (1999). Phenylalanine 138 in the second intracellular loop of human thromboxane receptor is critical for receptor-G-protein coupling. *Biochemical and Biophysical Research Communications*, *264*(1), 171–175.
Zocher, M., Fung, J. J., Kobilka, B. K., & Muller, D. J. (2012). Ligand-specific interactions modulate kinetic, energetic, and mechanical properties of the human beta2 adrenergic receptor. *Structure (London, England: 1993)*, *20*(8), 1391–1402.

CHAPTER TWELVE

From traveler to homebody: Which signaling mechanisms sponge larvae use to become adult sponges?

Ilya Borisenko[a,*], Olga I. Podgornaya[a,b,c], Alexander V. Ereskovsky[a,d]
[a]St. Petersburg State University, St. Petersburg, Russia
[b]Far Eastern Federal University, Vladivostok, Russia
[c]Institute of Cytology RAS, St. Petersburg, Russia
[d]Mediterranean Institute of Marine and Terrestrial Biodiversity and Ecology (IMBE), Aix Marseille University, CNRS, IRD, Avignon Université, Marseille, France
*Corresponding author: e-mail address: ilja.borisenko@gmail.com

Contents

Abstract

Cell-to-cell signaling is responsible for regulation of many developmental processes such as proliferation, cell migration, survival, cell fate specification and axis patterning. In this article we discussed the role of signaling in the metamorphosis of sponges with a focus on epithelial–mesenchymal transition (EMT) accompanying this event. Sponges (Porifera) are an ancient lineage of morphologically simple animals occupying a basal position on the tree of life. The study of these animals is necessary for understanding the origin of multicellularity and the evolution of developmental processes. Development of sponges is quite diverse. It finishes with the metamorphosis of a free-swimming larva into a young settled sponge. The outer surface of sponge larvae consists of a ciliated epithelial sheath, which ensures locomotion, while their internal structure varies from genus to genus. The fate of larval ciliated cells is the most intriguing aspect of metamorphosis. In this review we discuss the fate of larval ciliated cells, the processes going on in

Advances in Protein Chemistry and Structural Biology, Volume 116
ISSN 1876-1623
https://doi.org/10.1016/bs.apcsb.2019.02.002

cells during metamorphosis at the molecular level and the regulation of this process. The review is based on information about several sponge species with a focus on *Halisarca dujardini*, *Sycon ciliatum* and *Amphimedon queenslandica*. In our model sponge, *H. dujardini*, ciliated cells leave the larval epithelium during metamorphosis and migrate to the internal cell mass as amoeboid cells to be differentiated into choanocytes of the juvenile sponge. Ciliated cells undergo EMT and internalize within minutes. As EMT involves the disappearance of adherens junctions and as cadherin, the main adherens junction protein, was identified in the transcriptome of several sponges, we suppose that EMT is regulated through cadherin-containing adherens junctions between ciliated cells. We failed to identify the master genes of EMT in the *H. dujardini* transcriptome, possibly because transcription was absent in the sequenced stages. They may be revealed by a search in the genome. The master genes themselves are controlled by various signaling pathways. Sponges have all the six signaling pathways conserved in Metazoa: Wnt, TGF-beta, Hedgehog, Notch, FGF and NO-dependent pathways. Summarizing the new data about intercellular communication in sponges, we can put forward two main questions regarding metamorphosis: (1) Which of the signaling pathways and in what hierarchical order are involved in metamorphosis? (2) How is the organization of a young sponge related to that of the larva or, in other words, is there a heredity of axes between the larva and the adult sponge?

1. Introduction

Sponges (Porifera), which were among the first multicellular animals, are a sister group to all other Metazoa (Feuda et al., 2017). They are an ancient phyletic lineage of morphologically simple animals that diverged from other metazoans at least 700 Mya, well before the Cambrian explosion (Simion et al., 2017). Sponges occupy a key phylogenetic position, and the study of their molecular organization is a powerful tool for elucidating the first steps of the evolution of multicellularity. There is direct evidence of the origin of Metazoans from protists, and the first multicellular animals are sponges (Sebé-Pedrós, Degnan, & Ruiz-Trillo, 2017).

Although sponges are quite unlike other metazoans anatomically, their embryonic and larval development is surprisingly similar to that of higher animals. For instance, sponges express virtually all the morphogenetic movements observed during embryonic development of Eumetazoa (Ereskovsky, Renard, & Borchiellini, 2013). Yet there are relatively few omics data on sponges, and it is difficult to perform a meaningful comparative analysis with bilaterians.

The phylum Porifera is divided into four classes. The Calcarea are sponges with a calcareous skeleton. The Hexactinellida are syncytial glass sponges, whose skeleton consists of silicone dioxide. The Demospongiae are "common" sponges, and the Homoscleromorpha comprises sponge species possessing a basement membrane and metazoan-like cell junctions. Morphogenesis during embryonic development is highly diverse in different orders. Larvae, the only motile stage in the life cycle of sponges, are as diverse as the developmental types, with nine larval morphotypes being distinguished (Ereskovsky, 2010).

Our model organism, *Halisarca dujardini* Johnston, 1842 (Demospongiae) is a small sponge encrusting stones or algae in the sublittoral of boreal and Arctic seas (Fig. 1A). The embryonic development and metamorphosis of *H. dujardini* is well documented (Chen, 1976; Ereskovsky & Gonobobleva, 2000; Lévi, 1956; Metschnikoff, 1879; Schulze, 1877) (Fig. 1B and C). Transcriptomes of adults and larvae as well as a draft of genome sequenced with Illumina technology are available (Borisenko, Adamski, Ereskovsky, & Adamska, 2016). Although metamorphosis of numerous sponges was described at morphological level, the only study of this process with cell-type specific protein markers was performed on *H. dujardini* (Mukhina, Kumeiko, Podgornaya, & Efremova, 2006). Keeping in mind all information available on *H. dujardini* including data on morphology, transcriptome, and protein markers of larval cell types, we tried to analyze and compare the known facts about molecular mechanisms of sponge larva metamorphosis. In this review we focused on the signaling mechanisms involved in metamorphosis since the major morphogenetic events such as changes of the cell shape, rearrangement and migration of cells and changes in the gene expression profiles should be related to cell-to-cell communication.

2. Embryonic development

Three model species of sponges were considered in the review: *Halisarca dujardini*, *Sycon ciliatum* and *Amphimedon queenslandica*. Developmental mode is remarkably different in each of these species.

H. dujardini is a viviparous species. Embryos develop inside temporary embryonic capsules formed from dedifferentiated choanocytes (Korotkova & Apalkova, 1975) (Fig. 1B and C). Cleavage is complete, equal and polyaxial (Fig. 1C(b–c)) (Ereskovsky, 2010). A single-layered coeloblastula with a small cavity is formed, and outer cells become columnar in shape and ciliated.

The larvae start to move a few days before leaving the parent sponge. The walls of the embryonic capsule soon fuse with those of the aquiferous canal, and larvae escape through the osculum. Released larvae continue to

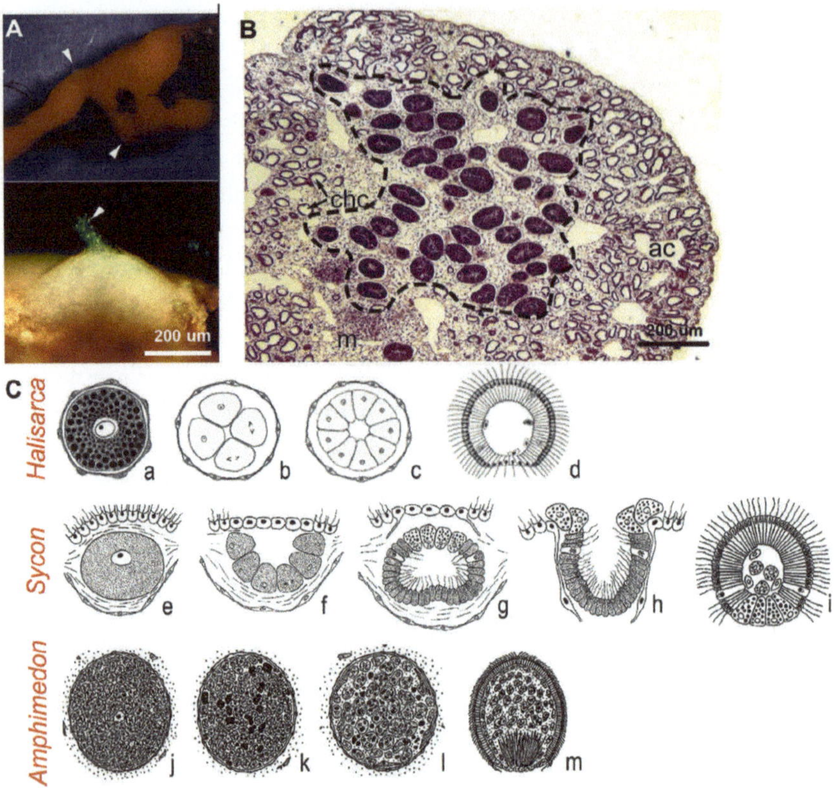

Fig. 1 (A and B). Sponge *Halisarca dujardini* from the White Sea. (A) *Top*—adult sponge on brown algae *Fucus vesiculosus*. Sponges have a canal system, through which they pump water. Water enters the sponge through pores, flows along the canals to choanocyte chambers and exits through large openings (oscula). Oscula are marked with arrowheads. *Bottom*—Rhagon, a young sponge just after metamorphosis. It has only a few choanocyte chambers and one osculum. (B) A sponge with developing larvae (a group of larvae is outlined with a dotted line). The aquiferous system in the central and the basal part of the sponge is destroyed, and normal tissue organization remains only at the periphery. ac, aquiferous canal, chc, choanocyte chambers, m, mesohyl. (C) Developmental stages of sponges with described developmental signalings from classes Demospongiae (*Halisarca* and *Amphimedon*) and Calcarea (*Sycon*; a scheme). a, e, j: egg; b, c: polyaxial cleavage; d: coeloblastula-like larva; f: incurvational cleavage; g, h: stomoblastula and its excurvation; j: amphiblastula larva; k: unequal and asynchronous cleavage; l: layers formation by delamination in morula; m: parenchymella larva. *Panel (C): Modified from Ereskovsky, A. V. (2010).* The Comparative Embryology of Sponges. *Dordrecht: Springer.*

swim freely by moving forward and rotating clockwise around the anterio-posterior axis. The larva of *H. dujardini* is rounded and measures about 120–130 μm in diameter. It is evenly ciliated, except at the posterior pole, which is ciliated less densely (Fig. 3B). Its external cells are connected with belt desmosome-like junctions (zonula adhaerens) in their apical parts (Gonobobleva, 2007; Gonobobleva & Ereskovsky, 2004a, 2004b).

The larvae of *S. ciliatum* develop in the mesohyl between the choanocyte and the pinacocyte layer (Fig. 1C(e–h); Franzen, 1988; Leys & Eerkes-Medrano, 2005). Early cleavage is equal and highly stereotypic. The first four blastomeres form a rhomboid arranged in the plane defined by the choanocyte and the pinacocyte layers. Since the third cleavage the mitotic spindle turns to divide daughter cells laterally and in the direction of choanoderm (Fig. 1C(f–g)). This results in the formation of a cup-shaped stomoblastula embryo with an opening directed to the choanoderm. By the end of cleavage and differentiation, the embryo is composed of three cell types: large, granular, non-ciliated macromeres adjacent to the choanocytes; smaller and more numerous micromeres, which have cilia pointing into the embryonic cavity; and four cruciform cells. The embryo then undergoes excurvation (inversion) through stomoblastula opening (Fig. 1C(h)). As a result of this process, during which the embryo is turned inside out, the cilia are positioned on the outer larval surface and the larva is moved to the choanocyte chamber. During metamorphosis, the larva settles on the anterior pole; within minutes the ciliated cells of the anterior half undergo epithelial–mesenchymal transition (EMT) and form the inner cell mass. In contrast, the macromeres maintain their epithelial organization, completely enclose the micromeres, and become the pinacocytes of the forming juvenile.

The embryos of *A. queenslandica* also develop in brooding chambers in follicular capsules (Fig. 1C(j); Adamska et al., 2007; Leys & Degnan, 2002). At the early stages of cleavage, chaotically arranged micromeres and macromeres of different size are formed (Fig. 1C(k)). By the end of cleavage, the morula (solid blastula) already contains micro- and macromeres, amoeboid cells, sclerocytes, ciliated cells and pigment cells (Fig. 1C(l)). All these cells are mixed until positional signals guide them to their definitive place in the larvae. Cleavage is followed by the formation of cell layers due to differential cell movements in the embryo. The main results of these movements are the formation of an outer ciliated cell layer by ciliated cells and an accumulation of pigment cells and spicules, referred to as the pigment spot, at the posterior pole of the embryo (Fig. 1C(m)). The outer ciliated cells will provide locomotion of the larva, while the pigment spot will transform into a pigment ring, a circular arrangement of pigmented cells with a photosensory function. By this stage the embryo elongates along the anterio-posterior axis, the pigment ring marking the posterior pole. Many of the developmental genes expressed in the embryo are closely related to the pigment ring.

3. Morphology of the metamorphosis

Metamorphosis is a process in the postembryonic period during which the larva develops into a juvenile. Fundamental morphological and physiological changes occurring during metamorphosis involve a reorganization of the basic body plan.

The most intriguing aspect of metamorphosis is the mechanism of morphogenetic movement, which results in the formation of the main definitive structures. Delage proposed the hypothesis of the "inversion of the germ layers" in sponges late in the 19th century (Delage, 1892). He described an inward migration of superficial ciliated cells during metamorphosis in some Demospongiae larvae and suggested that these cells formed the choanoderm. "Inversion" meant that external larval cells formed internal gut-like structures of an adult sponge. Delage's observations have been confirmed and disproved several times. The fate of ciliated cells and the formation of definitive tissues in sponges is still debated (Efremova, 1997; Ereskovsky, 2010; Ereskovsky & Dondua, 2006; Leys, 2004; Leys & Ereskovsky, 2006; Sogabe, Nakanishi, & Degnan, 2016). However, it has been demonstrated that the fate of ciliated cells depends on the larval organization (reviewed in Ereskovsky, 2010). This means that no single interpretation of their fate during metamorphosis can be proposed.

The larvae of *H. dujardini* attach to the substrate with their anterior pole and flatten along the anterio-posterior axis. The ciliated cells of the posterior pole form the exopinacoderm (Fig. 2A–C). Those of the anterior hemisphere flatten parallel to the substrate (Fig. 2B and C). An inner conglomerate, a

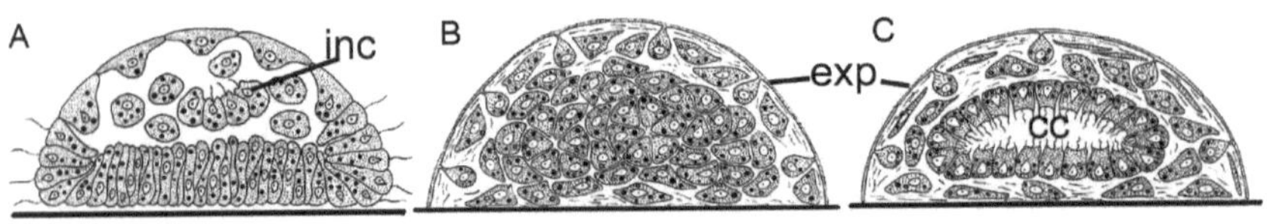

Fig. 2 Schematic drawing of the main stages of metamorphosis in *Halisarca dujardini*. (A) First stage, immediately after the attachment of the disphaerula and the destruction of the internal cavity (inc); cells of larval posterior pole (pci) begin to flatten to form exopinacoderm. (B) Pupa with developing exopinacoderm. (C) Early rhagon, which has a single choanocyte chamber (cc) and the exopinacoderm (exp), but lacks the basopinacoderm. *From Gonobobleva, E., & Ereskovsky, A. (2004a). Metamorphosis of the larva of Halisarca dujardini (Demospongiae, Halisarcida).* Bulletin de l'Institut Royal Des Sciences Naturelles de Belgique—Bulletin van Het Koninklijk Belgisch Instituut Voor Natuurwetenschappen, 74, *101–115, reproduced with permission of l'Institut Royal des Sciences Naturelles de Belgique.*

mass of ciliated cells of the anterior pole, the internal cells and the ciliated cells of the internal larval cavity, is formed below the exopinacoderm (Fig. 2B). By the end of metamorphosis, the cells of this conglomerate differentiate into choanocytes, endopinacocytes, basopinacocytes and mesohyl cells (Gonobobleva & Ereskovsky, 2004a). Calcium signaling plays an important role in initiation of metamorphosis in another demosponge, *Amphimedon queenslandica* (Nakanishi, Stoupin, Degnan, & Degnan, 2015).

Cytological markers have been successfully used to describe the differentiation of some cells of sponge larvae (Amano & Hori, 1993, 1996, 2001; Ereskovsky, Konjukov, & Willenz, 2007; Ereskovsky, Tokina, Bézac, & Boury-Esnault, 2007; Gaino, Baldacconi, & Corriero, 2007; Usher & Ereskovsky, 2005). Vital fluorescent non-specific tracer CMFDA was used to label larval peripheral cells (Leys & Degnan, 2002). Elegant experiments combining EdU-labeling of newly synthesized DNA with lipophilic tracer CM-DiI were performed on *A. queenslandica*, the first demosponge whose genome was sequenced (Sogabe et al., 2016). The authors followed the fate of ciliated cells to adult choanocytes and showed that they transformed into mesenchymal pluripotent stem cells resembling archaeocytes. Some of these EdU-labeled cells differentiated in choanocytes, but other, unlabeled cells were also involved in the formation of choanocyte chambers. These results indicate that multiple lineages of larval cells usually contribute to the formation of individual choanocyte chambers during metamorphosis, contradicting previous reports that in other sponge species choanocyte chambers form clonally (Nakanishi et al., 2015).

Molecular markers are now available for sclerocytes (cells producing the mineral skeleton) of *Ephydatia fluviatilis* and *Sycon ciliatum* (Funayama, Nakatsukasa, Kuraku, et al., 2005; Voigt et al., 2017) and archaeocytes of *E. fluviatilis* (Funayama, 2010). Annexin was shown to be a specific choanocyte marker, and archaeocytes were found to be the precursors of choanocytes during asexual reproduction of the freshwater sponge *E. fluviatilis* (Funayama, Nakatsukasa, Hayashi, & Agata, 2005). However, few of these molecules seem to be expressed in a lineage-restricted manner: homologs of *piwi* and *annexin* are present in *H. dujardini*, but they are not specifically expressed in, respectively, archaeocytes and choanocytes (GenBank accession numbers MH006695 for *piwiA*, MH107849 for *piwiB* and MH107850 for *annexin*). In general, the range of the available markers is very limited. This means that sponge cell lineages, especially those involved in development and reproduction, are poorly understood.

4. The fate of ciliated cells during metamorphosis of *H. dujardini*

The metamorphosis of *H. dujardini* was described using light and electron microscopy (Ereskovsky, Konjukov, & Willenz, 2007; Ereskovsky, Tokina, et al., 2007; Gonobobleva & Ereskovsky, 2004a, 2004b; Lévi, 1956). However, given considerable changes in the morphology of ciliated cells during metamorphosis, molecular markers are obviously required to trace their fate precisely. Intercellular junctions in sponges are calcium-dependent, and sponge tissues dissociate to cells in calcium-free media. A suspension of isolated sponge cells can be easily obtained by placing tissues in artificial seawater without Ca^{2+} supplied with a chelator such as EDTA (Ereskovsky, Konyukov, & Tokina, 2010; Ereskovsky, Konjukov, & Willenz, 2007; Ereskovsky, Tokina, et al., 2007; Mukhina et al., 2006; Mukhina, Kumeiko, Podgornaya, & Efremova, 2007). The larvae of *H. dujardini* were dissociated, and the cell suspension was overlaid onto discontinuous density gradient made with Percoll solution and centrifuged. The gradient consisted of layers from 10% to 45% Percoll in calcium-free seawater, producing five fractions numbered from top to bottom of a tube (Fig. 3).

The separation of larval cell types (Fig. 3C) made it possible to determine the polypeptide composition of the fractions by SDS-PAGE. The protein composition of fraction I (ciliated cells) differed dramatically from the rest. There were four major bands in the range of 42–68 kDa. One lane with a molecular weight of 42 kDa probably corresponded to actin. Another double lane of ~55 kDa corresponded to α and β tubulin isoforms. A third band of unknown nature with an apparent molecular weight (M_r) of 68 kDa (antigen S, p68) was chosen for polyclonal antibody production. Anti-p68 (S antigen) antibodies produced in rabbit were shown to be specific to larval ciliated cells, and were used in double immunofluorescence together with commercial anti-tubulin antibodies to trace the fate of ciliated cells during metamorphosis.

Double staining of larvae with both antibodies showed that protein p68 (S) was present in granules in the body of ciliated cells but was absent in the cilium. Tubulin was present in the cilium, the basal body and, to a lesser extent, in the cortical cytoskeleton structures (Fig. 4A and D(a)). Thus, anti-S and anti-tubulin antibodies stained different compartments in ciliated cells.

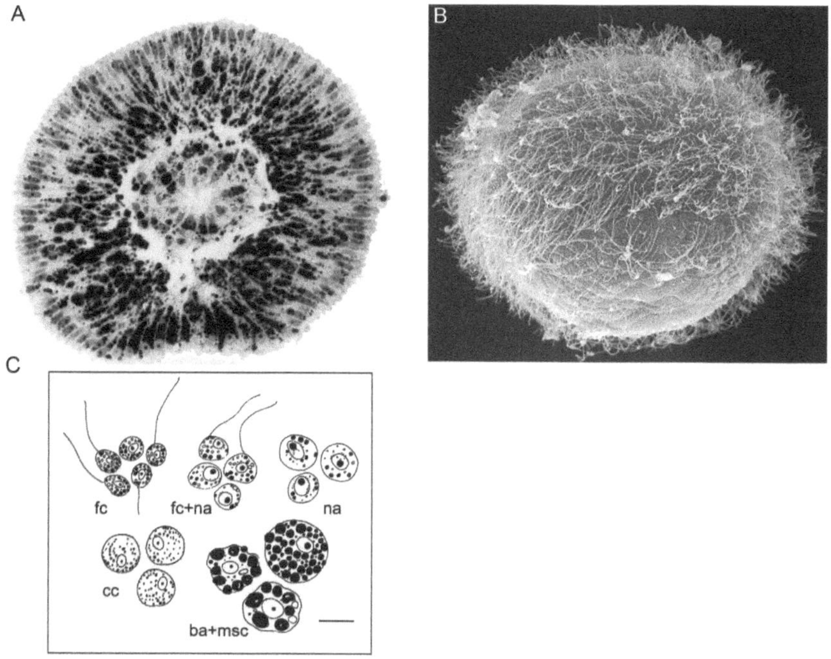

Fig. 3 Larvae of *H. dujardini*, their cell composition and the ability of cells to aggregate. (A) Semi-thin longitudinal section of the free-swimming disphaerula larva. ba: blastomere-like amoebocytes; ci: cilia; cil: ciliated cells; ich: internal ciliated cells cavity, msc: maternal spherulous cells; na: nucleolated amoebocytes; pci: posterior flagellated cells; *ap*, anterior pole; *pp*, posterior pole. Scale bar: 20 μm. (B) SEM of disphaerula larva. ci, cilia; pci, posterior flagellated cells; *ap*, anterior pole; *pp*, posterior pole. Scale bar: 20 μm. (C) Scheme of larval cell fractions of *H. dujardini* obtained by Percoll density gradient centrifugation. Scale bar: 5 μm.

Cilia visualized with anti-tubulin antibody were seen all over the larval surface except at the posterior pole, which was bare and lacked any specific fluorescence (Fig. 4A(a)). Such larvae settled on the substrate and flattened rapidly. In the narrow transparent monolayer of lateral ciliated cells attached to the substrate, the cilia appeared to be submerged into the cytoplasm, *i.e.*, were internalized (Fig. 4A(b–c)). Both anti-S and anti-T staining were confined to the internal cell mass at the middle sections of metamorphosing larvae (Fig. 4B). By 3 days post-settlement most of the ciliated cells joined the newly formed choanocyte chambers (Fig. 4C). S-specific labeling was seen in choanocytes but not in pinacocytes or amoebocytes.

At the morphological level choanocytes differ from ciliated cells by a prominent collar around cilia. S antigen (p68) was concentrated in the apical part of the choanocyte and visualized as rings or curves outlining the apical

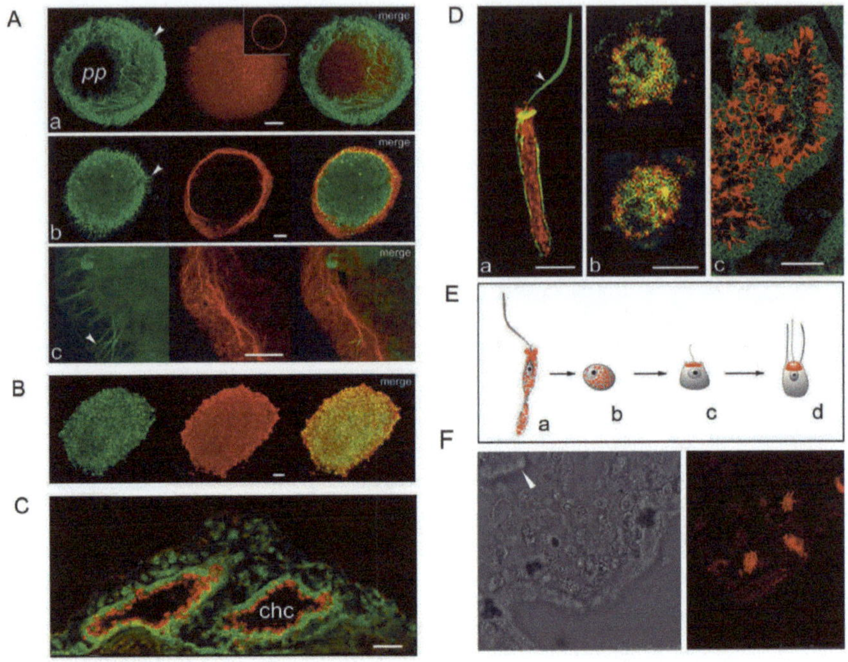

Fig. 4 Free-swimming and metamorphosing larva of *H. dujardini* double-stained with anti-tubulin and anti-p68 antibodies. Sequential steps of larval settlement (A), metamorphosis (B and C), transformation of ciliated cell (D and E) and adult sponge (F). (A) Larval settlement. Laser confocal scanning microscopy of whole mount larvae. a: Free-swimming larvae before settlement; top optical section and midsection (inset). Here and elsewhere green staining indicates tubulin, and red staining indicates S antigen. Arrow points to cilia. b: Larvae 40 min after contact with substrate (glass). c: Peripheral part of flattened larvae. Colors and designations are the same as above. Internalized cilia on **c** are indicated by arrow. (B) Larvae 6 h after contact with substrate. Ciliated cells are submerged into the larva. Scale bar for all images: 20 μm. (C) Young sponge 3 days after settlement, immunostained paraffin section. chc—choanocyte chamber formed *de novo*. Note the absence of S antigen staining in young pinacoderm. Scale bar 10 μm. (D) Transformation of ciliated cell (a) by inward migration (b) into the choanocyte chamber (c) of a definitive sponge. Scale bar 2 μm for a, b and 20 μm for **c.** (E) Scheme based on original image (D). a: Ciliated larval cells, b: ciliated cells inside larva, c: choanocyte of young sponge, d: choanocyte of definitive sponge. (F) Adult sponge with non-motile larva (correspond to Fig. 2B). a: DIC microscopy; b: immunofluorescence with anti-S. Pinacocytes are stained but note the absence of staining in cells of the larva before the release (white arrow). Scale bar: 20 μm.

border (Fig. 4D(c)). Hence, ciliated cells from the anterio-lateral surface of the larva internalized during metamorphosis, and then the cells of the internal cell mass with the same immunoreactivity (S/tubulin-positive) participated in the formation of choanocyte chambers in the juvenile.

Some of the larval ciliated cells appear to be the source of internal cells in juvenile *H. dujardini*. A very unusual polypeptide composition of the ciliated cells, with few prominent major proteins, could be a sign of their terminal differentiation, but no evidence of exfoliation of ciliated cells or their phagocytosis by any other cells during the short time of metamorphosis has yet been found (Gonobobleva & Ereskovsky, 2004b). Alternatively, a unique protein composition may be associated with the preparation of ciliated cells for a precipitous change of their fate. The S protein may be meant for choanocytes, which later become able to synthesize their own protein. The localization of p68 (S) must be checked by electron microscopy and its identity, by mass-spectrometry or Edman degradation procedure. The transcriptome of the adult sponge is available (Borisenko et al., 2016), and since the adult sponge can produce S antigen (Fig. 4(F)), we may be fairly sure that the S transcript can be found. Once this has been done, we may be able to trace specific cell-type transcription. By now we have proved its specificity for ciliated cells and juvenile choanocytes at the protein level.

Transdifferentiation of cells is common in sponges. For instance, it occurs during regeneration and the formation of primmorphs (Borisenko, Adamska, Tokina, & Ereskovsky, 2015; Ereskovsky et al., 2015; Lavrov, Bolshakov, Tokina, & Ereskovsky, 2018; Lavrov & Kosevich, 2016). Light and electron microscopic data demonstrate that choanoblasts can develop in parenchymella from ciliated cells and archaeocytes in some genera (*Mycale*, *Hamigera*) (Borojevič & Lévi, 1965; Boury-Esnault, 1976), only from archaeocytes in some other taxa (*Baikalospongia*, *Microciona*) (Efremova & Efremov, 1979; Kaltenbach, Kuhns, Simpson, & Burger, 1999), or from an internal population of archaeocytes and larval choanocytes (Spongillidae) (Weissenfels, 1989). A recent study of another demosponge, *A. queenslandica*, has shown that multiple larval cell types contribute to choanocyte chambers of the juvenile (Sogabe et al., 2016). So, while the fate of ciliated cells has been elucidated in *H. dujardini*, it has to be studied in other species with the use of a board range of markers.

Our approach could be extended to the search for new protein markers with the separation of adult cell types and powerful techniques such as 2D-DIGE. Different cell types of the adult sponge clearly possessed specific proteins (Fig. 5, arrows). Proteins in these spots could be identified after excision by mass-spectrometry and compared with the assembled transcriptome of the adult sponge. Then the specific synthesis could be traced by the whole-mount RNA–RNA *in situ* hybridization (Borisenko et al., 2016).

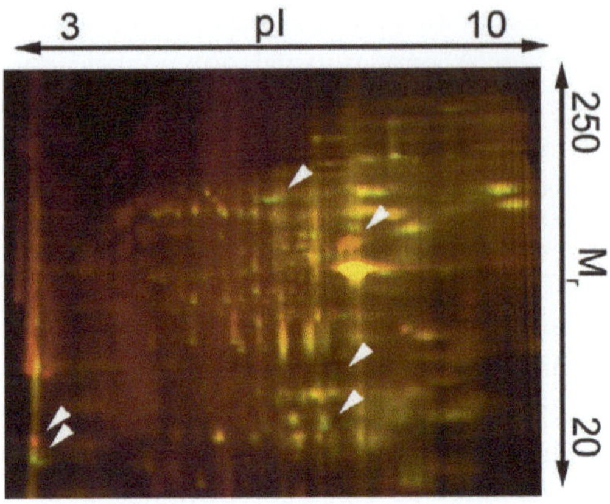

Fig. 5 Difference gel electrophoresis (DIGE; according Alban et al., 2003) revealed cell-specific protein markers. Cell suspension of an adult sponge was separated by centrifugation in density gradient, and proteins were extracted from choanocyte- and archaeocyte-enriched fractions with urea-containing solution. Each protein mixture was covalently labeled with reactive cyanine dye: choanocyte proteins were labeled with Cy3 (green), and archaeocyte proteins, with Cy5 (red). Then protein mixtures were pooled and underwent 2D electrophoresis by isoelectrofocusing and SDS-PAGE. After scanning the gel with lasers of different wavelength, we could recognize proteins specific for choanocytes (green), for amoebocytes (red), or common proteins (yellow).

5. A journey from the outside inward: How ciliated cells become detached

Antero-lateral ciliated cells leave the epithelium of the larva and migrate to the internal cell mass as amoeboid cells to be differentiated into choanocytes and/or other cell types of the juvenile sponge. Intercellular junctions have been found between larval ciliated cells, choanocytes and pinacocytes (Amano & Hori, 1996; Bergquist & Green, 1977; Boury-Esnault, Ereskovsky, Bézac, & Tokina, 2005; Evans, 1977; Ereskovsky et al., 2009; Ereskovsky & Tokina, 2004; Gonobobleva & Ereskovsky, 2004b; Gonobobleva & Maldonado, 2009). In our opinion, this indicates the presence of epithelial–mesenchymal transitions (EMT): ciliated cells break free from junctions and leave the polarized sheath to become amoeboid mesenchymal-like cells of the internal mass.

The loss of E-cadherin-mediated adherens junctions is the key event of EMT (Huang, Guilford, & Thiery, 2012). There are numerous proteins with cadherin repeats in their extracellular parts, but unlike true cadherins they have neither delta- nor beta-catenin binding domains in the cytoplasmic part. Proteins containing cadherin repeats have long been known, but the presence of true epithelial cadherin remained a mystery. The presence of E-cadherin has been shown for sponges from all classes (Belahbib et al., 2018). We identified one cadherin homolog in *H. dujardini* transcriptome, which contained multiple cadherin repeats, a transmembrane helix and a cytoplasmic tail with domains for binding intercellular components of junction, delta-catenin (p120 protein) and beta-catenin. The sequence also contains seven EGF-like domains and two laminin domains, which is characteristic of invertebrate cadherin (Fig. 6). So, *H. dujardini*, similarly to other sponges (Belahbib et al., 2018), also has true cadherin, which is the main component of adherens junctions serving as a cell-to-cell lock in epithelial sheath.

The repression of E-cadherin expression is followed by the disappearance of adherens junctions after the start of EMT (Huang et al., 2012). EMT is controlled by the master genes *snai1/snai2*, *twist* and *ZEB* (Lamouille, Xu, & Derynck, 2014). Their expression induces EMT in many processes associated with development, regeneration and diseases. Snai and ZEB are zinc finger transcription factors. Snai has a SNAG domain and sites for binding LATS2, KPNB1, NOTCH1 and PARP1 proteins; ZEB possesses an atypical homeobox and\or Glu-rich C-end. There are many transcripts for zinc finger DNA-binding proteins in *H. dujardini* transcriptome but none of them has the characters typical of Snai and ZEB. A similar situation observed for Twist. It is a basic helix-loop-helix (bHLH) transcription factor (TF), and *H. dujardini* has a lot of bHLH TFs, but they have only the DNA-binding domain in common with Twist. We failed to identify the master genes of EMT in *H. dujardini* transcriptome, possibly because transcription was absent during the sequenced staged. They may be revealed by a search in the genome.

Master genes themselves are controlled by various signaling pathways, mainly TGF-beta, FGF and Wnt. Finally, the process of internalization of flagellated cells during metamorphosis is under control of intercellular signals.

6. Signaling pathways in sponges

Cell-to-cell communications regulate many processes during development and adulthood of metazoans. During embryogenesis intercellular signals are the third main informational input to developmental process after maternal factors and zygotic genome. Key processes such as proliferation, cell migration and survival, cell fate specification and axis patterning are

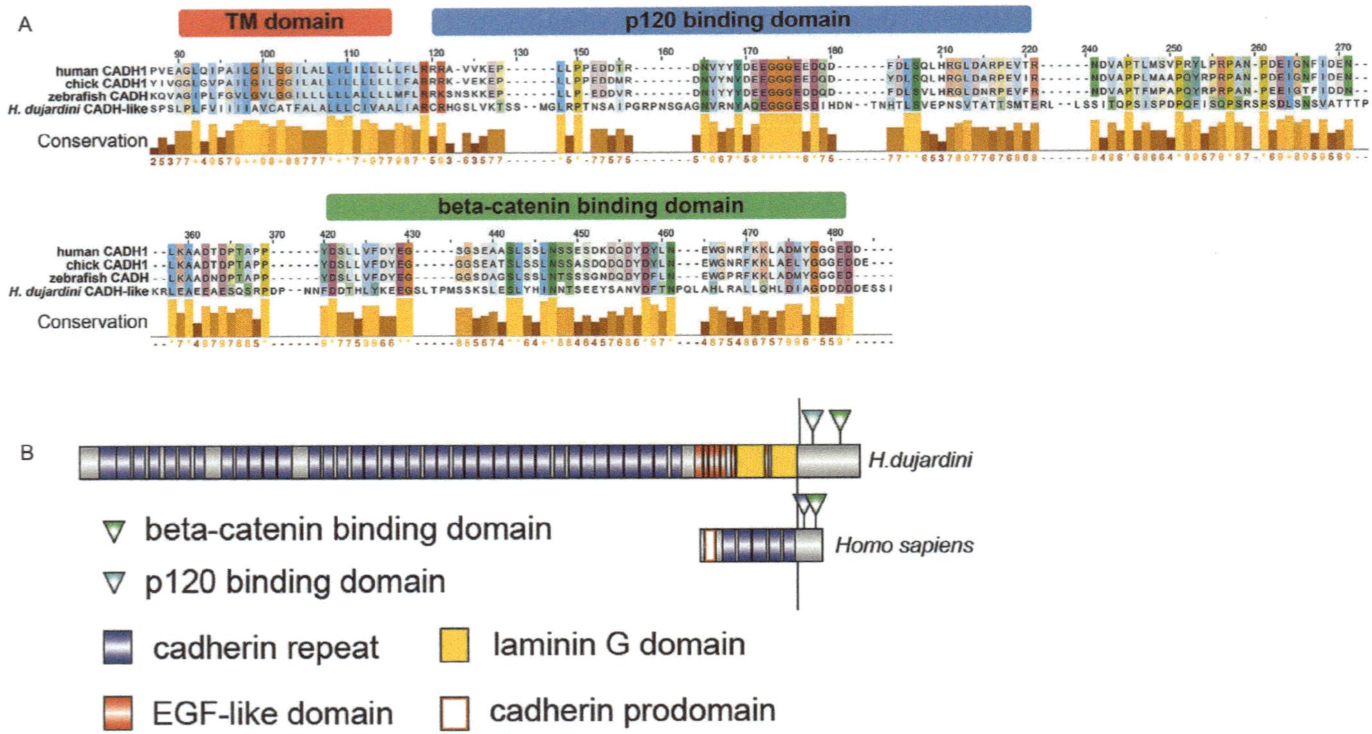

Fig. 6 Comparison of E-cadherin structure in sponge *H. dujardini* and bilateria. (A) Alignment of the cytoplasmic parts of E-cadherins in humans, chick, zebrafish and *H. dujardini*. Conserved domains responsible for interaction with delta-catenin (p120) and beta-catenin, present in the intracellular part of E-cadherin molecule in *H. dujardini*, anchor adherens junctions to the cytoskeleton. (B) Domain organization of E-cadherins in humans and sponges. Invertebrates cadherins are characterized by multiplication of conservative cadherin repeats and the presence of laminin G-like domains, while lacking the prodomain, a short N-terminal sequence releasing with proteolytic maturation. Sequence from *H. dujardini* was identified in transcriptome with local BLAST (option tblastn) and domains identified with SMART and InterPro 70.0 (EMBL-EBI); GenBank accession number is MH107848.

managed by cell-to-cell signaling. These processes are also regulated through intercellular communication. Six signaling pathways are conserved across Metazoa: Wnt, TGF-beta, Hedgehog, Notch, FGF and NO-dependent. Here we summarize facts and assumptions about each of these pathways in sponges.

Wnt signaling is the best studied pathway in sponges due its probable role in axial patterning, especially in early branching Metazoa. Representatives of all the four poriferan classes are studied for Wnt ligands. Syncytial sponges, Hexactinellida, have no Wnt ligands although they have all the other components of this pathway (Schenkelaars et al., 2017; Windsor Reid et al., 2018). Distribution of ligands and receptors is quite different even between genera. For instance, *A. queenslandica* (Demospongiae) has two Wnt ligands, three Frizzled receptors, a co-receptor LRP5/6, a secondary messenger Disheveled, beta-catenin, cytoplasmic beta-catenin destruction complex (GSK3, APC, Axin), nuclear factors TCF and Groucho (Adamska et al., 2010). Freshwater sponge *Ephydatia muelleri* has three ligands and three Frizzled receptors (Windsor Reid et al., 2018). In our model demosponge, *H. dujardini*, there are ten ligands and five receptors (Borisenko et al., 2015). Cytoplasmic scaffold proteins Axin and APC have similar features in both species. The beta-catenin binding site not found in Axins, and most of the conserved domains except armadillo repeats are not found in APCs. Biochemical studies on poriferan beta-catenin demonstrate that intermolecular interactions during cytoplasmic signal transduction can be mediated by highly diverged sequences (Nichols, Dirks, Pearse, & King, 2006; Schippers & Nichols, 2018). This may be a reason why such domains cannot be recognized by ordinary bioinformatic methods.

The champion in Wnts complexity, a calcisponge *Sycon ciliatum*, contains 21 ligands. In addition to ligands, four receptors and all core intracellular components except Axin are described in *S. ciliatum* (Leininger et al., 2014). Sponges from the genus *Oscarella* (Homoscleromorpha) have from 3–8 ligands (Lapébie et al., 2009; Leininger et al., 2014). The presence of other pathway components has not been ascertained.

Phylogenetic analysis cannot place poriferan Wnt in any of the established subfamilies of metazoan ligands (Adamska et al., 2010; Lapébie et al., 2009; Leininger et al., 2014). Phylogeny inference made with sponge sequences only demonstrates the presence of Porifera-specific subfamilies (Windsor Reid et al., 2018). In demosponges Wnts form three subfamilies (named PorWntA-C), in homoscleromorphs, two families (WntI-II), an in calcisponges, 21 families (CWntA-U). Notably, the nine of the ten *H. dujardini* ligands cannot be classified within any of these families.

Wnt ligands are expressed in embryos, larvae and adult sponges. The expression of some Wnt ligands was shown in adult and larvae of *H. dujardini* (Fig. 7A and B). Two ligands, *HdWntD* and *HdWntE*, were expressed in the oscular tube, while *HdWntG* transcripts were observed in exopinacocytes.

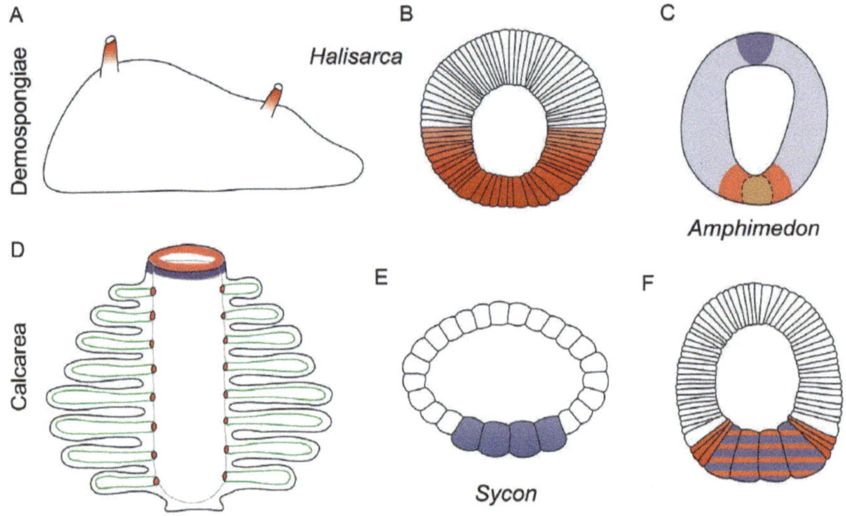

Fig. 7 Summarized data about axial-related expression of *wnt* and *TGF-beta* in three model sponges from Demospongiae (A–C) and Calcarea (D and F) classes. Wnt ligands are red and TGF-beta is blue; brightness depends on expression intensity. For embryo and larvae anterior end is oriented at top and posterior at bottom. (A and B) Main expression patterns of *HdWnt* is in oscular tube at adult *H. dujardini,* and posterior half of larva. Some *wnt*'s are expressed in basal part of sponge, and in exopinacoderm in "salt and pepper" pattern as well (data not shown). Expression patterns of *TGF-beta* are unknown in *Halisarca* yet. (C) expression of *AqWnt* and *AqTGFb* in *A. queenslandica* embryo at the pigment spot formation stage. (D) Adult *Sycon ciliatum* sponge demonstrate oscular expression of *wnt* and *TGF-beta* ligands; two *wnt*'s expresses in proximal part of choanocyte chambers as well. (E) Only *TGF-beta* transcript presented at stomoblastula stage, in aboral macromeres; *wnt* expression is not detected. (F) expression of both ligands at posterior pole of larva, with broader pattern of wnt. *Panels (A) and (B): Modified from Borisenko, I., Adamski, M., Ereskovsky, A., & Adamska, M. (2016). Surprisingly rich repertoire of Wnt genes in the demosponge Halisarca dujardini.* BMC Evolutionary Biology, 16, *123; Kozin, V. V., Borisenko, I. E., & Kostyuchenko, R. P. (2019). Establishment of the Axial Polarity and Cell Fate in Metazoa via Canonical Wnt Signaling: New Insights from Sponges and Annelids.* Biological Bulletin, 46*(1), 12; Panel (C): Modified from Adamska, M., Degnan, S. M., Green, K. M., Adamski, M., Craigie, A., Larroux, C., & Degnan, B. M. (2007). Wnt and TGF-β Expression in the Sponge Amphimedon queenslandica and the Origin of Metazoan Embryonic Patterning.* Plos One, 2*(10), e1031; Panel (D)–(F): Modified from Leininger, S., Adamski, M., Bergum, B., Guder, C., Liu, J., Laplante, M., et al. (2014). Developmental gene expression provides clues to relationships between sponge and eumetazoan body plans.* Nature Communications, 5, 3905. *https://doi.org/10.1038/ncomms4905.*

In addition, a set of *Wnt* is expressed in the posterior hemisphere of the larvae: *HdWntL* at the posterior pole, *HdWntK* from the equator to the latest lateral flagellated cells, and *HdWntJ* in a belt-like pattern at the equator (Borisenko et al., 2015). In *A. queenslandica AqWntA* is first expressed in micromeres throughout the embryo in a diffuse manner. During the development these cells migrate into the inner part of the embryo, then to the outer cell layer, and eventually turn up at the further posterior pole of the larva (Fig. 7C; Adamska et al., 2007, 2010).

Unlike *Amphimedon*, in *S. ciliatum* most *Wnt* expressed in oocytes and early cleavage stages, though transcripts disappeared by the end of embryonic development and before larvae formation. Expression domains of three *Wnt* are associated with the posterior half of larvae (Fig. 7E–F). Two out of five *Frizzled* receptors expressed in micromeres, at the anterior pole of the embryo. Such combinatorial expression of different ligands and receptors allow diversifying results from use of limited set of signaling molecules. Another set of *Wnt* is expressed in staggered domains in the osculum of the adult sponge, and in the opening of the choanocyte chamber (Leininger et al., 2014). In *S. ciliatum* the choanocyte chamber is the module of axis determined by water flow.

Expression of *Wnt* in ostia, incurrent opening of aquiferous canals, is described in adult *Oscarella lobularis* (Homoscleromorpha). This sponge has a true epithelium, and ostia form by epithelial invagination of the exopinacoderm. Localization of *Wnt* expression is characteristic of epithelial appendage formation in eumetazoans such as *Hydra* or *Nematostella* tentacles and wings in *Drosophila* (Lapébie et al., 2009).

The data on the diversity of Wnt pathway components, their phylogeny and expression allow us to make two important conclusions. (1) Sponges acquired their Wnt ligands independently of Cnidaria + Bilateria. It is still unclear whether the small number of ligands (two or three) in some species is a consequence of secondary loss or an ancestral condition. Lineage-specific expansion has definitely occurred in Calcarea. (2) Wnt pathway is likely to be involved in axes specification in Porifera. The sets of Wnt ligands expressed in a pole-dependent manner (the posterior half of the larva, the osculum, the opening of the choanocyte chamber). An opposite viewpoint is based on the study of *Wnt* expression in juvenile *E. muelleri* (Windsor Reid et al., 2018). It demonstrated the expression of *Wnt* ligands in different cell types not restricted to any region of the sponge. Although no expression of any *Wnt* in the osculum was shown, pharmacological alteration of this pathway increased the number of oscula upon activation and decreased upon inhibition (Windsor & Leys, 2010; Windsor Reid et al., 2018). So, we tend to think that in sponges Wnt is implicated in axial patterning rather than cell specification, as the first metazoan mechanism of anterio-posterior determination.

The second key player in axial patterning is the transforming growth factor (*TGF-beta*) *signaling* pathway. Core components of this pathway are extracellular ligands, transmembrane receptors and transcription factors SMADs. The superfamily of TGF-beta ligands encompasses many families of proteins, the main families being TGF-beta *sensu stricto*, BMPs, Nodals and Activins. Receptors are single pass serine-threonine kinases of two classes, TGFbRI or TGFbRII. Upon binding of the ligand, type II receptors activate type I receptors, and they phosphorylate SMADs, which leads to transcriptional activation. The patterning of the dorso-ventral axis during the development of Bilateria starts with the expression of BMP and Chordin (a soluble inhibitor of TGF-beta ligands) at the opposite sides of the embryo. At later stages TGF-beta signaling is involved in numerous developmental processes, from germ layer specification to the development of organs and tissues (Kitisin et al., 2007). This pathway is a novelty of Metazoa since neither ligand nor receptor molecules are found outside the animal kingdom (Srivastava et al., 2010).

A. queenslandica genome contains eight *TGF-beta* genes, of which two belong to TGF-beta *sensu stricto* and six fall into divergent clades such as GDF or DVR (Srivastava et al., 2010). Phylogenetic analysis of these diverged genes group suggests an independent expansion of ligands. There are ligand receptors of both types. The expression of one of the ligands is described in the development. Similarly to one of the *Wnt* genes, this *TGF-beta* is expressed in some micromeres from early cleavage stages. It is dynamically expressed throughout the outer cell layer of the embryo, and then in a concentric pattern around the posterior pole and at the anterior pole of embryo (Fig. 7C). In contrast to bilaterians and anthozoans, where TGF-beta patterns the axis perpendicular to the one patterned by Wnt signals, in *A. queenslandica* Wnt and TGF-beta are expressed along the same embryonic axis (Adamska et al., 2007).

A calcisponge *S. ciliatum* has 22 *TGF-beta* genes, one of which falls into TGF-beta *sensu stricto* family and remaining sequences cannot be ascribed to any defined eumetazoan families (Leininger et al., 2014). Several ligands are expressed uniformly in oocytes and during cleavage, while the expression of four *TGFs* is restricted to the macromeres located at the posterior pole of the embryo. The most prominent region of *TGF-beta* expression in the adult sponge is the apical region: seven ligands are expressed at a significantly higher level in the apical part of sponge bearing the osculum (Fig. 7D). Thus, the expression of TGFs is similar to that of Wnt in *S. ciliatum*, being predominantly posterior in the larvae and apical in the adults. *SMAD* genes are expressed either in the same cells or in the cells adjacent to those expressing *TGF-beta* ligands. In contrast to ligands, downstream components of pathway are expressed more uniformly along the apico-basal axis of the adult sponge.

A search in *H. dujardini* transcriptome revealed eight TGF-beta ligands. All of them have a structure typical of TGF-beta: a signaling peptide at the N-terminal end, a large latency-associated peptide, the site of protease furin cleavage, and the C-terminal signal peptide with cysteines in conserved positions. Phylogenetic analysis revealed some similarities with other sponges: three sequences fall into the TGF-beta family *sensu stricto*, while the other five ligands do not belong to any eumetazoan families (Borisenko, Podlevskikh, Prakhovnik, & Ereskovsky, in press). Intriguing feature of TGF-beta is absent of properly TGF-beta ligands out of dueterostomes and ctenophores, when genome screening in *C. elegans*, *Drosophila* or *Nematostella* did not demonstrate presence of the ligands from this class (Matus et al., 2006; Pang, Ryan, Baxevanis, & Martindale, 2011; Srivastava et al., 2010).

Many developmental processes in Bilateria involve *Hedgehog pathway* in a classical morphogen mode, acting in a concentration-dependent manner (Jiang & Hui, 2008). The Hedgehog ligand, a secreted protein, undergoes autocatalytic cleavage following acetylation and binds to Patched receptor on the cell surface. Hedgehog-competent signal cells express a zinc finger transcription factor Cubitus interruptus (Ci), which undergoes cleavage and releases a peptide (CiR) translocating in the nucleus and acting as a transcriptional repressor. The binding of Hedgehog to Patched allows the suppression of Ci cleavage; a full-size Ci works as a transcriptional activator. Thus, the level and the set of expressed target genes depend on the ligand concentration. In this way Hedgehog participates in many patterning processes such as limb formation in vertebrates and the establishment of segment polarity in *Drosophila*. However, in some cases this pathway acts as short-range signaling, *e.g.*, in case of the induction of the neural floor plate by the notochord in vertebrates.

Hedgehog genes have not been identified outside cnidarians and bilaterians but a Hedgling homolog was found in a choanoflagellate *Monosiga brevicolis* (King et al., 2008). The expression of two Hedgehog genes has been analyzed in *Nematostella vectensis* (Matus, Magie, Pang,

Martindale, & Thomsen, 2008). One of them is expressed in the pharyngeal ectoderm, while the other one is expressed in the endoderm; the receptor Patched is expressed only in the endoderm. This pattern suggests that both short- and long-range Hedgehog-mediated pathways may function in Cnidaria. Neither the ligand nor any other key components of this pathway were found in the ctenophore *Mnemiopsis leidyi* (Ryan et al., 2013). As at least some components of this pathway are present in choanoflagellates, its loss in Ctenophora seems to have been was secondary and independent (Adamska, 2015).

A typical Hedgehog protein contains an N-terminal Hedge domain and a C-terminal Hog domain. Hog domain separates due to autocatalytic cleavage. Both these domains are found in *A. queenslandica* genome but in different genes. The key domain, Hedge, is a part of a large transmembrane protein Hedgling also containing the von Willebrand domain, cadherin repeats, immunoglobulin and EGF domains. *Hedgling* expresses during late embryonic development and is associated with the formation of the sensory pigment ring at the posterior pole of the embryo. It is unknown whether Hedgling proteins undergo modifications and/or cleavage. If they do not, this transmembrane protein could only signal across adjacent cells through direct contact of cell surfaces, in contrast to the true Hedgehog. The genome of *A. queenslandica* contains several other components of the Hedgehog pathway but not the receptor Patched (Richards & Degnan, 2009; Srivastava et al., 2010). It is un clear whether Hedgling transmits intracellular signals through the canonical Hedgehog pathway or through an unidentified mechanism (Adamska et al., 2007).

Growth factors family encompasses *fibroblast growth factors (FGF)*, platelet-derived growth factors (PDGF), vascular endothelial growth factors (VEGF) and epidermal growth factors (EGF). All of these proteins were described as growth factors for defined cell types and act through the same type of receptors, although ligands do not demonstrate similarity of sequences. FGF bind to receptor tyrosine kinase that initiates a series of phosphorylation events, the tyrosine kinase cascade, resulting in activation of transcription factors. Unlike other signalings, there is no exact transcription factor activated by FGF pathway, and the output of the pathway activation depends on the combination of ligand, receptor and cellular content.

FGF signaling is involved in various developmental processes in Bilateria, from axes specification to tissue induction and organ patterning (Thisse & Thisse, 2005). Another conserved role of FGF is boundary formation as shown in *Hydra* (Sudhop et al., 2004). Nevertheless, a true FGF protein is the novelty of Bilateria + Cnidaria: no FGF ligand was found in *Trichoplax*,

A. queenslandica, *Mnemiopsis* or *Monosiga* genomes although FGF-like proteins with a similar domain organization were identified in many Metazoa including sponges (King et al., 2008; Ryan et al., 2013; Srivastava et al., 2008, 2010). Receptor tyrosine kinases are broadly distributed from Protista to Bilateria, and some of them display a weak similarity to the intracellular portion of the FGF receptors (Manning, Young, Miller, & Zhai, 2008). The intracellular signal transduction proteins are all present in *Monosiga*. This demonstrates a gradual assembly of the pathway, from signal transduction cascade utilized by single cell eukaryotes, through addition of receptors allowing more sophisticated sensing of the neighboring cells in the colony, to the emergence of a highly effective morphogen in the ancestor of Cnidaria + Bilateria (Adamska, 2015).

All signaling events mentioned above are mediated by secreted soluble proteins. In the *Delta/Notch signaling* pathway both the ligand (Delta or Serrate) and the receptor (Notch) are transmembrane proteins, and the signaling event means close contact of two neighbor cells. Upon binding of the ligand to the receptor, several proteolytic modifications occur at the ligand-receptor complex resulting in the production of Notch intracellular domain, which is translocated to the nucleus and acts as a transcription activator. This pathway appears to be the evolutionarily the oldest one: all the participants are identified in Monosiga genome except ligands. In choanoflagellates Delta/Notch might be involved in sensing environmental stimuli rather than intercellular signaling (Richards & Degnan, 2009).

Delta/Notch signaling is implicated in the development of many organs in Bilateria, and is crucial for segmentation in Vertebrata (Andersson, Sandberg, & Lendahl, 2011). In Cnidaria it is involved in the boundary formation during budding and affects oogenesis and the differentiation of nematoblasts and neurons (Käsbauer et al., 2007; Marlow, Roettinger, Boekhout, & Martindale, 2012; Matveev, Shaposhnikova, & Podgornaya, 2007; Münder et al., 2010). In sponges, Delta/Notch pathway components and their developmental expression are described in *A. queenslandica* (Gazave et al., 2009; Richards & Degnan, 2012; Richards et al., 2008). Five different Delta ligands, one Notch receptor and all core components were identified. The expression of *Notch* and *Delta* genes indicates that they are involved in specification of putative sensory cells, with *Notch* expression being broad and the five *Delta* genes displaying dynamic, unique patterns. The domain composition of the identified genes, as well as their expression patterns, indicate that the pathway operates in a manner similar to that in Cnidaria + Bilateria, and is involved in evolutionarily conserved processes such as cell-type specification and differentiation.

Nitric oxide (NO) pathway is another signaling mechanism usually mentioned in the context of blood vessel physiology or differentiation of endothelial cells. NO is an ancient signal molecule working *via* the activation of its primary cytoplasmic receptor, soluble guanylyl cyclase, or by direct covalent modification of target proteins such as *S*-nitrosylation and nitration. Data about its participation in early development are just beginning to accumulate although the role of NO in metamorphosis has been described for many phyla of invertebrates (Annona et al., 2017; Tomankova, Abaffy, & Sindelka, 2017).

Sponges have an almost full repertoire of neuronal genes needed to form functional neural scaffolds (Francis et al., 2017; Sakarya et al., 2007). Nevertheless, they have no neuronal structures. This is especially strange in case of sponge larvae, which have no neurons but are photosensitive and can coordinate ciliary beating for directional driving. It was elegantly demonstrated that NO is an activator of larva metamorphosis in *A. queenslandica* (Ueda et al., 2016). An inhibition of the NO pathway suppresses metamorphosis whereas an addition of endogenous NO activates it. In this case NO pathway works upstream of (or together with) ERK (extracellular signal-regulated *kinase*) pathway, one of the many receptor tyrosine kinase cascades, and the expression of NO synthase increases in localized cell populations when the larva becomes competent.

We can see that metazoans have an almost full toolkit for body patterning and the establishment of different cell types. Until new data about intercellular communication in sponges have been obtained, we can put forward two main questions regarding metamorphosis: (1) which of the signaling pathways and in what hierarchical order are involved in metamorphosis? (2) How is the body axis determined after metamorphosis and is it the same axis as in the larva?

7. Conclusions and perspective

During metamorphosis sponge larvae lose ciliary cells and this ability to locomotion. The similarity of larval ciliated cells with the choanocytes of the adult sponge gave rise to the classical hypothesis of the "inversion of the germ layers" in sponges (Delage, 1892). Now that we have evidence from different sponges, we know that the fate of ciliated cells is species-specific. At the same time, ciliated cells often leave their place in the epithelial-like sheath and migrate inside during metamorphosis, *i.e.,* undergo EMT. Here we tried to show the connection between this process and the signaling pathways present in sponges through molecules involved in intercellular junctions. While proteins managing EMT were not found in transcriptomes of adults or larvae of *H. dujardini*, the this process in sponges remains mysterious. At the same time, sponges have most of the components of the six main signaling pathways. There is evidence that in sponges Wnt and TGF-beta signaling pathways are implicated in axial patterning rather than cell specification, and these pathways were the first metazoan mechanism of anterio-posterior determination. The link between the axis of larvae and adult sponge looks highly likely as many signaling molecules are expressed at the posterior larval pole and in the osculum of the adult sponge. However, it is still unclear whether expression of signaling molecules in adult sponge marks the axis heredity from the larva, or it is *de novo* established after metamorphosis. The developmental regulation pathways may be traced with the help of transcriptomes of sponge development stages, which are available only for a couple of species. The application of traditional techniques to simple animals at the key root position of Metazoa is a necessary step for understanding developmental events at the molecular level.

Acknowledgments

The study was financially supported by the Russian Science Foundation (grants 17-14-01089 in part of bioinformatics analysis), by the Russian Foundation for the Basic Research (grant 18-34-00398 in part of TGF-beta study) and the "Molecular and Cell Biology" program of the Presidium of the Russian Academy of Sciences (grant 01.2.01457147). We gratefully appreciate the help provided at the White Sea Biological Station of the Zoological Institute of the Russian Academy of Sciences and at Belomorskaya Educational and Research Station of the St Petersburg State University. We thank the Research Resource Center for Molecular and Cell Technologies of St Petersburg State University for technical support, and Natalia Lentsman for improving the English.

References

Adamska, M. (2015). Developmental signalling and emergence of animal multicellularity. In I. Ruiz-Trillo & A. M. Nedelcu (Eds.), *Vol. 2. Evolutionary transitions to multicellular life* (pp. 425–450). Dordrecht: Springer Netherlands.

Adamska, M., Degnan, S. M., Green, K. M., Adamski, M., Craigie, A., Larroux, C., et al. (2007). Wnt and TGF-β expression in the sponge Amphimedon queenslandica and the origin of metazoan embryonic patterning. *PLoS One*, *2*(10), e1031.

Adamska, M., Larroux, C., Adamski, M., Green, K., Lovas, E., Koop, D., et al. (2010). Structure and expression of conserved Wnt pathway components in the demosponge Amphimedon queenslandica: Wnt pathway components in Amphimedon queenslandica. *Evolution & Development*, *12*(5), 494–518.

Alban, A., Olu, D. S., Lennart, B., Christian, A., Erik, S., Steve, L., et al. (2003). A novel experimental design for comparative two-dimensional gel analysis: Two-dimensional difference gel electrophoresis incorporating a pooled internal standard. *Proteomics*, *3*(1), 36–44.

Amano, S., & Hori, I. (1993). Metamorphosis of calcareous sponges II. Cell rearrangement and differentiation in metamorphosis. *Invertebrate Reproduction and Development, 24*(1), 13–26.

Amano, S., & Hori, I. (1996). Transdifferentiation of larval flagellated cells to choanocytes in the metamorphosis of the demosponge Haliclona permollis. *The Biological Bulletin, 190*, 161–172.

Amano, S., & Hori, I. (2001). Metamorphosis of coeloblastula performed by multipotential larval flagellated cells in the calcareous sponge Leucosolenia laxa. *The Biological Bulletin, 200*(1), 20–32.

Andersson, E. R., Sandberg, R., & Lendahl, U. (2011). Notch signaling: Simplicity in design, versatility in function. *Development (Cambridge, England), 138*(17), 3593–3612.

Annona, G., Caccavale, F., Pascual-Anaya, J., Kuratani, S., De Luca, P., Palumbo, A., et al. (2017). Nitric oxide regulates mouth development in amphioxus. *Scientific Reports*, 7(1), 8432.

Belahbib, H., Renard, E., Santini, S., Jourda, C., Claverie, J.-M., Borchiellini, C., et al. (2018). New genomic data and analyses challenge the traditional vision of animal epithelium evolution. *BMC Genomics, 19*(1), 393.

Bergquist, P. R., & Green, C. R. (1977). An ultrastructural study of settlement and metamorphosis in sponge larvae. *Cahiers de Biologie Marine, 18*, 289–302.

Borisenko, I. E., Adamska, M., Tokina, D. B., & Ereskovsky, A. V. (2015). Transdifferentiation is a driving force of regeneration in Halisarca dujardini (Demospongiae, Porifera). *PeerJ, 3*, e1211.

Borisenko, I., Adamski, M., Ereskovsky, A., & Adamska, M. (2016). Surprisingly rich repertoire of Wnt genes in the demosponge Halisarca dujardini. *BMC Evolutionary Biology, 16*, 123.

Borisenko. I., Podlevskikh. A., Prakhovnik. S. & Ereskovsky. A., TGF-beta components and their expression in demosponge *Halisarca dujardini*, *Development Genes and Evolution*, in press.

Borojevič, R., & Lévi, C. (1965). Morphogenese experimentale d'une Eponge a partir de cellules de la larvt nageante dissociee. *Zeitschrift Fur Zellforschung Und Mikroskopische Anatomie, 68*, 57–69.

Boury-Esnault, N. (1976). Ultrastructure de la larve parenchymella d'Hamigera hamigera (Schmidt) (Demosponge, Poecilosclerida). Origine des cellules grises. *Cahiers de Biologie Marine, 17*, 9–20.

Boury-Esnault, N., Ereskovsky, A., Bézac, C., & Tokina, D. (2005). Larval development in the Homoscleromorpha (Porifera, Demospongiae). *Invertebrate Biology, 122*(3), 187–202.

Chen, W. T. (1976). Reproduction and speciation in Halisarca. In F. W. Harrison & R. R. Cowden (Eds.), *Aspects of sponge biology* (pp. 113–139). New York, San Francisco, London: Academic Press.

Delage, Y. (1892). Embryogenese des eponges silicieuses. *Arch Z Exp Gen, 10*, 345–498.

Efremova, S. M. (1997). Once more on the position among Metazoa—gastrulation and germinal layers of sponges. In A. V. Ereskovsky, H. Keupp, & R. Kohring (Eds.), *Modern problems of Poriferan biology* (pp. 7–15). Berlin: Freie Universitat.

Efremova, S. M., & Efremov, V. I. (1979). Proliferation cellulaire chez la larve nageante de l' éponge d'eau douce: Baikalospongia bacillifera (Dybowski). In C. Lévi & N. Boury-Esnault (Eds.), *Vol. 291. Biologie des Spongiaires* (pp. 59–65). Paris: Colloq Int CNRS.

Ereskovsky, A. V. (2010). *The comparative embryology of sponges*. Dordrecht: Springer.

Ereskovsky, A. V., Borchiellini, C., Gazave, E., Ivanisevic, J., Lapébie, P., Perez, T., et al. (2009). The Homoscleromorph sponge Oscarella lobularis, a promising sponge model in evolutionary and developmental biology. *BioEssays, 31*(1), 89–97.

Ereskovsky, A. V., Borisenko, I. E., Lapébie, P., Gazave, E., Tokina, D. B., & Borchiellini, C. (2015). Oscarella lobularis (Homoscleromorpha, Porifera) regeneration: Epithelial morphogenesis and metaplasia. *PLoS One, 10*(8), e0134566.

Ereskovsky, A. V., & Dondua, A. K. (2006). The problem of germ layers in sponges (Porifera) and some issues concerning early metazoan evolution. *Zoologischer Anzeiger–A Journal of Comparative Zoology, 245*(2), 65–76.
Ereskovsky, A. V., & Gonobobleva, E. L. (2000). New data on embryonic development of Halisarca dujardini Johnston, 1842 (Demospongiae, Halisarcida). *Zoosystema, 22*(2), 355–368.
Ereskovsky, A. V., Konjukov, P., & Willenz, P. (2007). Experimental metamorphosis of Halisarca dujardini larvae (Demospongiae, Halisarcida): Evidence of flagellated cell totipotentiality. *Journal of Morphology, 268*(6), 529–536.
Ereskovsky, A. V., Konyukov, P. Y., & Tokina, D. B. (2010). Morphogenesis accompanying larval metamorphosis in Plakina trilopha (Porifera, Homoscleromorpha). *Zoomorphology, 129*(1), 21–31.
Ereskovsky, A. V., Renard, E., & Borchiellini, C. (2013). Cellular and molecular processes leading to embryo formation in sponges: Evidences for high conservation of processes throughout animal evolution. *Development Genes and Evolution, 223*(1–2), 5–22.
Ereskovsky, A., & Tokina, D. (2004). Morphology and fine structure of the swimming larvae of Ircinia oros (Porifera, Demospongiae, Dictyoceratida). *Invertebrate Reproduction and Development, 45*(2), 137–150.
Ereskovsky, A. V., Tokina, D. B., Bézac, C., & Boury-Esnault, N. (2007). Metamorphosis of cinctoblastula larvae (Homoscleromorpha, porifera). *Journal of Morphology, 268*(6), 518–528.
Evans, C. W. (1977). The ultrastructure or larvae from the marine sponge Halichondria moorei Bergquist (Porifera, Demospongiae). *Cahiers de Biologie Marine, 18*(4), 427–433.
Feuda, R., Dohrmann, M., Pett, W., Philippe, H., Rota-Stabelli, O., & Lartillot, N. (2017). Improved modeling of compositional heterogeneity supports sponges as sister to all other animals. *Current Biology, 27*(24), 3864–3870.e4.
Francis, W. R., Eitel, M., Vargas, S., Adamski, M., Haddock, S. H., & Krebs, S. (2017). The genome of the contractile demosponge Tethya wilhelma and the evolution of metazoan neural signalling pathways. *BioRxiv*. https://doi.org/10.1101/120998.
Franzen, W. (1988). Oogenesis and larval development of Scypha ciliata (Porifera, Calcarea). *Zoomorphology, 107*(6), 349–357.
Funayama, N. (2010). The stem cell system in demosponges: Insights into the origin of somatic stem cells: Stem cell system in demosponges. *Development, Growth & Differentiation, 52*(1), 1–14.
Funayama, N., Nakatsukasa, M., Hayashi, T., & Agata, K. (2005). Isolation of the choanocyte in the fresh water sponge, Ephydatia fluviatilis and its lineage marker, Ef annexin. *Development, Growth & Differentiation, 47*(4), 243–253.
Funayama, N., Nakatsukasa, M., Kuraku, S., Takechi, K., Dohi, M., & Iwabe, N. (2005). Isolation of Ef silicatein and Ef lectin as molecular markers Sclerocytes and cells involved in innate immunity in the freshwater sponge Ephydatia fluviatilis. *Zoological Science, 22*(10), 1113–1122.
Gaino, E., Baldacconi, R., & Corriero, G. (2007). Post-larval development of the commercial sponge Spongia officinalis L. (Porifera, Demospongiae). *Tissue & Cell, 39*(5), 325–334.
Gazave, E., Lapébie, P., Richards, G. S., Brunet, F., Ereskovsky, A. V., & Degnan, B. M. (2009). Origin and evolution of the notch signalling pathway: An overview from eukaryotic genomes. *BMC Evolutionary Biology, 9*, 249.
Gonobobleva, E. L. (2007). Basal apparatus formation in external flagellated cells of *Halisarca dujardini* larvae (Demospongiae: Halisarcida) in the course of embryonic development. In *Porifera research: Biodiversity, innovation and sustainability. Rio de Janeiro: Museu nacional* (pp. 345–351). Rio de Janeiro: Universidade Federal do Rio de Janeiro.
Gonobobleva, E., & Ereskovsky, A. (2004a). Metamorphosis of the larva of Halisarca dujardini (Demospongiae, Halisarcida). *Bulletin de l'Institut Royal Des Sciences Naturelles de Belgique–Bulletin van Het Koninklijk Belgisch Instituut Voor Natuurwetenschappen, 74*, 101–115.

Gonobobleva, E., & Ereskovsky, A. (2004b). Polymorphism in free-swimming larvae of Halisarca dujardini (Demospongiae, Halisarcida). *Bollettino dei Musei e degli istituti Biologici dell'Università di Genova, 68*, 349–356.

Gonobobleva, E., & Maldonado, M. (2009). Choanocyte ultrastructure in Halisarca dujardini (Demospongiae, Halisarcida). *Journal of Morphology, 270*(5), 615–627.

Huang, R. Y.-J., Guilford, P., & Thiery, J. P. (2012). Early events in cell adhesion and polarity during epithelial-mesenchymal transition. *Journal of Cell Science, 125*(19), 4417–4422.

Jiang, J., & Hui, C.-C. (2008). Hedgehog signaling in development and cancer. *Developmental Cell, 15*(6), 801–812.

Kaltenbach, J. C., Kuhns, W. J., Simpson, T. L., & Burger, M. M. (1999). Intense concanavalin a staining and apoptosis of peripheral flagellated cells in larvae of the marine sponge *Microciona prolifera*: Significance in relation to morphogenesis. *Biological Bulletin, 197*(2), 271–273.

Käsbauer, T., Towb, P., Alexandrova, O., David, C. N., Dall'armi, E., & Staudigl, A. (2007). The notch signaling pathway in the cnidarian Hydra. *Developmental Biology, 303*(1), 376–390.

King, N., Westbrook, M. J., Young, S. L., Kuo, A., Abedin, M., & Chapman, J. (2008). The genome of the choanoflagellate Monosiga brevicollis and the origin of metazoans. *Nature, 451*(7180), 783–788.

Kitisin, K., Saha, T., Blake, T., Golestaneh, N., Deng, M., Kim, C., et al. (2007). Tgf-Beta signaling in development. *Science's STKE: Signal Transduction Knowledge Environment, 2007*(399), cm1.

Korotkova, G. P., & Apalkova, L. V. (1975). Oogenesis of the Barents Sea sponge Halisarca dujardini. In I. B. Tokin (Ed.), *Comparative and experimental morphology of the sea organisms* (pp. 9–26). Academy of Sciences USSR. (in Russian) Apatity.

Lamouille, S., Xu, J., & Derynck, R. (2014). Molecular mechanisms of epithelial–mesenchymal transition. *Nature Reviews Molecular Cell Biology, 15*(3), 178–196.

Lapébie, P., Gazave, E., Ereskovsky, A., Derelle, R., Bézac, C., & Renard, E. (2009). WNT/β-catenin signalling and epithelial patterning in the Homoscleromorph sponge Oscarella. *PLoS One, 4*(6), e5823.

Lavrov, A. I., Bolshakov, F. V., Tokina, D. B., & Ereskovsky, A. V. (2018). Sewing up the wounds: The epithelial morphogenesis as a central mechanism of calcaronean sponge regeneration. *Journal of Experimental Zoology. Part B, Molecular and Developmental Evolution, 330*(6–7), 351–371.

Lavrov, A. I., & Kosevich, I. A. (2016). Sponge cell reaggregation: Cellular structure and morphogenetic potencies of multicellular aggregates. *Journal of Experimental Zoology Part A: Ecological Genetics and Physiology, 325*(2), 158–177.

Leininger, S., Adamski, M., Bergum, B., Guder, C., Liu, J., & Laplante, M. (2014). Developmental gene expression provides clues to relationships between sponge and eumetazoan body plans. *Nature Communications, 5*, 3905. https://doi.org/10.1038/ncomms4905.

Lévi, C. (1956). *Étude des halisarca de roscoff: Embryologie et systématique des démosponges*. 7, (pp. 3–181). Travaux de La Station Biologique Roscoff N. S.

Leys, S. P. (2004). Gastrulation in sponges. In C. Stern (Ed.), *Gastrulation* (pp. 23–31). Cold Spring Harbor, New York: Cold Spring Harbor Laboratory Press.

Leys, S. P., & Degnan, B. M. (2002). Embryogenesis and metamorphosis in a haplosclerid demosponge: Gastrulation and transdifferentiation of larval ciliated cells to choanocytes. *Invertebrate Biology, 121*(3), 171–189.

Leys, S. P., & Eerkes-Medrano, D. (2005). Gastrulation in calcareous sponges: In search of Haeckel's Gastraea. *Integrative and Comparative Biology, 45*(2), 342–351.

Leys, S. P., & Ereskovsky, A. V. (2006). Embryogenesis and larval differentiation in sponges. *Canadian Journal of Zoology, 84*(2), 262–287.

Manning, G., Young, S. L., Miller, W. T., & Zhai, Y. (2008). The protist, Monosiga brevicollis, has a tyrosine kinase signaling network more elaborate and diverse than found in any known metazoan. *Proceedings of the National Academy of Sciences of the United States of America, 105*(28), 9674–9679.

Marlow, H., Roettinger, E., Boekhout, M., & Martindale, M. Q. (2012). Functional roles of notch signaling in the cnidarian Nematostella vectensis. *Developmental Biology, 362*(2), 295–308.

Matus, D. Q., Magie, C. R., Pang, K., Martindale, M. Q., & Thomsen, G. H. (2008). The hedgehog gene family of the cnidarian, Nematostella vectensis, and implications for understanding metazoan hedgehog pathway evolution. *Developmental Biology, 313*(2), 501–518.

Matus, D. Q., Pang, K., Marlow, H., Dunn, C. W., Thomsen, G. H., & Martindale, M. Q. (2006). Molecular evidence for deep evolutionary roots of bilaterality in animal development. *Proceedings of the National Academy of Sciences of the United States of America, 103*(30), 11195–11200.

Matveev, I. V., Shaposhnikova, T. G., & Podgornaya, O. I. (2007). A novel Aurelia aurita protein mesoglein contains DSL and ZP domains. *Gene, 399*(1), 20–25.

Metschnikoff, E. (1879). Spongiologische Studien. *Zeitschrift für Wissenschaftliche Zoologie, 32*, 349–387.

Mukhina, Y. I., Kumeiko, V. V., Podgornaya, O. I., & Efremova, S. M. (2006). The fate of larval flagellated cells during metamorphosis of the sponge Halisarca dujardini. *The International Journal of Developmental Biology, 50*, 533–541.

Mukhina, Y. I., Kumeiko, V. V., Podgornaya, O. I., & Efremova, S. M. (2007). The events of metamorphosis in the demosponge Halisarca dujardini Johnston, 1842, studied with immunocytochemical method. In M. R. Custódio, G. Lôbo-Hajdu, E. Hajdu, & G. Muricy (Eds.), *Porifera research: Biodiversity, innovation and sustainability* (pp. 483–490). Rio de Janeiro: Museu Nacional.

Münder, S., Käsbauer, T., Prexl, A., Aufschnaiter, R., Zhang, X., Towb, P., et al. (2010). Notch signalling defines critical boundary during budding in Hydra. *Developmental Biology, 344*(1), 331–345.

Nakanishi, N., Stoupin, D., Degnan, S. M., & Degnan, B. M. (2015). Sensory flask cells in sponge larvae regulate metamorphosis via calcium signaling. *Integrative and Comparative Biology, 55*(6), 1018–1027.

Nichols, S. A., Dirks, W., Pearse, J. S., & King, N. (2006). Early evolution of animal cell signaling and adhesion genes. *Proceedings of the National Academy of Sciences, 103*(33), 12451–12456.

Pang, K., Ryan, J. F., Baxevanis, A. D., & Martindale, M. Q. (2011). Evolution of the TGF-β signaling pathway and its potential role in the ctenophore, Mnemiopsis leidyi. *PLoS One, 6*(9), e24152.

Richards, G. S., & Degnan, B. M. (2009). The dawn of developmental signaling in the metazoa. In *Vol. 74. Cold spring harbor symposia on quantitative biology* (pp. 81–90). Cold Spring Harbor Laboratory Press.

Richards, G. S., & Degnan, B. M. (2012). The expression of Delta ligands in the sponge Amphimedon queenslandica suggests an ancient role for notch signaling in metazoan development. *EvoDevo, 3*(1), 15.

Richards, G. S., Simionato, E., Perron, M., Adamska, M., Vervoort, M., & Degnan, B. M. (2008). Sponge genes provide new insight into the evolutionary origin of the neurogenic circuit. *Current Biology: CB, 18*(15), 1156–1161.

Ryan, J. F., Pang, K., Schnitzler, C. E., Nguyen, A.-D., Moreland, R. T., & Simmons, D. K. (2013). The genome of the ctenophore Mnemiopsis leidyi and its implications for cell type evolution. *Science, 342*(6164), 1242592.

Sakarya, O., Armstrong, K. A., Adamska, M., Adamski, M., Wang, I.-F., & Tidor, B. (2007). A post-synaptic scaffold at the origin of the animal kingdom. *PLoS One*, *2*(6), e506.
Schenkelaars, Q., Pratlong, M., Kodjabachian, L., Fierro-Constain, L., Vacelet, J., & Bivic, A. L. (2017). Animal multicellularity and polarity without Wnt signaling. *Scientific Reports*, 7(1), 15383.
Schippers, K. J., & Nichols, S. A. (2018). Evidence of signaling and adhesion roles for β-catenin in the sponge Ephydatia muelleri. *Molecular Biology and Evolution*, *35*(6), 1407–1421. https://doi.org/10.1093/molbev/msy033.
Schulze, F. E. (1877). Untersuchungen über den Bau und Entwicklung der Spongien. Die Gattung Halisarca. *Zeitschrift für Wissenschaftliche Zoologie*, *28*, 1–48.
Sebé-Pedrós, A., Degnan, B. M., & Ruiz-Trillo, I. (2017). The origin of Metazoa: A unicellular perspective. *Nature Reviews Genetics*, *18*(8), 498–512. https://doi.org/10.1038/nrg.2017.21.
Simion, P., Philippe, H., Baurain, D., Jager, M., Richter, D. J., Di Franco, A., et al. (2017). A large and consistent Phylogenomic dataset supports sponges as the sister group to all other animals. *Current Biology*, *27*(7), 958–967. https://doi.org/10.1016/j.cub.2017.02.031.
Sogabe, S., Nakanishi, N., & Degnan, B. M. (2016). The ontogeny of choanocyte chambers during metamorphosis in the demosponge Amphimedon queenslandica. *EvoDevo*, 7(1), 6. https://doi.org/10.1186/s13227-016-0042-x.
Srivastava, M., Begovic, E., Chapman, J., Putnam, N. H., Hellsten, U., Kawashima, T., et al. (2008). The Trichoplax genome and the nature of placozoans. *Nature*, *454*(7207), 955–960. https://doi.org/10.1038/nature07191.
Srivastava, M., Simakov, O., Chapman, J., Fahey, B., Gauthier, M. E. A., Mitros, T., et al. (2010). The Amphimedon queenslandica genome and the evolution of animal complexity. *Nature*, *466*(7307), 720–726. https://doi.org/10.1038/nature09201.
Sudhop, S., Coulier, F., Bieller, A., Vogt, A., Hotz, T., & Hassel, M. (2004). Signalling by the FGFR-like tyrosine kinase, Kringelchen, is essential for bud detachment in Hydra vulgaris. *Development (Cambridge, England)*, *131*(16), 4001–4011. https://doi.org/10.1242/dev.01267.
Thisse, B., & Thisse, C. (2005). Functions and regulations of fibroblast growth factor signaling during embryonic development. *Developmental Biology*, *287*(2), 390–402. https://doi.org/10.1016/j.ydbio.2005.09.011.
Tomankova, S., Abaffy, P., & Sindelka, R. (2017). The role of nitric oxide during embryonic epidermis development of Xenopus laevis. *Biology Open*, *6*(6), 862–871. https://doi.org/10.1242/bio.023739.
Ueda, N., Richards, G. S., Degnan, B. M., Kranz, A., Adamska, M., Croll, R. P., et al. (2016). An ancient role for nitric oxide in regulating the animal pelagobenthic life cycle: Evidence from a marine sponge. *Scientific Reports*, *6*(1), 37546. https://doi.org/10.1038/srep37546.
Usher, K., & Ereskovsky, A. (2005). Larval development, ultrastructure and metamorphosis in Chondrilla australiensis Carter, 1873 (Demospongiae, Chondrosida, Chondrillidae). *Invertebrate Reproduction and Development*, *47*(1), 51–62. https://doi.org/10.1080/07924259.2005.9652146.
Voigt, O., Adamska, M., Adamski, M., Kittelmann, A., Wencker, L., & Wörheide, G. (2017). Spicule formation in calcareous sponges: Coordinated expression of biomineralization genes and spicule-type specific genes. *Scientific Reports*, 7, 45658.https://doi.org/10.1038/srep45658.
Weissenfels, N. (1989). *Biologie und Mikroskopishe Anatomie der Süßwassershwämme (Spongillidae)*. Stuttgart, New York: Fisher.

Windsor, P. J., & Leys, S. P. (2010). Wnt signaling and induction in the sponge aquiferous system: Evidence for an ancient origin of the organizer: Polarity in sponges. *Evolution & Development*, *12*(5), 484–493. https://doi.org/10.1111/j.1525-142X.2010.00434.x.

Windsor Reid, P. J., Matveev, E., McClymont, A., Posfai, D., Hill, A. L., & Leys, S. P. (2018). Wnt signaling and polarity in freshwater sponges. *BMC Evolutionary Biology*, *18*(1), 12. https://doi.org/10.1186/s12862-018-1118-0.

Further reading

Giard, A. M. (1873). Contribution a l'histoire naturelle des synascidies. *Archives de Zoologie Expérimentale et Générale*, *2*, 481–514.

Kozin, V. V., Borisenko, I. E., & Kostyuchenko, R. P. (2019). Establishment of the axial polarity and cell fate in Metazoa via canonical Wnt signaling: New insights from sponges and annelids. *Biological Bulletin*, *46*(1), 12.

Kolibri 360 GmbH
Ferdinand-Jühlke-Straße 7
99095 Erfurt, Deutschland
produktsicherheit@kolibri360.de